Discrete-Time Dynamics of Structured Populations and Homogeneous Order-Preserving Operators

Mathematical Surveys and Monographs

Volume 281

Discrete-Time Dynamics of Structured Populations and Homogeneous Order-Preserving Operators

Horst R. Thieme

AMERICAN MATHEMATICAL SOCIETY
Providence, Rhode Island

2020 *Mathematics Subject Classification.* Primary 37N25, 39Axx, 47G10, 47H07, 92D25; Secondary 28A33, 28C15, 46E27, 47J10, 47N60.

For additional information and updates on this book, visit
www.ams.org/bookpages/surv-281

Library of Congress Cataloging-in-Publication Data

Names: Thieme, Horst R., 1948- author.
Title: Discrete-time dynamics of structured populations and homogeneous order-preserving operators / Horst R. Thieme.
Description: Providence, Rhode Island : American Mathematical Society, [2024] | Series: Mathematical surveys and monographs, 0076-5376 ; volume 281 | Includes bibliographical references and index.
Identifiers: LCCN 2024002996 | ISBN 9781470474652 (paperback) | ISBN 9781470477349 (ebook)
Subjects: LCSH: Population biology–Mathematical models. | Discrete-time systems. | Difference equations. | Integral operators. | Vector spaces. | AMS: Dynamical systems and ergodic theory – Applications – Dynamical systems in biology. | Difference and functional equations – Difference equations. | Operator theory – Integral, integro-differential, and pseudodifferential operators – Integral operators. | Operator theory – Nonlinear operators and their properties – Monotone and positive operators on ordered Banach spaces or other ordered topological vector spaces. | Biology and other natural sciences – Genetics and population dynamics – Population dynamics (general). | Measure and integration – Classical measure theory – Spaces of measures, convergence of measures. | Measure and integration – Set functions and measures on spaces with additional structure – Set functions and measures on topological spaces (regularity of measures, etc.). | Functional analysis – Linear function spaces and their duals – Spaces of measures. | Operator theory – Equations and inequalities involving nonlinear operators – Nonlinear spectral theory, nonlinear eigenvalue problems. | Operator theory – Miscellaneous applications of operator theory – Applications in chemistry and life sciences.
Classification: LCC QH352 .T44 2024 | DDC 577.8/80151111–dc23/eng/20240310
LC record available at https://lccn.loc.gov/2024002996

DOI: https://doi.org/10.1090/surv/281

∞ The paper used in this book is acid-free and falls within the guidelines
established to ensure permanence and durability.
Visit the AMS home page at `https://www.ams.org/`

10 9 8 7 6 5 4 3 2 1 29 28 27 26 25 24

To Ruth and Reese
and in memory of Adelheid

Contents

Preface

A fundamental question in the theory of discrete and continuous time population models concerns the conditions for the extinction or persistence of populations. This question is addressed by persistence theory ([**33, 68, 187, 226**] and the references therein). Since some time, it has been recognized that, if the dynamics of the population are mathematically captured by continuous or discrete semiflows and if these semiflows have first order approximations, the spectral radii of certain bounded linear operators (better known as basic reproduction numbers) act as thresholds between population extinction and persistence [**12,44,50,62,63,118–121,150,196,214**].

Work on continuous (positively) homogeneous operators (of degree one) by John Mallet-Paret and Roger Nussbaum [**156,157**] suggested to me that the spectral radii of these more general operators can play a similar role in population models that include the mating between two sexes. I thank Roger Nussbaum [**8, 145, 146, 167–170**] for guiding me through some of the literature on homogeneous operators. Indeed, as we will see, in discrete-time models of structured populations with two sexes, the spectral radius of an order-preserving homogeneous operator that is the first order approximation of the population turnover operator acts as threshold between population extinction and population persistence. Discrete-time models (difference equations) are chosen because the theory of homogeneous operators applies more readily than in continuous-time models ((partial, delay, integro) differential equations). For the role of homogeneous operators in differential equations models, see [**107,108,120,153,221,222**].

Work on the material in this monograph began as a series of papers [**199–202**], some of them coauthored by Wen Jin and/or Hal Smith [**127–130,187,188**], whom I gratefully acknowledge, and the dissertation by Wen Jin [**126**]. The mathematics play out in the framework of normed vector spaces, usually ordered by a closed cone. I thank Ulrich Krause for walking me through his book [**138**] which got me interested in some geometric aspects (Sections 2.1.3 and 2.6). I always wondered whether there are interesting ordered normed vector spaces that are not complete themselves or are even ordered by a cone that is not complete. I got my answer from a talk by Piotr Gwiazda at the Fields Institute in Toronto and in subsequent discussions with him, Karolina Kropielnicka and Anna Marciniak-Czochra ([**34,72, 97–102**], Chapter 3): The cone of nonnegative Borel measures on a separable metric space is complete under the flat (aka dual bounded Lipschitz) norm if and only if the underlying metric space is complete ([**113**, Thm.3.8], Section 3.2.4). This has largely influenced how the theory of ordered normed vector spaces is presented in this monograph and has led to the concept of a serially complete additive subset and a monotonically complete cone of a normed vector space (Sections 2.1.2 and 2.1.5). Chapter 2 can be used as a text for a course on ordered normed vector spaces.

I thank Odo Diekmann and Eugenia Franco for carefully reading [**203**] and motivating me to bring it into its present form (the original one is partially covered in Chapter 19). I thank Gaël Raoul for carefully reading [**205**] and convincing me that Feller kernels have no good backward interpretation (see Section 18.1.1). Fei Cao read a part of the monograph during a reading course at Arizona State University and did a couple of the exercises. My late colleagues John McDonald and Neil Weiss at ASU gave me a complimentary copy of their Real Analysis book [**159**], which turned out to be a wonderful resource I refer to frequently. I am grateful to the anonymous reviewers for their constructive comments and to my editor Dr. Ina Mette for encouragement that extended over more than a decade.

I thank my wife Adelheid for her love and support over the many years I worked on this monograph [**191**]; sadly, the love of my life did not live to see the book's completion. I dedicate it to her memory and to my daughters Ruth and Reese (Clara) in gratitude for their tolerance.

Finally, praise to the Holy Spirit, source of knowledge and persistence, for the completion of this book.

CHAPTER 1

Introduction

Many animal and plant populations have yearly cycles with reproduction occurring once a year during a relatively short period. They also carry population structures which may be due to spatial distribution, age or rank structure, or degree of maturity.

1.1. Discrete-time population models

It seems appropriate to model such populations by discrete-time structured models in the form of the recursion (or difference equation),

$$x_n = F(x_{n-1}), \qquad n \in \mathbb{N}, \quad x_0 \in X_F. \tag{1.1}$$

The population structure is encoded in the closed subset $X_F \ni 0$ of a normed vector space X over $\mathbb{R}$, and $F(0) = 0$.

(See Definitions A.19 and A.13.)

The vector x_n describes the structural distribution of the population in year n while $F : X_F \to X_F$ formulates the rule how the structural distribution in a given year follows from the structural distribution of the previous year; the zero vector represents the extinction state of the population. The norm $\|x_n\|$ is some measure of the population size in year n. F is called the *(yearly) population turnover operator.* The recursive equation (1.1) is solved by

$$x_n = F^n(x_0), \quad n \in \mathbb{N}, \tag{1.2}$$

where F^n is the n fold composition of F with itself (nth iterate, nth power). We set $F^0(h_0) = h_0$. $\{F^n; n \in \mathbb{Z}_+\}$ is the *discrete semiflow* induced by F [**187, 226**].

A fundamental question (Chapter 14) is as to whether the population dies out, $\|x_n\| \to 0$ as $n \to \infty$, or whether it persists uniformly weakly.

DEFINITION 1.1. *The population* persists uniformly weakly *if the following holds for the solutions (x_n) of (1.1):*

> *There is some $\epsilon > 0$ (independent of (x_n)) such that, if $x_0 \neq 0$, then $\|x_n\| \geq \epsilon$ for infinitely many $n \in \mathbb{N}$.*

In order to address this question, we assume that X_F is a *(positively) homogeneous subset* of X:

$$x \in X_F,\ \alpha \in \mathbb{R}_+ \implies \alpha x \in X_F. \tag{1.3}$$

Let X_B be the set of those $x \in X_F$ such that the directional derivative at 0 in the direction of x exists,

$$\lim_{\mathbb{R}_+ \ni b \to 0} \frac{1}{b} F(bx) = \partial F(0, x) =: B(x). \tag{1.4}$$

Since $0 \in X_F$ and $F(0) = 0$, $0 \in X_B$ and $B(0) = 0$. If $X_B = X_F$, $\partial F(0, \cdot)$ is called the *Gateaux derivative* of F at 0.

It is easy to see that $X_B \subseteq X_F$ is a homogeneous subset of X and the operator $B : X_B \to X$ is (positively) *homogeneous* (of degree one) (Theorem 14.1).

DEFINITION 1.2. *An operator $B : X_B \to X$, $X_B \subseteq X$, is* homogeneous *if X_B is a homogeneous subset of X and,*

$$x \in X_B, \alpha \in \mathbb{R}_+ \implies B(\alpha x) = \alpha B(x).$$

Since we rarely consider homogeneity in a different sense, B with this property is simply called *homogeneous*. B is a *first order approximation of F* in a weak sense, and we will need B to be a first order approximation in a stronger sense (Section 14.2) with which we do not want to burden the reader quite yet. We call B the *basic population turnover operator* because it approximates the turnover operator at low population densities.

The operator norm of a homogeneous operator $B : X_B \to X_B$ on a homogeneous subset X_B of X is defined as

$$\|B\| := \sup \big\{ \|B(x)\|;\ x \in X_B,\ \|x\| \le 1 \big\}, \tag{1.5}$$

and B is called *bounded* if this supremum exists.

LEMMA 1.3. *Assume that there are $\delta > 0$ and $c > 0$ such $F : X_F \to X_F$ satisfies $\|F(x)\| \le c\|x\|$ for all $x \in X_F$ with $\|x\| \le \delta$ and $B(x) = \partial F(0, x)$, $x \in X_B$. Then the homogeneous operator $B : X_B \to X$ is bounded.*

1.2. The spectral radius of a bounded homogeneous operator

If $B : X_B \to X$, $X_B \subseteq X$, is a bounded homogeneous operator, then

$$\|B(x)\| \le \|B\|\ \|x\|, \qquad x \in X_B. \tag{1.6}$$

If $B : X_B \to X_B$, this formula implies that the powers (iterates) $B^n : X_B \to X_B$ of a homogeneous bounded B are bounded homogenous operators and $\|B^n\| \le \|B\|^n$ for all $n \in \mathbb{N}$.

The *spectral radius* of a bounded homogeneous $B : X_B \to X_B \subseteq X$ is defined by the *Gelfand* formula [**93**]

$$\mathbf{r}(B) = \inf_{n \in \mathbb{N}} \|B^n\|^{1/n} = \lim_{n \to \infty} \|B^n\|^{1/n}. \tag{1.7}$$

The last equality is shown in the same well-known way as for a bounded linear everywhere-defined map. See Chapter 5 for more information.

The name "spectral radius" is motivated by the following fact. Let B be a bounded linear map on X. If $X_\mathbb{C}$ is the complexification of X and $B_\mathbb{C}$ the extension of B to $X_\mathbb{C}$ (see Section 5.4), then

$$\mathbf{r}(B) = \sup\{|\lambda|; \lambda \in \sigma^\mathbb{C}(B)\}, \tag{1.8}$$

where $\sigma^\mathbb{C}(B) \subseteq \mathbb{C}$ denotes the spectrum of $B_\mathbb{C}$ [**93, 167**].

Equation (1.8) is partially preserved for bounded homogeneous B. Recall that, for linear bounded everywhere-defined B, eigenvalues are special elements of the spectrum of B.

If X_B is a homogeneous subset of X, then $0 \in X_B$. See (1.3). We use the notation

$$\dot{X}_B = X_B \setminus \{0\}, \qquad \dot{X}_F = X_F \setminus \{0\}. \tag{1.9}$$

For homogeneous $B : X_B \to X_B$, we call $\lambda \in (0,\infty)$ an *eigenvalue* of B and $v \in \dot{X}_B$ an *associated eigenvector* of B if $B(v) = \lambda v$.

PROPOSITION 1.4. *Let $B : X_B \to X_B$ be homogeneous and bounded, $\lambda \in (0,\infty)$, $v \in \dot{X}_B$ and $B(v) = \lambda v$.*

Then $\lambda \le \mathbf{r}(B)$.

PROOF. Since B is homogeneous, we can assume that $\|v\| = 1$. By induction, $\lambda^n v = B^n(v)$ for all $n \in \mathbb{N}$. Since B^n is homogeneous and bounded, $\lambda^n \le \|B^n\|\,\|v\| \le \|B^n\|$ and $\lambda \le \|B^n\|^{1/n}$ for all $n \in \mathbb{N}$. The assertion now follows from (1.7). □

The Gelfand formula for the spectral radius is used for restrictions of bounded positive linear operators to a cone by Bonsall [**27**] under the name "partial spectral radius" and somewhat later by Nussbaum [**167**] under the name "cone spectral radius." Mallet-Paret and Nussbaum [**156, 157**] use the Gelfand formula for homogeneous bounded operators on a cone under the name "Bonsall cone spectral radius." But since the Gelfand formula also makes sense on homogeneous sets (which concept includes the vector space), there seems to be good reason to simply say "spectral radius." While a much richer theory can be developed for order-preserving homogeneous operators on a cone, upper semicontinuity of the spectral radius (Section 1.4) can be shown for homogeneous operators in a normed vector space without any order under relatively mild conditions (Chapter 7).

If B has an interpretation as a basic population turnover operator, $\mathbf{r}(B)$ is called the *basic population turnover number* and is denoted by $\mathcal{T}_0$. $\mathcal{T}_0$ has also been called *inherent population growth rate* [**44**] or *(population) growth factor* [**150**].

1.3. Preview of extinction and persistence results

The following results which highlight the role of the basic turnover number $\mathcal{T}_0 = \mathbf{r}(B)$ as threshold parameter between population extinction and persistence hold under additional assumptions, which we do not mention here (Chapter 14). We first consider the subthreshold case $\mathcal{T}_0 < 1$.

THEOREM 1.5. *Let $F, B : X_F \to X_F$ and let B be homogeneous and bounded and B an appropriate first order approximation of F, $\mathcal{T}_0 = \mathbf{r}(B) < 1$. Then the extinction state 0 is locally asymptotically stable in the following sense:*

For each $\alpha \in (\mathcal{T}_0, 1)$, there exist some $\delta_0 > 0$ and $M \ge 1$ such that for all solutions (x_n) of (1.1) $\|x_n\| \le M\alpha^n \|x\|$ for all $n \in \mathbb{N}$ and all $x_0 \in X_F$ with $\|x_0\| \le \delta_0$.

See Theorem 14.13 for the precise formulation. We turn to the superthreshold case $\mathcal{T}_0 > 1$.

THEOREM 1.6. *Let $F, B : X_F \to X_F$ and B be bounded homogeneous and let B be an appropriate first order approximation of F, $\mathcal{T}_0 = \mathbf{r}(B) > 1$.*

Then, under appropriate additional assumptions, the population persists uniformly weakly:

There exists some $\epsilon > 0$ such that for all solutions (x_n) of (1.1) with $x_0 \in X_F$ and $x_0 \ne 0$ we have $\|x_n\| \ge \epsilon$ for infinitely many $n \in \mathbb{N}$.

See Theorem 14.27 for the precise formulation.

In the threshold case $\mathcal{T}_0 = 1$, stability of the extinction state or even extinction of the population can be shown, but the conditions are quite restrictive and technical. See Section 14.5 and Theorems 14.23 and 14.24.

For these results, whether in the subthreshold, threshold, or superthreshold case, it will be not enough to assume that X_F is a closed homogeneous subset of X; rather X_F needs to be a closed cone. Therefore, cones, wedges and ordered vector spaces are studied in Chapter 2. Special care is given to being clear about when completeness of X or X_F are really needed. See motivation ($\diamondsuit_3$) below. Also, not much can be done without assuming that B is order-preserving (Chapter 4).

The previous theorems are known if B can be extended to a bounded linear map on X and B is the Frechet derivative of F at 0 [**187**, **226**].

There are at least three motivations to consider the more general situation of a bounded homogeneous order-preserving operator.

($\diamondsuit_1$) The first motivation is mathematical: The Gateaux derivative at the zero vector, (1.4), is homogeneous but not linear, and homogeneous operators are not Frechet differentiable at 0 unless they are linear. Also see Section 14.1.

($\diamondsuit_2$) The second, biological, motivation are two-sex population models which often use homogeneous mating (or pair formation) functions (Section 4.5, [**83**, **106**–**109**, **115**, **117**, **118**, **130**, **161**, **202**] and the references therein).

Let $\varphi(f, m)$ be the amount of offspring produced by f females and m males via mating (ignoring competition for resources). Assume that

$$\varphi(\alpha f, \alpha m) = \alpha\varphi(f, m), \qquad f, m, \alpha \in \mathbb{R}_+.$$

This means that if both the numbers of females and males double (or triple), the number of mated pairs also doubles (or triples).

This type of mating functions leads to homogeneous first order approximations of the population turnover operator (Chapters 15, 17, and 18). Males and females may have different survival probabilities, the sex ratio at birth may not be one, the sexes may move differently in their habitat [**162**], fighting between males may spread diseases more readily [**11**] (to mention a few reasons while one would like to model females and males separately).

($\diamondsuit_3$) The third motivation are structural population distributions which are best described by measures μ on a metric space S (Chapters 3, 13 and 19) with $\mu(T)$ representing the number of individuals whose structural characteristic lies in the Borel subset T of S. So, the population state space of interest is the ordered vector space of real measures on the Borel σ-algebra of S, denoted by $\mathcal{M}(S)$, ordered by the cone of nonnegative measures $\mathcal{M}_+(S)$. The variation norm (Section 3.1) makes this vector space a Banach space but is too strong to provide the required compactness of the basic turnover operator B on $X_F = \mathcal{M}_+(S)$ in Theorem 1.6 even if B can be extended to a bounded linear operator on X. A suitable alternative is the flat norm aka dual bounded Lipschitz norm (Section 3.2). Important linear basic turnover operators defined on all of X are compact and continuous on X_F, though they are not continuous on X (Chapters 13 and 19). As a trade off, the flat norm rarely makes

$X = \mathcal{M}(S)$ a Banach space and makes $\mathcal{M}_+(S)$ only complete if the metric space S can be chosen to be complete.

Theorems 1.5 and 1.6 highlight the importance of the spectral radius as threshold parameter separating population extinction and population persistence. Estimates, approximations, and equivalent characterizations of the spectral radius are therefore important (Chapters 5 and 6). This involves lower and upper Collatz–Wielandt numbers [**87**] and Collatz–Wielandt bounds and also Collatz–Wielandt spectral radii (Sections 5.5.1, 5.5.2, and 6.2).

Uniform weak versus uniform persistence. A stronger and more desirable concept than uniform weak persistence (Definition 1.1) is uniform persistence.

DEFINITION 1.7. *The population* persists uniformly *if the following holds for the solutions* (x_n) *of* (1.1)*:*

There is some $\epsilon > 0$ *(independent of* (x_n)*) such that, if* $x_0 \neq 0$*, then* $\|x_n\| \geq \epsilon$ *for all but finitely many* $n \in \mathbb{N}$*.*

It may be worth-while to write down the two definitions in formal logic.
Uniform persistence:

$$\exists \epsilon > 0\, \forall x_0 \neq 0\, \exists m \in \mathbb{N}\, \forall n \in \mathbb{N}\, \big(n \geq m \implies \|x_n\| \geq \epsilon\big).$$

Uniform weak persistence:

$$\exists \epsilon > 0\, \forall x_0 \neq 0\, \forall m \in \mathbb{N}\, \exists n \in \mathbb{N}\, \big(n \geq m \wedge \|x_n\| \geq \epsilon\big).$$

Verbally, in both concepts, there is some level $\epsilon > 0$ for the population size that does not depend on the initial state of the population.

If the population persists uniformly, after some time (which depends on the initial state of the population) the population size stays above the level ϵ.

If the population persists uniformly weakly, the population size may become arbitrarily small at times but always bounces back to or above the level ϵ.

If one is sufficiently familiar with the notions of limit superior and limit inferior of real functions or sequences (Appendix A.3.1)[**159**, Def.2.5], there is yet another way of formulating the concept of uniform weak and uniform persistence. Recall that if (x_n) is a solution of $x_n = F(x_{n-1}), n \in \mathbb{N}$, then $x_n = F^n(x_0)$ where F^n is the nth iterate or power of F.

- The population persists uniformly if there is some $\epsilon > 0$ such that, for all $x_0 \in \dot{X}_F$, $\liminf_{n\to\infty} \|F^n(x_0)\| \geq \epsilon$.
- The population persists uniformly weakly if there is some $\epsilon > 0$ such that, for all $x \in \dot{X}_F$, $\limsup_{n\to\infty} \|F^n(x_0)\| \geq \epsilon$.

See Exercise A.3.2. We will use the limsup liminf formulations in the rest of the book, e.g., Section 14.9. See also Appendix A.3.1.

Uniform persistence requires much stronger assumptions than uniform weak persistence. Compare Theorems 14.27 and 14.32. There are well-established routes [**92**], [**130**], Section 5 and Theorem 7.2, and [**187**], Sections 4.1 and 4.2, of getting from uniform weak to uniform persistence. These routes do not depend on whether the basic turnover operator B is a bounded homogeneous order preserving operator on X_F or whether it can be extended to a bounded linear positive map on all of X. Therefore, we mostly concentrate on uniform weak persistence (Section 14.9.1), but give a taste of uniform persistence in Section 14.9.2. As for the case studies, we work out uniform persistence for the less familiar case that the population state space is the cone of nonnegative real measures on a metric space.

We mention that there is also a direct route to uniform persistence ([**154, 226**] and the references therein and in [**187**]).

1.4. Continuous dependence of the spectral radius on its operator

The threshold property of the spectral radius (between extinction and persistence of populations) outlined in Section 1.3 raises interest in the question whether the spectral radius of a bounded homogeneous operator B on X_B continuously depends on B. In other words: If the basic turnover operator is only slightly changed, does the same holds for the basic turnover number, $\mathcal{T}_0$? In mathematical formulation:

> If (B_k) is a sequence of bounded homogeneous operators on X_B and $\|B_k - B\| \to 0$ as $k \to \infty$, then $\mathbf{r}(B_k) \to \mathbf{r}(B)$ as $k \to \infty$.

Here $\|\cdot\|$ is the operator norm for bounded homogeneous operators on X_B as defined in (1.5).

The definition of the spectral radius, (1.7), involves taking an infimum and upper semicontinuity of functions is preserved under taking infima (Lemma A.52). Thus, it is not surprising that upper semicontinuity of the spectral radius,

> if (B_k) is any sequence of bounded homogeneous operators on X_B with $\|B_k - B\| \to 0$ as $k \to \infty$, then
>
> $$\limsup_{k\to\infty} \mathbf{r}(B_k) \le \mathbf{r}(B),$$

can be proved under rather mild extra conditions (Theorem 7.1).

Conditions for lower semicontinuity,

> if (B_k) is any sequence of bounded homogeneous operators on X_B with $\|B_k - B\| \to 0$ as $k \to \infty$, then
>
> $$\liminf_{k\to\infty} \mathbf{r}(B_k) \ge \mathbf{r}(B),$$

are much harder to find (Chapter 10); actually lower semicontinuity may fail even if X_B is an additive homogeneous subset of the normed vector space X and the homogeneous B is compact, additive and continuous on X_B. See [**146**] and Section 10.4.

Instead of finding conditions for the lower semicontinuity of the spectral radius directly, we try to find them for the lower Collatz–Wielandt bound. The definition of the Collatz–Wielandt bound, (5.30), involves taking a supremum, an operation known to preserve lower semicontinuity of functions (Lemma A.53). Eventually, the upper semicontinuity of the spectral radius is combined with the lower semicontinuity of the lower CW bound, which requires some more conditions (see next section).

1.5. The spectral radius as (lower) eigenvalue

In Theorem 1.6, the link between the assumption $\mathbf{r}(B) > 1$ and the persistence statement is provided by $\mathbf{r}(B) =: \mathcal{T}_0$ to be an eigenvalue (or lower eigenvalue) of the homogeneous operator B associated with an eigenvector

$$v \in X_B, \quad \|v\| = 1, \quad B(v) = \mathcal{T}_0\, v,$$

(Chapter 9), or with a homogeneous eigenfunctional

$$\theta : X_B \to \mathbb{R}_+, \quad \theta(B(x)) = \mathcal{T}_0\, \theta(x), \quad x \in X_B,$$

(Chapter 11).

If $\theta : X_B \to \mathbb{R}_+$ is an eigenfunctional of B associated with $\mathcal{T}_0$, then, by induction,

$$\theta(B^n(x)) = \mathcal{T}_0^n \theta(x), \qquad x \in X_B.$$

Assume that $\theta(x) > 0$ is related to the size of a population with structural distribution $x \in X_B$ and that the basic turnover operator B were to coincide with the turnover operator F such that $\theta(B^n(x))$ were the population size after n years. Then the population would die out geometrically fast if $\mathcal{T}_0 < 1$ and would persist (in the form of geometric growth) if $\mathcal{T}_0 > 1$.

However, typically $B \neq F$, but these behaviors somewhat carry over to F (Section 1.3).

The existence of eigenvectors or homogeneous eigenfunctionals of B associated with $\mathcal{T}_0 = \mathbf{r}(B)$ occurs if B has sufficiently strong compactness properties (Section 9.1) or order boundedness properties ([**8, 200, 201**], Section 9.2, and Chapter 11) and/or the cone X_B has particular properties [**199**].

Some of the proofs rely on perturbation arguments similar to the ones used for bounded linear positive operators and on the monotonicity (Section 6.5) and the upper semicontinuity of the spectral radius (as a function of its operator) under reasonably general assumptions (Chapter 7).

As much as we can, we will avoid assuming that X_B is a complete subset of X (Definition A.27). The reason is that we want to consider population models in the cone $\mathcal{M}_+(S)$ of nonnegative measures with the flat norm, which is only complete if S is a complete metric space (see Section 1.3, ($\diamondsuit_3$) and Section 3.2.4). The flat norm is used to make the operator B compact (Sections 13.5 and A.1.8).

In turn, if there is an eigenvector of B associated with $\mathcal{T}_0 = \mathbf{r}(B)$, this helps to show that the spectral radius is also lower semicontinuous function of B by relating the spectral radius of B to the lower Collatz–Wielandt bound of B.

This is why the chapters on upper and lower semicontinuity (Chapters 7 and 10) of the spectral radius are separated by the chapter on eigenvectors (Chapter 9).

A tool that is very effective for the spectral analysis of bounded linear operators but works only partially for homogeneous operators are the resolvents which only work from one side (Chapter 8). This is enough to show that the spectral radius of an order-bounded homogeneous order-preserving operator is an eigenvalue associated with a positive bounded eigenfunctional (Chapter 11). However, this functional may not be continuous (Section 11.6).

1.6. Turnover versus reproduction number

While the basic turnover number $\mathcal{T}_0$, the spectral radius of the basic turnover operator B, is the threshold parameter between population extinction and persistence (Section 1.3), it may have little tangible biological interpretation except in special cases (Section 18.6). If $B = H + Q$ with bounded linear positive operators H and Q, $\mathbf{r}(Q) < 1$, the operator

$$H(\mathbb{I} - Q)^{-1} = H \sum_{n=0}^{\infty} Q^n \tag{1.10}$$

can be interpreted as *next generation operator* and its spectral radius, $\mathcal{R}_0$, as *basic reproduction number*. See Section 12.1. Under suitable assumptions (Section 12.2, [**50, 150**], [**216**, Thm.5.1]), the basic turnover number $\mathcal{T}_0$ and the basic reproduction number $\mathcal{R}_0$ have the same threshold behavior: They are simultaneously strictly smaller than one, equal to one, or strictly larger than one. Typically, $\mathcal{R}_0$ has a more tangible biological interpretation than $\mathcal{T}_0$ and can sometimes be even explicitly expressed in terms of the model parameters [**12, 50, 128, 150**]. Unfortunately, if H is an order-preserving continuous homogeneous operator, the relation between $\mathcal{T}_0$ and $\mathcal{R}_0$ seems to hold only under quite technical conditions [**207**]. An exception is the special case that H is a rank-one operator, $H(x) = \phi(x)u$, with u being a nonzero vector in cone and ϕ an order-preserving continuous homogeneous functional (Section 12.3, [**207**]). An application to a rank-structured population with mating can be found in Chapter 16.

1.7. Case studies in population dynamics

Theorems 1.5 and 1.6 are applied to a number of mathematical case studies. This includes the simplest possible discrete-time two-sex population model (Section 15.1) and a simple hybrid-time (periodically impulsive) population model with mating (Section 15.2).

As for structured populations, we first study a rank-structured population, with the state-space chosen as the sequence space ℓ^1_+ (Chapter 16). This is an example where we can derive a concrete basic reproduction number rather than the abstract basic turnover number as a threshold parameter to separate extinction from persistence (Chapter 12).

Later, we concentrate on spatial structure and consider a hybrid-time (periodically impulsive) model where females and males move by diffusion during the year and mate once a year during a very short mating season (Chapter 17). We also consider nonlocal spatial spread (where space can be interpreted in a very wide sense) for one-sex and two-sex populations (Chapters 18 and 19). Special consideration is given to the case where state transitions are modeled by Feller kernels (Section 13.3).

Let the structural characteristics of a population be represented by a metric space S. If the structure is given by spatial location, a point $s \in S$ is a location in the habitat S of the population. For a primer on metric spaces, see Appendix A.1.

Let $\mathcal{B}$ be the σ-algebra of Borel subsets of S (Section 3.2). A function $\kappa : \mathcal{B} \times S \to \mathbb{R}_+$ is called a *Feller kernel* if $\kappa(\cdot, s)$ is a finite measure on $\mathcal{B}$ for all $s \in S$ and $\int_S f(t)\kappa(dt, s)$ is a continuous function of $s \in S$ for all $f \in C^b(S)$, the space of bounded continuous real-valued functions on S. It follows that $\kappa(T, \cdot)$ is Borel measurable for each $T \in \mathcal{B}$.

If $\kappa(S, s) \in [0, 1]$ for all $s \in S$, the Feller kernel has the following forward interpretation:

$\kappa(T, s)$ is the probability that an individual that is at location $s \in S$ at the beginning at the year is alive and located in the Borel set T at the end of the year. The forward interpretation is associated with a linear positive map on the space of real-valued Borel measures on S, $\mathcal{M}(S)$,

$$(A\mu)(T) = \int_S \kappa(T, s)\mu(ds), \qquad \mu \in \mathcal{M}(S), \quad T \in \mathcal{B}. \tag{1.11}$$

If $\mathcal{M}(S)$ is equipped with the variation norm, A is a bounded linear map. If $\mathcal{M}(S)$ is equipped with the flat norm, A map not be bounded, but its restriction A_+ to the cone $\mathcal{M}_+(S)$ of nonnegative Borel measures is bounded. See Chapter 13.

By definition, a Feller kernel also induces a map A_* on $C^b(S)$, the Banach space of bounded continuous real-valued functions on S, with the supremum norm,

$$(A_*f)(s) = \int_S f(t)\kappa(dt,s), \qquad f \in C^b(S), \quad s \in S. \tag{1.12}$$

The operators A and A_* are dual to each other,

$$\int_S f\, d(A\mu) = \int_S (A_*f)\, d\mu, \qquad f \in C^b(S), \quad \mu \in \mathcal{M}(S). \tag{1.13}$$

A and A_* have the same spectral radius $\mathcal{T}_0$ which we define to be the spectral radius of the Feller kernel κ, $\mathcal{T}_0 = \mathbf{r}(\kappa)$.

A_* is a bounded linear operator on $C^b(S)$ that maps $C^b_+(S)$ into itself, the closed cone of nonnegative bounded continuous functions. A_* can be extended to a bounded linear operator on $M^b(S)$, the Banach space of bounded measurable functions. This makes it possible to use Feller kernels to formulate density models for two-sex population models (Chapter 18) on $C^b_+(S)$ or $M^b_+(S)$. The basic turnover operator takes the form

$$(B\chi_T)(s) = \varphi\big(s, \kappa_1(T,s), \kappa_2(T,s)\big), \qquad s \in S, \quad T \in \mathcal{B}. \tag{1.14}$$

Here χ_T is the characteristic (indicator) function of the Borel subset T of S, $\chi_T(s) = 1$ if $s \in T$, and $\chi_T(s) = 0$ if $s \in S \setminus T$. For each $s \in S$, $\varphi(s, f, m)$ is the amount of offspring produced at location s by f females and m males ignoring competition for resources but not competition for mates. Cf. ($\diamondsuit_2$) in Section 1.3. As for dispersal, $\kappa_j(T,s)$ describes how females ($j = 1$), and males ($j = 2$) move from the set T to the point s. The expression in (1.14) relates the number of neonates (newborn individuals) at location s in the present year to the number of neonates in the set T in the previous year (again by ignoring competition or cooperation in using resources).

The class of these models intersects with integro-difference equations [**116, 125, 134, 135, 148, 151, 152, 162, 165, 171, 192, 221**] and integral projection models [**74, 78, 79, 89**]. To reduce the technicality of the presentation, only semelparous populations are considered, where individuals reproduce only once in their lifetime and, as far as population persistence goes, only uniform weak persistence is established (for uniform persistence see [**205**]). Unisex models for iteroparous populations, in which individuals reproduce several times during their lifespan, are considered in [**206**] where also uniform persistence is established. Differently from the state space $\mathcal{M}_+(S)$, the Feller kernel has no good biological interpretation for $C^b_+(S)$ but is still a handy mathematical tool. See Section 18.1.1.

It seems quite cumbersome to formulate two-sex population models on $\mathcal{M}_+(S)$ that can be analyzed by the methods developed in this monograph (see [**101**, Sec.6.4] and Section 19.5 for attempts); so mainly unisex population models are considered on the cone $\mathcal{M}_+(S)$ of the ordered vector space $\mathcal{M}(S)$ which will be endowed with either the variation norm or the flat norm (See ($\diamondsuit_3$) in Section 1.3 and Chapter 19). The models cover both semelparous and iteroparous populations, and both uniform weak and uniform persistence are established. A more general model framework is considered in [**204**].

For continuous-time population models on the space of measures we refer to **[72]** and the references therein. Following Section 1.5, we will look for and use an eigenfunction $f \in \dot{C}^b_+(S)$ and an eigenmeasure $\mu \in \dot{\mathcal{M}}_+(S)$ of the kernel κ, i.e., eigenvectors of A_* and A, respectively,

$$\mathcal{T}_0\, f(s) = \int_S f(t)\kappa(dt, s), \qquad s \in S,$$

$$\mathcal{T}_0\, \mu(T) = \int_S \kappa(T, s)\mu(ds), \qquad T \in \mathcal{B}.$$

1.8. Other aspects of population dynamics

In this monograph, we investigate the extinction and persistence of populations (in discrete-time models), but we do not explore the many ways in which the populations persist in the framework of discrete-time models (recursive equations): at stable equilibrium, in stable periodic or quasiperiodic oscillations or even chaotically, in competition or cooperation with other populations, or in adaptive evolution. See **[6, 7]** **[42]** – **[49]** **[53]** **[54]** **[58]** **[64]** **[65]** **[66]** **[67]** **[103]** **[104]** **[138]** **[160]** for a small sample of references. Other models do not only study the persistence of a population but also the persistence of a disease that afflicts it **[12, 80, 155, 176]**. We also do not investigate the effects of forcing **[89]**.

Of course, there are many population models formulated in continuous time, some of them with homogeneous nonlinearities. For homogeneous differential equations, see **[35, 83, 84, 107–109, 115, 117, 120]** for time-autonomous and **[106, 219]** for periodically forced differential equations; see **[219]** (and the references therein) for periodically forced homogeneous partial differential equations. For continuous-time population models using the cone of nonnegative measures, see **[72]** and the references therein.

CHAPTER 2

Cones and ordered vector spaces

While homogeneous operators can be considered in any normed vector space, a much richer theory can be developed if there also is an order structure given by a cone.

Let X be a vector space over $\mathbb{R}$.

A nonempty subset W of X is called *homogeneous*, if $\alpha x \in W$ for all $\alpha \in \mathbb{R}_+$ and $x \in W$.

Any homogeneous subset contains the vector 0. If W is a homogeneous subset of X, we use the notation

$$\dot{W} = W \setminus \{0\}. \tag{2.1}$$

X and $\{0\}$ are obvious homogeneous subsets of X, and so are all other linear subspaces.

2.1. Wedges, cones, and associated orders

While some results just hold under the assumption of homogeneity (Theorem 5.7 and Section 7.1), mostly an order structure needs to be assumed for a satisfying theory.

2.1.1. Convex and additive subsets. A subset W of X is called *convex*, if $\alpha x + (1-\alpha)y \in W$ for all $x, y \in W$ and $\alpha \in (0,1)$.

Geometrically, this means that a line segment is contained in W whenever its endpoints are elements of W.

W is called *additive* if $x + y \in W$ for all $x, y \in W$. A homogeneous subset of X is convex if and only if it is additive.

2.1.2. Serially complete additive subsets. Let X be a normed vector space (Appendix A.19).

An additive subset W of X is called *serially complete* if any series $\sum_{n=1}^{\infty} x_n$ with $x_n \in W$ for all $n \in \mathbb{N}$ and $\sum_{n=1}^{\infty} \|x_n\|$ converging in $\mathbb{R}$ converges in W.

THEOREM 2.1 ([**159**, Prop.9.3]). *A normed vector space is a Banach space if and only if it is serially complete.*

COROLLARY 2.2. *Let W be a closed additive subset of a Banach space X. Then W is serially complete.*

Wedges and cones (see next) that are serially complete will play an important role in this monograph. See Sections 2.1.5, 2.3.1, 9.2, and 12.2.1. An important example of an ordered normed vector space that is not a Banach space but has a serially complete closed cone is the vector space of real measures on a metric space with the flat norm and the cone of nonnegative finite measures. If the metric space is not complete, even the cone is not complete but still serially complete. See Section 3.2.

2.1.3. Wedges and cones. $W \subseteq X$ is called a *wedge* if it is convex and homogeneous. A wedge W is called a *cone* if $W \cap (-W) = \{0\}$, i.e., $x = 0$ is the only vector $x \in X$ such that x and $-x$ are elements in W.

X is a wedge and $\{0\}$ is a cone. If $x \in X$, $\{\alpha x; \alpha \in \mathbb{R}_+\}$ is a cone.

DEFINITION 2.3. *A cone W is called* Archimedean *if, for any $x \in W$ and any $y \in X$, $x + ny \in W$ for all $n \in \mathbb{N}$ implies that $y \in W$.*

A subset L in X is called a line *if there exist $x \in X$ and $y \in \dot{X}$ such that $L = \{x + \alpha y; \alpha \in \mathbb{R}\}$.*

If X is a normed vector space over $\mathbb{R}$ and a cone (wedge) W is a closed subset of X, W is called a closed cone (wedge).

A closed wedge is Archimedean. The next result gives a geometric interpretation of an Archimedean cone.

LEMMA 2.4 ([**138**]). *A wedge that contains no lines is a cone. An Archimedean cone contains no lines.*

PROOF. Let W be a wedge that does not contain any lines. Let $\pm x \in W$. Since W is homogeneous, W contains the set $\{\alpha x; \alpha \in \mathbb{R}\}$. Since W contains no lines, $x = 0$ and W is a cone.

Now let W be an Archimedean cone. Assume that W contains the set $\{x + \alpha y; \alpha \in \mathbb{R}\}$ for some $x, y \in X$. Then $x \in W$ and $x + n(\pm y) \in W$ for all $n \in \mathbb{N}$. Since W is an Archimedean cone, $\pm y \in W$ and $y = 0$. So W does not contain any lines. □

2.1.4. Partial orders. If $W \subseteq X$ is a cone, the definition

$$x \leq y \iff y - x \in W \tag{2.2}$$

provides a partial *order* $\leq$ on X. We also write $y \geq x$ for $x \leq y$. This order is compatible with addition and with the multiplication by positive real numbers. We recover

$$W = \{x \in X; x \geq 0\}.$$

A real vector space X has many cones. When we talk about an *ordered vector space* we typically have a specific cone in mind which induces the order. This cone is denoted by X_+, and we will talk about it as the *order cone* in order to single it out from the other cones contained in X.

X is called an *ordered Banach space* if X is an ordered normed vector space which is complete.

REMARK 2.5. If X is a vector space of real-valued functions, X_+ typically is the cone of nonnegative functions.

If nothing else is said, $\mathbb{R}$ is considered the ordered Banach space with the absolute value as norm and $\mathbb{R}_+ = [0, \infty)$ as order cone.

2.1.5. Monotonically complete subsets. The next concept is an example of singling out the order cone.

DEFINITION 2.6. *If X is an ordered normed vector space with order cone X_+, a subset S of X_+ is called* monotonically complete *if any increasing Cauchy sequence in S (increasing with respect to the partial order induced by X_+) has a limit in S.*

REMARK 2.7. Let X be an ordered normed vector space with order cone X_+. If X_+ is monotonically complete, then every decreasing Cauchy sequence in X converges.

PROOF. Let (x_n) be a decreasing Cauchy sequence in X. Define $y_n = x_1 - x_n$ for all $n \in \mathbb{N}$. Then (y_n) is an increasing Cauchy sequence in X_+ and has a limit $y \in X_+$ according to Definition 2.6. Then $x_n \to x_1 - y$. □

Recall the concept of serial completeness in Section 2.1.2.

THEOREM 2.8. *Let X be an ordered normed vector space with order cone X_+ and let W be an additive subset of X. Then the following holds:*

(a) *If W is a monotonically complete subset of X_+, then W is serially complete.*
(b) *If W is a serially complete subset of X with $X_+ \subseteq W$, then W is monotonically complete.*

PROOF. Let X be an additive subset of X.

(a) Let W be also a monotonically complete subset of X_+ and (x_n) a sequence in W such that $\sum_{n=1}^{\infty} \|x_n\|$ converges in $\mathbb{R}$. Let (y_m) be the sequence of partial sums $y_m = \sum_{n=1}^{m} x_n$. Since $x_n \in W \subseteq X_+$ and by the triangle inequality, (y_m) is an increasing Cauchy sequence in W which converges in W by assumption.

(b) The proof follows the one of Theorem 2.1 given in [**159**, Prop.9.3].

Assume that $X_+ \subseteq W \subseteq X$ and that W is a serially complete additive set. Let (y_m) be an increasing Cauchy sequence in W. For convergence, it is sufficient that (y_m) has a convergent subsequence with limit in W.

Since (y_m) is a Cauchy sequence, we can find a strictly increasing sequence of numbers (m_j) in $\mathbb{N}$ such that $\|y_{m_{j+1}} - y_{m_j}\| \leq 2^{-j}$ for all $j \in \mathbb{N}$. Set

$$z_k = \sum_{j=1}^{k} [y_{m_{j+1}} - y_{m_j}], \qquad k \in \mathbb{N}.$$

Since (y_m) is increasing, $y_{m_{j+1}} - y_{m_j} \in X_+ \subseteq W$ and $z_k \in W$ because W is an additive set. By construction, since W is serially complete, (z_k) converges to some $z \in W$. By telescoping, $z_k = y_{m_{k+1}} - y_{m_1}$. So (y_{m_k}) converges with limit $z + y_{m_1} \in W$ and (y_m) has a converging subsequence in W. □

COROLLARY 2.9. *Let X be an ordered normed vector space with order cone X_+. Then X_+ is monotonically complete if and only if it is serially complete.*

An example of an ordered normed vector space that is not a Banach space but has a serially complete closed cone is the vector space of real measures on a metric space with the flat norm and the cone of nonnegative finite measures. If the metric space is not complete, even the cone is not complete but still serially complete. See Section 3.2.

An example of an ordered normed vector space with a closed cone that is not serially complete is the vector space of bounded Lipschitz continuous real-valued functions on $[0, 1]$ with the supremum norm and the cone of nonnegative bounded Lipschitz continuous functions. See Section 2.7.1.

2.1.6. Operators between ordered normed vector spaces. Let X and Y be vector spaces and $B : X_B \to Y$ with $X_B \subseteq X$.

DEFINITION 2.10. *B is called* additive *if $B(x+z) = B(x) + B(z)$ for all $x, z \in X_B$ with $x + z \in X_B$.*

B is called homogeneous if $B(\alpha x) = \alpha B(x)$ for all $\alpha \in \mathbb{R}_+$ and $x \in X_B$ with $\alpha x \in X_B$.

REMARK 2.11. Let X and Y be normed vector spaces over $\mathbb{R}$ and W be a closed wedge contained in X. If $B : W \to Y$ is additive and continuous, then B is homogeneous.

PROOF. Let $x \in W$. Since B is additive and W a wedge, $nx \in W$ and $(1/n)x \in W$ and $B(nx) = nB(x)$ for all $n \in \mathbb{N}$ and $B(x) = B(n(1/n)x = nB((1/n)x)$ for all $n \in \mathbb{N}$. So $B(qx) = qB(x)$ for all $q \in \mathbb{Q}_+$. Since B is continuous and $\mathbb{Q}_+$ dense in $\mathbb{R}_+$, $\alpha x \in W$ and $B(\alpha x) = \alpha B(x)$ for all $\alpha \in \mathbb{R}_+$. □

DEFINITION 2.12. *Let Y be an ordered vector space with order cone Y_+.*

- *B is called* superadditive *if $B(x+z) \geq B(x) + B(z)$ for all $x, z \in X_B$ with $x + z \in X_B$.*
- *B is called* subadditive *if $B(x+z) \leq B(x) + B(z)$ for all $x, z \in X_B$ with $x + z \in X_B$.*

  Let X_B be a convex subset of X.
- *B is called a* convex operator *if X_B is convex and $B(\alpha x + (1-\alpha)z) \leq \alpha B(x) + (1-\alpha)B(z)$ for all $x, z \in X_B$ and $\alpha \in (0,1)$.*
- *B is called a* concave operator *if X_B is convex and $B(\alpha x + (1-\alpha)z) \geq \alpha B(x) + (1-\alpha)B(z)$ for all $x, z \in X_B$ and $\alpha \in (0,1)$.*

REMARK 2.13. If X_B is a wedge and B is homogeneous, superadditivity (subadditivity) of B is equivalent to concavity (convexity) of B.

DEFINITION 2.14. *Let X and Y be ordered normed vector spaces with order cones X_+ and Y_+. We use the same symbols $\|\cdot\|$ and $\leq$ for the norms on X and Y and for the orders induced by X_+ on X and by Y_+ on Y, respectively.*

- *B is called* order-preserving *if $B(x) \leq B(z)$ for all $x, z \in X_B$ with $x \leq z$.*
- *B is called* order-preserving on *$Z \subseteq X_B$ if $B(x) \leq B(z)$ for all $x, z \in Z$ with $x \leq z$.*

EXAMPLE 2.15. Let X and Y be as in Definition 2.14 and $B : X_B \to Y$ with $X_+ \subseteq X_B \subseteq X$. If B is superadditive and $B(X_+) \subseteq Y_+$, then B is order-preserving.

PROOF. Let $x, z \in X_B$ and $x \leq z$. By (2.2), $z - x \in X_+ \subseteq X_B$. Since B is superadditive and $B(z-x) \in Y_+$, $B(x) \leq B(x) + B(z-x) \leq B\big(x + (z-x)\big) = B(z)$. □

2.1.7. Infima, suprema, and lattices. Let X be an ordered vector space with order cone X_+ and Z be a subset of X.

A point x in X is called a *lower bound* for Z if

$$x \leq z, \qquad z \in Z. \tag{2.3}$$

A point $\tilde{x} \in X$ is called a *largest lower bound* for the set Z if $\tilde{x}$ is a lower bound for Z and $\tilde{x} \geq x$ for any lower bound x of Z. A largest lower bound $\tilde{x}$ for Z is uniquely determined and also called the *infimum* of Z,

$$\tilde{x} = \inf Z. \tag{2.4}$$

A point x in X is called an *upper bound* for Z if

$$x \geq z, \qquad z \in Z. \tag{2.5}$$

A point $\tilde{x} \in X$ is called a *least upper bound* for Z if $\tilde{x}$ is an upper bound for Z and $\tilde{x} \leq x$ for any upper bound x of Z.

A least upper bound $\tilde{x}$ for Z is uniquely determined and also called the *supremum* of Z,

$$\tilde{x} = \sup Z. \tag{2.6}$$

Recall the concept of a monotonically complete subset (Definition 2.6).

THEOREM 2.16. *Let X be an ordered normed vector space with closed order cone X_+ and let S be a monotonically complete subset of X_+. Then any increasing Cauchy sequence in S has a supremum in S.*

PROOF. Let (x_n) be an increasing Cauchy sequence in S. Since S is monotonically complete, (x_n) has a limit x in S. For any $n \geq m$ in $\mathbb{N}$, $x_n \geq x_m$. Since X_+ is closed, taking the limit as $n \to \infty$ yields $x \geq x_m$ for any $m \in \mathbb{N}$. So x is an upper bound for (x_n). Let $y \in X$ be an upper bound for (x_n). Then $y \geq x_n$ for all $n \in \mathbb{N}$. Since X_+ is closed, taking the limit as $n \to \infty$ yields $y \geq x$. So x is the least upper bound of S. □

Let S be a subset of X. S is called an inf-*semilattice* [8] (or *minihedral* [**135**]) if

$$x \wedge y = \inf\{x, y\} \tag{2.7}$$

exist and are elements of S for all $x, y \in S$.

By induction, finite subsets of an inf-semilattice S have infima that are elements of S.

S is called a sup-*semilattice* if

$$x \vee y = \sup\{x, y\} \tag{2.8}$$

exist and are elements of S for all $x, y \in S$. S is called a *lattice* if $x \wedge y$ and $x \vee y$ exist and are elements of S for all $x, y \in S$.

By induction, finite subsets of sup-semilattices S have suprema that are elements of S.

X is a *lattice* if $x \vee y$ exist for all $x, y \in X$. Since $x \wedge y = -((-x) \vee (-y))$, also $x \wedge y$ exist for all x, y in a lattice X.

Equivalently, X is a lattice if for any $x \in X$ the *absolute value*

$$|x| = x \vee (-x) = \sup\{x, -x\} \tag{2.9}$$

exists. Since $0 = x + (-x) \leq 2|x|$, $|x| \in X_+$. The following connections hold if X is a lattice:

$$x \vee y = (1/2)(x + y + |x - y|), \qquad x \wedge y = (1/2)(x + y - |x - y|). \tag{2.10}$$

EXAMPLE 2.17. Let $X = \mathbb{R}^S$, the vector space of functions $f : S \to \mathbb{R}$ where S is an arbitrary nonempty set. Choose the cone $X_+ = \mathbb{R}_+^S$ of functions $f : S \to \mathbb{R}_+$. This is an Archimedean cone. The partial ordering induced by X_+ is the pointwise ordering $f \le g$ if and only if $f(s) \le g(s)$ for all $s \in S$.

$\mathbb{R}^S$ is a lattice: If $f, g \in \mathbb{R}^S$,

$$(f \wedge g)(s) = \inf\{f(s), g(s)\}, \quad (f \vee g)(s) = \sup\{f(s), g(s)\}, \quad |f|(s) = |f(s)|.$$

An ordered normed vector space X that also is a lattice is called a *normed lattice* if

$$x, y \in X, |x| \le |y| \implies \|x\| \le \|y\|. \tag{2.11}$$

A normed lattice that is complete is called an *Banach lattice*. The order cone X_+ of a normed lattice is closed (Exercise 2.1.2).

In a normed lattice, $|\cdot| : X \to X$ and $\wedge, \vee : X^2 \to X$ are continuous.

A normed lattice X is called an *abstract M-space* if

$$\|x \vee y\| = \max\{\|x\|, \|y\|\}, \qquad x, y \in X_+. \tag{2.12}$$

It is called an *abstract L-space* if

$$\|x + y\| = \|x\| + \|y\|, \qquad x, y \in X_+. \tag{2.13}$$

For more information on lattices, see [**9, 10, 181**].

EXAMPLE 2.18. Let $X = B(S)$, the vector space of bounded functions $f : S \to \mathbb{R}$ which is a linear subspace of $\mathbb{R}^S$ and becomes a Banach space with the supremum norm $\|f\|_\infty = \sup_{s \in S} |f(s)|$. Choose the cone X_+ of nonnegative bounded functions, $X_+ = B(S) \cap \mathbb{R}_+^S = B_+(S)$ which is a closed cone and thus complete. $B(S)$ inherits the lattice structure from $\mathbb{R}^S$ and is a Banach lattice and an abstract M-space. $B_+(S)$ is a solid cone, its open interior is formed by the positive functions f that are bounded away from 0, $\inf f(S) > 0$.

We will meet an example for an abstract L-space in Chapter 3.

Exercises.

2.1.1. Let X be a vector lattice and $x, y \in X$. Show that $|x + y| \le |x| + |y|$ and $\big||x| - |y|\,\big| \le |x - y|$.

2.1.2. Let X be a normed vector lattice and $x, y \in X$. Show:

(a) $\big\||x| - |y|\,\big\| \le \|x - y\|$.
Hint: Use Exercise 2.1.1 and (2.11).

(b) The order cone X_+ is closed.

2.1.3. Let X be a normed vector space and a lattice. Show that (2.12) implies (2.11).

2.1.8. Cones generated by convex sets. Convex subsets of a real vector space generate wedges (though not every cone is generated this way). We study how properties of the convex set are inherited by the generated wedge. Recall Sections 2.1.1 and 2.1.3.

THEOREM 2.19. *Let X be a real vector space and Z be a convex subset of X. Define $W = \{\alpha x; x \in Z, \alpha \in \mathbb{R}_+\}$. The following hold:*

(a) *W is a wedge, called the* wedge generated by *the convex set Z in X.*

(b) *If Z does not contain the zero vector of X, W is a cone.*

If X is a normed vector space, the following also hold:

(c) *If $Z \not\ni 0$ is bounded and closed, then W is closed.*
(d) *If $Z \not\ni 0$ is bounded and complete in addition, then W is complete.*

The convex set Z is called a base of the wedge W that it generates [**10**, Sec.1.7] provided that $\alpha \in \mathbb{R}_+$ and $z \in Z$ are uniquely determined by $w \in W$ with $w = \alpha z$. See [**10**, Sec.1.7] and [**138**, Sec.3.3] for more information.

PROOF. (a) Obviously, W is a homogeneous set. To show that W is additive, let $w_1, w_2 \in W$. Then $w_j = \alpha_j z_j$ with $z_j \in Z$ and $\alpha_j \in \mathbb{R}_+$, $j = 1, 2$. We can assume that $\alpha_1 + \alpha_2 > 0$. Since Z is convex,

$$\frac{1}{\alpha_1 + \alpha_2}(w_1 + w_2) = \frac{\alpha_1}{\alpha_1 + \alpha_2} z_1 + \frac{\alpha_2}{\alpha_1 + \alpha_2} z_2 \in Z.$$

So $w_1 + w_2 \in W$.

(b) Assume that the zero vector of X is not an element of Z. To show that W is a cone, assume that $w, -w \in W$ and $w \neq 0$. By definition of W, $w = \alpha z$ and $-w = \beta y$ for $y, z \in Z$ and $\alpha, \beta \in \mathbb{R}_+ \setminus \{0\}$. So, $0 = \alpha z + \beta y$. Since Z is convex,

$$0 = \frac{\alpha}{\alpha + \beta} z + \frac{\beta}{\alpha + \beta} y \in Z.$$

Our claim follows by contradiction.

(c) Let us now assume that Z is bounded and closed in addition. To show that W is closed, let (w_n) be a sequence in W and $x \in X$ and $w_n \to x$. There exist sequences (α_n) in $\mathbb{R}_+$ and (z_n) in Z such that $w_n = \alpha_n z_n$. We claim that (α_n) is bounded. If not, there is a subsequence $\alpha_{n_j} \to \infty$ for $j \to \infty$. Since (w_n) is bounded, $z_{n_j} \to 0$ contradicting that the closed bounded set Z does not contain the zero vector. Since (α_n) is bounded, it contains a convergent subsequence $\alpha_{n_j} \to \alpha$ with $\alpha \in \mathbb{R}_+$. If $\alpha = 0$, $w_{n_j} \to 0$ because (z_n) is bounded, and the proof is finished. So $\alpha > 0$. Then $z_{n_j} \to (1/\alpha)x$ as $j \to \infty$. Since Z is closed, $(1/\alpha)x \in Z$ and $x \in W$.

(d) Let us now assume that Z is bounded and complete in addition. Let (w_n) be a Cauchy sequence in W. It is sufficient that (w_n) has a convergent subsequence. By definition, $w_n = \alpha_n z_n$ with $\alpha_n \in \mathbb{R}_+$ and $z_n \in Z$. We claim that (α_n) is bounded. If not, there is a subsequence $\alpha_{n_j} \to \infty$ for $j \to \infty$. Since (x_n) is bounded, $z_{n_j} \to 0$ contradicting that the complete bounded set Z does not contain the zero vector. Since (α_n) is bounded, it contains a convergent subsequence $\alpha_{n_j} \to \alpha$ with $\alpha \in \mathbb{R}_+$. If $\alpha = 0$, $x_{n_j} \to 0$ because (z_n) is bounded, and the proof is finished. So $\alpha > 0$. This implies that (z_{n_j}) is a Cauchy sequence in the complete set Z and has a limit $z \in Z$. So $w_{n_j} \to \alpha z \in W$ as $j \to \infty$. □

Exercises.

2.1.4. Consider the real vector space $\mathbb{R}^\Omega$ of real-valued function on a nonempty set Ω. Let I be the interval $[0, 1]$ and I^Ω be the set of functions $f : \Omega \to I$.

(a) Show that I^Ω is convex subset of $\mathbb{R}^\Omega$.
(b) Show that the set $B(\Omega) \cap \mathbb{R}^\Omega_+$ of bounded nonnegative functions is the wedge generated by I^Ω in $\mathbb{R}^\Omega$.

2.2. Normal and (fully) regular cones

The order cones in ordered normed vector spaces are called normal or regular respectively, if there is a particular interplay between the order induced by the cone and the topology induced by the norm.

2.2.1. Normal cones. The following result for normed vector spaces is well-known [**135**, Sec.1.2].

THEOREM 2.20. *Let X be a normed vector space with wedge X_+. Then the following three properties are equivalent:*

(i) *X_+ is* normal*: There exists some $\delta > 0$ such that $\|x+z\| \geq \delta$ whenever $x \in X_+$, $z \in X_+$ and $\|x\| = 1 = \|z\|$.*
(ii) *The norm is semi-monotonic: There exists some $M \geq 0$ such that $\|x\| \leq M\|x+z\|$ for all $x, z \in X_+$.*
(iii) *There exists some $\tilde{M} \geq 0$ such that $\|x\| \leq \tilde{M}\|y\|$ whenever $x \in X$, $y \in X_+$, and $y+x$ and $y-x$ are elements in X_+.*

REMARK 2.21. By (i), any normal wedge is a cone. We have formulated these equivalences for wedges to make clear that any of the properties (i), (ii), (iii) makes a wedge a cone.

In a normed vector lattice, the cone X_+ is normal (see (2.11)).

Further, X_+ is normal if the norm is order-preserving on X_+.

PROOF OF THEOREM 2.20. (iii) $\Rightarrow$ (ii): Suppose (iii) holds and choose $\tilde{M}$ accordingly. Let $x, z \in X_+$. Set $y = x + z$. Then $y \in X_+$ and $y - x = z \in X_+$ and $y + x \in X_+$. By (iii), $\|x\| \leq \tilde{M}\|y\| = \tilde{M}\|x+z\|$.

(ii) $\Rightarrow$ (i): Suppose (ii) holds and choose M accordingly. Let $x, z \in X_+$ and $\|x\| = 1 = \|z\|$. Then $1 = \|x\| \leq M\|x+z\|$. So $\|x+z\| \geq 1/M$ and X_+ is normal.

(i) $\Rightarrow$ (iii): Suppose that X_+ is a normal wedge, but that (iii) does not hold. Then, from negating (iii), for every $n \in \mathbb{N}$ there exist $x_n \in X$ and $y_n \in X_+$ such that $-y_n \leq x_n \leq y_n$ and $\|x_n\| > n\|y_n\|$. This implies that $x_n \neq 0$ and $y_n \neq 0$. Since X_+ is a cone,

$$g_n = \frac{x_n}{\|x_n\|} + \frac{y_n}{n\|y_n\|} = \frac{x_n + y_n}{\|x_n\|} + \Big(\frac{1}{n\|y_n\|} - \frac{1}{\|x_n\|}\Big) y_n \in X_+$$

and

$$h_n = \frac{y_n}{n\|y_n\|} - \frac{x_n}{\|x_n\|} = \Big(\frac{1}{n\|y_n\|} - \frac{1}{\|x_n\|}\Big) y_n + \frac{y_n - x_n}{n\|x_n\|} + \in X_+.$$

Further $g_n \neq 0$ and $h_n \neq 0$. Since X_+ is normal, there exists some $\delta > 0$ such that

$$\Big\|\frac{g_n}{\|g_n\|} + \frac{h_n}{\|h_n\|}\Big\| \geq \delta.$$

Notice that

$$\begin{aligned}\frac{g_n}{\|g_n\|} + \frac{h_n}{\|h_n\|} &= \frac{g_n + h_n}{\|g_n\|} + \Big(\frac{1}{\|h_n\|} - \frac{1}{\|g_n\|}\Big) h_n \\ &= \frac{2y_n}{n\|y_n\|\|g_n\|} + \Big(\frac{\|g_n\| - \|h_n\|}{\|g_n\|\|h_n\|}\Big) h_n.\end{aligned}$$

By the triangle inequality and reverse triangle inequality,

$$\begin{aligned}\Big\| \frac{g_n}{\|g_n\|} + \frac{h_n}{\|h_n\|} \Big\| &\le \frac{2}{n\|g_n\|} + \frac{\big|\|g_n\| - \| - h_n\|\big|}{\|g_n\|} \\ &\le \frac{1}{\|g_n\|}\Big(\frac{2}{n} + \|g_n + h_n\|\Big) \le \frac{1}{\|g_n\|}\frac{4}{n}.\end{aligned}$$

Since $\|g_n\| \ge 1 - \frac{1}{n}$, the right-hand side converges to 0 yielding a contradiction. □

Exercises.

2.2.1. Let X_+ be the normal order cone of an ordered normed vector space X. Show: There exists some $c > 0$ such that $\|y\| \le c(\|x\| + \|z\|)$ for all $x, y, z \in X$ with $x \le y \le z$.

2.2.2 (Sandwich theorem). Let X_+ be the normal order cone of an ordered normed vector space X. Let $x \in X$ and (x_n), (y_n), (z_n) be sequence in X with $x_n \le y_n \le z_n$ for all $n \in \mathbb{N}$ and $x_n \to x$ and $z_n \to x$ as $n \to \infty$.

Show that $y_n \to x$ as $n \to \infty$.

Hint: First consider $x = 0$.

2.2.3. Let c_0 be the space of sequences $(x_n)_{n=0}^\infty$ such that $x_n \to 0$ as $n \to \infty$, with the supremum norm. Let

$$L = \{(x_n) \in c_0; x_0 \ge \Big(\sum_{n=1}^\infty x_n^2\Big)1/2.$$

Show that L is a closed normal cone of c_0.

Hint: For normality, check property (i) of Theorem 2.20.

2.2.2. Normal points. Let X be an ordered normed vector space with order cone X_+. Let W be a homogeneous subset contained in X_+. Then $u \in X_+$ is called a *normal point* for W if $\{\|x\|; x \in W, x \le u\}$ is a bounded subset in $\mathbb{R}$. Cf. [**168**].

Recall the concept of serial completeness in Section 2.1.2.

THEOREM 2.22. *Let X be an ordered normed vector space with closed order cone X_+. Let W be a serially complete cone contained in X_+ such that each $u \in W$ is a normal point for W. Then there exists some $c > 0$ such that $\|x\| \le c\|y\|$ for all $x, y \in W$ with $x \le y$.*

PROOF. Suppose that such a $c > 0$ does not exist. Then, for any $n \in \mathbb{N}$, there exist $x_n, y_n \in W$ such that $x_n \le y_n$ and $\|x_n\| \ge 4^n \|y_n\|$. Set

$$v_n = \frac{y_n}{2^n\|y_n\|} \quad \text{and} \quad u_n = \frac{x_n}{2^n\|y_n\|}, \qquad n \in \mathbb{N}.$$

Then $u_n, v_n \in W$ and $u_n \le v_n$ and $\|v_n\| = 2^{-n}$ and $\|u_n\| \ge 2^n$ for all $n \in \mathbb{N}$. Since W is a serially complete cone, the series $v = \sum_{n=1}^\infty v_n$ converges in W, $\|v\| \le 1$. Since X_+ is a closed cone that contains W, $u_n \le v$ for all $n \in \mathbb{N}$. So $v \in W$ is not a normal point for W. □

COROLLARY 2.23. *Let X be an ordered normed vector space with closed order cone X_+.*

(a) *If X_+ is a normal cone, then every point in X_+ is a normal point for X_+.*
(b) *If X_+ is serially complete and every point in X_+ is a normal point for X_+, then X_+ is a normal cone.*

Recall Theorem 2.20 for the proof and see Example 2.43 for a cone that is not normal though all its elements are normal points.

2.2.3. A convergence principle. The following convergence principle, which has been distilled from the proof of [**8**, Thm.5.2], will be applied several times.

PROPOSITION 2.24. *Let X be an ordered normed vector space with closed order cone. Let $S \subseteq X^{\mathbb{N}}$ be a set of sequences with terms in X with the property that with $(x_n) \in S$ also all subsequences $(x_{n_j}) \in S$. Assume that every increasing (decreasing) sequence $(x_n) \in S$ has a convergent subsequence. Then every increasing (decreasing) sequence $(x_n) \in S$ converges.*

PROOF. Let $(x_n) \in S$ be increasing (the decreasing case is similar). Then there exist a subsequence (x_{n_j}) and some $x \in X$ such that $x_{n_j} \to x$. Suppose that (x_n) does not converge to x. Then there exist some $\epsilon > 0$ and a strictly increasing sequence (k_i) in $\mathbb{N}$ such that $\|x_{k_i} - x\| \geq \epsilon$ for all $i \in \mathbb{N}$. By assumption, $(x_{k_i}) \in S$; further it is an increasing sequence. So, after choosing a subsequence, $x_{k_i} \to y$ for some $y \neq x$. Fix n_j. If i is sufficiently large, $k_i \geq n_j$ and $x_{n_j} \leq x_{k_i}$. Taking the limit $i \to \infty$ yields $x_{n_j} \leq y$ because the cone is closed. Now we let $j \to \infty$ and obtain $x \leq y$. By symmetry, $y \leq x$, a contradiction. $\square$

2.2.4. Regular and fully regular cones. The order cone X_+ of an ordered normed vector space is called *regular* if any decreasing sequence in X_+ converges. X_+ is called *fully regular* if any increasing bounded sequence in X_+ converges.

The norm of X is called *additive* on X_+ if $\|x+z\| = \|x\|+\|z\|$ for all $x, z \in X_+$.

PROPOSITION 2.25. *Let X_+ be the order cone of an ordered normed vector space X. Then X_+ is regular if and only if any increasing sequence (y_n) in X_+ converges for which there is some $z \in X_+$ with $y_n \leq z$ for all $n \in \mathbb{N}$.*

PROOF. Assume that X_+ is regular and (y_n) an increasing sequence in X_+ and $z \in X_+$ such that $y_n \leq z$ for all $n \in \mathbb{N}$. Let $x_n = z - y_n$. Then (x_n) is a decreasing sequence in the regular cone X_+ and converges to some $x \in X$. Set $y = z - x \in X$. Then $y_n = z - x_n \to z - x = y$.

Now assume that any increasing sequence (y_n) in X_+ converges for which there is some $z \in X_+$ with $y_n \leq z$ for all $n \in \mathbb{N}$.

Let (x_n) be a decreasing sequence in X_+. Set $y_n = x_1 - x_n$. Then (y_n) is an increasing sequence in X_+ and $y_n \leq x_1$ for all $n \in \mathbb{N}$. By assumption, $y_n \to y$ with some $y \in X$. Then $x_n = x_1 - y_n \to x_1 - y$. $\square$

THEOREM 2.26. *Let X be an ordered normed vector space with closed order cone X_+.*

If X_+ is fully regular, then X_+ is normal.

If X_+ is regular and serially complete, then X_+ is normal.

PROOF. Suppose that X_+ is not normal. By Theorem 2.20 (i), there exist sequences (x_n) and (y_n) in X_+ with norm one such that $\|x_n + y_n\| \leq 2^{-n}$ for all $n \in \mathbb{N}$. Set

$$v_{2n} = \sum_{j=1}^{n} (x_n + y_n), \quad v_{2n+1} = v_{2n} + x_{n+1}, \quad n \in \mathbb{N}.$$

Then $\|v_{2n}\| \leq 1$, $\|v_{2n+1}\| \leq 2$ and $v_{2n+1} \leq v_{2(n+1)}$ for all $n \in \mathbb{N}$. So (v_n) is an increasing bounded sequence.

If X_+ is fully regular, (v_n) converges. But $\|v_{2n+1} - v_{2n}\| = 1$ for all $n \in \mathbb{N}$, a contradiction.

Now assume that X_+ is regular and serially complete. Then the sequence (v_{2n}) converges to some $z \in X_+$ and $v_n \leq z$ for all $n \in \mathbb{N}$. Since X_+ is regular, (v_n) converges by Proposition 2.25, yielding the same contradiction as before. □

THEOREM 2.27. *Let X be an ordered normed vector space with closed order cone X_+. If X_+ is fully regular and every point in X_+ is a normal point for X_+, then X_+ is regular.*

PROOF. Let (x_n) be a decreasing sequence in X_+. Set $y_n = x_1 - x_n$ and $y_n \in X_+$ and $y_n \leq x_1$ for all $n \in \mathbb{N}$. Since x_1 is a normal point for X_+, (y_n) is bounded. It is also an increasing sequence. Since X_+ is fully regular, $y_n \to y$ for some $y \in X_+$. Since X_+ is closed, $y \leq x_1$. Then $x_n \to x_1 - y \in X_+$. □

THEOREM 2.28. *Let X be an ordered normed vector space with cone X_+. If X_+ is serially complete with additive norm, then X_+ is fully regular.*

PROOF. Let (x_n) be an increasing sequence in X_+ such that there is some $c > 0$ such that $\|x_n\| \leq c$ for all $n \in \mathbb{N}$. Define $y_n = x_{n+1} - x_n$ for $n \in \mathbb{N}$. Then $y_n \in X_+$ and

$$\sum_{n=j}^{m} y_n = x_{m+1} - x_j, \qquad m \geq j. \tag{2.14}$$

Since the norm is additive on X_+,

$$\sum_{n=1}^{m} \|y_n\| = \Big\|\sum_{n=1}^{m} y_n\Big\| = \|x_{m+1} - x_1\| \leq 2c, \qquad m \in \mathbb{N},$$

and the bounded increasing sequence $(\sum_{n=1}^{m} \|y_n\|)_{m\in\mathbb{N}}$ converges in $\mathbb{R}_+$. Since X_+ is serially complete, the sequence $(\sum_{n=1}^{m} y_n)_{m\in\mathbb{N}}$ converges in X_+. By (2.14), (x_n) converges in X_+. □

DEFINITION 2.29. *An element $u \in X_+$ is called a* regular point *for X_+ if every increasing sequence (x_n) in X_+ with $x_n \leq u$ for all $n \in \mathbb{N}$ converges.*

Notice that X_+ is regular if all points in X_+ are regular points of X_+ (Exercise 2.2.5).

PROPOSITION 2.30. *Let X be an ordered normed vector space with cone X_+. Then every regular point of X_+ is a normal point for X_+.*

PROOF. Let u be a regular point of X_+ that is not a normal point for X_+. By Section 2.2.2, for any $n \in \mathbb{N}$, there exists some $x_n \in X_+$ with $x_n \leq u$ and $\|x_n\| \geq 4^n$. Define

$$y_m = \sum_{n=1}^{m} 2^{-n} x_n, \qquad m \in \mathbb{N}.$$

Then (y_m) is an increasing sequence in X_+ and $y_m \leq u$ for all $n \in \mathbb{N}$. Since u is a regular point in X_+, (y_m) converges and is a Cauchy sequence. Hence $y_m - y_{m-1} = 2^{-m} x_m \to 0$. But $\|2^{-m} x_m\| \geq 2^m$, a contradiction. □

Exercises.

2.2.4. Let X be an ordered normed vector space with closed order cone X_+ and $u \in X_+$. Then u a regular point of X_+ if and only if every decreasing sequence (y_n) with $y_n \in X_+$ and $y_n \leq u$ for all $n \in \mathbb{N}$ converges.

2.2.5. Let X_+ be the closed order cone of an ordered normed vector space X. Show that X_+ is a regular cone if and only if every point $x \in X_+$ is a regular point for X_+.

2.3. Generating, total and non-flat wedges

Let X be a real vector space and W a wedge in X. Then

$$W - W = \{w - v; v, w \in W\} \tag{2.15}$$

is a linear subspace of X, and the following fruitful concepts have been developed around it.

- W is called *generating* if $X = W - W$.

  Now let W be a wedge in a normed vector space X with norm $\|\cdot\|$.
- W is called *total* if $W - W$ is dense in X.
- W is called *solid* if W contains an interior point.
- W is called *non-flat* [**136**, Sec.1.8] if there exists some $c \geq 1$ such that for any $x \in W - W$ there exist $v, w \in W$ such that $x = v - w$ and $\|v\| + \|w\| \leq c\|x\|$.
- W is called *flat* if there is a bounded sequence (x_n) in $W - W$ such that, for all sequences (v_n) and (w_n) in W with $x_n = v_n - w_n$ for all $n \in \mathbb{N}$, at least one of the sequences (v_n) or (w_n) is unbounded.

Bonsall [**27**] says that $W - W$ has the strict bounded decomposition property if W is non-flat.

One easily checks the following relation between the last two concepts.

PROPOSITION 2.31. *Let W be a wedge in a normed vector space X. Then W is non-flat if and only if it is not flat.*

THEOREM 2.32. *Let W be a solid wedge in a normed vector space X. Then W is generating and non-flat.*

PROOF. Fix an interior point u of W and some $\epsilon > 0$ such that $u \pm \frac{\epsilon}{\|x\|} x \in W$ for all $x \in \dot{X}$. Then $x = v - w$ with

$$v = \frac{1}{2}\Big(\frac{\|x\|}{\epsilon} u + x\Big) \in W, \qquad w = \frac{1}{2}\Big(\frac{\|x\|}{\epsilon} u - x\Big) \in W.$$

By the triangle inequality, $\|v\| \leq \|x\| \frac{1}{2}\Big(\frac{\|u\|}{\epsilon} + 1\Big)$ and the same estimate holds for w. □

Later we will show that a generating wedge in a Banach space is also non-flat (Corollary 2.38).

THEOREM 2.33. *Let X be an ordered vector space with order cone X_+ and a vector lattice. Then X_+ is a generating cone.*

If X is a normed vector space and $\|\,|x|\,\| \leq \|x\|$ for all $x \in X$, X_+ is a non-flat cone.

PROOF. For each $x \in X$, $x = x_+ - x_-$ with $x_+ = (1/2)(|x| + x) \in X_+$ and $x_- = (1/2)(|x| - x) \in X_+$. Then

$$\|x_+\| \leq (1/2)\big(\| \, |x| \, \| + \|x\|\big) \leq \|x\|$$

and, similarly, $\|x_-\| \leq \|x\|$. □

Later, we will meet two examples for flat cones:

The cone of nonnegative increasing bounded sequences in the Banach space of converging sequences c with the supremum norm is total and flat (Section 2.4.2).

If S is a metric space that is not uniformly discrete (Definition A.3), the nonnegative measures on the Borel sets of S form a flat separating cone in the incomplete normed space of real measures with the flat norm (Corollary 3.11).

2.3.1. The span of a wedge as normed vector space. Let W be a wedge in the normed vector space X. The following construction is well-known from the literature [**18, 167, 179**]. We will put it to new use [**203**, Sec.7.5], in particular if the wedge is serially complete (Section 2.1.2) but not complete. An important example is the cone of nonnegative measures in the space of real measures on a non-complete metric space if the space of measures is endowed with the flat norm (Section 3.2.4).

Let $\tilde X$ be the vector space $W - W$ in (2.15) and define a seminorm on $\tilde X$ by

$$\|x\|^\sim = \inf \big\{ \|v\| + \|w\|; x = v - w, v, w \in W \big\}. \tag{2.16}$$

We have

$$\|x\| \leq \|x\|^\sim, \quad x \in \tilde X, \qquad \|x\| = \|x\|^\sim, \quad x \in W. \tag{2.17}$$

The inequality follows from $\|v - w\| \leq \|v\| + \|w\|$. It shows that $\| \cdot \|^\sim$ is a norm rather than a seminorm. For $x \in W$, $x = x - 0$ and so $\|x\| = \|x\|^\sim$.

REMARK 2.34. The inequality in (2.17) shows that W is a generating non-flat wedge under $\| \cdot \|^\sim$. Further, if W is closed under $\| \cdot \|$, then it is closed under $\| \cdot \|^\sim$.

The equality in (2.17) shows that W is normal under $\| \cdot \|^\sim$ if it is normal under $\| \cdot \|$.

Recall the concept of serial completeness introduced in Section 2.1.2.

THEOREM 2.35.

(a) *If W is serially complete under $\| \cdot \|$, $\tilde X$ is a Banach space under $\| \cdot \|^\sim$.*
(b) *$\| \cdot \|^\sim$ is equivalent to $\| \cdot \|$ on $\tilde X$ if and only if W is a non-flat wedge of X under $\| \cdot \|$.*
(c) *If $\tilde X$ is complete under $\| \cdot \|$, W is a non-flat wedge.*
(d) *If W is a serially complete non-flat wedge, then $\tilde X$ is complete under the original norm $\| \cdot \|$.*
(e) *If W is complete under $\| \cdot \|$, then W is closed in the Banach space $(\tilde X, \| \cdot \|^\sim)$.*

PROOF. (a) We use Theorem 2.1 and show that $\tilde X$ is serially complete under $\| \cdot \|^\sim$.

Let (x_k) be a sequence in $\tilde X$ such that $\sum_{n=1}^\infty \|x_n\|^\sim$ converges in $\mathbb{R}$. For any $n \in \mathbb{N}$, there exist $v_n, w_n \in W$ such that $x_n = v_n - w_n$ and $\|v_n\| + \|w_n\| < \|x_n\|^\sim + 2^{-n}$. Since W is serially complete, $v = \sum_{n=1}^\infty v_n$ and $w = \sum_{n=1}^\infty w_n$ converge in W with respect to $\|\cdot\|$. Since $\|\cdot\|$ and $\|\cdot\|^{\sim}$ coincide on W, these series also converge with respect to $\|\cdot\|^{\sim}$. So $\sum_{n=1}^{\infty} x_n = \sum_{n=1}^{\infty}(v_n - w_n)$ converges with respect to $\|\cdot\|^{\sim}$.

(b) This statement follows directly from the definition of the norm $\|\cdot\|^{\sim}$ and the concept of a non-flat wedge.

(c) By (2.17), the identity is continuous from $(\tilde{X}, \|\cdot\|^{\sim})$ to $(\tilde{X}, \|\cdot\|)$. Assume that $(\tilde{X}, \|\cdot\|)$ is complete. By part (a), W is serially complete under $\|\cdot\|$ and $\tilde{X}$ is a Banach space under $\|\cdot\|^{\sim}$. Then the identity is continuous from $(\tilde{X}, \|\cdot\|)$ to $(\tilde{X}, \|\cdot\|^{\sim})$ by the open mapping theorem [**223**, II.5], and the two norms are equivalent which implies that W is non-flat by part (b).

(d) This follows from combining (a) and (b). □

REMARK 2.36. Theorem 2.35 (a) also follows from the proof of [**18**, L.2.2]. Though the statement of [**18**, L.2.2] assumes that the normed vector space X is complete, the proof only uses that the wedge W is serially complete.

EXAMPLE 2.37. Let c be the Banach space of converging real sequences with supremum norm and W be the closed cone of nonnegative increasing bounded sequences. Then $W - W = bc$ and the norm on bc given by (2.16) is the variation norm. See Exercise 2.4.4.

The next result, which is well-known, follows from Theorem 2.35, part (c).

COROLLARY 2.38 ([**136**, Thm.1.5]). *Let X be a Banach space and W a generating wedge in X. Then W is non-flat. In particular, there exists some $c > 0$ such that for all $x \in X$ there exist $v, w \in W$ such that $x = v - w$ and $\|v\| + \|w\| \le c\|x\|$.*

Exercises.

2.3.1. Prove Proposition 2.31.

2.3.2. Prove Remark 2.34.

2.3.3. Prove Theorem 2.35 (b) and (e).

2.3.4. Prove Corollary 2.38.

2.3.2. A normed vector space with many cones. Let $X = C_0(0,1]$, the Banach space of continuous functions $f : [0,1] \to \mathbb{R}$ with $f(0) = 0$, endowed with the supremum norm. Let X_+ be the cone of continuous functions $f : [0,1] \to \mathbb{R}_+$ with $f(0) = 0$, a closed cone. Notice that $C_0(0,1]$ is a linear subspace of the ordered vector space $\mathbb{R}^{[0,1]}$.

$C_0(0,1]$ also inherits the lattice structure from $\mathbb{R}^{[0,1]}$ and is a Banach lattice (Section 2.1.7).

For $\alpha \ge 0$, let X_α be the cone of $f \in X_+$ such that $x^{-\alpha} f(x)$ is an increasing function of $x \in (0,1]$. X_0 is the cone of increasing function. We have $X_+ \supseteq X_0 \supseteq X_\alpha \supseteq X_\beta$ for all $\beta > \alpha > 0$.

Other cones are formed by the $f \in X_+$ that are convex on $[0,1]$ or by the f that are concave on $[0,1]$. Notice that any convex function in $C_{0+}(0,1]$ is increasing.

All these cones are closed and consequently complete. Since X_+ is normal, all these cones are normal.

A cone that is not closed is the cone X_{pol} of polynomials $f(s) = \sum_{j=1}^{n} \alpha_j s^j$ with coefficients $\alpha_j \ge 0$, $X_{pol} \subseteq X_{con}$.

The cone $X_+ = C_{0+}(0,1]$ of nonnegative functions is generating and non-flat, but not solid. The cone X_0 of nonnegative increasing functions is not generating because the difference of two such functions is of bounded variation. X_0 is a flat cone: Let $f_n(s) = \sin(2n\pi s)$, $n \in \mathbb{N}$. Then, for any $g_n, h_n \in X_0$ with $f_n = g_n - h_n$, $\|g_n\|_\infty \to \infty$ and $\|h_n\|_\infty \to \infty$ as $n \to \infty$.

The cone X_{pol} is total. Recall that any $f \in C[0,1]$ can be uniformly approximated by Bernstein polynomials [**223**, 0.2]

$$p_n(s) = \sum_{j=0}^{n} \binom{n}{j} f\Big(\frac{j}{n}\Big) s^j (1-s)^{n-j}.$$

Notice that $p_n(0) = 0$ if $f \in C_0(0,1]$. Each such polynomial can be written as the difference of polynomials in standard form with nonnegative coefficients and zero-value at 0.

Since, for all $\alpha \in [0,1]$, $X_0 \supseteq X_\alpha \supseteq X_1 \supseteq X_{con} \supseteq X_{pol}$, X_α are not generating but total.

The cone $C_{0+}(0,1]$ is neither regular nor fully regular as shown by the sequence (f_n) with $f_n(t) = t^n$.

If we consider X_0 as order cone of $C_0(0,1]$, then it is both regular and fully regular. Since X_0 is normal, we only need to show that X_0 is fully regular (Theorem 2.27). Let (f_n) be a sequence in X_0 that is increasing with respect to the order induced by X_0 and $(\|f_n\|)$ is bounded. Then $(f_n(1))$ is an increasing bounded sequence in $\mathbb{R}_+$ and converges. Let $n \geq m$. Then $f_n - f_m$ is an increasing nonnegative function and, for all $t \in [0,1]$,

$$0 = f_n(0) - f_m(0) \leq f_n(t) - f_m(t) \leq f_n(1) - f_m(1).$$

This shows that (f_n) is a Cauchy sequence with respect to the supremum norm and converges in $C_{0+}(0,1]$. The limit is also in X_0. Since $X_\alpha \subseteq X_0$, also X_α is regular and fully regular if considered as order cone in $C_{0+}(0,1]$.

2.4. The space of sequences of bounded variation: a choice of cones

Since we have introduced the concept of a normal point (Section 2.2.2) in addition to the concept of a normal cone (Section 2.2.1), it may be appropriate to present examples of cones that have many normal points but are not normal themselves. This section is similar to [**199**, Sec.2.2] but (with some clarifications) has been adapted to the format of this monograph.

Well-known are the Banach sequence spaces ℓ^∞, c, c_0 of bounded sequences, converging sequences and sequences converging to 0 with the supremum norm and the space ℓ^1 of summable sequences with the sum-norm [**73**, IV.2.3]. The cones of c_0 and ℓ^1, formed by the respective nonnegative sequences, are regular (and thus normal), and the cones of ℓ^∞ and c are solid and normal.

The following example for an ordered Banach space whose cone is not normal follows a suggestion by Wolfgang Arendt. A sequence (x_n) in $\mathbb{R}^\mathbb{N}$ is called of *bounded variation* if the following series converges

$$|x_1| + \sum_{n=1}^{\infty} |x_{n+1} - x_n| =: \|(x_n)\|_{bv}. \tag{2.18}$$

The sequences of bounded variation form a vector space, bv, over $\mathbb{R}$ with $\|\cdot\|_{bv}$ being a norm called the *variation-norm* [**73**, IV.2.8]. Notice

$$\ell^1 \subseteq bv \subseteq c, \qquad \begin{cases} \|x\|_\infty \leq \|x\|_{bv}, & x \in bv, \\ \|x\|_{bv} \leq 2\|x\|_1, & x \in \ell_1. \end{cases} \tag{2.19}$$

We mention that these sequence spaces are employed as state spaces for rank-structured two-sex populations in [**130**, Sec.9], [**199**, Sec.14], [**207**, Sec.6.2].

LEMMA 2.39. *The space bv with the variation-norm is a Banach space.*

This follows from (2.19) and the fact that ℓ^∞ is complete under the sup-norm. Other properties of bv are listed in the Table IV.A, p.375, in [**73**, IV.15]. Also see [**73**, IV.13.11].

The space bv contains all constant sequences and $\|(x_n)\|_{bv} = |x_1| = \|(x_n)\|_\infty$ if (x_n) is a constant sequence. More generally, all monotone bounded nonnegative sequences are of bounded variation.

LEMMA 2.40. *If (x_n) is a nonnegative increasing bounded sequence, then (x_n) is of bounded variation and $\|(x_n)\|_{bv} = \|(x_n)\|_\infty = \lim_{n\to\infty} x_n$.*

If (x_n) is a nonnegative decreasing sequence, then (x_n) is of bounded variation and $\|(x_n)\|_{bv} = 2x_1 - \lim_{n\to\infty} x_n$.

There are several cones we can consider in bv. The one we are going to consider first is the cone of nonnegative sequences of bounded variation, bv_+. Others are the cone of nonnegative increasing bounded sequences and the cone of nonnegative decreasing sequences.

The cone of nonnegative sequences of bounded variation is not normal.

PROPOSITION 2.41. *bv_+ is generating, solid, but not normal (and not regular and not fully regular). X is a lattice and the lattice operations are continuous: If $x = (x_j), y = (y_j) \in bv$, then $|x| = (|x_j|) \in bv$ and $\big\| |x| \big\|_{bv} \le \|x\|_{bv}$ with strict inequality being possible. However, $\big\| |x| - |y| \big\|_{bv} > \|x - y\|_{bv}$ is possible.*

PROOF. The continuity of the lattice operations can be seen from the following inequality: For any $m \in \mathbb{N}$ and any $x = (x_n)$ and $y = (y_n)$ in bv,

$$\begin{aligned}\big\| |x| - |y| \big\|_{bv} &= \big| |x_1| - |y_1| \big| + \sum_{j=1}^{\infty} \big| |x_{j+1}| - |y_{j+1}| - |x_j| + |y_j| \big| \\ &\le \big||x_1| - |y_1|\big| + \sum_{j=1}^{m} \big||x_{j+1}| - |y_{j+1}|\big| + \sum_{j=1}^{m} \big||x_j| - |y_j|\big| \\ &\quad + \sum_{j=m+1}^{\infty} \big| |x_{j+1}| - |x_j| \big| + \sum_{j=m+1}^{\infty} \big| |y_{j+1}| - |y_j| \big|.\end{aligned}$$

By the inverse triangle inequality,

$$\big\| |x| - |y| \big\|_{bv} \le (2m+1)\|x - y\|_\infty + \sum_{j=m+1}^{\infty} \big|x_{j+1} - x_j\big| + \sum_{j=m+1}^{\infty} \big|y_{j+1} - y_j\big|.$$

By the triangle inequality,

$$\begin{aligned}\big\||x| - |y|\big\|_{bv} &\le (2m+1)\|x - y\|_\infty + 2\sum_{j=m+1}^{\infty} \big|x_{j+1} - x_j\big| + 2\|x - y\|_{bv} \\ &\le (2m+3)\|x - y\|_{bv} + 2\sum_{j=m+1}^{\infty} \big|x_{j+1} - x_j\big|.\end{aligned}$$

This implies that $|\cdot|$ is continuous at x.

As an example for $\big\||x|-|y|\big\|_{bv} > \|x-y\|_{bv}$ choose $x_1 = 0$, $2 = x_2 = -y_1$ and $y_2 = 1$, $x_n = x_2$ and $y_n = y_2$ for $n \geq 3$. Then $x - y = (2, 1, 1, \cdots)$ and

$$\|x - y\|_{bv} = 2 + |1-2| = 3, \qquad \big\||x| - |y|\big\|_{bv} = 2 + |2 - 1 + 2 = 5. \qquad \square$$

Every monotone nonnegative bounded sequence that is bounded away from zero is in the interior of bv_+ as can be seen from (2.19). The space of Lipschitz continuous functions with Lipschitz norm is an example of an ordered normed vector space and lattice where the cone of nonnegative functions contains no nonzero normal points and the lattice operations are not continuous [**156**, p.535]. See Section 2.7.2.

That bv_+ is not normal follows from the following result that characterizes the normal points. Since bv_+ is complete, bv_+ is then also not regular and not fully regular (Theorem 2.26).

THEOREM 2.42. *The following are equivalent for $x = (x_n) \in bv_+$:*

(i) *x is a normal point for bv_+,*
(ii) *$x \in \ell_1$,*
(iii) *x is a regular point for bv_+.*

PROOF. (i) $\Rightarrow$ (ii): Assume that $x = (x_n) \in bv_+$ is a normal point for bv_+. Let $m \in \mathbb{N}$ and (y^m) be the sequence given as $y_n^m = 0$ if $n > m$, $y_n^m = 0$ if n is even and $y_n^m = x_n$ if n is odd and $n \leq m$.

Further let (z^m) be the sequence given as $z_n^m = 0$ if $n > m$, $y_n^m = 0$ if n is odd and $y_n^m = x_n$ if n is even and $n \leq m$.

Then $(y^m), (z^m) \in bv_+$ and $y^m, z^m \leq x$. Since x is a normal point for bv_+ (Section 2.2.2), there exists some$c > 0$ such that $\|y^m\|_{bv} \leq c$ and $\|z^m\|_{bv} \leq c$ for all $m \in \mathbb{N}$. This implies that $\sum_{n=1}^m |x_n| \leq 2c$ for all $m \in \mathbb{N}$ and so $x \in \ell^1$, $\sum_{n=1}^\infty |x_n| \leq 2c$.

(ii) $\Rightarrow$ (iii): Let $x = (x_j) \in \ell^1$ and (x^n) be an increasing sequence in ℓ^1_+ with $x^n \leq x$, $x^n = (x_j^n)_{j\in\mathbb{N}}$. Then $x_j^n \leq x_j$ for all $j, n \in \mathbb{N}$ and all sequences $(x_j^n)_{n\in\mathbb{N}}$ are increasing. So there exists some $y = (y_j)$ in ℓ^1_+ such that, for each $j \in \mathbb{N}$, $x_j^n \to y_j$ as $n \to \infty$, $y \leq x$. By the dominated convergence theorem, $x^n \to y$ in ℓ^1 and so also in bv. Recall (2.19). This shows that x is a regular point for X_+ (Definition 2.29).

(iii) $\Rightarrow$ (i): In general, each regular point of a cone X_+ is a normal point for X_+ (Proposition 2.30). $\square$

EXAMPLE 2.43. The cone ℓ^1_+ in ℓ^1 with the variation norm is not normal though all points in ℓ^1_+ are normal elements of bc. It is regular, but not fully regular. It is generating and non-flat, though neither it is solid nor is the normed vector space complete.

PROOF. Let x^m be the sequence where $x_j^m = 0$ for $j > 2m$, $x_j^m = 0$ for all odd indices and $x_j^m = 1$ otherwise. Let u^m be the sequence where $u_j^m = 0$ for $j > 2m$ and $u_j^m = 1$ otherwise. Then $\|u^m\|_{bv} = 2$ and $\|x^m\|_{bv} = 2m$. For all $m \in \mathbb{N}$, $x^m \leq u^m$ but $\|x^m\|_{bv} = m\|u^m\|_{bv}$. So ℓ^1_+ is not normal under the variation norm. Theorem 2.26 implies that ℓ^1_+ is not fully regular under the variation norm.

To show that ℓ^1_+ is regular under the variation norm, let (x_n) be a decreasing sequence in ℓ^1_+. Since ℓ^1_+ is regular under the sum-norm, there exists some $x \in \ell^1_+$ with $\|x_n - x\|_1 \to 0$. By (2.19), $\|x_n - x\|_{bv} \to 0$. $\square$

Corollary 2.44. *bv_+ under the variation norm is complete and not normal, but contains the dense set ℓ^1_+ of normal points.*

2.4.1. The cone of nonnegative increasing bounded sequences makes bv a Banach lattice under the variation norm. Let W denote the cone of nonnegative increasing bounded sequences. W is a generating cone for bv. In fact, W makes bv a lattice (Section 2.1.7). If $x = (x_n) \in bv$, its absolute value $|x| = \sup_W\{x, -x\}$ is given by

$$|x|_1 = |x_1|, \quad |x|_n = |x_1| + \sum_{j=1}^{n-1} |x_{j+1} - x_j|, \quad n \in \mathbb{N}, \quad n \geq 2. \tag{2.20}$$

Here $\sup_W$ is to signal that the supremum is taken with respect to the order induced by W.

If bv is endowed with the bounded variation norm which makes bv a Banach space, then W makes bv a Banach lattice,

$$\big\| |x| \big\|_{bv} = \|x\|_{bv}, \qquad x \in bv, \tag{2.21}$$

as can be seen from (2.20) and (2.18). In particular, W is a normal closed non-flat cone under the variation norm.

2.4.2. The cone of nonnegative increasing bounded sequences is a flat cone for bv or c with the supremum norm. The space bv is not complete but rather dense in the space c under the supremum norm. However, the cone W of nonnegative bounded increasing sequences is complete under the supremum norm.

To show that W is flat in bc or c with the supremum norm, consider the sequences

$$x_\alpha = \big((-1)^n \alpha^n\big)_{n\in\mathbb{N}}, \qquad \alpha \in (0,1), \tag{2.22}$$

as elements in bc and in c. The supremum norms of these sequences are smaller than 1, $\|x_\alpha\|_\infty = \alpha \leq 1$.

Suppose that W is non-flat under the supremum norm (Section 2.3). By definition, there exists some $M > 0$ and $y_\alpha \in W$ and $z_\alpha \in W$ such that $x_\alpha = y_\alpha - z_\alpha$ and $\|y_\alpha\|_\infty \leq M\|x_\alpha\|_\infty \leq M$ and $\|z_\alpha\|_\infty \leq M\|x_\alpha\|_\infty \leq M$.

Let $\leq_W$ be the order induced by the cone W. Then $x_\alpha \leq_W y_\alpha$ and $-x_\alpha \leq_W z_\alpha$ which implies $|x_\alpha| \leq_W y_\alpha + z_\alpha$ where x_α is given by (2.20). Since $\|v\|_\infty \leq \|v+w\|_\infty$ for all $v, w \in W$,

$$2M \geq \|y_\alpha + z_\alpha\|_\infty \geq \big\| |x_\alpha| \big\|_\infty.$$

By (2.20) and (2.22),

$$\begin{aligned} 2M &\geq 1 + \sum_{j=1}^{\infty} |(-1)^{j+1}\alpha^{j+1} - (-1)^j \alpha^j| \\ &= 1 + \sum_{j=1}^{n} \alpha^j (1+\alpha) = 1 + \frac{\alpha(1+\alpha)}{1-\alpha} = \frac{1+\alpha^2}{1-\alpha} \xrightarrow{\alpha\to 1} \infty, \end{aligned}$$

a contradiction because this inequality holds for all $\alpha \in (0,1)$.

Exercises.

2.4.1. Show that $bv \subseteq c$, but $c_0 \not\subseteq bv$. Actually, there is a nonnegative sequence that converges to 0 but is not of bounded variation, $x_n = 0$ if n is odd and $x_n = 2/n$ if n is even.

2.4.2. Consider the map $T : \mathbb{R}^N \to \mathbb{R}^N$ defined by

$$(Tx)_n = \left\{ \begin{array}{ll} 0, & n \text{ odd} \\ x_{n/2}, & n \text{ even} \end{array} \right\} \quad x = (x_n) \in \mathbb{R}^{\mathbb{N}},$$

i.e., $T(x_n) = (0, x_1, 0, x_2, 0, x_3, \cdots)$. Obviously, T is linear and injective and $T(\mathbb{R}^{\mathbb{N}}_+) \subseteq \mathbb{R}^{\mathbb{N}}_+$.

(a) Show that $Tx \in bv$ if and only if $x \in \ell^1$ and $\|Tx\|_{bv} = \sum_{n=1}^\infty |x_n| = \|x\|_1$ for $x = (x_n) \in \ell_1$.

(b) $T(c_c) \subseteq c_0$ and $\|T(x)\|_\infty = \|x\|_\infty = \sup_{n\in\mathbb{N}} |x_n|$ for all $x = (x_n) \in \ell_1$ and $\lim_{n\to\infty} (Tx)_n = \lim_{n\to\infty} x_n = 0$ for $x = (x_n) \in c_0$.

(c) $T(c_c \setminus \ell^1) \subseteq c_0 \setminus bv$.

2.4.3. Show Lemma 2.39.

2.4.4. Let c be the Banach space of converging real sequences with supremum norm and W be the cone of nonnegative increasing bounded sequences.

Show:

(a) W is closed under the supremum norm and thus complete.

(b) $W - W = bc$ and the norm on bc given by (2.16) is the variation norm. See Example 2.3.3.

Hint: For $x = (x_n) \in bc$ define $|x| = (|x|_n)$ with $|x|_1 = |x_1|$ and $|x|_n = |x_1| + \sum_{j=2}^n |x_j - x_{j-1}|$. Notice that $x = x_+ - x_-$ with $x_+ = (1/2)(|x| + x)$ and $x_- = (1/2)(|x| - x)$ and $x_+, x_- \in W$.

2.5. Monotone half-norm and companion norm

Every ordered normed vector space X also carries an order-preserving half-norm [cf. [**17**], [**28**, L.4.1], [**136**, (4.2)], [**199**, Sec.4]] which we call the *(monotone) companion half-norm* (called the *canonical half-norm* in [**17**]). This half-norm is particularly useful if the order cone X_+ is closed which we assume in this section.

PROPOSITION 2.45. *The (monotone) companion half-norm $\psi : X \to \mathbb{R}_+$ is defined by*

$$\psi(x) = \inf\{\|x + z\|; z \in X_+\} = d(x, -X_+) \tag{2.23}$$

$$= \inf\{\|y\|; y \in X, x \leq y\}, \qquad x \in X. \tag{2.24}$$

The following hold:

(a) *ψ is homogeneous and order-preserving on X: For all $x, y \in X$ with $x \leq y$ and for all $\alpha \geq 0$, $\psi(\alpha x) = \alpha\psi(x)$ and $\psi(x) \leq \psi(y)$.*

(b) *ψ is subadditive on X ($\psi(x + y) \leq \psi(x) + \psi(y)$, $x, y \in X$),*

$$|\psi(x) - \psi(y)| \leq \|x - y\|, \qquad x, y \in X.$$

(c) *For $x \in X$, $\psi(x) = 0$ if and only if $x \in -X_+$. In particular ψ is strictly positive: $\psi(x) > 0$ for all $x \in \dot{X}_+$.*

(d) *X_+ is normal if and only if there exists some $\delta > 0$ such that $\delta\|x\| \le \psi(x)$ for all $x \in X_+$.*

(e) *If the original norm $\|\cdot\|$ is order-preserving on X_+, then $\|x\| = \psi(x)$ for all $x \in X_+$.*

Here $d(x, -X_+)$ denotes the distance of x from $-X_+$.

Proof. The functional ψ inherits positive homogeneity from the norm and is order-preserving by (2.24). For all $x \in X$, $x \le x$ and so $\|x\| \ge \psi(x)$.

Most of the other properties follow from (2.23) and the assumption that X_+ is a cone. See Exercises 2.5.2 and 2.5.4. Since ψ is subadditive,

$$|\psi(x) - \psi(y)| \le \psi(x-y) \le \|x-y\|.$$

If $-x \in X_+$, then $x \le 0$ and $\psi(x) \le \|0\| = 0$.

In turn, let $x \in X$ and $\psi(x) = 0$. By (2.24), there exists a sequence (y_n) in X with $\|y_n\| \to 0$ and $y_n \ge x$ for all $n \in \mathbb{N}$. Then $y_n - x \in X_+$. Since X_+ is closed, $-x = \lim_{n\to\infty}(y_n - x) \in X_+$.

The strict positivity of ψ follows from $X_+ \cap (-X_+) = \{0\}$.

(d) Assume that X_+ is normal. By Theorem 2.20 (ii), there exists some $c > 0$ such that $\|x\| \le c\|y\|$ whenever $x, y \in X_+$ and $x \le y$. By (2.24), $\|x\| \le c\psi(x)$ for all $x \in X_+$. Set $\delta = 1/c$.

The converse follows from ψ being order-preserving and $\psi(x) \le \|x\|$ for all $x \in X_+$. □

X_+ is closed under the monotone half-norm.

Proposition 2.46. *Let (x_n) be a sequence in X_+ and $x \in X$ and $\psi(x_n - x) \to 0$ as $n \to \infty$. Then $x \in X_+$.*

Proof. By (2.23), there exists a sequence (y_n) in X_+ such that $\|x_n + y_n - x\| \to 0$. Since $x_n + y_n \in X_+$ and X_+ is closed under $\|\cdot\|$, $x \in X_+$. □

Theorem 2.47. *Let (x_n) and (y_n) be sequences in X such that $x_n \le y_n$ for all $n \in \mathbb{N}$. Let $x, y \in X$ and $\psi(y_n - y) \to 0$ and $\psi(x - x_n) \to 0$. Then $x \le y$.*

Proof. Set $z_n = y_n - x_n$ and $z = y - x$. Then $z_n \in X_+$. Since ψ is subadditive on X by Proposition 2.45 (b),

$$\psi(z_n - z) = \psi(y_n - y + x - x_n) \le \psi(y_n - y) + \psi(x - x_n) \to 0.$$

By Proposition 2.46, $z \in X_+$ and $y \ge x$. □

The functional ψ induces a monotone norm on X which is equivalent to the original norm if and only if X_+ is a normal cone.

Theorem 2.48 (cf. [**136**, Thm.4.4]). *With ψ from Proposition 2.45, define*

$$\sharp x \sharp = \max\{\psi(x), \psi(-x)\}, \qquad x \in X.$$

Then $\sharp \cdot \sharp$ is a norm on X with the following properties.

- *$\sharp x \sharp \le \|x\|$ for all $x \in X$, $\sharp x \sharp = \psi(x)$ if $x \in X_+$, and $\sharp x \sharp = \psi(-x)$ if $-x \in X_+$.*
- *$\sharp \cdot \sharp$ is order-preserving on X_+: $\sharp x \sharp \le \sharp y \sharp$ for all $x, y \in X_+$ with $x \le y$. Moreover, for all $x, y, z \in X$ with $x \le y \le z$,*

$$\sharp y \sharp \le \max\{\sharp x \sharp, \sharp z \sharp\}. \tag{2.25}$$

- *X_+ is closed under $\sharp \cdot \sharp$.*
- *X_+ is a normal cone if and only if the norm $\sharp \cdot \sharp$ is equivalent to the original norm.*
- *If the original norm $\| \cdot \|$ is order-preserving on X_+, then $\|x\| = \sharp x \sharp$ for all $x \in X_+ \cup (-X_+)$.*

PROOF. The properties of ψ imply that $\sharp \alpha x \sharp = |\alpha| \, \sharp x \sharp$ for all $\alpha \in \mathbb{R}$, $x \in X$, and that $\sharp \cdot \sharp$ is subadditive (Exercise 2.5.1).

By Proposition 2.45, $\psi(x) \le \|x\|$ and $\psi(-x) \le \| - x\| = \|x\|$ and so $\sharp x \sharp \le \|x\|$.

To prove (2.25), let $x, y, z \in X$ and $x \le y \le z$. Then $y \le z$ and $-y \le -x$. Since ψ is order-preserving on X,

$$\psi(y) \le \psi(z) \le \sharp z \sharp, \qquad \psi(-y) \le \psi(-x) \le \sharp x \sharp.$$

(2.25) now follows from the definition of $\sharp \cdot \sharp$ as do most of the remaining assertions.

The closedness of X_+ under $\sharp \cdot \sharp$ follows from Proposition 2.46.

That $\| \cdot \|$ and $\sharp \cdot \sharp$ are equivalent norms if the cone is normal is shown in [**136**, Thm.4.4]. See Lemma 2.49. The converse follows from Theorem 2.20 (ii). □

LEMMA 2.49. *Let X_+ be normal and $c > 0$ such that $\|z\| \le c\|y\|$ for all $z, y \in X_+$ with $y - z \in X_+$. Then $\|x\| \le (2c+1)\sharp x \sharp$ for all $x \in X$.*

PROOF. Let $u, x, v \in X$ and $u \le x \le v$. Then $x - u \in X_+$ and $v - x \in X_+$ and $v - u \ge x - u$. By the triangle inequality and the normality of the cone,

$$\|x\| \le \|x - u\| + \|u\| \le c\|v - u\| + \| - u\| \le c\|v\| + (c+1)\| - u\|.$$

We keep u fixed for a moment, $u \le x$. Then

$$\|x\| - (c+1)\| - u\| \le c\|v\|, \qquad v \ge x.$$

By (2.24),

$$\|x\| - (c+1)\| - u\| \le c\psi(x).$$

Hence

$$\|x\| - c\psi(x) \le (c+1)\| - u\|, \qquad u \le x.$$

This is equivalent to

$$\|x\| - c\psi(x) \le (c+1)\|w\|, \qquad w \ge -x.$$

Again by (2.24),

$$\|x\| - c\psi(x) \le (c+1)\psi(-x).$$

By definition of $\sharp \cdot \sharp$,

$$\|x\| - c\sharp x \sharp \le (c+1)\sharp x \sharp,$$

and $\|x\| \le (2c+1)\sharp x \sharp$. □

DEFINITION 2.50. *The norm $\sharp \cdot \sharp$ is called the (monotone) companion norm on the ordered normed vector space X.*

REMARK 2.51. Let X be an ordered Banach space and let the order cone X_+ be not normal. Then X is not complete under the companion norm.

PROOF. Suppose that X is also complete under the companion norm. Since the identity map is continuous from X with the original norm to X with the companion norm by Theorem 2.48, the norms are equivalent by the open mapping theorem [**223**, II.5], and X_+ is normal by Theorem 2.48. □

COROLLARY 2.52. *Let X_+ be a normal cone. Then there exists some $c \geq 0$ such that $\|y\| \leq c \max\{\|x\|, \|z\|\}$ for all $x, y, z \in X$ with $x \leq y \leq z$.*

PROOF. Let $\sharp \cdot \sharp$ be the monotone companion norm from Theorem 2.48, which is equivalent to the original norm because X_+ is normal. Choose $c \geq 0$ such that $\sharp x \sharp \leq \|x\| \leq c \sharp x \sharp$ for all $x \in X$. Let $x \leq y \leq z$. By Theorem 2.48,

$$\|y\| \leq c\sharp y\sharp \leq c \max\{\sharp x\sharp, \sharp z\sharp\} \leq c \max\{\|x\|, \|z\|\}. \qquad \square$$

COROLLARY 2.53 (Squeezing theorem [**136**, Thm.4.3]). *Let X_+ be a normal cone. Let $y \in X$ and (x_n), (y_n), (z_n) be sequences in X with $x_n \leq y_n \leq z_n$ for all $n \in \mathbb{N}$ and $x_n \to y$ and $z_n \to y$. Then $y_n \to y$.*

PROOF. Notice that $x_n - y \leq y_n - y \leq z_n - y$. By Corollary 2.52, with some $c \geq 0$ that does not depend on n,

$$\|y_n - y\| \leq c \max\{\|z_n - y\|, \|x_n - y\|\} \to 0. \qquad \square$$

EXAMPLE 2.54. (a) Let $X = \mathbb{R}$ with the absolute value. Then the monotone companion norm is also the absolute value and the companion half-norm is the positive part, $\psi(x) = \max\{x, 0\}$ for all $x \in \mathbb{R}$.

(b) Let $X = \mathbb{R}^2$ with the maximum norm and $X_+ = \mathbb{R}^2_+$. Then the monotone companion norm is also the maximum norm.

(c) Let $X = \mathbb{R}^2$ with $\|\cdot\|$ being either the Euclidean norm or the sum norm and $X_+ = \mathbb{R}^2_+$. Then $\sharp x\sharp = \|x\|$ if $x \in X_+ \cup (-X_+)$ while $\sharp x \sharp$ is the maximum norm of x otherwise.

EXAMPLE 2.55. Let X be a normed vector lattice (Section 2.1.7, [**181**, II.5]). Since $x \leq x^+$, $\psi(x) \leq \|x^+\|$. Let $x \leq y$. Then $x^+ \leq y^+$, and $\|x^+\| \leq \|y^+\| \leq \|y\|$. Hence $\|x^+\| \leq \psi(x)$. In combination,

$$\psi(x) = \|x^+\|, \qquad x \in X.$$

Further $\psi(-x) = \|(-x)^+\| = \|x^-\|$. So

$$\sharp x\sharp = \max\{\|x^+\|, \|x^-\|\}, \qquad x \in X,$$

and $\sharp x\sharp = \|x\|$ for all $x \in X_+ \cup (-X_+)$. If X is an abstract M-space (Section 2.1.7, [**181**, II.7]),

$$\sharp x\sharp = \|x^+ \vee x^-\| = \||x|\| = \|x\|.$$

We turn to ordered Banach spaces the cone of which may be not normal.

EXAMPLE 2.56. Let $X \subseteq \tilde{X}$ where $X, \tilde{X}$ are ordered normed vector spaces with norms $\|\cdot\|$ and $\|\cdot\|^\sim$ and $\|x\| \geq \|x\|^\sim$ for all $x \in X$. Let $\psi, \tilde{\psi}$ be the respective monotone companion half-norms. Then $\psi(x) \geq \tilde{\psi}(x)$ for all $x \in X$.

Assume that $(\tilde{X}, \|\cdot\|^\sim)$ is an abstract M-space. By the previous example,

$$\sharp x\sharp \geq \|x\|^\sim, \qquad x \in X.$$

Now assume that there exists some $u \in X$ with $\|u\| \leq 1$ and $|x| \leq \|x\|^\sim u$ for all $x \in X$. Then $\psi(\pm x) \leq \|x\|^\sim$ and $\sharp x\sharp \leq \|x\|^\sim$. In combination,

$$\sharp x\sharp = \|x\|^\sim, \qquad x \in X.$$

EXAMPLE 2.57. Let us determine the monotone companion (half-) norm in a concrete case where the cone is not normal. To this end, we revisit the space bv of sequences of bounded variation with the variation-norm (Section 2.4).

Recall that $bv \subseteq \ell^\infty$ and $\|x\|_{bv} \geq \|x\|_\infty$ for all $x \in bv$. ℓ^∞ is an abstract M space under the sup-norm. We apply the previous example with u being the sequence all the terms of which are 1. Then $\|u\|_{bv} = 1$ and $|x| \leq \|x\|_\infty u$. We obtain that $\sharp x \sharp = \|x\|_\infty$ for all $x \in bv$.

DEFINITION 2.58. *Let X be a real vector space. A functional $\psi : X \to \mathbb{R}_+$ is called a* half-norm *if*

(a) *ψ is subadditive: $\psi(x+y) \leq \psi(x) + \psi(y)$ for all $x, y \in X$.*
(b) *ψ is homogeneous: $\psi(\alpha x) = \alpha\psi(x)$ for all $x \in X$ and $\alpha \in \mathbb{R}_+$.*
(c) *If $x \in X$ and $\psi(x) = 0 = \psi(-x)$, then $x = 0$.*

Exercises.

2.5.1. Let X be a real vector space and $\psi : X \to \mathbb{R}_+$ a half-norm. Define $\sharp x \sharp = \max\{\psi(x), \psi(-x)\}$, $x \in X$. Show $\sharp \cdot \sharp$ is a norm on X.

2.5.2. Let X be a normed vector space and W a nonempty homogeneous subset of X. Define $\psi_W(x) = \inf\{\|x+w\|; w \in W\} = d(x, -W)$ for $x \in X$. Show:

(a) $\psi_W : X \to \mathbb{R}_+$ is homogeneous.
(b) If $-x \in W$, then $\psi_W(x) = 0$. If W is closed and $\psi_W(x) = 0$, then $-x \in W$.
(c) $\psi_W(x) \leq \|x\|$ for all $x \in X$.

2.5.3. Let X be normed vector space and W a nonempty convex subset of X. Then the distance function $\psi_W : X \to \mathbb{R}$ defined by

$$\psi_W(x) = d(x, -W) = \inf\{\|x+y\|; y \in W\}, \qquad x \in X,$$

is convex.

2.5.4. Let X be normed vector space and W a nonempty subset of X. Assume that W is additive, i.e., $x + y \in W$ for any $x, y \in W$.

Define the distance function $\psi_W : X \to \mathbb{R}$ by

$$\psi_W(x) = d(x, -W) = \inf\{\|x+y\|; y \in W\}, \qquad x \in X,$$

Show:

(a) ψ_W is subadditive, i.e., $\psi_W(x+y) \leq \psi_W(x) + \psi_W(y)$ for all $x, y \in X$.
(b) $\psi_W(x) \leq \psi_W(x+w)$ for all $x \in X$ and $w \in W$.

2.6. How orders create norms

Let X be an ordered vector space with order cone X_+.

DEFINITION 2.59. *Let $x \in X$ and $u \in X_+$. Then x is called* u-bounded *if there exists some $c > 0$ such that $-cu \leq x \leq cu$. If x is u-bounded, we define*

$$\|x\|_u = \inf\{c > 0; -cu \leq x \leq cu\}. \tag{2.26}$$

If $x \in X$, $u \in X_+$ and x is not u-bounded, we define $\|x\|_u = \infty$.

The set of u-bounded elements in X (called u-measurable in [**135**, Sec.1.2.1]*) is denoted by X_u,*

$$X_u = \{x \in X; \|x\|_u < \infty\}, \tag{2.27}$$

and u is called an order unit for X_u. *X_u is called the* (principal) ideal *generated by u in* [**10**, Sec.2.7].

Two elements v and u in X_+ are called comparable *if v is u-bounded and u is v-bounded, i.e., if there exist $\epsilon, c > 0$ such that*

$$\epsilon u \le v \le cu. \tag{2.28}$$

If v and u are comparable, we sometimes say that v is u-comparable or u is v-comparable.

Comparability is an equivalence relation for elements of X_+, and we write

$$u \sim v \tag{2.29}$$

if u and v are comparable. Notice that $X_u = X_v$ if and only if $u \sim v$.

A subset Y of X is called u-bounded *if there exists some $c > 0$ such that*

$$-cu \le x \le cu, \qquad x \in Y, \tag{2.30}$$

(equivalently, if $\{\|x\|_u; x \in Y\}$ is a bounded subset of $\mathbb{R}$).

If X is a vector space of real-valued functions f on a set S,

$$\|f\|_u = \sup\Big\{\frac{|f(x)|}{u(x)}; x \in S, u(x) > 0\Big\}. \tag{2.31}$$

2.6.1. The normed linear space of u-bounded vectors. One easily checks that X_u is a linear subspace of X and $\|\cdot\|_u$ is a seminorm on X_u, which is order-preserving on $X_u \cap X_+$. Further, $u \in X_u$. If $u \in \dot{X}_+$, $\|u\|_u = 1$. The following discussion has been inspired by [**138**]. See also [**10**, Sec.2.7].

PROPOSITION 2.60. *Let $u \in X_+$ and $x \in X$. Then $\|x\|_u = 0$ if and only if X_+ contains the line $\{u + \alpha x; \alpha \in \mathbb{R}\}$.*

So, $\|\cdot\|_u$ is a norm on X_u for all $u \in X_+$ if and only if X_+ does not contain any lines.

PROOF. The following chain of equivalences holds for $x \in X$ and $u \in X_+$:

$$\begin{aligned} & \|x\|_u = 0 \\ \iff & \forall \epsilon > 0 : -\epsilon u \le x \le \epsilon u \\ \iff & \forall \epsilon > 0 : \epsilon u \pm x \in X_+ \\ \iff & \forall \alpha > 0 : u \pm \alpha x \in X_+ \\ \iff & \{u + \alpha x; \alpha \in \mathbb{R}\} \subseteq X_+. \end{aligned}$$

□

So, in the following, we assume that X does not contain any lines. Then $\|\cdot\|_u$ is a norm on the linear subspace X_u, called the *order norm* with respect to u, u-norm for short.

We will look for conditions under which X_u, under this norm, is an ordered normed vector space with cone $X_+ \cap X_u$ which is closed, normal, generating and non-flat, and has nonempty interior. Recall Definition 2.3.

LEMMA 2.61. *The following statements are equivalent.*

(i) *X_+ contains no lines and, for all $u \in \dot{X}_+$, the cone X_+ is closed under the u-norm in the following extended sense:*
   If (x_n) is a sequence in X_+ and $x \in X$ such that $x_n - x \in X_u$ for all $n \in \mathbb{N}$ and $\|x_n - x\|_u \to 0$, then $x \in X_+$.
(ii) *X_+ is an Archimedean cone of X.*

PROOF. Assume (ii). Further, assume that X_+ contains the set $\{u+\alpha x; \alpha \in \mathbb{R}\}$ with some $x \in X$. Then X_+ contains $u + nx$ and $u + n(-x)$ for all $n \in \mathbb{N}$. Since X_+ is Archimedean, $x \in X_+$ and $-x \in X_+$. Since X_+ is a cone, $x = 0$. So X_+ contains no lines.

Let $u \in \dot{X}_+$. To show that X_+ is closed under the u-norm in our extended sense, let (x_n) be a sequence in X_+ and $x \in X$ such that $x_n - x \in X_u$ for all $n \in \mathbb{N}$ and $\|x_n - x\|_u \to 0$. Let $\epsilon > 0$. Then there exists some $n \in \mathbb{N}$ such that $x - \epsilon u \leq x_n \leq x + \epsilon u$. So $x + \epsilon u \in X_+$. Since this hold for any $\epsilon > 0$, $nx + u \in X_+$ for all $n \in \mathbb{N}$. Since X_+ is Archimedean (Section 2.1.3), $x \in X_-$.

Now assume (i). To show that X_+ is Archimedean, let $u \in X_+$ and $x \in X$ and $u + nx \in X_+$ for all $n \in \mathbb{N}$.

If $u = 0$, then $x \in X_+$. If $u \in \dot{X}_+$, $\|\cdot\|_u$ is a norm on X_u and X_+ is closed under the u-norm in our extended sense.

Since X_+ is homogeneous, $x_n := x + (1/n)u \in X_+$ and $x_n - x \in X_u \cap X_+$ for all $n \in \mathbb{N}$ and $\|x_n - x\|_u \to 0$. Since X_+ is closed under the u-norm in our extended sense, $x \in X_+$. □

Since we want the cones $X_+ \cap X_u$ to be closed, we better assume that X_+ is Archimedean.

COROLLARY 2.62. *If X_+ is an Archimedean cone, X_+ contains no lines and $X_u \cap X_+$ is a closed cone of X_u under the u-norm.*

Another benefit of this assumption is the following inequality which makes manipulations with the u-norm much more comfortable.

LEMMA 2.63. *Assume that the cone X_+ is Archimedean. Then*

$$-\|x\|_u u \leq x \leq \|x\|_u u, \qquad x \in X_u. \tag{2.32}$$

We mentioned the normality of the cone $X_u \cap X_+$ (Exercise 2.6.2). Actually, the following holds.

LEMMA 2.64. *Let $x, u \in X_+$ and $x \in X_u$. If $y \in X$ satisfies $-x \leq y \leq x$, then $y \in X_u$ and, if X_+ is Archimedean, $\|y\|_u \leq \|x\|_u$.*

PROPOSITION 2.65. *Let X_+ be Archimedean, $u, v \in \dot{X}_+$ and u be v-bounded. Then $X_u \subseteq X_v$ and $\|x\|_v \leq \|x\|_u \|u\|_v$ for all $x \in X_u$.*

PROOF. Since u is v-bounded, $u \leq \|u\|_v v$. Let $x \in X_u$. By Lemma 2.63, $\pm x \leq \|x\|_u u \leq \|x\|_u \|u\|_v v$. So $x \in X_v$ and $\|x\|_v \leq \|x\|_u \|u\|_v$. □

We present the following characterization of the completeness of X_u. Recall Section 2.1.7.

THEOREM 2.66. *Let X_+ contain no lines and $u \in \dot{X}_+$. Consider the following three statements concerning the u-norm.*

(i) *Every increasing Cauchy sequence in $X_u \cap X_+$ has a supremum in X.*
(ii) *$X_u \cap X_+$ is monotonically complete.*
(iii) *X_u is complete.*

Then (i) ⇒ (ii) ⇒ (iii).

If X_+ is Archimedean (or, more weakly, $X_+ \cap X_u$ is closed), then all three statement are equivalent.

PROOF. (i) $\Rightarrow$ (ii): Let (x_n) be an increasing Cauchy sequence in $X_+ \cap X_u$. By (i), $x = \sup_{n\in\mathbb{N}} x_n$ exists, $x \in X$. Since $x \geq x_1 \in X_+$, $x \in X_+$. Let $\epsilon > 0$. Then there exists some $N \in \mathbb{N}$ such that $\|x_n - x_m\|_u \leq \epsilon$ for all $n, m \in \mathbb{N}$ with $n, m \geq N$. If $n > m > N$,

$$x_m - \epsilon u \leq x_n \leq x_m + \epsilon u.$$

Since the supremum is an upper bound, $x_m - \epsilon u \leq x$. Since (x_n) is increasing, $x_n \leq x_m + \epsilon u$ for all $n \in \mathbb{N}$. Since the supremum is the least upper bound, $x \leq x_m + \epsilon u$. Since $x_m \in X_u$, $x \in X_u$ and $\|x - x_m\|_u \leq \epsilon$ for all $m > N$. This shows that $X_+ \cap X_u$ is monotonically complete.

(ii) $\Rightarrow$ (iii): Assume that $X_u \cap X_+$ is monotonically complete under the u-norm. Let (x_n) be a Cauchy sequence in X_u under the u-norm. Then (x_n) is u-bounded and there exists some $c > 0$ such that $x_n \geq -cu$ for all $n \in \mathbb{N}$. Set $y_n = x_n + (c+1)u$. Then (y_n) is a Cauchy sequence with respect to the u-norm and

$$y_n \geq u, \qquad n \in \mathbb{N}. \tag{2.33}$$

It is sufficient to show that (y_n) converges with respect to the u-norm. For this, it is sufficient that (y_n) has a convergent subsequence with respect to the u-norm. So we can assume that $\|y_{n+1} - y_n\|_u \leq 2^{-(n+1)}$ for all $n \in \mathbb{N}$ because the Cauchy sequence (y_n) has a subsequence with this property. We have $y_{n+1} \geq y_n - 2^{-(n+1)}u$. Set $z_n = y_n - 2^{-n}u$. By (2.33), $z_n \in X_+$ for all $n \in \mathbb{N}$. For all $n \in \mathbb{N}$,

$$z_{n+1} = y_{n+1} - 2^{-(n+1)}u \geq y_n - 2 \cdot 2^{-(n+1)}u = z_n.$$

So (z_n) is an increasing sequence in X_+ and a Cauchy sequence with respect to the u-norm. Since $X_+ \cap X_u$ is monotonically complete with respect to the u-norm, (z_n) converges with respect to the u-norm and so does (y_n).

Now assume that X_+ is Archimedean.

(ii) $\Rightarrow$ (i): Let (x_n) be an increasing Cauchy sequence in $X_u \cap X_+$. Since $X_u \cap X_+$ is monotonically complete, (x_n) has a limit $x \in X_+$. Since $X_+ \cap X_u$ is closed because X_+ is Archimedean (Corollary 2.62), $x \geq x_n$ for all $n \in \mathbb{N}$. Let y be an upper bound for (x_n). Again since $X_+ \cap X_u$ is closed, $y \geq x$. So x is the supremum of (x_n).

(iii) $\Rightarrow$ (ii): This is obvious. $\square$

EXAMPLE 2.67. Recall $X = \mathbb{R}^\Omega$, the vector space of functions $f : \Omega \to \mathbb{R}$ where Ω is an arbitrary nonempty set. Choose the cone $X_+ = \mathbb{R}_+^\Omega$ of functions $f : \Omega \to \mathbb{R}_+$. The partial ordering induced by X_+ is the pointwise ordering $f \leq g$ if and only if $f(x) \leq g(x)$ for all $x \in \Omega$.

Let $u \in X_+$. Then X_u is the space of functions f for which $\{f(x)/u(x); x \in \Omega, u(x) > 0\}$ is a bounded subset of $\mathbb{R}$ and $\|f\|_u$ is given by (2.31). X_u is complete under the u-norm as one can see from Theorem 2.66. Indeed, any increasing u-bounded sequence (f_n) in X_u has a supremum in $\mathbb{R}^\Omega$, namely

$$(\sup f_n)(x) = \sup_{n\in\mathbb{N}} f_n(x), \qquad x \in \Omega.$$

Notice that, for any $x \in \Omega$, $\{f_n(x); n \in \mathbb{N}\}$ is a bounded subset of $\mathbb{R}$.

If u is a positive constant function, $X_u = B(\Omega, \mathbb{R})$, the ordered Banach space of bounded real-valued functions on Ω with the supremum norm, and the u-norm is the supremum norm, $\|\cdot\|_u = \|\cdot\|_\infty$.

EXAMPLE 2.68. Let Ω be a metric space with metric d and $X = C(\Omega)$, the vector space of continuous real-valued functions on Ω. Since $C(\Omega)$ is a linear subspace of $\mathbb{R}^\Omega$, it inherits the cone $C_+(\Omega) = C(\Omega) \cap \mathbb{R}^\Omega_+$ of nonnegative continuous functions. $C(\Omega)$ also inherits the lattice structure. This is most easily seen by noticing that $|f| \in C(\Omega)$ if $f \in C(\Omega)$.

Let again u be a positive constant function, then $X_u = C^b(\Omega)$ the ordered Banach space of bounded continuous functions.

Exercises.

2.6.1. Let X be an ordered vector space with order-cone X_+ and $u \in \dot{X}_+$. Show that X_u is a linear subspace of X and $\|\cdot\|_u$ a seminorm on X_u.

2.6.2. Prove Lemma 2.64.

2.6.3. Let X be an ordered vector space with order cone X_+. Show that $X_+ \cap X_u$ under the u-norm is a solid, generating and non-flat cone. For the latter, if $x \in X$, for $x = v - w$ with $v, w \in X_u \cap X_+$, choose $v = (1/2)(\|x\|_u u + x)$ and $w = (1/2)(\|x\|_u u - x)$.

2.6.4. Prove that $\mathbb{R}^S_u$, where $u \in \mathbb{R}^S_+$, is complete (see Examples 2.17 and 2.67).

2.6.5. Let X be a vector lattice with order cone X_+ and $u \in \dot{X}_+$. Assume that X_+ contains no lines. Show that X_u under the u-norm is an abstract M-space.

2.6.2. X_u when X already has a norm. If the ordered vectored space X already carries a norm, one would like to know how it relates to the u-norms. We assume that the order cone X_+ is closed with respect to that norm. This implies that X_+ is Archimedean.

We write

$$X_{u+} = X_+ \cap X_u, \qquad \dot{X}_{u+} = X_{u+} \setminus \{0\}.$$

Recall the monotone companion norm $\sharp \cdot \sharp$ (Theorem 2.48 and Definition 2.50).

LEMMA 2.69. *Let $u \in \dot{X}_+$. Then the following hold:*

(a) $\sharp x \sharp \le \|x\|_u \, \sharp u \sharp = \|x\|_u \, d(u, -X_+), \qquad x \in X_u.$

(b) *The following are equivalent:*
   - (i) *u is a normal point for X_+ under $\|\cdot\|$ (Section 2.2.2).*
   - (ii) *There exists some $c \ge 0$ such that $\|x\| \le c\|x\|_u$ for all $x \in X_{u+}$.*
   - (iii) *There exists some $c \ge 0$ such that $\|x\| \le c\|x\|_u$ for all $x \in X_u$.*

(c) *If X_+ is solid and u is in the interior of X_+ with respect to $\|\cdot\|$, then $X = X_u$, $d(u, X \setminus X_+) > 0$, and*

$$\|x\| \ge \|x\|_u \, d(u, X \setminus X_+), \qquad x \in X.$$

(d) *In turn, if $X_u = X$ and there exists some $\epsilon > 0$ such that $\|x\| \ge \epsilon\|x\|_u$ for all $x \in X$, then u is an interior point of X_+ with respect to $\|\cdot\|$.*

(e) *$X = X_u$ and $\|\cdot\|$ and $\|\cdot\|_u$ are equivalent norms if and only if X_+ is solid and u is both an interior and a normal point for X_+ with respect to $\|\cdot\|$.*

PROOF. (a) By (2.32), if $x \in X_u$,

$$x, -x \le \|x\|_u u.$$

Since the companion half-norm ψ is order-preserving on X (Section 2.5),

$$\psi(x) \le \|x\|_u \psi(u), \qquad \psi(-x) \le \|x\|_u \psi(u),$$

and so $\sharp x\sharp \le \|x\|_u \psi(u)$.

(b) Obviously (iii) $\Rightarrow$ (ii) $\Rightarrow$ (i).

Let u be normal point for X_+. Then there exists some $\tilde{c} > 0$ such that $\|y\| \le \tilde{c}$ for all $y \in X_+$ with $y \le u$. For $x \in X_u$, $\pm x \le \|x\|_u u$ by (2.26). If $x \ne 0$ in addition,

$$0 \le \frac{1}{2\|x\|_u} x + \frac{1}{2} u \le u.$$

So

$$\Big\| \frac{1}{2\|x\|_u} x + \frac{1}{2} u \Big\| \le \tilde{c}.$$

By the triangle inequality,

$$\Big\| \frac{1}{2\|x\|_u} x \Big\| \le \frac{1}{2} \|u\| + \tilde{c}.$$

This implies (iii).

(c) Let u be an interior element of X_+. Then $d(u, X \setminus X_+) > 0$. For any $\delta \in \big(0, d(u, X \setminus X_+)\big)$ we have $u \pm \frac{\delta}{\|x\|} x \in X_+$ for all $x \in \dot{X}$. So $\pm x \le \frac{\|x\|}{\delta} u$ and so $x \in X_u$ and $\|x\|_u \le \frac{\|x\|}{\delta}$. Since this holds for any $\delta \in \big(0, d(u, X \setminus X_+)\big)$, it also holds for $\delta = d(u, X \setminus X_+)$.

(d) Assume that $X_u = X$ and there exists some $\epsilon > 0$ such that $\|x\| \ge \epsilon \|x\|_u$ for all $x \in X$. By (2.32),

$$\pm x \le \|x\|_u u \le (1/\epsilon) \|x\| u.$$

Hence $u \pm \frac{\epsilon}{\|x\|} x \in X_+$ for all $x \in \dot{X}$. This implies that u is an interior point of X_+.

(e) This follows from combining (b), (c) and (d). □

The functional $\|\cdot\|_u$ is lower semicontinuous in the following sense:

PROPOSITION 2.70. *If (x_n) is a sequence in X_u and $x \in X$ and $x_n \to x$ with respect to the original norm (or its monotone companion norm) and $\liminf_{n\to\infty} \|x_n\|_u < \infty$, then $x \in X_u$ and $\|x\|_u \le \liminf_{n\to\infty} \|x_n\|_u$.*

PROOF. Set $\beta = \liminf_{n\to\infty} \|x_n\|_u$. For any $k \in \mathbb{N}$, there exists some $n_k \ge k$ in $\mathbb{N}$ such that $\|x_{n_k}\|_u \le \beta + (1/k)$. By (2.32), $\pm x_{n_k} \le (\beta + (1/k))u$. Since X_+ is closed under its original norm (and its monotone companion norm by Theorem 2.48), we can take the limit as $k \to \infty$ and obtain $\pm x \le \beta u$. □

The next result should be compared to a result due to Andô [**14**] [**10**, Thm.2.60].

THEOREM 2.71. *Let X be an ordered normed vector space with closed order cone X_+ and $u \in \dot{X}_+$. Assume that any u-bounded increasing Cauchy sequence in X_+ has a limit.*

Then the following statements are equivalent:

(i) *u is a normal point for X_+ (Section 2.2.2).*
(ii) *$X_u \cap X_+$ is monotonically complete under the u-norm.*
(iii) *X_u is complete under the u-norm.*

PROOF. (i) $\Rightarrow$ (ii): Let (x_n) be an increasing sequence in $X_u \cap X_+$ that is a Cauchy sequence with respect to the u-norm. Then (x_n) is u-bounded. Since u is a normal point for X_+, by Lemma 2.69 (b), (x_n) is a Cauchy sequence with respect to the original norm and thus has a limit $x \in X_+$ by the overall assumption of the theorem. Let $\epsilon > 0$. Then there exists some $N \in \mathbb{N}$ such that $-\epsilon u \leq x_n - x_m \leq \epsilon u$ for all $n, m \in \mathbb{N}$ with $n, m \geq N$. Since X_+ is closed, we can take the limit as $n \to \infty$ and obtain $-\epsilon u \leq x - x_m \leq \epsilon u$ for all $m \in \mathbb{N}$ with $m \geq N$. So $\|x_m - x\|_u \leq \epsilon$ for all $m \in \mathbb{N}$ with $m \geq N$ and $\|x_m - x\|_u \to 0$ as $m \to \infty$.

(ii) $\Rightarrow$ (iii): This follows from Theorem 2.66.

(iii) $\Rightarrow$ (i): Let X_u be complete under the u-norm. Endow the vector space

$$V = \{(x, x); x \in X_u\}$$

with the norm $\|(x, x)\|_V = \|x\| + \|x\|_u$.

To show that V is complete under this norm, consider a Cauchy sequence $((x_n, x_n))$ in V. It is sufficient to show that this sequence has a converging subsequence. (x_n) is a Cauchy sequence in X under the original norm $\|\cdot\|$ and in X_u under the u-norm. By the same construction as in the proof of Theorem 2.66 (ii) $\Rightarrow$ (iii), after choosing a subsequence, we can assume that there is an increasing sequence (z_n) in $X_u \cap X_+$ which is both a Cauchy sequence with respect to the u-norm and the original norm $\|\cdot\|$ such that $z_n - x_n \to 0$ with respect to both norms. By our overall assumption that every increasing u-bounded sequence that is a Cauchy sequence with respect to $\|\cdot\|$ converges, there exists some $x \in X$ such that $\|z_n - x\| \to 0$. Since X_+ is closed, $x \in X_+$. Since X_u is complete under the u-norm, there exists some $z \in X_+ \cap X_u$ such that $\|z_n - z\|_u \to 0$. For any $\epsilon > 0$, there exists some $N \in \mathbb{N}$ such that $z - \epsilon u \leq z_n \leq z + \epsilon u$ for all $n \in \mathbb{N}$ with $n \geq N$. Since the cone X_+ is closed, taking the limit as $n \to \infty$ yields $z - \epsilon u \leq x \leq z + \epsilon u$ for arbitrary $\epsilon > 0$ and so $z = x$. This implies $\|x_n - x\| \to 0$ and $\|x_n - x\|_u \to 0$ and $\|(x_n, x_n) - (x, x)\|_V \to 0$.

This completes the proof that V is complete.

The map $(x, x) \mapsto x$ from the Banach space V to the Banach space X_u with the u-norm is bijective and continuous and has a continuous inverse map by the Open Mapping Theorem. So there exists some $c > 0$ such that $\|x\| + \|x\|_u \leq c\|x\|_u$ for all $x \in X_u$. In particular, $\|x\| \leq c$ for all $x \in X_+$ with $x \leq u$. So u is a normal point for X_+ (Section 2.2.2). □

An element $u \in \dot{X}_+$ is called a *uniform order unit* for X_+ if there is some $c = c_u > 0$ such that

$$x \leq c\|x\|u, \qquad x \in X_+. \tag{2.34}$$

We say that X_+ has a uniform order unit if there exists a uniform order unit in X_+ for X_+.

The following is immediate from Definition 2.59.

LEMMA 2.72. *Let $u \in \dot{X}_+$ be a* uniform order unit *for X_+.*

Then any $x \in X_+$ is an element of X_u and there is some $c > 0$ such that $\|x\|_u \leq c\|x\|$ for all $x \in X_+$.

PROPOSITION 2.73. *Let X be an ordered normed vector space with closed order cone X_+. Then $u \in \dot{X}_+$ is an interior point of X_+ if and only if u is a uniform order unit for X_+ and X_+ is generating and non-flat.*

PROOF. One direction follows from Theorem 2.32.

Assume that u is a uniform order unit for X_+ and X_+ is a non-flat generating closed cone of X. Let $x \in X$. According to the definitions in Section 2.3, there exists some $\tilde{c} > 0$ that is independent of x and some $y, z \in X_+$ such that $x = y - z$ and $\|y\| + \|z\| \le \tilde{c}\|x\|$. Since u is a uniform order unit for X_+, there is some $c_u > 0$ such that $y \le c_u\|y\|u$ and $z \le c_u\|z\|u$. So $y, z \in X_u$ and $\|y\|_u \le c_u\|y\|$ and $\|z\|_u \le c_u\|z\|$. Since X_u is a vector space, $x \in X_u$ and $\|x\|_u \le c_u\tilde{c}\|x\|$. By Lemma 2.69 (d), u is an interior point of X_+. □

2.7. The ordered vector space of (Lipschitz) continuous functions

If S is an arbitrary nonempty set, we have met the ordered vector space $\mathbb{R}^S$ of real-valued functions on S and the ordered Banach space $B(S)$ which is a subspace of $S^{\mathbb{R}}$ (Examples 2.17 and 2.18).

Let S be a metric space with metric d (Appendix A.1). Then we have interesting ordered subspaces of $B(S)$.

- A function $f : S \to \mathbb{R}$ is called *lower semicontinuous* at $s \in S$ if for any $\epsilon > 0$ there exists some $\delta > 0$ such that $f(t) > f(s) - \epsilon$ for all $t \in S$ with $d(t, s) < \delta$.
- It is called *upper semicontinuous* at $s \in S$ if for any $\epsilon > 0$ there exists some $\delta > 0$ such that $f(t) < f(s) + \epsilon$ for all $t \in S$ with $d(t, s) < \delta$.
- f is *continuous at* $s \in S$ if and only if it is lower and upper semicontinuous at s.

See Appendix A.4 for more information.

The set of continuous functions $f : S \to \mathbb{R}$, denoted by $C(S)$, forms an ordered linear subspace and a sublattice of $\mathbb{R}^S$ (Example 2.17). The set of bounded continuous functions $f : S \to \mathbb{R}$, denoted by $C^b(S)$, forms an ordered closed subspace and a sublattice of $B(S)$ which is a Banach lattice in its own right. The lattice property follows from (2.10)

The cones are formed by the respective nonnegative functions in these spaces and denoted by $C_+(S)$ and $C_+^b(S)$. $C_+^b(S)$ is a solid cone, its open interior is formed by the positive continuous functions f that are bounded away from 0, $\inf f(S) > 0$. $C_+^b(S)$ will be the population state space for spatially distributed semelparous two-sex populations in Chapter 18.

DEFINITION 2.74. *A function $f : S \to \mathbb{R}$ is called* Lipschitz (continuous) *if there is some $\Lambda \in \mathbb{R}_+$ such that $|f(s) - f(t)| \le \Lambda d(s,t)$ for all $s, t \in S$.*

Λ is called a Lipschitz constant for f. THE Lipschitz constant of a Lipschitz continuous function is defined as

$$\Lambda(f) = \sup\Big\{\frac{|f(s) - f(t)|}{d(s,t)}; s, t \in S, s \neq t\Big\}. \tag{2.35}$$

One readily checks that the set $\mathrm{LC}(S)$ of real-valued Lipschitz continuous functions is a linear subspace of $C(S)$ and so inherits the pointwise order structure with the Archimedean cone $\mathrm{LC}(S) \cap \mathbb{R}_+^S = \mathrm{LC}_+(S)$. $\Lambda(f) = 0$ if and only if the function f is constant. So Λ is only a semi-norm on $\mathrm{LC}(S)$ (Definition A.19).

Examples of elements in $\mathrm{LC}_+(S)$ can be constructed from the metric d. Recall the distance from a point $x \in S$ to a subset T of S,

$$d(s, T) = \inf\{d(s, t); t \in T\}, \tag{2.36}$$

$$|d(s,T) - d(t,T)| \le d(s,t), \tag{2.37}$$

$$d(s,T) = 0 \iff s \in \bar{T}. \tag{2.38}$$

See Definition A.8 and Lemma A.9.

Further, recall the open δ-neighborhood of T,

$$U_\delta(T) = \{s \in S; d(s,T) < \delta\} = \bigcup_{s \in T} U_\delta(s). \tag{2.39}$$

(Exercise A.1.4) and the closed δ-neighborhood of T,

$$\bar{U}_\delta(T) = \{s \in S; d(s,T) \le \delta\}. \tag{2.40}$$

While $\overline{U_\delta(T)} \subseteq \bar{U}_\delta(T)$, equality may not hold necessarily (it holds if S is a normed vector space and T is a singleton set).

The following notation for spaces of real-valued functions on S is used: $C(S)$ continuous, $C^b(S)$ continuous bounded, UC(S) uniformly continuous, UC$^b(S)$ uniformly continuous and bounded, $C_c(S)$ continuous with compact support, $C_0(S)$ continuous and vanishing at infinity, LC(S) Lipschitz continuous, LC$^b(S)$ Lipschitz continuous and bounded.

A function $f : S \to R$ has *compact support* if there is a compact subset K of S such that $f(s) = 0$ for all $s \in S \setminus K$.

A function $f : S \to \mathbb{R}$ *vanishes at infinity* if, for any $\epsilon > 0$, there exists a compact subset K of S such that $|f(s)| < \epsilon$ for all $s \in S \setminus K$.

See Definition A.10 for the concept of a compact set and Appendix A.1.2 for upcoming concepts of metric topology.

LEMMA 2.75. *Let $T, \tilde{T}$ be two subsets of S the closures of which are disjoint. Then the function $f : S \to \mathbb{R}_+$,*

$$f(s) = \frac{d(s,\tilde{T})}{d(s,T) + d(s,\tilde{T})}, \qquad s \in S, \tag{2.41}$$

separates T and $\tilde{T}$, i.e., $f \in C^b(S)$, $0 \le f(s) \le 1$ for all $s \in S$, $f(s) = 1$ for all $s \in T$ and $f(s) = 0$ for all $s \in \tilde{T}$ [**190**, p.59].

If $\delta > 0$ and $\tilde{T} = S \setminus U_\delta(T)$ or $T = S \setminus U_\delta(\tilde{T})$, then f is Lipschitz continuous,

$$|f(s) - f(t)| \le (1/\delta) d(s,t), \qquad s,t \in S.$$

Further, f is Lipschitz continuous if one of T and $\tilde{T}$ is compact and the other closed.

PROOF. We show the statement concerning the Lipschitz continuity for the case that $\tilde{T} = S \setminus U_\delta(T)$. The other case follows by symmetry. We first claim that

$$d(x,T) + d(x, S \setminus U_\delta(T)) \ge \delta, \qquad x \in S.$$

Suppose not. Then there exists some $y \in T$ and $z \in S \setminus U_\delta(T)$ such that

$$\delta > d(x,y) + d(x,z) \ge d(y,z).$$

Since $y \in T$, this implies that $z \in U_\delta(T)$, a contradiction.

Set $g(s,t) = \frac{s}{s+t}$ for $s,t \ge 0$, $s + t > 0$. Then

$$|g(s,t) - g(\tilde{s},\tilde{t})| \le \frac{\max\{|s-\tilde{s}|, |t-\tilde{t}|\}}{s+t}.$$

(2.37) implies the assertion.

Now let T or $\tilde{T}$ be compact. Our previous consideration shows that f is Lipschitz continuous, if $d(x,T)+d(x,\tilde{T})$ is bounded away from 0. Suppose not. Then there exists a sequence (x_n) in S such that $d(x_n,T)\to 0$ and $d(x_n,\tilde{T})\to 0$. By symmetry, we can assume that T is compact. Then there exists some $x\in T$ such that $x_n\to x$ after choosing subsequences. By (2.37), $d(x,\tilde{T})=0$. By (2.38), $x\in\tilde{T}$, a contradiction. □

PROPOSITION 2.76. *Let T be a closed subset of S. Then there exists a decreasing sequence of Lipschitz continuous functions (f_n) with values between 0 and 1 such that $f_n\to\chi_T$ pointwise.*

Let T be a open subset of S. Then there exists an increasing sequence of Lipschitz continuous functions (f_n) with values between 0 and 1 such that $f_n\to\chi_T$ pointwise.

PROOF. Let T be closed. For $n\in\mathbb{N}$, set

$$f_n(x)=\frac{d\big(x,S\setminus U_{1/n}(T)\big)}{d(x,T)+d\big(x,S\setminus U_{1/n}(T)\big)},\qquad x\in S.$$

Let T be open. Then $S\setminus T$ is closed and there exists a decreasing sequence (f_n) with values between 0 and 1 that converges pointwise towards $\chi_{S\setminus T}=\chi_S-\chi_T$. Set $g_n=\chi_S-f_n$. □

Since there seem to be so many Lipschitz continuous functions, one may wonder under which conditions there is a bounded uniformly continuous real-valued function which is not Lipschitz continuous.

PROPOSITION 2.77. *Assume that the metric space S has an accumulation point (Definition A.12). Then there exists some $g:S\to\mathbb{R}_+$ which is uniformly continuous and bounded but not Lipschitz continuous.*

PROOF. Let $s\in S$ be an accumulation point of S. Define $h,g:S\to\mathbb{R}_+$ by $h(t)=\min\{1,d(s,t)\}$ and $g(t)=\sqrt{h(t)}$. Then $h\in\mathrm{LC}^b_+(S)$, and g is uniformly continuous on S as composition of two uniformly continuous functions and is bounded. Since s is an accumulation point of S, there exists a sequence (s_n) in S such that $0<d(s_n,s)$ for all $n\in\mathbb{N}$ and $s_n\to s$ as $n\to\infty$. Without loss of generality, we can assume that $d(s_n,s)<1$ for all $n\in\mathbb{N}$. Then

$$\frac{|g(s_n)-g(s)|}{d(s_n,s)}=\frac{\sqrt{h(s_n)}}{d(s_n,s)}=\frac{1}{\sqrt{d(s_n,s)}}\to\infty,\qquad n\to\infty.$$ □

2.7.1. The space of bounded Lipschitz functions under the supremum norm. Again, we choose u as a positive constant function on S (which is trivially Lipschitz) and consider the ordered normed vector space of u-bounded Lipschitz-continuous functions (Section 2.6).

This is the same as the ordered normed vector space $\mathrm{LC}^b(S)$ of bounded Lipschitz continuous functions with the supremum norm and the cone $\mathrm{LC}^b_+(S)$ of non-negative functions in $\mathrm{LC}^b(S)$. This cone is normal and solid and thus generating and non-flat (Theorem 2.32).

$\mathrm{LC}^b(S)$ under the supremum norm is a normed vector lattice. This is seen most easily from (2.10) by observing that, for any Lipschitz continuous function f, also its absolute value $|f|$ is Lipschitz continuous and $\Lambda(|f|)\le\Lambda(f)$.

However, if there exists a function $f : S \to \mathbb{R}_+$ which is bounded and uniformly continuous but not Lipschitz continuous (like $f(x) = \sqrt{|x|}$ for $S = [0, 1]$), the cone $\mathrm{LC}^b_+(S)$ is not monotonically complete under the supremum norm.

This follows from the fact that every nonnegative bounded uniformly continuous function can be uniformly approximated by an increasing sequence of nonnegative bounded Lipschitz continuous functions.

For a bounded function $f : S \to \mathbb{R}$, define the sequence (f_n) of functions $f_n : S \to \mathbb{R}$ by

$$f_n(x) = \inf\{f(y) + nd(x, y);\ y \in S\}, \qquad x \in S, \quad n \in \mathbb{N}, \tag{2.42}$$

[**9**, Thm.3.13].

PROPOSITION 2.78. *(f_n) is an increasing sequence of bounded Lipschitz continuous functions (with n being a Lipschitz constant of f_n) such that* $\inf f \le f_n \le f$ *for all $n \in \mathbb{N}$. If $f \ge 0$, then $0 \le f_n \le f$. The following convergence results hold.*

(a) *If $f \in \mathrm{UC}^b(S)$, $f_n \to f$ uniformly on S.*
(b) *If $f \in C^b(S)$, $f_n \to f$ uniformly on compact subsets of S.*
(c) *If f is locally Lipschitz continuous on S, then, for any compact set K in S, there exists some $\delta > 0$ such that $f_n \to f$ uniformly on $U_\delta(K)$.*
(d) *If $f \in C_{0+}(S)$, so is f_n and $f_n \to f$ uniformly on S*
(e) *If f is lower semi-continuous, $f_n \to f$ pointwise on S.*
(f) *If $n \in \mathbb{N}$ and $|f(x) - f(y)| \le nd(x, y)$ for all $x, y \in S$, then $f_n = f$.*

This proposition will also be used in the proof of Theorem 3.5 which concerns the convergence of measures.

PROOF. To see that f_n is well-defined and bounded, we notice that

$$f(y) + nd(x, y) \ge -\|f\|_\infty, \qquad x, y \in S.$$

Hence, for each $x \in S$, the set $\{f(y)+nd(x, y); y \in S\}$ is bounded below, by $-\|f\|_\infty$, and has an infimum. So $f_n(x)$ is defined and $f_n(x) \ge -\|f\|_\infty$. Further, for each $x \in S$,

$$f_n(x) \le f(x) + nd(x, x) = f(x).$$

This shows that f_n is bounded and $\|f_n\|_\infty \le \|f\|_\infty$.

Let $n \in \mathbb{N}$, $x, z \in S$. Then, for all $y \in S$, by the triangle inequality,

$$f_n(x) \le f(y) + nd(x, y) \le f(y) + n(d(x, z) + d(z, y)).$$

So, for all $y \in S$,

$$f(y) + nd(z, y) \ge f_n(x) - nd(x, z).$$

This implies that $f_n(x) - nd(x, z)$ is a lower bound for $\{f(y) + nd(z, y); y \in S\}$. Since the infimum is the largest lower bound,

$$f_n(z) \ge f_n(x) - nd(x, z),$$

and so

$$f_n(x) - f_n(z) \le nd(x, z).$$

We switch x and z, and obtain

$$f_n(z) - f_n(x) \le nd(z, x) = nd(x, z).$$

This implies $|f_n(z) - f_n(x)| \le nd(x, z)$. So f_n is Lipschitz continuous with n being a Lipschitz constant.

(a) Suppose f is uniformly continuous and that $f_n(x)$ does not converge to $f(x)$ uniformly for $x \in S$ as $n \to \infty$. Then there exists some $\epsilon > 0$ such that for all $k \in \mathbb{N}$ there exist $n_k > k$ and $x_k \in S$ such that

$$|f_{n_k}(x_k) - f(x_k)| \geq \epsilon.$$

Notice that the f_{n_k} do not necessarily form a subsequence of (f_n) as we have not chosen (n_k) to be strictly increasing. Since $f_{n_k}(x_k) \leq f(x_k)$,

$$f(x_k) \geq f_{n_k}(x_k) + \epsilon, \qquad k \in \mathbb{N}.$$

By definition of the infimum, there exist $y_k \in S$ such that

$$f(x_k) \geq f(y_k) + n_k d(x_k, y_k) + \epsilon/2.$$

Since f is bounded and $n_k \to \infty$, $d(x_k, y_k) \to 0$ as $k \to \infty$. Since f is uniformly continuous, $f(x_k) - f(y_k) \to 0$ by the sequential characterization of uniform continuity. This leads to the contradiction $0 \geq \epsilon/2 > 0$.

(b) If f is continuous and K a compact set in S, to show uniform convergence on K, we have $x_k \in K$ in the proof of part (a) and $x_k \to x \in K$ after choosing subsequences. Then $y_k \to x$ and the continuity of f at x provides the contradiction.

(c) Let f be locally Lipschitz continuous and K a compact set in S. Then there exists some $\delta > 0$ such that f is Lipschitz continuous on $U_{2\delta}(K)$. In proving uniform convergence on $U_\delta(K)$, the sequence (x_k) is in $U_\delta(K)$ and y_k turns out to be in $U_{2\delta}(K)$ for all large $k \in \mathbb{N}$. The Lipschitz continuity of f on $U_{2\delta}(K)$ now provides the contradiction.

(e) Assume that $x \in S$, f is lower semicontinuous at x and $f_n(x)$ does not converge to $f(x)$ as $n \to \infty$. The proof works as in (a) with $x_k = x$ for all $k \in \mathbb{N}$. We then find $y_k \to x$ such that

$$f(x) \geq f(y_k) + \epsilon/2, \qquad k \in \mathbb{N}.$$

Since f is lower semicontinuous at x, $\liminf_{k\to\infty} f(y_k) \geq f(x)$, a contradiction. □

REMARK 2.79. Let T be a subset of S and $f : T \to \mathbb{R}$ be Lipschitz continuous with $\Lambda > 0$ being a Lipschitz constant. Then $\tilde{f} : S \to \mathbb{R}$ defined by

$$\tilde{f}(s) = \inf\{f(t) + \Lambda d(s,t); t \in T\}, \qquad s \in S,$$

extends f from T to S and is Lipschitz continuous with Λ being a Lipschitz constant.

COROLLARY 2.80. *Every continuous function f on S with compact support can be uniformly approximated by bounded Lipschitz continuous functions f_n with compact support such that $|f_n| \leq |f|$.*

We combine Proposition 2.77 and Proposition 2.78 (a).

COROLLARY 2.81. *If the metric space S has an accumulation point, the cone $\mathrm{LC}_+^b(S)$ is not monotonically complete under the supremum norm.*

DEFINITION 2.82. *Let $\mathcal{F}$ be a set of functions $f : S \to \mathbb{R}$ and $s \in S$. $\mathcal{F}$ is called* equicontinuous *at s if for any $\epsilon > 0$ there exists some $\delta > 0$ such that $|f(t) - f(s)| < \epsilon$ for all $f \in \mathcal{F}$ and all $t \in S$ with $d(t,s) < \delta$. $\mathcal{F}$ is called equicontinuous on S if it is equicontinuous at all $s \in S$.*

$\mathcal{F}$ is called equibounded *if there exists some $c > 0$ such that $|f(s)| \leq c$ for all $s \in S$ and all $f \in \mathcal{F}$.*

A similar proof as for Proposition 2.78 (b) shows the following result which will be used in the proof of Proposition 3.15 which concerns the convergence of measures. Cf. [**203**, L.6.9].

LEMMA 2.83. *Let $\mathcal{F}$ be an equicontinuous and equibounded family of functions $f : S \to \mathbb{R}_+$. For each $f \in \mathcal{F}$, let $([f]_k)$ be the approximation (2.42). Then, for each $s \in S$, $\sup_{f\in\mathcal{F}} \big|f(s) - [f]_k(s)\big| \to 0$ as $k \to \infty$.*

PROOF. Let $s \in S$ and suppose that $\sup_{f\in\mathcal{F}} \big|f(s) - [f]_k(s)\big| \to 0$ as $k \to \infty$ does not hold. Since $f(s) \geq [f]_k(s)$, there exists some $\epsilon > 0$ and a sequence (k_n) in $\mathbb{N}$ such that $k_n \to \infty$ as $n \to \infty$ and $\sup_{f\in\mathcal{F}}(f(s) - [f]_{k_n}(s)) > \epsilon$ for all $n \in \mathbb{N}$. Then there exist $f_n \in \mathcal{F}$ such that

$$f_n(s) > [f_n]_{k_n}(s) + \epsilon, \qquad n \in \mathbb{N}.$$

By (2.42), for $[f_n]_{k_n}$ replacing f_n, there exists a sequence $(t_n) \in S$ such that

$$f_n(s) > f_n(t_n) + k_n d(s, t_n) + \epsilon, \qquad n \in \mathbb{N}.$$

Since $k_n \to \infty$ and $\{f_n\}$ is equibounded, $d(s, t_n) \to 0$ as $n \to \infty$. Since $\{f_n; n \in \mathbb{N}\} \subseteq \mathcal{F}$ is equicontinuous, $f_n(s) - f_n(t_n) \to 0$ as $n \to \infty$ which implies $0 \geq \epsilon$, a contradiction. □

2.7.2. The space of bounded Lipschitz functions under the bounded Lipschitz norm. Alternatively, we can endow the ordered vector space of bounded Lipschitz functions with the bounded Lipschitz norm

$$\|f\|_\flat = \max\big\{\|f\|_\infty, \Lambda(f)\big\}, \qquad f \in \mathrm{LC}^b_+(\mathbb{R}). \tag{2.43}$$

Equivalently, we could use any norm on $\mathbb{R}^2$ instead of the maximum norm. It is left as an exercise that $\mathrm{LC}^b(S)$ is an ordered Banach space under the bounded Lipschitz norm. The cone of nonnegative functions is still solid, generating and non-flat, but no longer normal in general (Exercise 2.7.3).

Further, the lattice operations are no longer continuous.

Recall the definition of a normal point (Section 2.2.2).

PROPOSITION 2.84. $\mathrm{LC}^b_+(\mathbb{R})$ *contains no normal point for itself.*

PROOF. Let f be a nonzero nonnegative bounded Lipschitz function on $\mathbb{R}$. Then there exists some $x_0 \in \mathbb{R}$ such that $f(x_0) \geq (1/2)\|f\|_\infty$. Further, there exists some $\Lambda > 0$ such that

$$|f(y) - f(x)| \leq \Lambda|y - x|, \qquad x, y \in \mathbb{R}.$$

With $y = 0$,

$$f(x) \geq f(0) - \Lambda|x|, \qquad x \in \mathbb{R}.$$

Set

$$f_n(x) = \big[f(0) - n|x|\big]_+, \qquad x \in \mathbb{R}, \quad n \in \mathbb{N}.$$

Here $[z]_+ = \max\{z, 0\}$ is the positive part of $z \in \mathbb{R}$. For all $n \in \mathbb{N}$, f_n is Lipschitz continuous with Lipschitz constant n. Since $f(x) \geq 0$ for all $x \in \mathbb{R}$, for all $n > \Lambda$, $0 \leq f_n(x) \leq f(x)$ for all $x \in \mathbb{R}$ and $\|f_n\|_\flat \geq \Lambda(f_n) = n$.

So f is not a normal point for $\mathrm{LC}^b_+(\mathbb{R})$ (Section 2.2.2). □

In order to show that the lattice operations are not continuous, we follow [**156**] in considering $f_s(t) = t - s$ for $t, s \in \mathbb{R}$. Then, for all $t, s, r \in \mathbb{R}$,

$$|f_s(t) - f_r(t)| = |t - s - (t - r)| = |s - r|,$$

and so $\|f_s - f_r\|_\infty \le |s - r|$. Further,

$$\frac{|(f_s - f_r)(t) - (f_s - f_r)(\tilde{t})|}{|t - \tilde{t}|} = \frac{|(t - s) - (t - r) - (\tilde{t} - s) + (\tilde{t} - r)|}{|t - \tilde{t}|} = 0,$$

and so $\Lambda(f_r - f_s) = 0$ and $\|f_s - f_r\|_\flat = \|f_r - f_s\|_\infty = |r - s|$.

Now, if $r \ne s$,

$$\Big|(|f_r| - |f_s|)(s) - (|f_r| - |f_s|)(r)\Big| = \Big||f_r(s)| + |f_s(r)|\Big| = 2|s - r|,$$

and so $\Lambda(|f_r| - |f_s|) = 2$ for all $s, r \in \mathbb{R}$, $s \ne r$.

PROPOSITION 2.85. *Consider* $\mathrm{LC}^b(S)$ *with the bounded Lipschitz norm* $\|\cdot\|_\flat$ *under which the cone typically is not normal. Then the associated monotone companion norm* $\sharp \cdot \sharp_\flat$ *is the supremum norm.*

This follows from Example 2.56 and the observation that $\mathrm{LC}^b(S)$ under the supremum norm is an abstract M space.

The next result [**101**, Prop.4.2] will be used in the proof of Proposition 3.23.

PROPOSITION 2.86. *Let* (S, d) *be separable. Then any bounded sequence in* $\mathrm{LC}^b(S)$ *has a subsequence that converges to some* $f \in \mathrm{LC}^b(S)$ *pointwise on* S *and uniformly on every totally bounded subset* T *of* S.

For the concepts of separable and totally bounded sets see Definition A.31 and A.23.

PROOF. Let (f_n) be a bounded sequence in $\mathrm{LC}^b(S)$. We can assume that $\|f_n\|_\flat \le 1$ for all $n \in \mathbb{N}$. By the Arzela-Ascoli theorem, after choosing a subsequence, $f_n \to f$ pointwise with $f \in \mathrm{LC}^b(S)$ [**159**, Thm.8.5] and $\|f\|_\flat \le 1$. Indeed, for all $n \in \mathbb{N}$,

$$|f_n(s) - f_n(t)| \le d(s,t), \qquad s, t \in S$$

By taking the limit $n \to \infty$,

$$|f(s) - f(t)| = \lim_{n\to\infty} |f_n(s) - f_n(t)| \le d(s,t), \qquad s, t \in S.$$

Let T be a totally bounded subset of S. Let $\epsilon > 0$. Then there exist $s_1, \dots, s_m \in S$ such that $T \subseteq \bigcup_{j=1}^m U_\epsilon(s_j)$. Let $s \in T$. Then $d(s, s_j) < \epsilon$ for some $j \in \{1, \dots, m\}$. For all $n \in \mathbb{N}$,

$$\begin{aligned} |f_n(s) - f(s)| \le& |f_n(s) - f_n(s_j)| + |f_n(s_j) - f(s_j)| + |f(s_j) - f(s)| \\ \le& 2d(s, s_j) + |f_n(s_j) - f(s_j)| \le 2\epsilon + |f_n(s_j) - f(s_j)|. \end{aligned}$$

Since this hold for all $s \in T$, for all $n \in \mathbb{N}$,

$$\sup_{s\in T} |f_n(s) - f(s)| \le 2\epsilon + \sup_{j=1}^{m} |f_n(s_j) - f(s_j)|.$$

Since $f_n \to f$ pointwise,

$$\limsup_{n\to\infty} \sup_{s\in T} |f_n(s) - f(s)| \le 2\epsilon.$$

Since $\epsilon > 0$ has been arbitrary, $\limsup_{n\to\infty} \sup_{s\in T} |f_n(s) - f(s)| = 0$. ☐

Exercises.

2.7.1. Show that $\mathrm{LC}^b(S)$ is an ordered Banach space under the bounded Lipschitz norm $\|\cdot\|_\flat$. Show that the order cone of nonnegative bounded Lipschitz functions is solid.

2.7.2. Consider the Banach space of bounded uniformly continuous real-valued functions on the metric space S, $X = \mathrm{UC}^b(S)$, with the supremum norm. Show that the set of nonnegative bounded Lipschitz continuous functions $\mathrm{LC}^b_+(S)$ is a total cone in $\mathrm{UC}^b(S)$.

2.7.3. Let $S = [0,1]$ with the metric induced by the absolute value. Show that $\mathrm{LC}^b_+(S)$ is not a normal cone in $\mathrm{LC}^b(S)$ with the bounded Lipschitz norm by considering the function $f_n(s) = s^n$, $n \in \mathbb{N}$, $s \in S$.

2.7.4. For $s \in \mathbb{R}_+$ and $n \in \mathbb{N}$, let $f(s) = \sqrt{s}$ and $f_n(s) = \sqrt{(1/n) + s}$. Show: $f_n(s) \geq f_{n-1}(s)$ for all $n \in \mathbb{N}$ and $s \in \mathbb{R}_+$, f_n is Lipschitz on $\mathbb{R}_+$ with $\Lambda(f_n) \to \infty$, f is not Lipschitz on $\mathbb{R}_+$, and $f_n \to f$ uniformly on $\mathbb{R}_+$ as $n \to \infty$.

CHAPTER 3

The ordered vector space of real measures

Let S be a nonempty set and $\mathcal{B}$ a sigma-algebra on S. Let $\mathcal{M}(S)$ denote the vector space of all real-valued measures on $\mathcal{B}$ and $\mathcal{M}_+(S)$ the cone of all nonnegative measures.

For a structured population, S can be interpreted as the state space of individual characteristics where the state could be location, age, body size, etc. [**60**]. If $\mu : \mathcal{B} \to \mathbb{R}_+$ is a measure and $T \in \mathcal{B}$, $\mu(T)$ gives the number of individuals whose characteristic state is in the set T. With this interpretation, μ is the structural distribution of the population and $\mathcal{M}_+(S)$ the structural population state space [**60**]. Note that the properties of a sigma-algebra and a measure make immediate sense like additivity [**203**, intro]:

DEFINITION 3.1. *$\mathcal{B}$ is a σ-*algebra *on S and μ is a* measure *on $\mathcal{B}$ if the following hold:*

- *$\emptyset, S \in \mathcal{B}$ and $\mu(\emptyset) = 0$.*
- *If $T_1, T_2 \in \mathcal{B}$, then $T_1 \cup T_2 \in \mathcal{B}$, $T_1 \cap T_2 \in \mathcal{B}$, $T_1 \setminus T_2 \in \mathcal{B}$ and*

$$\begin{aligned} \mu(T_1 \cup T_2) + \mu(T_1 \cap T_2) &= \mu(T_1) + \mu(T_2), \\ \mu(T_1 \setminus T_2) + \mu(T_1 \cap T_2) &= \mu(T_1). \end{aligned} \tag{3.1}$$

- *If $(T_j)_{j\in\mathbb{N}}$ is a sequence in $\mathcal{B}$, then $\bigcup_{j\in\mathbb{N}} T_j \in \mathcal{B}$. If, in addition, $T_j \cap T_k = \emptyset$ for all $j, k \in \mathbb{N}$, $j \neq k$, then*

$$\mu\Big(\bigcup_{j\in\mathbb{N}} T_j\Big) = \sum_{j=1}^{\infty} \mu(T_j). \tag{3.2}$$

The property (3.2) *is called the σ-additivity of the measure μ. Actually,* (3.2) *implies* (3.1)*.*

Many books cover (aspects of) measure theory [**9, 22, 24, 70, 72, 73, 159**] and [**195**, App.B]. See the historic and bibliographic remarks in [**24**] for a stunning list of references. Measures as a modeling tool are highlighted in the special issues [**1, 2**]. See also [**203**] for more information and references.

In spite of this natural setting, mostly the special case is considered that $\mu(T) = \int_T f(x)\nu(dx)$, $T \in \mathcal{B}$, with some master measure ν and a density f. The structural dynamics of the population are then investigated in terms of densities [**57, 142**], perhaps because densities appear mathematically easier than measures.

Nevertheless, there is a substantial body of work in population or other dynamics in terms of measures. The reasons, among others, are [**203**]

- the conceptual naturalness of the setting [**2, 59–61, 90, 91, 110**],
- the phenomenon that, even if the initial conditions are given by densities, the solution can become measure-valued in finite time [**3, 193**] [**188**, Sec.9],

or as time tends to infinity [**4, 5, 31, 209**], or via some other limiting procedure [**82**],

- the inclusion of random structural transitions ([**114**] and the references therein) like random mutations [**4, 38, 39, 94**],
- yearly random spatial movements described by a measure kernel (Chapter 19 and Section 13.3).

In this book, as in [**72, 101, 113, 203, 204**] we want to consider the population state space $\mathcal{M}(S)$ as an ordered vector space with the cone $\mathcal{M}_+(S)$ as structural population state space in order to take advantage of the material we have developed so far and of the rich resources that are available in the literature [**135, 136, 138, 177, 178, 180, 181**], in particular the theory of homogeneous order-preserving operators and their spectral radius [**27, 130, 145, 156, 157, 167, 170, 199–202**].

$\mathcal{M}(S)$ is a vector lattice (Section 2.1.7) which is even order-complete: Each subset $\mathcal{N}$ of $\mathcal{M}(S)$ which has an lower (upper) bound has a infimum (supremum). For this, it is enough to provide an infimum (largest lower bound) to any subset $\mathcal{N}$ of $\mathcal{M}_+(S)$,

$$\mu(T) = \inf\Big\{\sum_{j=1}^{m} \mu_j(T_j)\Big\}, \qquad T \in \mathcal{B}, \tag{3.3}$$

where the infimum is taken over all $m \in \mathbb{N}$, all subsets $\{\mu_1, \dots, \mu_m\}$ of $\mathcal{N}$ and all subsets $\{T_1, \dots, T_m\}$ of $\mathcal{B}$ such that T is their disjoint union [**73**, III.7.5]. The same construction works for any subset $\mathcal{N}$ of $\mathcal{M}(S)$ with lower bound. If $\mathcal{N}$ has an upper bound, the infimum in (3.3) is replaced by a supremum.

The absolute value $|\mu|$ of a measure (in this context also called the variation of the measure) is given by $|\mu| = \sup\{\mu, -\mu\}$,

$$\begin{aligned} |\mu|(T) &= \sup\{\mu(U) - \mu(T \setminus U); \mathcal{B} \ni U \subseteq T\} \\ &= \sup\{|\mu(U)| + |\mu(T \setminus U)|; \mathcal{B} \ni U \subseteq T\} \\ &= \sup\Big\{\sum_{j=1}^{n} |\mu(T_j)|\Big\}, \end{aligned} \tag{3.4}$$

where the supremum is taken over all $n \in \mathbb{N}$ and subsets $\{T_1, \dots, T_n\}$ of $\mathcal{B}$ such that T is its disjoint union [**9**, Cor.10.54 and Thm.10.56].

The following result is a consequence of the Vitali-Hahn-Saks Theorem [**73**, III.7] [**223**, II.2].

THEOREM 3.2. *Let (μ_n) be a sequence in $\mathcal{M}(S)$ such that*

$$\lim_{n\to\infty} \mu_n(T) =: \mu(T)$$

exists for every $T \in \mathcal{B}$. Then $\mu \in \mathcal{M}(S)$.

COROLLARY 3.3. *Let (μ_n) be an increasing sequence in $\mathcal{M}_+(S)$ such that $(\mu_n(S))$ is a bounded sequence in $\mathbb{R}$. Then*

$$\mu(T) = \lim_{n\to\infty} \mu_n(T)$$

exists for every $T \in \mathcal{B}$ and provides a measure $\mu \in \mathcal{M}_+(S)$.

PROOF. Let $T \in \mathcal{B}$. Since $\mu_n \in \mathcal{M}_+(S)$, $\mu_n(T) \le \mu_n(S)$. So, $(\mu_n(T))$ is an increasing bounded sequence in $\mathbb{R}$ and has a limit. These limits provide a measure $\mu \in \mathcal{M}_+(S)$ by Theorem 3.2. □

3.1. Measures under the variation norm

The variation norm (also called total variation) on $\mathcal{M}(S)$ is defined by

$$\|\mu\|_\sharp = |\mu|(S), \qquad \mu \in \mathcal{M}(S), \tag{3.5}$$

where $|\mu|$ is the absolute value of μ defined by (3.4).

If $\mu \in \mathcal{M}_+(S)$, $\|\mu\|_\sharp = \mu(S)$. So the variation norm is additive and order-preserving on $\mathcal{M}_+(S)$. The variation norm makes $\mathcal{M}(S)$ a Banach lattice (Section 2.1.7), actually an abstract L-space, (2.13); in particular, $\mathcal{M}_+(S)$ is a non-flat generating cone: every real-valued measure μ can be written as the difference of its positive and negative variation, $\mu = \mu_+ - \mu_-$, and $\|\mu_\pm\|_\sharp \le \|\mu\|_\sharp$. By Corollary 3.3, the cone $\mathcal{M}_+(S)$ is fully regular, regular and monotonically complete (Section 2.2.4 and Definition 2.6).

The variation norm is equivalent to the supremum norm

$$\|\mu\|_\infty = \sup_{T \in \mathcal{B}} |\mu(T)| \tag{3.6}$$

and the two norms are equal on $\mathcal{M}_+(S)$.

3.2. Measures on a metric space under the flat norm

Let (S, d) be a metric space (Appendix A.1). $\mathcal{B}$ now denotes the *Borel* σ-algebra of S which is the smallest σ-algebra (Definition 3.1) that contains all open and closed sets (Definitions A.18 and A.13). The sets in the Borel σ-algebra are called *Borel sets*. In a metric space, the Borel σ-algebra is also the smallest σ-algebra for which all (bounded) continuous functions are measurable [**70**, Thm.7.1.1]. This second σ-algebra is often [**70**] but not always [**9**] called the *Baire-σ-algebra*. The following lemma guarantees that the two σ-algebras coincide in a metric space, however, for all common definitions of Baire sets. This section has been adapted from [**101**] to the format of this monograph.

LEMMA 3.4. *$\mathcal{B}$ is the smallest σ-algebra that contains all sets $f^{-1}(\{0\})$ where $f : S \to [0,1]$ and $|f(s) - f(t)| \le d(s,t)$ for all $s, t \in S$.*

The proof is a slight modification of the one of [**70**, Thm.7.1.1].

PROOF. For every closed set T in S, consider $f : S \to [0,1]$ given by $f(s) = \min\{1, d(s,T)\}$. By (2.38) and (2.37), f has the properties stated in the lemma and $T = f^{-1}(\{0\})$. $\square$

Recall the concepts of lower and upper semicontinuous functions (Appendix A.4).

THEOREM 3.5. *Let $\mu \in \mathcal{M}(S)$ and (μ_n) be a sequence in $\mathcal{M}_+(S)$ such that $\int_S h\mu_n \to \int_S h d\mu$ as $n \to \infty$ for every bounded Lipschitz continuous function $h : S \to \mathbb{R}_+$.*

Then the following hold for bounded $f : S \to \mathbb{R}$.

(a) *If f is lower semicontinuous, then* $\liminf_{n\to\infty} \int_S f d\mu_n \ge \int_S f d\mu$.

(b) *If f is upper semicontinuous, then* $\limsup_{n\to\infty} \int_S f d\mu_n \le \int_S f d\mu$.

(c) *If f is continuous, then* $\int_S f d\mu_n \to \int_S f d\mu$.

(d) *If U is an open subset of S, then* $\liminf_{n\to\infty} \mu_n(U) \geq \mu(U)$.

(e) *If V is a closed subset of S, then* $\limsup_{n\to\infty} \mu_n(V) \leq \mu(V)$.

(f) *(f) If T is a Borel set in S and $\mu(\partial T) = 0$, then* $\mu(T) = \lim_{n\to\infty} \mu_n(T)$.

PROOF. If $h : S \to \mathbb{R}$ is a bounded Lipschitz continuous function, $h = h_+ - h_-$ where $h_+ = (1/2)(|h| + h)$ and $h_- = (1/2)(|h| - h)$ are bounded nonnegative Lipschitz continuous functions. So

$$\int_S h d\mu_n \to \int_S h d\mu, \qquad n \to \infty, \tag{3.7}$$

for all bounded Lipschitz continuous functions $h : S \to \mathbb{R}$.

(a) Let $f : S \to \mathbb{R}$ be bounded and lower semicontinuous. By Proposition 2.78 (e), there exists an increasing sequence (f_m) of bounded Lipschitz continuous functions $f_m : S \to \mathbb{R}$ such that $f_m \to f$ pointwise and $\inf f \leq f_m \leq f$ for all $m \in \mathbb{N}$. For all $m, n \in \mathbb{N}$, since $\mu_n \in \mathcal{M}_+(S)$,

$$\int_S f d\mu_n \geq \int_S f_m d\mu_n.$$

By (3.7), for all $m \in \mathbb{N}$,

$$\liminf_{n\to\infty} \int_S f d\mu_n \geq \liminf_{n\to\infty} \int_S f_m d\mu_n = \int_S f_m d\mu.$$

Since $f_m \to f$ pointwise and $\inf f \leq f_m \leq f$ for all $m \in \mathbb{N}$, we can take the limit as $m \to \infty$ by the dominated convergence theorem and obtain (a).

(b) Let $f : S \to \mathbb{R}$ be bounded and upper semicontinuous. Apply (a) to the lower semicontinuous function $g : S \to \mathbb{R}$ defined by $g(s) = \sup f - f(s)$.

(c) If f is continuous, it is both lower and upper semicontinuous and (c) follows from (a) and (b).

(d)–(e) If U is open, χ_U is lower semicontinuous. If V is closed, χ_V is upper semicontinuous.

(f) Let T be a Borel set in S. By (d) and (e),

$$\mu(\breve{T}) \leq \liminf_{n\to\infty} \mu(\breve{T}) \leq \limsup_{n\to\infty} \mu(\bar{T}) \leq \mu(\bar{T}).$$

If $\mu(\partial T) = 0$, then $\mu(\breve{T}) = \mu(\bar{T})$, and equality holds. □

We next show that nonnegativity or real measures is preserved under weak forms of convergence.

THEOREM 3.6. *Let $\mu \in \mathcal{M}(S)$ and (μ_n) be a sequence in $\mathcal{M}_+(S)$ such that $\int_S h\mu_n \to \int_S h d\mu$ as $n \to \infty$ for every bounded Lipschitz continuous function $h : S \to \mathbb{R}_+$.*

Then $\mu \in \mathcal{M}_+(S)$.

PROOF. We have $\int_S f d\mu \geq 0$ first for all nonnegative bounded Lipschitz continuous functions and then all nonnegative bounded continuous functions (Proposition 2.78). By Proposition 2.76, $\mu(T) \geq 0$ for all closed and open sets T in S. Since $\mathcal{M}_+(S)$ is generating, there exist $\mu_\pm \in \mathcal{M}_+(S)$ with $\mu = \mu_+ - \mu_-$. Let $\delta > 0$ and T a Borel set. Since μ_+ is outer regular, [**22**, Cor.7.2.6], there exists an open superset T_2 of T in S such that $\mu_+(T_2 \setminus T) \le \delta$. Since μ_- is inner regular, there exists a closed subset T_1 such that $\mu_-(T \setminus T_1) \le \delta$. See [**9**, Sec.12.1]. In one view,

$$T_1 \subseteq T \subseteq T_2 \subseteq S, \qquad \mu_+(T_2 \setminus T) \le \delta, \qquad \mu_-(T \setminus T_1) \le \delta. \tag{3.8}$$

By Lemma 2.75, since T_1 and $S \setminus T_2$ are closed disjoint sets, there exists a bounded continuous function f with values between 0 and 1 such that $f = 1$ on T_1 and $f = 0$ on $S \setminus T_2$. Since $\int_S f d\mu \ge 0$ and $\mu = \mu_+ - \mu_-$,

$$\mu(T) \ge \mu_+(T) - \mu_-(T) - \int_S f d(\mu_+ - d\mu_-).$$

We reorganize,

$$\mu(T) \ge \mu_+(T) - \int_S f d\mu_+ + \int_S f d\mu_- - \mu_-(T).$$

Since $f = 0$ on $S \setminus T_2$ and $f \ge 0$ on S,

$$\mu(T) \ge \mu_+(T) - \int_{T_2} f d\mu_+ + \int_{T_1} f d\mu_- - \mu_-(T).$$

Since $f = 1$ on T_1 and $f \le 1$ on S,

$$\mu(T) \ge \mu_+(T) - \mu_+(T_2) + \mu_-(T_1) - \mu_-(T).$$

By (3.8),

$$\mu(T) \ge -\mu_+(T_2 \setminus T) - \mu_-(T \setminus T_1) \ge -2\delta.$$

Since this holds for any $\delta > 0$, $\mu(T) \ge 0$. □

For perspective, we present the following result for the variation norm.

THEOREM 3.7. *For all* $\mu \in \mathcal{M}(S)$,

$$\|\mu\|_\sharp = |\mu|(S) = \sup\Big\{\Big|\int_S f d\mu\Big|; f \in C^b(S), \|f\|_\infty \le 1\Big\}.$$

PROOF. By [**73**, IV.6.2], the map $\mu \mapsto \theta$ with $\theta(f) = \int_S f d\mu$, $f \in C^b(S)$, is an isometric isomorphism between the Banach space of regular additive set functions with the variation norm and the dual space of $C^b(S)$. The assertion now follows because every real measure on $\mathcal{B}$ is regular [**9**, Thm.12.5] [**73**, III.9.22] (Exercise 3.2.2). □

We introduce the following functional on $\mathcal{M}(S)$,

$$\begin{aligned} \|\mu\|_\flat &= \sup_{f \in \mathcal{F}} \Big|\int_S f d\mu\Big|, \\ \mathcal{F} &= \Big\{f \in [0,1]^S;\ |f(s) - f(t)| \le d(s,t) \text{ for } s, t \in S\Big\}. \end{aligned} \tag{3.9}$$

Recall that M^S denotes the set of functions from S to a set M. One readily checks that $\|\cdot\|_\flat$ is a seminorm on $\mathcal{M}(S)$ and

$$\|\mu\|_\flat \le \|\mu\|_\sharp, \qquad \mu \in \mathcal{M}(S). \tag{3.10}$$

To show that $\|\cdot\|_\flat$ is a norm, let $\|\mu\|_\flat = 0$. Then $\int_S f d\mu = 0$ for all bounded Lipschitz continuous functions. Since the indicator function of every closed or open set is the pointwise limit of bounded Lipschitz continuous functions (Proposition 2.76), $\mu(T) = 0$ for all closed or open subsets of S. So $\mu(T) = 0$ for all Borel sets T in S and $\|\cdot\|_\flat$ is a norm.

In the literature, definitions different from (3.9) are used [**70**, **99**, **100**, **102**, **213**] that lead to equivalent norms. For instance, $[0,1]^S$ is replaced by $[-1,1]^S$. For details see Section 3.2.6.

Sometimes in the literature [**70**, **166**], (3.9) is only defined for $\mu = P - Q$ where P and Q are probability measures providing a metric on the set of probability measures. The norm $\|\cdot\|_\flat$ (or an equivalent norm) is called the *flat* norm [**85**, Sec.4.1.12; **86**] or the *bounded Lipschitz* norm [**166**, p.63] or BL^* norm [**69**] or the *dual bounded Lipschitz* norm [**70**, p.394, 404] or the *Kantorovich–Rubinstein* norm [**24**, Sec.8.3]. To acknowledge the work in [**88**] and [**69**, **70**], it is also called *Fortet-Mourier* norm [**143**] or *Dudley* norm [**82**]. See Section 3.2.6 for more. We call $\|\cdot\|_\flat$ the *flat norm.*

All the definitions have in common that

$$\|\mu\|_\flat = \mu(S) = \|\mu\|_\sharp, \qquad \mu \in \mathcal{M}_+(S). \tag{3.11}$$

This implies that the flat norm is additive on $\mathcal{M}_+(S)$,

$$\|\mu + \nu\|_\flat = \|\mu\|_\flat + \|\nu\|_\flat, \qquad \mu, \nu \in \mathcal{M}_+(S). \tag{3.12}$$

Further, the norm is order-preserving on $\mathcal{M}_+(S)$, i.e., if $\mu, \nu \in \mathcal{M}_+(S)$ and $\mu \le \nu$, then $\|\mu\|_\flat \le \|\nu\|_\flat$.

The space of bounded Lipschitz continuous functions, $X = \mathrm{LC}^b(S)$ is an ordered Banach space under the norm

$$\|f\|_\flat = \max\left\{ \sup_{s\in S} |f(s)|, \sup\left\{\frac{|f(s) - f(t)|}{d(s,t)}; s, t \in S, s \neq t\right\}\right\}, \tag{3.13}$$

with the cone X_+ being formed by the nonnegative functions in $\mathrm{LC}^b(S)$. The function $f \equiv 1$ is an interior point of X_+, and X_+ is generating. In general, X_+ is not normal. Since the infimum and the supremum of two bounded Lipschitz functions f and g are again bounded and Lipschitz continuous, $\mathrm{LC}^b(S)$ is a vector lattice. However, the infimum and supremum do not depend continuously on f and g [**156**, p.535]. See Section 2.7.2.

Our preceding consideration shows that $\mathcal{M}(S)$ can be considered a linear subspace of X^* (the dual space of X) which is an ordered Banach space with the dual cone $X_+^* = \{x^* \in X^*; x^*x \ge 0, x \in X_+\}$.

There are many interesting duality considerations involving $\mathcal{M}(S)$ and $X = \mathrm{LC}^b(S)$ [**72**, Sec.1.3.] [**113**] into which we will not go here. Rather, for lean notation, we will formulate our results in terms of $\mathcal{M}(S)$ only. For other metrics on spaces of measures (or on the set of probability measures) than the one induced by the flat norm see [**24**, Sec.8.10(viii)] and [**70**, Sec.11.3].

Once we have endowed $\mathcal{M}(S)$ with the flat norm, we need to show that $\mathcal{M}_+(S)$ is closed under it. In the following, all topological notions concerning $\mathcal{M}(S)$ and $\mathcal{M}_+(S)$ are meant with respect to the flat norm unless it is explicitly said otherwise.

The next result basically follows from Theorem 3.6.

THEOREM 3.8. *$\mathcal{M}_+(S)$ is a generating, normal, closed, and completely regular cone in $\mathcal{M}(S)$.*

LEMMA 3.9. *Let (S,d) be a metric space. For $s \in S$, let δ_s denote the Dirac measure at s.*

Then $1 = \|\delta_s\|_\flat$ and, for $t, s \in S$,

$$\|\delta_s - \delta_t\|_\flat = \min\{1, d(s,t)\}.$$

Similar estimates are shown in [**113, 213**]. See Section 3.2.6. Recall that $\|\delta_s - \delta_t\|_\sharp = 2$ for all $s, t \in S$ with $s \neq t$. So, unless the metric space S is uniformly discrete (Definition A.3), the metric induced by flat norm on $\mathcal{M}_+(S)$ is not topologically equivalent to the metric induced by the variation norm on $\mathcal{M}_+(S)$. See Appendix A.1.7.

PROOF. By (3.11), $\|\delta_s\|_\flat = 1$, $s \in S$.

Let $s, t \in S$.

Consider $f \in \mathcal{F}$, see (3.9), i.e., $f : S \to [0,1]$ and $|f(s) - f(t)| \leq d(s,t)$ for all $s, t \in S$. Then

$$|\delta_s(f) - \delta_t(f)| = |f(s) - f(t)|.$$

On the one hand, since $0 \leq f \leq 1$,

$$|\delta_s(f) - \delta_t(f)| \leq 1.$$

On the other hand,

$$|\delta_s(f) - \delta_t(f)| \leq d(s,t).$$

By (3.9), $\|\delta_s - \delta_t\|_\flat \leq \min\{1, d(s,t)\}$.

To show the opposite inequality, with $s, t \in S$ given above, define

$$f(u) = \min\{1, d(u,s)\}, \qquad u \in S.$$

Then $0 \leq f \leq 1$. Further, by discussing the various cases,

$$|f(u) - f(v)| \leq |d(u,s) - d(v,s)| \leq d(u,v), \qquad u, v \in S.$$

So $f \in \mathcal{F}$. By (3.9),

$$\|\delta_s - \delta_t\|_\flat \geq |f(s) - f(t)| = f(t) = \min\{1, d(t,s)\}. \qquad \square$$

This lemma has important consequences for the relation between the metric on S and the flat norm.

COROLLARY 3.10. *If $U_1(0) \cap \mathcal{M}_+(S)$ is totally bounded, then S is totally bounded.*

Later, we will also show the converse (Theorem 3.32).

COROLLARY 3.11. *If S is not uniformly discrete (Definition A.3), then $\mathcal{M}_+(S)$ is a flat generating cone in the ordered normed vector space $\mathcal{M}(S)$ which is not complete: There exists a sequence $(\mu_n) \in \mathcal{M}(S)$ such that $\|\mu_n\|_\flat \to 0$ and, for all sequences $(\mu_n^\pm)_n$ in $\mathcal{M}_+(S)$ with $\mu_n = \mu_n^+ - \mu_n^-$, it follows that*

$$\liminf_{n\to\infty} \|\mu_n^+\|_\flat > 0 \text{ and } \liminf_{n\to\infty} \|\mu_n^-\|_\flat > 0.$$

PROOF. Since S is not uniformly discrete, there are sequences (s_n) and (t_n) in S such that $0 < d(s_n, t_n) \to 0$. Choose $\mu_n = \delta_{s_n} - \delta_{t_n}$. By Lemma 3.9, $\|\mu_n\|_\flat \leq d(s_n, t_n) \to 0$. Let $\mu_n^\pm \in \mathcal{M}_+(S)$ such that $\mu_n = \mu_n^+ - \mu_n^-$ for all $n \in \mathbb{N}$. Since $s_n \neq t_n$,

$$\|\mu_n^+\|_\flat = \mu_n^+(S) \geq \mu_n^+(\{s_n\}) \geq \mu_n(\{s_n\}) = 1 - 0 = 1.$$

Since $\mu_n^- - \mu_n^+ = \delta_{t_n} - \delta_{s_n}$, by symmetry, $\|\mu_n^-\|_\flat \geq 1$ for all $n \in \mathbb{N}$.

This shows that the generating cone $\mathcal{M}_+(S)$ is flat. Since any generating cone in an ordered Banach space is non-flat [**136**, Thm.1.5] (Corollary 2.38), $\mathcal{M}(S)$ is not complete. $\square$

THEOREM 3.12. *The following are equivalent:*

(i) $\mathcal{M}(S)$ *is complete.*
(ii) S *is uniformly discrete (Definition* A.3*).*
(iii) *The flat norm and the variation norm are equivalent on* $\mathcal{M}(S)$ *(Appendix* A.1.7*).*

PROOF. (i) $\Rightarrow$ (ii): This follows from Corollary 3.11.
(ii) $\Rightarrow$ (iii): This is shown in the proof of [**113**, Thm.3.11].
(iii) $\Rightarrow$ (i): This follows from $\mathcal{M}(S)$ being complete under the variation norm. $\square$

REMARK 3.13. We mention that it is impossible to find a norm $\|\cdot\|$ on $\mathcal{M}(S)$ with all of the following three properties:

(i) $\mathcal{M}(S)$ is complete under $\|\cdot\|$.
(ii) $\mathcal{M}_+(S)$ is closed under $\|\cdot\|$.
(iii) $\|\cdot\|$ is not equivalent to the variation norm.

PROOF. Use [**18**, Thm.2.9] and the fact that $\mathcal{M}(S)$ is an ordered Banach space under the variation norm. $\square$

Recall that $\mathcal{M}_+(S)$ is a Banach lattice under the variation norm. In particular, taking the infimum of two measures is a continuous operation. This does not hold under the flat norm.

COROLLARY 3.14. *Let* S *have an accumulation point. Then the operation* $(\mu,\nu) \to \mu \wedge \nu$ *is not continuous from* $\mathcal{M}_+(S)^2 \to \mathcal{M}_+(S)$ *under the flat norm.*

PROOF. Choose an accumulation point $s \in S$ and a sequence (s_n) in S such that $s \neq s_n \to s$. By Lemma 3.9, $\|\delta_{s_n} - \delta_s\|_\flat \to 0$, but $\delta_{s_n} \wedge \delta_s = 0$ and $\delta_s \wedge \delta_s = \delta_s$. $\square$

In the following, we focus on the cone $\mathcal{M}_+(S)$. Recall that all topological notions are meant with respect to the flat norm unless explicitly said otherwise.

Exercises.

3.2.1. Let S be a nonempty set with the discrete metric $d(s,t) = 1$ if $t, s \in S$ and $s \neq t$ and $d(s,s) = 0$ for all $s \in S$.

Then $\|\mu\|_\flat \geq \sup_{T \subseteq S} |\mu(T)|$ for all $\mu \in \mathcal{M}(S)$ and the flat norm is equivalent to the variation norm.

3.2.1. Convergence in $\mathcal{M}_+(S)$. Recall that all topological notions are meant with respect to the flat norm unless explicitly said otherwise. In the following, we focus on the cone $\mathcal{M}_+(S)$. The subsequent results may not hold in $\mathcal{M}(S)$. See Exercises 3.2.4 and 3.2.5.

Recall the definitions at the beginning of Section 2.7.

Recall Definition 2.82.

PROPOSITION 3.15. *Let* $\mathcal{F}$ *be an equicontinuous and equibounded family of functions* $f : S \to \mathbb{R}_+$. *Assume that* $\mu \in \mathcal{M}_+(S)$ *and* (μ_n) *is a sequence in* $\mathcal{M}_+(S)$ *such that* $\|\mu_n - \mu\|_\flat \to 0$ *as* $n \to \infty$.

Then $\int_S f d\mu_n \to \int_S f d\mu$ *as* $n \to \infty$ *uniformly for* $f \in \mathcal{F}$.

Cf. [**203**, Prop.6.10].

PROOF. We can assume that $0 \leq f(s) \leq 1$ for all $f \in \mathcal{F}$. For $f \in \mathcal{F}$, let (f_k) be the approximation in (2.42). Then $0 \leq f_k(s) \leq f(s) \leq 1$ for all $k \in \mathbb{N}$ and $f \in \mathcal{F}$. By Lemma 2.83,

$$h_k(s) := \sup_{f \in \mathcal{F}} |f(s) - f_k(s)| \to 0, \qquad k \to \infty, \quad s \in S.$$

Since $\{f - f_k; f \in \mathcal{F}\}$ is equicontinuous, h_k is continuous and thus integrable. Since $\mathcal{F}$ is equibounded, $h_k(s) \leq \sup_{f \in \mathcal{F}} f(s) =: c \in \mathbb{R}_+$. By the dominated convergence theorem,

$$\sup_{f \in \mathcal{F}} \Big(\int_S f d\mu - \int_S f_k d\mu \Big) = \sup_{f \in \mathcal{F}} \int_S |f - f_k| d\mu \leq \int_S h_k d\mu \to 0, \qquad k \to \infty.$$

Let $\epsilon > 0$. Then there exists $k \in \mathbb{N}$ such that

$$0 \leq \int_S f d\mu - \int_S f_k d\mu \leq \frac{\epsilon}{4}, \qquad f \in \mathcal{F}.$$

For all $n \in \mathbb{N}$,,

$$\int_S f d\mu \leq \int_S f_k d\mu + \frac{\epsilon}{4} = \int_S f_k d(\mu - \mu_n) + \int_S f_k d\mu_n + \frac{\epsilon}{4}.$$

Since k is a Lipschitz constant for f_k and $f_k \leq f$,

$$\int_S f d\mu \leq k\|\mu - \mu_n\|_\flat + \int_S f d\mu_n + \frac{\epsilon}{4}, \qquad n \in \mathbb{N}.$$

Recall (3.9). Since $\|\mu - \mu_n\|_\flat \to 0$ as $n \to \infty$, there exists some $N = N_k \in \mathbb{N}$ such that $\|\mu - \mu_n\| \leq \epsilon/(4k)$ for all $n \geq N_k$,

$$\int_S f d\mu \leq \int_S f d\mu_n + \frac{\epsilon}{2}, \qquad n \geq N_k, \quad f \in \mathcal{F}. \tag{3.14}$$

Since $\mathcal{F}$ is equibounded, there exists some $c > 0$ such that $f(s) \leq c$ for all $s \in S$ and all $f \in \mathcal{F}$. Define $g_f(s) = c - f(s)$ for all $s \in S$ and all $f \in \mathcal{F}$. Then $\{g_f; f \in \mathcal{F}\}$ is an equicontinuous and equibounded family in $C_+^b(S)$. By (3.14), applied to $\{g_f\ f \in \mathcal{F}\}$ instead of $\mathcal{F}$, there exists some $N \in \mathbb{N}$ such that, for all $f \in \mathcal{F}$,

$$c\mu(S) - \int_S f du = \int_S g_f d\mu \leq \int_S g_f d\mu_n + \frac{\epsilon}{2} \leq c\mu_n(S) - \int_S f d\mu_n + \frac{\epsilon}{2}, \qquad n \geq N.$$

We regroup,

$$\int_S f d\mu \geq \int_S f d\mu_n - \frac{\epsilon}{2} + c(\mu(S) - \mu_n(S)), \qquad n \geq N, \quad f \in \mathcal{F}.$$

Since $\mu_n(S) \to \mu(S)$ as $n \to \infty$, we find some $\check{N} \in \mathbb{N}$ such that

$$\int_S f d\mu \geq \int_S f d\mu_n - \epsilon, \qquad n \geq \check{N}, \quad f \in \mathcal{F}.$$

We combine the last inequality with (3.14) and obtain $\int_S f d\mu_n \to \int_S f d\mu$ uniformly for $f \in \mathcal{F}$. $\square$

We aim at finding conditions under which a converse of Theorem 3.5 (c) holds.

DEFINITION 3.16. *A subset $\mathcal{N}$ of $\mathcal{M}_+(S)$ is called* tight *if for any $\epsilon > 0$ there exists a compact subset K of S such that $\mu(S \setminus K) < \epsilon$ for all $\mu \in \mathcal{N}$.*

A single measure $\mu \in \mathcal{M}(S)$ is called tight if $\{|\mu|\}$ is tight.

The set of tight measures is denoted by $\mathcal{M}^t(S)$ and the set of nonnegative tight measures is denoted by $\mathcal{M}^t_+(S)$. $\mathcal{M}^t(S)$ is a linear subspace of $\mathcal{M}(S)$ and $\mathcal{M}^t_+(S)$ is a generating cone of $\mathcal{M}^t(S)$.

S is called universally tight *if every $\mu \in \mathcal{M}_+(S)$ is tight.*

Universally tight metric spaces are called Radon spaces in [**13**].

Since, in a metric space, a set T is compact if and only if it is totally bounded and complete (Theorem A.30), we also consider the following concepts [**69**] [**113**] [**215**, Sec.1.3].

DEFINITION 3.17. *A subset $\mathcal{N}$ of $\mathcal{M}_+(S)$ is called* pre-tight *if for any $\epsilon > 0$ there exists a closed totally bounded subset K of S such that $\mu(S \setminus K) < \epsilon$ for all $\mu \in \mathcal{N}$.*

A single measure $\mu \in \mathcal{M}(S)$ is called pre-tight if $\{|\mu|\}$ is pre-tight.

A measure $\mu \in \mathcal{M}(S)$ is called separable *if there exists a separable Borel subset K of S such that $|\mu|(S \setminus K) = 0$.*

The linear subspace of separable measures is denoted by $\mathcal{M}^s(S)$ and the cone of nonnegative separable measures by $\mathcal{M}^s_+(S)$.

S is called universally pre-tight *if every $\mu \in \mathcal{M}_+(S)$ is pre-tight, i.e., $\mathcal{M}_+(S) = \mathcal{M}^s_+(S)$.*

To show that the universal tightness does not change under topologically equivalent metrics (Appendix A.1.7) we note the following [**215**, L.1.3.2].

PROPOSITION 3.18. *Let $\mu \in \mathcal{M}_+(S)$. The following are equivalent:*

(i) *μ is pre-tight.*
(ii) *For every $\epsilon > 0$, there exists a separable Borel set $U \subseteq S$ such that $\mu(S \setminus U) < \epsilon$.*
(iii) *μ is separable.*

PROOF. Since every totally bounded set is separable, (i) implies (ii). (ii) implies (iii) because countable unions of separable sets are separable (Exercise A.1.7).

The proof of (iii) $\Rightarrow$ (i) follows the one of [**22**, Thm.7.3.3]. Since μ is separable, there exists a separable Borel set $T \subseteq S$ such that $\mu(S \setminus T) = 0$. Let $\{x_k; k \in \mathbb{N}\}$ be a subset of T such that T is contained in its closure. Then $T \subseteq \bigcup_{k\in\mathbb{N}} \bar{U}_\delta(x_k)$ for all $\delta > 0$. Since $\mu \in \mathcal{M}_+(S)$ and $\mu(S) = \mu(T)$,

$$\mu(S) = \lim_{\ell\to\infty} \mu\Big(\bigcup_{k=1}^{\ell} \bar{U}_\delta(x_k)\Big), \qquad \delta > 0.$$

Let $\epsilon > 0$. For each $n \in \mathbb{N}$, there exists some $k_n \in \mathbb{N}$ such that

$$\mu(D_n) \ge \mu(S) - \frac{\epsilon}{2^n}, \qquad D_n = \bigcup_{k=1}^{k_n} \bar{U}_{1/n}(x_k).$$

Each D_n is closed and so is $D = \bigcap_{n=1}^\infty D_n$,

$$\mu(S \setminus D) = \mu\Big(\bigcup_{n\in\mathbb{N}} \mu(S \setminus D_n)\Big) \le \sum_{n=1}^\infty \mu(S \setminus D_n) \le \sum_{n=1}^\infty \frac{\epsilon}{2^n} = \epsilon.$$

Since $D \subseteq D_n$ for all $n \in \mathbb{N}$ and D_n is the union of finitely many closed neighborhoods with radius $1/n$, D is totally bounded. $\square$

REMARK 3.19. Universally tight metric spaces are universally pre-tight, and complete universally pre-tight metric spaces are universally tight.

Separable metric spaces are universally pre-tight. Separable complete metric spaces are universally tight (Ulam's Theorem [**70**, 7.1.4]).

PROPOSITION 3.20. *Let (μ_n) be a sequence in $\mathcal{M}_+(S)$ and $\mu \in \mathcal{M}_+(S)$ and $\|\mu_n - \mu\|_\flat \to 0$. Then the following hold:*

(i) *If (μ_n) is tight, then μ is tight.*
(ii) *If (μ_n) is pre-tight, then μ is pre-tight.*
(iii) *If all μ_n are separable, then μ is separable.*

PROOF. (i) follows from Proposition 3.5 (d).

(iii). For each $n \in \mathbb{N}$, since μ_n is separable, there exists a separable Borel set set T_n in S such that $\mu_n(S \setminus T_n) = 0$. Let T be the closure of the union of the T_n. Then T is separable and $\mu_n(S \setminus T) = 0$ for all $n \in \mathbb{N}$. By Proposition 3.5 (d), $\mu(S \setminus T) \le \liminf_{n\to\infty} \mu_n(S \setminus T) = 0$.

(ii) follows from (iii) and Proposition 3.18. $\square$

COROLLARY 3.21. *$\mathcal{M}_+^s(S)$ is a closed cone of $\mathcal{M}(S)$.*

REMARK 3.22. $\mathcal{M}_+^t(S)$ is a closed cone of $\mathcal{M}(S)$ if S is complete because then $\mathcal{M}_+^t(S) = \mathcal{M}_+^s(S)$. In general, $\mathcal{M}_+^t(S)$ is not a closed cone of $\mathcal{M}(S)$ because it is not closed. It will follow from Theorem 3.26 that $\mathcal{M}_+^t(S)$ is dense in $\mathcal{M}_+^s(S)$. Assume that $\mathcal{M}_+^t(S)$ is closed and S is separable. Then $\mathcal{M}_+^t(S) = \mathcal{M}_+(S)$. This implies that S is universally measurable, however, not all separable metric spaces are universally measurable ([**70**, Sec.11.5]).

We proceed on our characterization of convergence.

PROPOSITION 3.23. *Let (μ_n) be a sequence in $\mathcal{M}_+(S)$ and $\mu \in \mathcal{M}_+^s(S)$ such that $\int_S f\, d(\mu_n - \mu) \to 0$ as $n \to \infty$ for any Lipschitz continuous function $f : S \to [0,1]$.*

Then $\|\mu_n - \mu\|_\flat \to 0$ as $n \to \infty$.

PROOF. Assume that $\int_S f\, d(\mu_n - \mu) \to 0$ for any Lipschitz continuous function $f : S \to [0,1]$.

Since $\mu \in \mathcal{M}_+^s(S)$, μ is pre-tight by Proposition 3.18, and we can choose a sequence of totally bounded subsets (T_k) of S such that

$$\mu(S \setminus T_k) \to 0, \qquad k \to \infty. \tag{3.15}$$

Let

$$T \text{ be the closure of } \bigcup_{k\in\mathbb{N}} T_k. \tag{3.16}$$

T is a separable subset of S (Exercise A.1.7).

Suppose that $\|\mu_n - \mu\|_\flat \to 0$ is false. Then there exists some $\epsilon > 0$ and a sequence (f_n) of Lipschitz continuous functions with values between 0 and 1 and Lipschitz constants ≤ 1 such that

$$\Big| \int_S f_n\, d(\mu_n - \mu) \Big| \ge \epsilon. \tag{3.17}$$

Since T is separable, after choosing a subsequence, there is some $f : T \to \mathbb{R}$ such that $f_n \to f$ as $n \to \infty$ pointwise on T and uniformly on each set T_k (Proposition 2.86). The function f is also bounded and Lipschitz continuous on T with a Lipschitz constant ≤ 1. By Remark 2.79, we can extend f so that f is defined and Lipschitz continuous on S with a Lipschitz constant ≤ 1.

Let $\delta > 0$. By (3.15), choose some $k \in \mathbb{N}$ such that $\mu(S \setminus T_k) < \delta$.

We claim that there exists some $\eta > 0$ such that

$$\limsup_{n\to\infty} \sup_{t\in U_\eta(T_k)} |f_n(t) - f(t)| \le \delta. \tag{3.18}$$

Suppose not. Then, for any $j \in \mathbb{N}$, there exists some $n_j \in \mathbb{N}$ with $n_j > j$ such that

$$\sup_{t\in U_{1/j}(T_k)} |f_{n_j}(t) - f(t)| > \delta.$$

So there exists a sequence (s_j) in S such that $d(s_j, T_k) < 1/j$ and

$$|f_{n_j}(s_j) - f(s_j)| > \delta, \qquad j \in \mathbb{N}. \tag{3.19}$$

There exists a sequence (t_j) in T_k such that $d(s_j, t_j) < 1/j$.

Since $f_{n_j} \to f$ uniformly on T_k, $f_{n_j}(t_j) - f(t_j) \to 0$ as $j \to \infty$ and

$$\begin{aligned}|f_{n_j}(s_j) - f(s_j)| \le& |f_{n_j}(s_j) - f_{n_j}(t_j)| + |f_{n_j}(t_j) - f(t_j)| + |f(t_j) - f(s_j)|\\ \le& 2d(s_j, t_j) + |f_{n_j}(t_j) - f(t_j)| \to 0, \qquad j \to \infty,\end{aligned}$$

a contradiction to (3.19). This proves (3.18).

For $\eta > 0$ in (3.18), let g be the Lipschitz continuous function from Lemma 2.75 which takes values between 0 and 1 and $g \equiv 1$ on T_k and $g \equiv 0$ on $S \setminus U_\eta(T_k)$. For all $n \in \mathbb{N}$,

$$\begin{aligned}\Big|\int_S f_n\, d(\mu_n - \mu)\Big| \le& \Big|\int_S f_n(1-g)\, d\mu_n\Big| + \Big|\int_S (f_n - f) g\, d\mu_n\Big|\\ &+ \Big|\int_S fg\, d(\mu_n - \mu)\Big| + \Big|\int_S (f - f_n) g\, d\mu\Big| + \Big|\int_S f_n(g-1)\, d\mu\Big|.\end{aligned}$$

For all $n \in \mathbb{N}$, since $0 \le f_n, f, g \le 1$ and $g \equiv 0$ on $S \setminus U_\eta(T_k)$,

$$\begin{aligned}\Big|\int_S f_n\, d(\mu_n - \mu)\Big| \le& \int_S (1-g)\, d\mu_n + \sup_{s\in U_\eta(T_k)} |f_n(s) - f(s)|(\mu_n(S) + \mu(S))\\ &+ \Big|\int_S fg\, d(\mu_n - \mu)\Big| + \int_S (1-g)\, d\mu.\end{aligned}$$

Since fg and $1 - g$ are Lipschitz continuous with values between 0 and 1, by assumption

$$\begin{aligned}&\limsup_{n\to\infty} \Big|\int_S f_n\, d(\mu_n - \mu)\Big|\\ &\quad\le 2\int_S (1-g) d\mu + \limsup_{n\to\infty} \sup_{s\in U_\eta(T_k)} |f_n(s) - f(s)|\, 2\mu(S).\end{aligned}$$

Since $g \equiv 1$ on T_k and $\mu(S \setminus T_k) < \delta$ and (3.18) holds,

$$\limsup_{n\to\infty} \Big|\int_S f_n\, d(\mu_n - \mu)\Big| \le 2\delta + 2\delta\, \mu(S).$$

Since this holds for any $\delta > 0$, we obtain

$$\limsup_{n\to\infty}\Big|\int_S f_n\, d(\mu_n-\mu)\Big| = 0,$$

which contradicts (3.17). □

We combine Theorem 3.5 and Proposition 3.23 to the following result.

THEOREM 3.24. *Let (μ_n) in $\mathcal{M}_+(S)$ and $\mu \in \mathcal{M}_+^s(S)$. Then the following are equivalent:*

(i) $\|\mu_n - \mu\|_\flat \to 0$,
(ii) $\int_S f d(\mu_n - \mu) \to 0$ *for all continuous functions* $f \in C^b(S)$,
(iii) $\int_S f d(\mu_n - \mu) \to 0$ *for all Lipschitz continuous functions* $f : S \to [0,1]$.

Cf. [**70**, Thm.11.3.3].

The next result states that a converging sequence of tight measures with tight limit measure forms a tight set of measures.

THEOREM 3.25 (Le Cam, 1957). *Let (μ_n) be a sequence in $\mathcal{M}_+^t(S)$ and μ an element of $\mathcal{M}_+^t(S)$. Assume that*

$$\int_S f d\mu_n \to \int_S f d\mu, \qquad n\to\infty, \quad f \in C_+^b(S).$$

Then $\{\mu_n; n\in\mathbb{N}\}\cup\{\mu\}$ is a tight set of measures.

PROOF. This is [**70**, Thm.11.5.3] if all μ_n are probability measures. We can assume that μ is not the zero measure. Otherwise, $\mu_n(S) \to 0$ as $n \to \infty$ and the statement is valid.

Without loss of generality, we can assume that all μ_n are not the zero measure. Set

$$\nu(T) = \frac{\mu(T)}{\mu(S)}, \qquad \nu_n(T) = \frac{\mu_n(T)}{\mu_n(S)}, \quad n \in \mathbb{N},\ T \in \mathcal{B}.$$

Then

$$\int_S f d\nu_n \to \int_S f d\nu, \qquad n\to\infty, \quad f \in C_+^b(S).$$

By [**70**, Thm.11.5.3], $\{\nu_n; n\in\mathbb{N}\}\cup\{\nu\}$ is a tight set of measures. Since $\{\mu_n(S); n \in \mathbb{N}\}$ is bounded, $\{\mu_n; n\in\mathbb{N}\}\cup\{\mu\}$ is a tight set of measures. □

3.2.2. Dense subsets. Linear combinations of point measures are dense with respect to the flat norm.

THEOREM 3.26. *The set of measures $\sum_{j=1}^n q_j\delta_{s_j}$, $n\in\mathbb{N}$, $q_j \in \mathbb{Q}_+$ and $s_j \in S$, is dense in $\mathcal{M}_+^s(S)$.*

The set of measures $\displaystyle\sum_{j=1}^n q_j\delta_{s_j}$, $n\in\mathbb{N}$, $q_j\in\mathbb{Q}$ *and* $s_j \in S$, *is dense in* $\mathcal{M}^s(S)$.

If (S,d) is separable, then $\mathcal{M}(S)$ is separable.

This result is one part of [**113**, Thm.3.9]. For a proof, see [**101**, Thm.4.19] [**72**, Thm.1.37]. This result will be used in Proposition 13.21.

Next we show that Lebesgue absolutely continuous measures are dense in $\mathcal{M}(\mathbb{R}^n)$ with the flat norm.

Let $\mu \in \mathcal{M}(\mathbb{R}^N)$. Let $h \in C_+^b(\mathbb{R}^N)$ such that $h(x) = 0$ if $\|x\| \geq 1$ and $\int_{\mathbb{R}^N} h(x)dx = 1$. It is well known that h can be chosen in $C_0^\infty(\mathbb{R}^N)$ [**189**, Lem.A.1]. For $\epsilon > 0$, we define functions

$$h_\epsilon(x) = \epsilon^{-N} h(\epsilon^{-1}x), \qquad x \in \mathbb{R}^N. \tag{3.20}$$

Notice that

$$\int_{\mathbb{R}^N} h_\epsilon(x)dx = 1 \tag{3.21}$$

and

$$\int_{\mathbb{R}^N} \|x\| h_\epsilon(x)dx = \epsilon \int_{\mathbb{R}^N} \|y\| h(y)dy \leq \epsilon \int_{\mathbb{R}^N} h(y)dy = \epsilon. \tag{3.22}$$

Further we define $g_\epsilon : \mathbb{R}^N \to \mathbb{R}$ by

$$g_\epsilon(x) = \int_{\mathbb{R}^N} h_\epsilon(x - z)\mu(dz), \qquad x \in \mathbb{R}^N. \tag{3.23}$$

Let μ_ϵ be the Lebesgue absolutely continuous measure with density g_ϵ.

THEOREM 3.27. *For all $\epsilon > 0$, $\|\mu_\epsilon - \mu\|_\flat \leq \epsilon\|\mu\|$. The Lebesgue absolutely continuous measures with densities that have bounded derivatives of any order are dense in $\mathcal{M}(\mathbb{R}^N)$.*

PROOF. Let $f : \mathbb{R}^N \to [0,1]$ and $|f(x) - f(y)| \leq \|x - y\|$ for all $x, y \in \mathbb{R}^N$. Then

$$\int_{\mathbb{R}^N} f(x)\mu_\epsilon(dx) = \int_{\mathbb{R}^N} f(x)g_\epsilon(x)dx = \int_{\mathbb{R}^N} f(x)\Big(\int_{\mathbb{R}^N} h_\epsilon(x - z)\mu(dz)\Big)dx.$$

We change the order of integration and perform a change of variables,

$$\begin{aligned}\int_{\mathbb{R}^N} f(x)\mu_\epsilon(dx) &= \int_{\mathbb{R}^N} \Big(\int_{\mathbb{R}^N} f(x)h_\epsilon(x - z)dx\Big)\mu(dz)\\ &= \int_{\mathbb{R}^N} \Big(\int_{\mathbb{R}^N} f(y + z)h_\epsilon(y)dy\Big)\mu(dz).\end{aligned}$$

By (3.21),

$$\begin{aligned}&\Big|\int_{\mathbb{R}^N} f(x)\mu_\epsilon(dx) - \int_{\mathbb{R}^N} f(z)\mu(dz)\Big|\\ &\quad= \Big|\int_{\mathbb{R}^N} \Big(\int_{\mathbb{R}^N} [f(y + z) - f(z)]h_\epsilon(y)dy\Big)\mu(dz)\Big|\\ &\quad\leq \int_{\mathbb{R}^N} \Big(\int_{\mathbb{R}^N} |f(y + z) - f(z)|h_\epsilon(y)dy\Big)|\mu|(dz)\\ &\quad\leq \int_{\mathbb{R}^N} \Big(\int_{\mathbb{R}^N} \|y\|h_\epsilon(y)dy\Big)|\mu|(dz) \leq \epsilon|\mu|(\mathbb{R}^N).\end{aligned}$$

In the last inequality, we have used (3.22). □

If S is an open subset of $\mathbb{R}^N$, we only obtain a density result for $\mathcal{M}_+(S)$. For $\epsilon > 0$, define $g_\epsilon : S \to \mathbb{R}$ by

$$g_\epsilon(x) = \int_S h_\epsilon(x - z)\mu(dz), \qquad x \in \mathbb{R}^N,$$

where h_ϵ is given by (3.20).

THEOREM 3.28. *Let S be an open subset of $\mathbb{R}^N$ and $\mu \in \mathcal{M}_+(S)$. Let (ϵ_n) be a sequence with $0 < \epsilon_n \to 0$ and let μ_n be the Lebesgue absolutely continuous measures in $\mathcal{M}_+(S)$ with densities g_{ϵ_n} (restricted to S).*

Then $\|\mu_n - \mu\|_\flat \to 0$ as $n \to \infty$.

The measures with nonnegative Lebesgue densities that have bounded derivatives of any order are dense in $\mathcal{M}_+(S)$.

PROOF. Since S is the subset of the separable normed vector space $\mathbb{R}^N$, it is separable itself. By Proposition 3.18, μ is pre-tight.

Let $f : S \to [0,1]$ with $|f(x) - f(y)| \le \|x - y\|$ for all $x, y \in S$. By Proposition 3.23, it is sufficient to show $\int_S f d\mu_n \to \int_S f d\mu$.

To prove this, we try to apply Theorem 3.27.

To this end, we extend μ to a measure in $\tilde{\mu} \in \mathcal{M}_+(\mathbb{R}^N)$ by setting $\tilde{\mu}(T) = \mu(T \cap S)$ for all Borel sets T in $\mathbb{R}^N$.

We also extend f to a function $\tilde{f} : \mathbb{R}^N \to [0,1]$ with $|\tilde{f}(s) - \tilde{f}(t)| \le \|s - t\|$ for all $s, t \in \mathbb{R}^N$ (Remark 2.79). Let $\tilde{\mu}_n$ be the Lebesgue absolutely continuous measures in $\mathcal{M}_+(\mathbb{R}^N)$ with densities g_{ϵ_n}. By Theorem 3.27,

$$\int_S f d\mu = \int_{\mathbb{R}^N} \tilde{f} d\tilde{\mu} = \lim_{n\to\infty} \int_{\mathbb{R}^N} \tilde{f}(s) g_{\epsilon_n} ds.$$

The assertion follows if $\int_{\mathbb{R}^N \setminus S} \tilde{f}(s) g_{\epsilon_n}(s) ds \to 0$ as $n \to \infty$. Since $\tilde{f}$ takes values between 0 and 1, it is sufficient to show $\int_{\mathbb{R}^N \setminus S} g_{\epsilon_n}(s) ds \to 0$ as $n \to \infty$. Changing the order of integration, these integrals equal $\int_S v_n d\mu$ with

$$v_n(z) = \int_{\mathbb{R}^N \setminus S} h_{\epsilon_n}(x - z) dx.$$

By the dominated convergence theorem, $\int_S v_n d\mu \to 0$ if $v_n(z) \to 0$ for all $z \in S$ and there is some $c > 0$ such that $v_n(z) \le c$ for all $z \in S$. By (3.20),

$$v_n(z) = \int_{\mathbb{R}^N \setminus S} \epsilon_n^{-N} h(\epsilon_n^{-1}(x - z)) dx.$$

Since S is open, any $z \in S$ has a positive distance $\delta > 0$ from $\mathbb{R}^N \setminus S$. Recall that $h(y) = 0$ if $\|y\| \ge 1$. So $v_n(z) = 0$ if $\epsilon_n < \delta$, and $v_n(z) \to 0$ as $n \to \infty$ for all $z \in S$.

Finally

$$v_n(z) \le \int_{\mathbb{R}^N} \epsilon_n^{-N} h(\epsilon_n^{-1}(x - z)) dx = 1.$$

This finishes the proof. □

Exercises.

3.2.2. Let $\mu \in \mathcal{M}_+(S)$. Show that μ has the following *regularity property* [**73**, III.V.10]: For all $T \in \mathcal{B}$,

$$\mu(T) = \sup \{\mu(V); V \subseteq T, V \text{ closed}\} = \inf \{\mu(U); U \supseteq T, U \text{ open}\}.$$

Hint: Define $\tilde{\mathcal{B}}$ as the set of sets $T \in \mathcal{B}$ which have the property above and show that $\tilde{\mathcal{B}}$ is a σ-algebra and contains all closed sets [**9**, Thm.12.5].

3.2.3. Let $\mu \in \mathcal{M}_+^t(S)$. Show that μ has the following extended *tightness property*: For any $T \in \mathcal{B}$ and $\epsilon > 0$, there exists some compact subset K of S such that $K \subseteq T$ and $\mu(T \setminus K) < \epsilon$.

Hint: Use Exercise 3.2.2.

3.2.4. Consider the compact metric space $S = [0,1]$ with the metric induced by the absolute value and the sequence of measures

$$\mu_n = n^{3/4}(n+1)\big(\delta_{1/n} - \delta_{1/(n+1)}\big), \qquad n \in \mathbb{N}.$$

(a) Show that $\|\mu_n\|_\flat = n^{-1/4} \to 0$ as $n \to \infty$ (Lemma 3.9).
(b) Consider the sequence of Lipschitz functions $f_n : S \to [0,1]$, $f_n(s) = ns$ for $s \in [0,1/n]$ and $f_n(s) = 1$ for $s \in [1/n, 1]$.
Show that $\int_S f_n d\mu_n \to \infty$ and in particular $\|\mu_n\|_\sharp \to \infty$.
(c) Consider the uniformly continuous function $f : S \to [0,1]$ with $f(s) = \sqrt{s}$. Show that $\int_S f d\mu_n \to \infty$ as $n \to \infty$.

3.2.5. Consider $S = (0,1)$ with the metric induced by the absolute value and $f : S \to [0,1]$, $f(s) = |\sin(1/s)|$, $s \in (0,1)$.

Notice that $f \in C_+^b(S)$. Find a sequence of measures $\mu_n \in \mathcal{M}(S)$ such that $\|\mu_n\|_\flat \to 0$ as $n \to \infty$ but $\int_S f d\mu_n \to \infty$ for all $n \in \mathbb{N}$.

Hint: Choose sequences (s_n) and (t_n) in S such that $1/s_n = 2n\pi$ and $1/t_n = 2n\pi + (\pi/2)$ for all $n \in \mathbb{N}$ and set $\mu_n = n\big(\delta_{t_n} - \delta_{s_n}\big)$ (see Lemma 3.9).

3.2.6. Set S be a metric space. Let $\mathcal{N}$ be a subset of $\mathcal{M}(S)$ such that $\Big\{ \int_S f d\mu;\ \mu \in \mathcal{N}\Big\}$ is a bounded subset of $\mathbb{R}$ for all $f \in C^b(S)$. Then $\{\|\mu\|_\sharp; \mu \in \mathcal{N}\}$ is a bounded subset of $\mathbb{R}$.

Hint: Use the proof of Theorem 3.7 and the uniform boundedness principle.

3.2.3. Compact sets in $\mathcal{M}_+(S)$. One of our main motivations of considering the flat norm consists in the characterization of sets with compact closure.

THEOREM 3.29 ([**70**, Thm.11.5.4], [**72**, Thm.1.55]). *Let (S,d) be a complete separable space and $\mathcal{N}$ be a set of probability measures on S. Then $\mathcal{N}$ has compact closure if and only if $\mathcal{N}$ is tight.*

Recall Definition 3.16. Davies [**52**] constructs a separable metric space S (a countable union of compact sets) and a set of tight probability measures $\mathcal{N}$ that is compact but not tight. But one direction of Theorem 3.29 holds in more generality.

THEOREM 3.30. *Let (μ_n) be a tight sequence in $\mathcal{M}_+(S)$ such that $(\mu_n(S))$ is bounded. Then (μ_n) has a converging subsequence (with the limit measure being tight as well).*

This theorem, for a sequence of probability measures, is an exercise in [**70**]. We sketch a proof for the convenience of the reader. An alternative proof would build on Theorem 3.29 [**101**, Th.4.22].

PROOF. Let (μ_n) be a tight sequence in $\mathcal{M}_+(S)$ such that $(\mu_n(S))$ is bounded. After choosing a subsequence, there exists some $\mu \in \mathcal{M}(S)$ such that $\int_S f d\mu_n \to \int_S f d\mu$ for all $f \in C^b(S)$ [**72**, Thm.G.28] [**133**, Thm.13.29]. Then $\mu \in \mathcal{M}_+(S)$ by Theorem 3.6. By Proposition 3.20, $\mu \in M_+^s(S)$, and, by Theorem 3.24, $\|\mu_n - \mu\|_\flat \to 0$ as $n \to \infty$. □

THEOREM 3.31. *If S is compact, every bounded closed subset of $\mathcal{M}_+(S)$ is compact and $\mathcal{M}_+(S)$ is locally compact (Definition* A.1.8 *(e)).*

PROOF. Let S be compact. Then $\mathcal{M}_+(S)$ is tight and the claim follows from Theorem 3.30. □

THEOREM 3.32. *The following are equivalent:*

(a) *S is totally bounded.*
(b) *$\bar{U}_1(0) \cap \mathcal{M}_+(S)$ is totally bounded.*
(c) *Every bounded subset of $\mathcal{M}_+(S)$ is totally bounded.*
(d) *0 has a totally bounded neighborhood in $\mathcal{M}_+(S)$.*

PROOF. (b) $\Rightarrow$ (a) has been shown in Corollary 3.10.

(c) $\Rightarrow$ (b) is immediate.

(b) $\Rightarrow$ (c) follows from the fact that every bounded subset of $\mathcal{M}_+(S)$ is contained in $c(\bar{U}_1(0) \cap M_+(S))$ for some $c > 0$ and that multiplication by a scalar is uniformly continuous.

By the same token, (d) implies (b) while the converse is obvious.

(a) $\Rightarrow$ (b). Let $(\tilde{S}, \tilde{d})$ be a completion of (S, d), i.e., there exists an isometry $J : S \to \tilde{S}$ such that $J(S)$ is dense in $\tilde{S}$. Since S is totally bounded, so is $J(S)$ and its closure $\tilde{S}$. Since $\tilde{S}$ is complete, $\tilde{S}$ is compact. By Theorem 3.31, every bounded closed subset of $\mathcal{M}_+(\tilde{S})$ is compact. Let (μ_n) be a bounded sequence in $\mathcal{M}_+(S)$. It is sufficient to show that (μ_n) has a subsequence that is a Cauchy sequence (Theorem A.28). Define a sequence $(\tilde{\mu}_n)$ in $\mathcal{M}_+(\tilde{S})$ by

$$\tilde{\mu}_n(\tilde{T}) = \mu_n(J^{-1}(\tilde{T})), \qquad n \in \mathbb{N}, \quad \tilde{T} \in \mathcal{B}(\tilde{S}).$$

Then $(\tilde{\mu}_n)$ is a bounded sequence in $\mathcal{M}_+(\tilde{S})$ and has a subsequence that converges and thus is a Cauchy sequence. Let $f : S \to [0, 1]$ and $|f(x) - f(y)| \le d(x, y)$. Define $\tilde{f} : J(S) \to [0, 1]$ by $\tilde{f}(J(x)) = f(x)$. Then $|\tilde{f}(\tilde{x}) - \tilde{f}(\tilde{y})| \le \tilde{d}(\tilde{x}, \tilde{y})$ for all $\tilde{x}, \tilde{y} \in J(S)$. Since $J(S)$ is dense in $\tilde{S}$, $\tilde{f}$ can be extended to $\tilde{S}$ and the extension maps into [0,1] and has 1 as a Lipschitz constant. Hence (μ_n) has a subsequence that is a Cauchy sequence with respect to the flat metric. $\square$

COROLLARY 3.33. *If S is not totally bounded (e.g., S complete but not compact), $\mathcal{M}_+(S)$ is not locally compact.*

3.2.4. Completeness of $\mathcal{M}_+(S)$. Completeness of S is necessary for the completeness of $\mathcal{M}_+(S)$. Cf. [**113**, Thm.3.8].

THEOREM 3.34. *If (S, d) is not complete, then there exists a Cauchy sequence (s_n) in S such that (δ_{s_n}), which is a Cauchy sequence in $M_+^t(S)$, does not converge to some $\mu \in \mathcal{M}_+(S)$.*

PROOF. Assume that (S, d) is not complete and let $(\tilde{S}, \tilde{d})$ be its completion. There exists an isometry $J : S \to \tilde{S}$ such that $J(S)$ is dense in $\tilde{S}$. Since S is not complete, there exists a Cauchy sequence (s_n) in S and some $\tilde{s} \in \tilde{S} \setminus J(S)$ such that $J(s_n) \to \tilde{s}$ as $n \to \infty$. By Lemma 3.9, (δ_{s_n}) is a Cauchy sequence with respect to the flat norm. Suppose that there exists some $\mu \in \mathcal{M}_+(S)$ such that $\|\delta_{s_n} - \mu\|_\flat \to 0$. In particular, $\mu(S) = 1$.

We define a Borel measure $\tilde{\mu} \in \mathcal{M}_+(\tilde{S})$ by

$$\tilde{\mu}(\tilde{T}) = \mu(J^{-1}(\tilde{T})), \qquad \tilde{T} \in \tilde{\mathcal{B}}. \tag{3.24}$$

Let $\epsilon > 0$. Since

$$\tilde{S} \setminus \{\tilde{s}\} = \bigcup_{n \in \mathbb{N}} (\tilde{S} \setminus U_{1/n}(\tilde{s})),$$

there exists a closed subset $\tilde{V}$ of $\tilde{S} \setminus \{\tilde{s}\}$ such that

$$\tilde{\mu}(\tilde{V}) > \tilde{\mu}(\tilde{S} \setminus \{\tilde{s}\}) - \epsilon. \tag{3.25}$$

By Lemma 2.75, there exists a Lipschitz continuous function $\tilde{f} : \tilde{S} \to [0,1]$ such that $\tilde{f} \equiv 1$ on $\tilde{V}$ and $\tilde{f}(\tilde{s}) = 0$. Let $f = \tilde{f} \circ J$. Then $f : S \to [0,1]$ is Lipschitz continuous. Since $\|\delta_{s_n} - \mu\|_\flat \to 0$ as $n \to \infty$, by (3.24),

$$\int_{\tilde{S}} \tilde{f} d\tilde{\mu} = \int_S f d\mu = \lim_{n\to\infty} f(s_n) = \lim_{n\to\infty} \tilde{f}(J(s_n)) = \tilde{f}(\tilde{s}) = 0.$$

Since $\tilde{f} \equiv 1$ on $\tilde{V}$,

$$0 = \tilde{\mu}(\tilde{V}) > \tilde{\mu}\big(\tilde{S} \setminus \{\tilde{s}\}\big) - \epsilon.$$

Here, the second inequality follows from by (3.25). Since this holds for any $\epsilon > 0$, $\tilde{\mu}(\tilde{S} \setminus \{\tilde{s}\}) = 0$. Since $J(S) \subseteq \tilde{S} \setminus \{\tilde{s}\}$,

$$0 = \tilde{\mu}(J(S)) = \mu(J^{-1}(J(S))) = \mu(S).$$

The last equality holds because J is injective. We have obtained the contradiction $0 = \mu(S) = 1$. □

See [**113**, Thm.3.8] and its proof for an alternative proof.

THEOREM 3.35. *If $\mathcal{M}_+(S)$ or $\mathcal{M}_+^s(S)$ or $\mathcal{M}_+^t(S)$ are complete, then (S,d) is complete.*

We work towards converses for $\mathcal{M}_+^s(S)$ and $\mathcal{M}_+^t(S)$ following the proof of [**70**, Thm.11.5.4] only using the subsequent lemma rather than the full relation between the flat norm and the Prokhorov metric [**70**, Sec.11.3]. Cf. [**101**, Sec.4.3] and, for the case that S is separable, [**72**, G.5]. Recall the concepts of the open and closed δ-neighborhoods of a set T, (2.39) and (2.40).

LEMMA 3.36. *For each $\delta \in (0,1]$, all $T \in \mathcal{B}$, and all $\mu, \nu \in \mathcal{M}_+(S)$,*

$$\mu(T) \le \nu(U_\delta(T)) + (1/\delta)\|\mu - \nu\|_\flat,$$
$$\mu(S \setminus U_\delta(T)) \le \nu(S \setminus T) + (1/\delta)\|\mu - \nu\|_\flat.$$

PROOF. Let $\delta \in (0,1]$, T be a Borel set in S, and $\mu, \nu \in \mathcal{M}_+(S)$. By Lemma 2.75, there exists a Lipschitz continuous function $f : S \to [0,1]$ with $1/\delta$ being a Lipschitz constant such that $f \equiv 1$ on T and $f \equiv 0$ on $S \setminus U_\delta(T)$. Then $\delta f : S \to [0,1]$ and δf has 1 as a Lipschitz constant and so

$$\Big| \int_S f d\mu - \int_S f d\nu \Big| \le (1/\delta)\|\mu - \nu\|_\flat.$$

Since $f \equiv 1$ on T,

$$\mu(T) \le \int_S f d\mu \le \int_S f d\nu + (1/\delta)\|\mu - \nu\|_\flat \le \nu(U_\delta(T)) + (1/\delta)\|\mu - \nu\|_\flat.$$

There also exists a Lipschitz continuous function $g : S \to [0,1]$ with $1/\delta$ being a Lipschitz constant such that $g \equiv 1$ on $S \setminus U_\delta(T)$ and $g \equiv 0$ on T. Then

$$\mu(S \setminus U_\delta(T)) \le \int_S g d\mu \le \int_S g d\nu + \frac{\|\mu - \nu\|_\flat}{\delta} \le \nu(S \setminus T) + \frac{\|\mu - \nu\|_\flat}{\delta}. \qquad \square$$

PROPOSITION 3.37. *Let $\mathcal{N} \subseteq \mathcal{M}_+^s(S)$ be a totally bounded set of pre-tight measures. Then $\mathcal{N}$ is pre-tight and, if S is complete, tight.*

PROOF. Let $\mathcal{N}$ be totally bounded under the flat norm. Let $\epsilon \in (0,1)$. By Definition A.23. there exists a finite subset $\tilde{\mathcal{N}}$ of $\mathcal{N}$ such that

$$\mathcal{N} \subseteq U_{\epsilon^2/4}(\tilde{\mathcal{N}}). \tag{3.26}$$

Here the neighborhood is taken within the normed vector space $\mathcal{M}(S)$ with the flat norm.

Since each measure in $\tilde{\mathcal{N}}$ is pre-tight, for any $\nu \in \tilde{\mathcal{N}}$ there exists a closed totally bounded subset K_ν of S such that $\nu(S \setminus K_\nu) < \epsilon/2$. Set $K = \bigcup_{\nu \in \tilde{\mathcal{N}}} K_\nu$. Since $\tilde{\mathcal{N}}$ is finite, K is a closed totally bounded subset of S. Further,

$$\nu(S \setminus K) \leq \frac{\epsilon}{2}, \qquad \nu \in \tilde{\mathcal{N}}. \tag{3.27}$$

Since K is a totally bounded subset of S, there exists a finite subset F of K such that $K \subseteq U_{\epsilon/2}(F)$. Here the neighborhood is taken in the metric space S. By (3.27), $\nu(S \setminus U_{\epsilon/2}(F)) < \epsilon/2$ for all $\nu \in \tilde{\mathcal{N}}$.

Notice that $U_{\epsilon/2}(U_{\epsilon/2}(F)) \subseteq U_\epsilon(F)$. By Lemma 3.36, for any $\mu \in \mathcal{M}_+(S)$ and any $\nu \in \tilde{\mathcal{N}}$,

$$\mu(S \setminus U_\epsilon(F)) \leq \nu(S \setminus U_{\epsilon/2}(F)) + (2/\epsilon)\|\mu - \nu\|_\flat < \epsilon/2 + (2/\epsilon)\|\mu - \nu\|_\flat.$$

By (3.26), for each $\mu \in \mathcal{N}$, there exists some $\nu \in \tilde{\mathcal{N}}$ such that $\|\mu - \nu\|_\flat < \epsilon^2/4$. Hence, for each $\epsilon \in (0,1)$ we have found a finite set F_ϵ in S such that

$$\mu(S \setminus U_\epsilon(F_\epsilon)) < \epsilon, \qquad \mu \in \mathcal{N}.$$

Now let $\delta \in (0,1]$ and set $\epsilon_n = \delta 2^{-n}$, $n \in \mathbb{N}$. Then, for each $n \in \mathbb{N}$, there exists a finite set F_n in S such that

$$\mu\big(S \setminus U_{\epsilon_n}(F_n)\big) < \epsilon_n, \qquad \mu \in \mathcal{N}.$$

Set $K_0 = \bigcap_{n \in \mathbb{N}} \bar{U}_{\epsilon_n}(F_n)$. Then, for all $\mu \in \mathcal{N}$,

$$\mu(S \setminus K_0) \leq \mu\Big(\bigcup_{n \in \mathbb{N}} (S \setminus U_{\epsilon_n}(F_n))\Big) \leq \sum_{n=1}^\infty \mu(S \setminus U_{\epsilon_n}(F_n)) \leq \sum_{n=1}^\infty \epsilon_n \leq \delta.$$

K_0 is totally bounded and closed, and thus $\mathcal{N}$ is pre-tight.

If S is a complete metric space, K_0 is compact and $\mathcal{N}$ is tight. □

PROPOSITION 3.38. *Let (μ_n) be a Cauchy sequence in $\mathcal{M}_+^s(S)$. Then $\{\mu_n\}$ is pre-tight. If S is complete, $\{\mu_n\}$ is tight and (μ_n) converges (to a tight measure).*

PROOF. Since (μ_n) is a Cauchy sequence of pre-tight measures, $\{\mu_n; n \in \mathbb{N}\}$ is a totally bounded subset of $\mathcal{M}_+(S)$ consisting of pre-tight measures. Since S is complete, (μ_n) is tight by Proposition 3.37. By Theorem 3.30, (μ_n) has a converging subsequence and thus converges itself. The limit measure is tight by Proposition 3.20. □

Proposition 3.38 and Theorem 3.35 can be combined as follows.

THEOREM 3.39 ([**113**, Thm.3.8]). *$\mathcal{M}_+^s(S)$ is complete if and only if S is complete.*

We have the following consequence from Definition 3.17 and Remark 3.19.

COROLLARY 3.40. *Let S be universally pre-tight (e.g., separable). Then $\mathcal{M}_+(S)$ is complete if and only if S is complete.*

3.2.5. Metrizable topological spaces. Metric spaces are special topological spaces (Exercise A.1.3).

DEFINITION 3.41. *Let S be a nonempty set. A* topology *τ on S is a set of subsets of S with the following properties:*

(a) $\emptyset, S \in \tau$.
(b) *Each arbitrary union of elements in τ is an element of τ.*
(c) *Each finite intersection of elements in τ is an element in τ.*

If τ is a topology on S, then (S, τ) is called a topological space.
See [**71**, Def.1.1], [**159**, Def.7.2].

A topological space S is called metrizable *if its topology consists of open sets as defined via a metric by Definition* A.18.

S is called completely metrizable *if that metric makes S a complete metric space.*

Two metrics d_1 and d_2 on a set S are called topologically equivalent *if they induce the same topology; equivalently, the identity is continuous from (X, d_1) to (X, d_2) and vice versa.*

They are called (Lipschitz) equivalent *if the identity is Lipschitz continuous from (X, d_1) to (X, d_2) and vice versa* [**21**, Def.2.10]. *See Appendix* A.1.7 *and the various exercises therein.*

A metric space is called topologically complete *if it is completely metrizable, i.e., complete under a topologically equivalent metric* [**70**, p.59] [**71**, XIV.2.2].

Since compactness only depends on the topology and not on the inducing metric (Exercise A.1.2), it depends only on the topology of S whether a subset of $\mathcal{M}_+(S)$ is tight or whether S is universally tight. Further $\mathcal{M}^s(S)$ does not depend on the inducing metric. See Definition 3.16.

THEOREM 3.42. *Assume that d_1 and d_2 are metrics on S and let $\|\cdot\|_{\flat_1}$ and $\|\cdot\|_{\flat_2}$ be the associated flat norms on $\mathcal{M}(S)$.*

Let $d_j^\flat$ be the metrics induced by $\|\cdot\|_{\flat_j}$ on $\mathcal{M}_+^s(S)$,

$$d_j^\flat(\mu,\nu) = \|\mu - \nu\|_{\flat_j}, \qquad \mu,\nu \in \mathcal{M}_+^s(S).$$

Then $d_1^\flat$ and $d_2^\flat$ are topologically equivalent on $\mathcal{M}_+^s(S)$ if and only if d_1 and d_2 are topologically equivalent on S.

PROOF. Let d_1 and d_2 be topologically equivalent on S. Then $C^b(S)$ is the same for d_1 and d_2. By Theorem 3.24, the identity is continuous between $(\mathcal{M}_+^s(S), d_1^\flat)$ and $(\mathcal{M}_+^s(S), d_2^\flat)$ in both directions, i.e., the metrics $d_1^\flat$ and $d_2^\flat$ are topologically equivalent on the cone $\mathcal{M}_+^s(S)$.

Now assume that the metrics $d_1^\flat$ and $d_2^\flat$ are topologically equivalent on the cone $\mathcal{M}_+^s(S)$. Let (s_n) be a sequence in S and $s \in S$ such that $d_1(s_n, s) \to 0$ as $n \to \infty$. By Lemma 3.9, $d_1^\flat(\delta_{s_n}, \delta_s) \to 0$ as $n \to \infty$. Since $d_1^\flat$ and $d_2^\flat$ are topologically equivalent, $d_2^\flat(\delta_{s_n}, \delta_s) \to 0$ as $n \to \infty$. Again by Lemma 3.9, $d_2(s_n, s) \to 0$ as $n \to \infty$. By symmetry, we can exchange the roles of d_1 and d_2 and obtain that d_1 and d_2 are topologically equivalent on S. $\square$

If d_1 and d_2 are topologically equivalent on S, the metrics $d_j^\flat$, $j = 1, 2$, may not be topologically equivalent on the vector space $\mathcal{M}^s(S)$; see Remark 13.5 for an example.

However, if d_1 and d_2 are equivalent, i.e., the identity is Lipschitz continuous between (S, d_1) and (S, d_2) in both directions, then also the two associated flat norms are equivalent to each other. The converse does not hold as it can happen that d_1 and d_2 are topologically equivalent but not equivalent and that the associated flat norms are equivalent, even the same.

EXAMPLE 3.43. Let d_1 be a metric on S and $d_2(s,t) = \min\{2, d_1(s,t)\}$. Then d_2 is a metric that is topologically equivalent to d_1. If d_1 is not bounded (like the metric induced by the absolute value on $\mathbb{R}$), d_2 is not equivalent to d_1. However, $\|\mu\|_{\flat_1} = \|\mu\|_{\flat_2}$ for all $\mu \in \mathcal{M}(S)$.

PROOF. Let $f : S \to [0,1]$. By (3.9), it is sufficient to show that the set $\mathcal{F}$ in this definition is the same for d_1 and d_2, i.e., for all $s, t \in X$,

$$|f(s) - f(t)| \le d_1(s,t) \iff |f(s) - f(t)| \le d_2(s,t).$$

Let $s, t \in S$. If $|f(s) - f(t)| \le d_2(s,t)$, then also $|f(s) - f(t)| \le d_1(s,t)$.

Now let $|f(s) - f(t)| \le d_1(s,t)$.

If $d_1(s,t) \le 2$, then $|f(s) - f(t)| \le d_2(s,t)$.

If $d_1(s,t) \ge 2$, then $|f(s) - f(t)| \le |f(s)| + |f(t)| \le 2 = d_2(s,t)$. □

As for compactness of subsets of $\mathcal{M}_+(S)$, we have the following result which implies a sufficient condition.

THEOREM 3.44. *Let S be a metrizable topological space and (μ_n) be a tight sequence in $\mathcal{M}_+(S)$ such that $(\mu_n(S))$ is bounded. Then there exist a subsequence (μ_{n_k}) and some tight $\mu \in \mathcal{M}_+(S)$ with the following property: If d is any metric that induces the topology of S, then (μ_{n_k}) converges to μ with respect to the flat norm associated with d.*

PROOF. Let d_0 be a specific metric that induces the topology of S. By Definition 3.16, (μ_n) is tight with respect to d_0. By Theorem 3.30, (μ_n) has a subsequence which converges with respect to the flat norm associated with d_0 and the limit measure μ is tight. Let d be any other metric that induces the topology of S. Then the subsequence converges also with respect to the flat norm associated with d by Theorem 3.24. □

If S is completely metrizable, there is a metric d which makes S a complete metric space. There may be another metric $\tilde{d}$ which also induces the topology of S but does not make S a complete metric space. By Theorem 3.39, $\mathcal{M}_+^s(S)$ is complete with respect to the flat norm associated with respect to d, but not with respect to the flat norm associated with respect to $\tilde{d}$.

EXAMPLE 3.45. Let (S, d) be a complete metric space and U an open strict subset of S (e.g., $S = \mathbb{R}$ and $U = (0,1)$). U will not be complete in general (for instance, if S is connected). However, a topologically equivalent metric d_U on U that makes U complete is given by

$$d_U(s,t) = d(s,t) + |f(s) - f(t)|, \qquad f(s) = \frac{1}{d(s, S \setminus U)}, \qquad s, t \in U.$$

Thus, U is completely metrizable (topologically complete).

By Corollary 3.40, the flat norm on $\mathcal{M}(U)$ associated with d_U would make $\mathcal{M}_+^s(U)$ complete, but the flat norm associated with the restriction of d to U will not in general.

More generally, by results by Alexandrov and Mazurkiewicz, a subset of a completely metrizable space S is completely metrizable itself if and only if it is a countable intersection of open subsets (i.e., a so-called G_δ subset) of S [**9**, Sec.3.7] [**70**, Thm.2.5.4].

Topological spaces that are separable and completely metrizable are called Polish [**70**, p.344].

Polish spaces are universally tight (Definition 3.16, Proposition 3.37).

All locally compact metric spaces (Definition A.1.8) are completely metrizable [**71**, XIV.2.3], and all separable locally compact metric spaces are Polish.

The set of irrational numbers with the absolute value inducing the metric is Polish [**70**, p.60] but not locally compact. The set of irrational numbers with the discrete metric is not separable but locally compact and complete (Exercise A.1.8).

The set of rational numbers with the absolute value inducing the metric is neither completely metrizable nor locally compact but it is universally tight.

3.2.6. Norms equivalent to the flat norm. Recall our definition of the flat norm, (3.9),

$$\|\mu\|_\flat = \sup\Big\{\Big|\int_S f d\mu\Big|;\ f\in[0,1]^S,\ \forall x,y\in S: |f(x)-f(y)|\le d(x,y)\Big\}, \qquad \mu\in\mathcal{M}(S). \tag{3.28}$$

[**166**] (called the *bounded Lipschitz norm* therein); we use M^S for the set of functions from the metric space S to a set M.

There are at least two norms on $\mathcal{M}(S)$ in the literature that are equivalent (Appendix A.1.7, Exercise A.1.9) to the flat norm [**101**]:

$$\|\mu\|_\diamond = \sup\Big\{\Big|\int_S f d\mu\Big|;\ f\in[-1,1]^S,\ \forall x,y\in S: |f(x)-f(y)|\le d(x,y)\Big\}, \tag{3.29}$$

[**85**, p.367] [**86**] (called the *flat norm* therein);

$$\|\mu\|_\bullet = \sup\Big\{\int_S f d\mu;\ f\in[-1,1]^S,\ \forall x,y\in S: |f(x)-f(y)|\le d(x,y)\Big\}, \tag{3.30}$$

[**24**, Sec.8.3] (called the *Kantorovich–Rubinstein norm therein*), [**97, 99, 102**]. For Kantorovich–Rubinstein Theorems see [**70**, Sec.11.8]. To gain further insight, let

$$\|f\|_\infty = \sup_{x\in S}|f(x)|, \qquad \Lambda(f) = \sup\Big\{\frac{|f(x)-f(y)|}{d(x,y)}; x,y\in S, x\ne y\Big\},$$

and let p be any norm on $\mathbb{R}^2$. Then, we can define a norm on $\mathrm{LC}^b(S)$ by

$$\|f\|_p = p(\|f\|_\infty, \|f\|_L)$$

and a dual norm on $\mathcal{M}(S)$ by

$$\|\mu\|_p^* = \sup\Big\{\Big|\int_S f d\mu\Big|; f\in \mathrm{LC}^b(S), \|f\|_p\le 1\Big\}. \tag{3.31}$$

The norm in (3.29) corresponds to $\|\cdot\|_p^*$ with p being the maximum norm on $\mathbb{R}^2$ and is also called *Fortet-Mourier norm* [**143**].

$\|\cdot\|_p^*$ with p being the sum norm on $\mathbb{R}^2$ is introduced in [**69**] [**70**, p.394], called BL^* norm in [**69**] and *dual bounded Lipschitz norm* in [**70**, p.404]. It is attributed to Fortet and Mourier [**88**] in [**70**, p.434] and called the *Dudley norm* in [**82**].

For different norms p on $\mathbb{R}^2$, the norms $\|\cdot\|_p$ are equivalent on $\mathrm{LC}^b(S)$, and so the norms $\|\cdot\|_p^*$ are equivalent on $\mathcal{M}(S)$.

Since we restrict ourselves to real-valued functions and measures, and $\|-f\|_p = \|f\|_p$, we can omit the absolute value in (3.31),

$$\|\mu\|_p^* = \sup\Big\{ \int_S f d\mu; f \in \mathrm{LC}^b(S), \|f\|_p \leq 1 \Big\}. \tag{3.32}$$

This has the consequence that $\|\cdot\|_\diamond$ and $\|\cdot\|_\bullet$ in (3.29) and (3.30) are equal.

LEMMA 3.46. *The norms $\|\cdot\|_\flat$ and $\|\cdot\|_\bullet$ are equivalent.*

PROOF. Let $f \in [0,1]^S$, $|f(x) - f(y)| \leq d(x,y)$ for all $x, y \in S$. By (3.30),

$$\Big| \int_S f d\mu \Big| = \max\Big\{ \int_S f d\mu, -\int_S f d\mu \Big\} = \max\Big\{ \int_S f d\mu, \int_S (-f) d\mu \Big\} \leq \|\mu\|_\bullet.$$

By (3.28), $\|\mu\|_\flat \leq \|\mu\|_\bullet$.

To see the opposite inequality, let $f \in [-1,1]^S$, $|f(x) - f(y)| \leq d(x,y)$ for all $x, y \in S$. Then $f = f_+ - f_-$ with

$$f_+ = \max\{f, 0\} = (1/2)(|f| + f), \qquad f_- = \max\{-f, 0\} = (1/2)(|f| - f).$$

One readily checks that $f_+, f_- \in [0,1]^S$ and $|f_\pm(x) - f_\pm(y)| \leq d(x,y)$ for all $x, y \in S$. So

$$\int_S f d\mu = \int_S f_+ d\mu - \int_S f_- d\mu \leq \Big| \int_S f_+ d\mu \Big| + \Big| \int_S f_- d\mu \Big| \leq 2\|\mu\|_\flat.$$

This implies $\|\mu\|_\bullet \leq 2\|\mu\|_\flat$. □

As mentioned before, $\|\mu\|_\flat = \mu(S)$ for all $\mu \in \mathcal{M}_+(S)$. This also holds for $\|\cdot\|_\diamond$ and $\|\cdot\|_\bullet$. It holds for $\|\cdot\|_p^*$ as well provided the norm p on $\mathbb{R}^2$ is order-preserving on $\mathbb{R}_+^2$ with respect to the natural, coordinate-wise, order and $\|(1,0)\|_p = 1$.

Which version of the flat norm should be chosen? Since we concentrate on positivity aspects in this book, it seems natural to choose a norm that only involves nonnegative bounded Lipschitz continuous functions like in the definition of $\|\cdot\|_\flat$, (3.28).

This has some consequences. Recall that, for the variation-norm,

$$\|\delta_s - \delta_t\|_\sharp = 2, \qquad s, t \in S, \quad s \neq t. \tag{3.33}$$

and for the flat norm,

$$\|\delta_s - \delta_t\|_\flat = \min\{1, d(s,t)\}, \qquad s, t \in S, \tag{3.34}$$

(Lemma 3.9). To show $\geq$, we used the function $f(x) = \min\{1, d(x,t)\}$. If one chooses one of the other versions of the flat norm, (3.31), for example with p being the max-norm on $\mathbb{R}^2$, one obtains

$$\|\delta_s - \delta_t\|_\diamond = \|\delta_s - \delta_t\|_p^* = \min\{2, d(s,t)\}, \qquad s, t \in S, \tag{3.35}$$

(Remark after Lemma 3.5) [**113**] [**72**, L.1.32].

If p is the sum-norm, $\|\delta_s - \delta_t\|_p^* = \frac{2d(x,y)}{2+d(x,y)}$ [**113**, L.3.5].

See [**101**, Sec.8] for more details.

Why is the flat norm called *flat*? We could not find anything in the literature to answer this question. It is a unwitting terminological coincidence that $\mathcal{M}_+(S)$ is a non-flat generating cone of $\mathcal{M}(S)$ with respect to the variation norm but generally lacks non-flatness with respect to the flat norm (Theorem 3.11).

CHAPTER 4

Homogeneous operators

Let X and Y be vector spaces over $\mathbb{R}$ and $B : X_B \to Y$ a map defined on a homogeneous subset X_B of X.

Then B is called *homogeneous* if $B(\alpha x) = \alpha B(x)$ for all $x \in X_B$ and $\alpha \in \mathbb{R}_+$.

Notice that B is homogeneous if and only if the graph of B, $\{(x, B(x)); x \in X_B\}$, is a homogeneous subset of $X \times Y$.

B is called a *homogeneous operator* from X to Y if B is a homogeneous map from some homogeneous subset X_B to Y.

EXAMPLE 4.1. Let the ordered vector space X with order cone X_+ be a lattice (Section 2.1.7). Then the operators $B, \tilde{B} : X_+^2 \to X_+$ defined by $B(x, y) = x \wedge y$ and $\tilde{B}(x, y) = x \vee y$ are homogeneous operator. See [**9**, Thm.8.6]. If X is a normed lattice, B and $\tilde{B}$ are continuous [**9**, Cor.8.7] (see (2.11)). The statements also hold if the operators are considered as maps from X^2 to X.

4.1. Operator norm for homogeneous operators

Let X and Y be normed vector spaces (with $\|\cdot\|$ denoting the norm on both) and $B : X_B \to Y$, $X_B \subseteq X$, be homogeneous. Then we define

$$\|B\| = \sup\big\{\|B(x)\|; x \in X_B, \|x\| \le 1\big\}, \tag{4.1}$$

and B is called *bounded* if the set in (4.1) is a bounded subset of $\mathbb{R}$.

By the homogeneity of B, if B is bounded,

$$\|B(x)\| \le \|B\|\,\|x\|, \qquad x \in X_B. \tag{4.2}$$

We call $\|B\|$ the *operator norm* of B. This can be justified if there is a fixed homogeneous subset $\check{X}$ and we let $H(\check{X}, Y)$ be the vector space of bounded homogeneous maps from $\check{X}$ to Y with addition and scalar multiplication be defined pointwise in the usual way. Then (4.1) defines a norm on $H(\check{X}, Y)$.

PROPOSITION 4.2. *Let $B : X_B \to Y$ be homogeneous. Then B bounded if and only if B is continuous at 0.*

PROOF. Assume that B is continuous at 0. Then there exists some $\delta > 0$ such that $\|B(z)\| = \|B(z) - B(0)\| \le 1$ for all $z \in X_B$ with $\|z\| \le \delta$. Let $x \in X_B$ with $\|x\| \le 1$. Set $y = \delta x$. Then $y \in X_B$ and $\|y\| \le \delta$. Since B and the norm are homogeneous, $1 \ge \|B(y)\| = \|B(\delta x)\| = \delta\|B(x)\|$.

So $\|B(x)\| \le 1/\delta$ for all $x \in X_B$ with $\|x\| \le 1$, and B is bounded.

The converse implication follows from (4.2). □

If order structures are assumed, some homogeneous operators are automatically bounded.

DEFINITION 4.3. *Let X be an ordered normed vector space with order cone X_+. Further, let $B : X_B \to Y$ be a homogeneous operator, $X_B \subseteq X$. Then a point $u \in X$ is called a* normal point *for B if there exists some $c > 0$ such that $\|B(x)\| \le c$ for all $x \in X_B$ with $x \le u$.*

EXAMPLE 4.4. Let Y be an ordered norm vector space with order cone Y_+ and $B : X_B \to Y_+$ be a homogeneous operator on a homogeneous subset X_B of X and B be order-preserving.

(a) If $u \in \dot{X}_B$ and $B(u)$ is a normal point for Y_+, then u is a normal point for B.
(b) If Y_+ is a normal cone, any point $u \in \dot{X}_B$ is a normal point for B.

PROOF. (a) Let $u \in \dot{X}_B$ such that $B(u)$ is a normal point for Y_+. Then there exists some $c > 0$ such that $\|y\| \le c$ for all $y \in Y_+$ with $y \le B(u)$ (Section 2.2.2).

Now let $x \in X_B$ such that $x \le u$. Since B is order-preserving, $B(x) \le B(u)$ and so $\|B(x)\| \le c$.

(b) This part follows from part (a). □

THEOREM 4.5. *Let X be an ordered normed vector space with closed order cone X_+. Let X_B be a serially complete cone contained in X_+ and $B : X_B \to Y$ be homogeneous.*

Assume that each $x \in \dot{X}_B$ is a normal point for B.

Then B is bounded.

See Section 2.1.2 for the notion of a serially complete cone.

PROOF. Suppose that B is not bounded. Then there exists a sequence (x_n) in X_B with $\|x_n\| = 1$ and $\|B(x_n)\| \ge n^4$ for all $n \in \mathbb{N}$. Since X_B is serially complete, $x = \sum_{n=1}^{\infty} n^{-2} x_n$ converges and is an element in X_B. Since X_+ is closed and $X_B \subseteq X_+$, $x \ge n^{-2} x_n$ for all $n \in \mathbb{N}$. Since x is a normal point for B and B is homogeneous, there exists some $c > 0$ such

$$c \ge \|B(n^{-2} x_n)\| = n^{-2} \|B(x_n)\| \ge n^2, \qquad n \in \mathbb{N},$$

a contradiction. □

EXAMPLE 4.6. Let

$$B(x) = \sum_{k=1}^{n} \theta_k(x) x_k, \qquad x \in X_B,$$

with $x_k \in \dot{X}_B$ and $\theta_k : X_B \to \mathbb{R}_+$ being homogeneous and order-preserving. Let $u \in X_B$ and $x \in X_B$ with $x \le u$. Then

$$\|B(x)\| \le \sum_{k=1}^{n} \theta_k(x) \|x_k\| \le \sum_{k=1}^{n} \theta_k(u) \|x_k\|.$$

Hence there exists some $c > 0$ such that $\|B(x)\| \le c$ for all $x \in X_B$ with $x \le u$. So every $u \in X_B$ is a normal point for B, and B is bounded provided X_B is a serially complete cone contained in X_+.

The next result follows from (4.2).

THEOREM 4.7. *Let X, Y, Z be normed vector spaces over $\mathbb{R}$ and $B : X_B \to Y$, $C : Y_C \to Z$ be homogeneous bounded operators defined on homogeneous subsets $X_B \subseteq X$ and $Y_C \subseteq Y$ such that $B(X_B) \subseteq Y_C$. Then the composition $C \circ B : X_B \to Z$ is defined, homogeneous, and bounded and $\|C \circ B\| \le \|C\|\,\|B\|$.*

4.1.1. A uniform boundedness principle.

THEOREM 4.8. *Let X be an ordered normed vector space with closed order cone X_+ and Z be an ordered normed vector space with a normal order cone Z_+. Let $X_{\mathcal{H}} \subseteq X_+$ be a serially complete cone and $\mathcal{H}$ be a family of homogeneous order-preserving operators $H : X_{\mathcal{H}} \to Z_+$.*

If $\{\|H(x)\|; H \in \mathcal{H}\}$ is bounded for each $x \in X_{\mathcal{H}}$, then $\{\|H\|; H \in \mathcal{H}\}$ is bounded.

Notice that each $H \in \mathcal{H}$ is bounded by Theorem 4.5 and Example 4.4 (b).

PROOF. Assume that $\{\|H(x)\|; H \in \mathcal{H}\}$ is bounded for each $x \in X_{\mathcal{H}}$, but $\{\|H\|; H \in \mathcal{H}\}$ is not bounded. Then, for every $n \in \mathbb{N}$, there exists some $H_n \in \mathcal{H}$ such that $\|H_n\| > n^4$ and some $x_n \in X_{\mathcal{H}}$ with $\|x_n\| = 1$ such that $\|H_n(x_n)\| > n^4$. Since $X_{\mathcal{H}}$ is serially complete, $x = \sum_{n=1}^{\infty} n^{-2} x_n$ converges and is an element in $X_{\mathcal{H}}$. Since X_+ is closed, $x \ge n^{-2} x_n$ for all $n \in \mathbb{N}$. Since each H_n is order-preserving and homogeneous,

$$H_n(x) \ge n^{-2} H_n(x_n), \qquad n \in \mathbb{N}.$$

Since Z_+ is normal, there exists some $c > 0$ such that

$$c\|H_n(x)\| \ge \|n^{-2} H_n(x_n)\| \ge n^2,$$

a contradiction. □

Exercises.

4.1.1. Prove Example 4.4.

4.2. Order-preserving operators and the companion norm

In the following let X and Y be ordered normed vector spaces with closed order cones X_+ and Y_+, respectively. We use the same symbols $\|\cdot\|$, ψ and $\sharp\cdot\sharp$ for the norms and the monotone companion (half-) norms on X and Y. Recall Section 2.5.

THEOREM 4.9. *Let $B : X_+ \to Y_+$ be bounded, homogeneous, and order-preserving. Then $\psi(B(x)) \le \|B\|\psi(x)$ for all $x \in X_+$. In particular, B is bounded with respect to the monotone companion norms and $\sharp B \sharp \le \|B\|$.*

PROOF. Since B is order-preserving, for $x \in X_+$,

$$\{y \in Y; B(x) \le y\} \supseteq \{B(z); x \le z \in X_+\}.$$

By definition of ψ,

$$\begin{aligned}\psi(B(x)) &\le \inf\{\|B(z)\|; z \in X_+, x \le z\} \le \inf\{\|B\|\,\|z\|; x \le z \in X_+\} \\ &= \|B\|\psi(x).\end{aligned}$$

□

THEOREM 4.10. *Let $B : X \to Y$ be bounded, linear and positive. Then B is bounded with respect to the monotone companion norm, $\psi(B(x)) \le \|B\|\psi(x)$ for all $x \in X$, and $\sharp B \sharp \le \|B\|$.*

PROOF. By a similar proof as for Theorem 4.9, since B is order-preserving, $\psi(B(x)) \leq \|B\|\psi(x)$ for all $x \in X$ and $\psi(-B(x)) = \psi(B(-x)) \leq \|B\|\psi(-x)$. The assertion now follows from $\sharp x\sharp = \max\{\psi(x), \psi(-x)\}$. □

Various concepts of continuity are preserved if one switches from the original norm to the monotone companion norm. We only look at the most usual concept here.

PROPOSITION 4.11. *Let $B : X \to Y$ be order-preserving and continuous at x with respect to the original norms. Then B is continuous at x with respect to the monotone companion norms.*

PROOF. Let $x \in X$ and B be continuous at x with respect to the original norms. Let $y \in X$ as well. For any $z \geq y - x$, $z \in X$, we have

$$B(y) - B(x) = B(x + y - x) - B(x) \leq B(x + z) - B(x).$$

By definition of the companion half-norm on Y (Proposition 2.45),

$$\psi(B(y) - B(x)) \leq \|B(x + z) - B(x)\|, \qquad y - x \leq z \in X.$$

Let $\epsilon > 0$. Since B is continuous at x with respect to the original norms, there exists some $\delta_+ > 0$ such that $\|B(x + z) - B(x)\| < \epsilon$ for all $z \in X$, $\|z\| < \delta_+$. Now let $y \in X$ and $\psi(y - x) < \delta_+$. By definition of the companion half-norm on X, there exists some $z \in X$ such that $y - x \leq z$ and $\|z\| < \delta_+$. Then $\|B(x + z) - B(x)\| < \epsilon$ and $\psi(B(y) - B(x)) < \epsilon$.

Also, for any $z \in X$ with $x - y \leq z$, we have

$$B(x) - B(y) = B(x) - B(x - (x - y)) \leq B(x) - B(x - z).$$

By definition of the companion half-norm on Y,

$$\psi(B(x) - B(y)) \leq \|B(x) - B(x - z)\|, \qquad z \in X, \quad x - y \leq z.$$

Let $\epsilon > 0$. Since B is continuous with respect to the original norms, there exists some $\delta_- > 0$ such that $\|B(x) - B(x - z)\| < \epsilon$ for all $z \in X$, $\|z\| < \delta_-$. Now let $y \in Y$ and $\psi(x - y) < \delta_-$. By definition of the companion half-norm on X, there exists some $z \in X$ such that $x - y \leq z$ and $\|z\| = \| - z\| < \delta_-$. Thus $\psi(B(x) - B(y)) < \epsilon$.

Set $\delta = \min\{\delta_+, \delta_-\}$ and $\sharp y - x\sharp < \delta$. Then $\psi(\pm(y - x)) < \delta_\pm$ and $\psi(\pm(B(y) - B(x)) < \epsilon$. This implies $\sharp B(y) - B(x)\sharp < \epsilon$. □

Recall that in $Y = \mathbb{R}$ with the absolute value, the monotone companion norm is also the absolute value and the companion half-norm is the positive part, $\psi(x) = \max\{x, 0\}$ for all $x \in \mathbb{R}$ (Example 2.54).

PROPOSITION 4.12. *Let $\phi : X_+ \to \mathbb{R}_+$ be homogeneous.*

(a) *If ϕ is bounded with respect to the monotone companion norm, then it is bounded with respect to the original norm and $\|\phi\| \leq \sharp\phi\sharp$.*
(b) *If ϕ is order-preserving and bounded with respect to the original norm, then it is bounded with respect to the monotone companion norm and $\|\phi\| = \sharp\phi\sharp$. Further $\phi(x) \leq \psi(x)\|\phi\|$ for all $x \in X_+$.*

PROOF. (a) Let $x \in X_+$. Since ϕ is bounded with respect to the monotone companion norm,

$$\phi(x) \leq \sharp\phi\sharp\, \sharp x\sharp \leq \sharp\phi\sharp\, \|x\|.$$

(b) This follows from part (a) and Theorem 4.9. □

PROPOSITION 4.13. *Let $\phi : X \to \mathbb{R}$ be linear.*

(a) *If ϕ is bounded with respect to the monotone companion norm, then it is bounded with respect to the original norm and $\|\phi\| \le \sharp\phi\sharp$.*
(b) *If ϕ is positive and bounded with respect to the original norm, then it is bounded with respect to the monotone companion norm and $\|\phi\| = \sharp\phi\sharp$ and $\phi(x) \le \|\phi\|\psi(x)$ for all $x \in X$.*

PROOF. (a) The proof is similar to the one for Proposition 4.12.

(b) This follows from part (a) and Theorem 4.10 and the fact that the monotone half-norm on $\mathbb{R}$ is given by the positive part. □

4.3. Duality for ordered normed vector spaces

If X is a normed vector space, the set X^* of bounded linear functionals $x^* : X \to \mathbb{R}$ is a Banach space under the operator norm, called the *dual space* of X. If X is an ordered normed vector space with order cone X_+, the set X_+^* of functionals in X^* with $x^*(X_+) \subseteq \mathbb{R}_+$ forms a closed wedge in X^* (Section 2.1.3), called the *dual wedge.*

By Proposition 4.13, if X_+ is a closed cone, X_+^* is also the dual wedge of linear functionals that are bounded with respect to the monotone companion norm $\sharp \cdot \sharp$ (Section 2.5). The respective norms induced on X^* are the same on X_+^*.

PROPOSITION 4.14. *Let X be an ordered normed vector space with closed order cone X_+. Let $x \in X$. Then there exists $x^* \in X_+^*$ such that $x^*x = \psi(x)$ and $\|x^*\| = \sharp x^*\sharp \le 1$. If $x \in X_+$, $\|x^*\| = 1$ can be achieved. Actually,*

$$\begin{aligned} \psi(x) &= \max\{x^*x; x^* \in X_+^*, \|x^*\| \le 1\}, \qquad x \in X, \\ \psi(x) &= \max\{x^*x; x^* \in X_+^*, \|x^*\| = 1\}, \qquad x \in X_+. \end{aligned} \tag{4.3}$$

Further [**136**, (4.3)], *for all $x \in X$,*

$$\sharp x\sharp = \max\{|x^*x|; x \in X_+^*, \|x^*\| = 1\} = \max\{|x^*x|; x \in X_+^*, \|x^*\| \le 1\}. \tag{4.4}$$

PROOF. Recall that X_+^* and its norm do not depend on whether we consider X with the original norm or its monotone companion norm (Proposition 4.13).

Let $u \in X$. By [**223**, IV.6], we find $x^* \in X^*$ with $x^*u = \psi(u)$ and

$$-\psi(-x) \le x^*x \le \psi(x), \qquad x \in X,$$

and so

$$|x^*x| \le \sharp x\sharp \le \|x\|, \qquad x \in X.$$

Since $\psi(-x) = 0$ for all $x \in X_+$ by Proposition 2.45, we have $x^* \in X_+^*$. By Proposition 4.13, $\sharp x^*\sharp = \|x^*\| \le 1$.

Now let $u \in X_+$. Then $x^*u = \psi(u) = \sharp u\sharp$ and so $\sharp x^*\sharp = 1$. Notice that we have proved $\le$ in (4.3).

Suppose that $\psi(u) > \psi(-u)$. Then there exists $x^* \in X_+^*$ with $\sharp x^*\sharp \le 1$ such that $x^*u = \psi(u) = \sharp u\sharp$. Hence $\sharp x^*\sharp = 1$.

If $\psi(u) < \psi(-u)$, there exists $x^* \in X_+^*$ with $\sharp x^*\sharp \le 1$ such that $x^*(-u) = \sharp u\sharp = |x^*u|$. Again $\sharp x^*\sharp = 1$.

If $\psi(u) = \psi(-u) > 0$, then $u, -u \notin X_+$ and we can make the same conclusion.

It remains the case $\psi(u) = \psi(-u) = 0$. But then $u = 0$, and all equalities hold trivially. So the $\le$ inequalities hold in (4.4). The $\ge$ inequalities follow from Proposition 4.13 (b). □

The rest of this section explores how properties of X_+ and properties of X_+^* are connected.

PROPOSITION 4.15. *The wedge X_+^* is a cone (called the* dual cone*) if and only if X_+ is a total cone, i.e., X is the closure of $X_+ - X_+$.*

PROOF. Assume that X_+ is total. Let $x^* \in X_+^*$ such that also $-x^* \in X_+^*$. Then x^* is zero on $X_+ - X_-$ and, since x^* is continuous, on the closure of this set which is X. So, x^* is the zero functional.

Suppose that X_+ is not total. The the closure of $X_+ - X_+$ is a closed linear subspace of X different from X. By a consequence of the Hahn–Banach theory, there is some nonzero functional $x^* \in X^*$ which is zero on $X_+ - X_+$, $x^* \in X_+^*$ and $-x^* \in X_+^*$. See [**223**, IV.6]. □

THEOREM 4.16. *Let X be an ordered normed vector space with generating non-flat cone X_+. Then the dual cone X_+^* is normal.*

For the concepts of generating (non-flat) or normal cones see Section 2.3 and 2.2.

PROOF. Since X_+^* is complete, it is sufficient to show that all points in X_+^* are normal points for X_+^* (Corollary 2.23).

Let $u^* \in X_+^*$. Let $x^* \in X_+^*$ with $x^* \le u^*$. By definition of the dual cone, $x^*x \le u^*x$ for all $x \in X_+$. Since X_+ is a generating non-flat cone, there exists some $c > 0$ such that for any $x \in X$ there exists $v, w \in X_+$ such that $x = v - w$ and $\|v\| + \|w\| \le c\|x\|$. Since x^* is linear and bounded,

$$|x^*x| \le x^*v + x^*w \le u^*v + u^*w \le \|u^*\|(\|v\| + \|w\|) \le c\|u^*\|\|x\|.$$

So u^* is a normal point for X_+^*. □

THEOREM 4.17. *Let X be an ordered normed vector space with order cone X_+. If X_+^* is generating, then X_+ consists of normal points for itself (Section 2.2.2).*

PROOF. Let $u \in X_+$ and assume that the set $\{x \in X_+; x \le u\}$ is not bounded. Then there exists a sequence (x_n) in X_+ with $x_n \le u$ and $\|x_n\| \ge n$ for all $n \in \mathbb{N}$. By one of the corollaries to the Hahn–Banach theorem, there exist $x_n^* \in X^*$ such that $x_n^* x_n \to \infty$ and $\|x_n^*\| = 1$. Since the wedge X_+^* is generating and closed and X^* is complete, X_+^* is non-flat by Corollary 2.38 and the x_n^* can be chosen in X_+^*. Then $x_n^* x_n \le x_n^* u \le \|u\|$, a contradiction. □

The converse does not hold in general. See Remark 4.20. The next result combines Corollary 2.23 with Theorem 4.17.

COROLLARY 4.18. *Let X be an ordered normed vector space with closed order cone X_+. If X_+^* is generating and X_+ is serially complete (Section 2.1.2), then X_+ is a normal cone.*

THEOREM 4.19. *Let X be an ordered normed vector space with closed cone X_+. If X_+ is normal, then the dual wedge X_+^* is generating.*

The proof follows the one in [**10**, Thm.2.26], which proves the more general result for locally convex spaces. Notice that X_+ does not need to be complete.

PROOF. Let X_+ be a normal cone. By Theorem 2.48, the monotone companion norm $\sharp\cdot\sharp$ is equivalent to the original norm. Let $f \in X^*$. To show that $f = f_+ - f_-$ for some $f_\pm \in X_+^*$, we can assume that

$$|f(x)| \leq \sharp x\sharp, \qquad x \in X. \tag{4.5}$$

We consider the ordered normed vector space

$$Z = X \times X \times \mathbb{R} \quad \text{with cone} \quad Z_+ = X_+ \times X_+ \times \mathbb{R}_+ \tag{4.6}$$

and with the norm of $z = (x, y, \alpha) \in Z$ being given by $\sharp x\sharp + \sharp y\sharp + |\alpha|$. We consider the linear subspace S of Z,

$$S = \{(x, -x, -f(x)); x \in X\}, \tag{4.7}$$

and the open convex set T of Z,

$$T = U \times U \times (-1, 1), \qquad U = \{x \in X; \sharp x\sharp < 1\}. \tag{4.8}$$

The aim is to separate the convex sets $S + T - (0, 0, 5)$ and Z_+ by a bounded linear nonzero functional $F : Z \to \mathbb{R}$,

$$F(z) \leq F(\tilde{z}) \quad \text{for all } z \in S + T - (0, 0, 5) \text{ and all } \tilde{z} \in Z_+,$$

[**9**, Thm.5.67] [**180**, II.9.1].

To this aim, we show that the sets $S + T - (0, 0, 5)$ and Z_+ do not have common points. Suppose they do.

Then there exist $x, y, u, v \in X$ and $\alpha, \beta \in \mathbb{R}$ such that

$$(x, u, \alpha) \in S, \quad (y, v, \beta) \in T, \quad -1 < \alpha + \beta - 5 < 1$$

and

$$x + y \in X_+, \quad u + v \in X_+, \quad \alpha + \beta - 5 \geq 0.$$

By definition of S, (4.7), $u = -x$ and $\alpha = -f(x)$. By definition of T, (4.8), $\sharp y\sharp < 1$, $\sharp v\sharp < 1$ and $-1 < \beta < 1$. So $v - x \in X_+$ and $x \leq v$ and $-x \leq y$ and $0 \leq -f(x) + \beta - 5 < 1$.

We use the monotone companion half-norm,

$$\psi(x) \leq \psi(v) \leq \sharp v\sharp < 1, \qquad \psi(-x) \leq \psi(y) \leq \sharp y\sharp < 1,$$

and so $\sharp x\sharp < 1$. By (4.5), $|f(x)| \leq \sharp x\sharp < 1$. So,

$$0 \leq -f(x) + \beta - 5 < 1 + 1 - 5 = -3 < 0, \tag{4.9}$$

a contradiction.

This shows that the convex sets $S + T - (0, 0, 5)$ and Z_+ have an empty intersection. Notice that the open set T has the interior point $(0, 0, 0)$ which is also an element of S. So, $-(0, 0, 5)$ is an interior point of $S + T - (0, 0, 5)$.

By a separation theorem for convex sets [**9**, Thm.5.67] [**180**, II.9.1], there exists a bounded linear nonzero functional F on Z such that

$$F(z) \leq F(\tilde{z}) \quad \text{for all } z \in S + T - (0, 0, 5) \text{ and all } \tilde{z} \in Z_+. \tag{4.10}$$

Since Z_+ contains the zero vector, $F(z) \leq 0$ for all $z \in S + T - (0, 0, 5)$.

Since S is a linear subspace of Z and T contains the zero vector, by (4.10), $F(nz - (0, 0, 5)) \leq 0$ for all $n \in \mathbb{Z}$ and $z \in S$. So $nF(z) \leq F(0, 0, 5)$ for all $n \in \mathbb{Z}$ and $z \in S$. This implies $F(z) = 0$ for all $z \in S$. Further, $F(0, 0, 5) \geq 0$ and so $F(0, 0, 1) \geq 0$.

Let $\tilde{z} \in Z_-$. Then $n\tilde{z} \in Z_+$ for all $n \in \mathbb{N}$ and so $nF(\tilde{z}) = F(n\tilde{z}) \geq F\big(-(0, 0, 5)\big)$ for all $n \in \mathbb{N}$ by (4.10) because $-(0, 0, 5) \in S + T - (0, 0, 5)$. This implies $F(\tilde{z}) \geq 0$.

By definition of S, (4.7), and $F(z) = 0$ for all $z \in S$, for all $x \in X$ we have $0 = F(x, -x, -f(x))$ and so

$$F(x,0,0) - F(0,x,0) = f(x)F(0,0,1), \qquad x \in X. \tag{4.11}$$

Suppose that $F(0,0,1) = 0$. By (4.10), $F(T) \subseteq -\mathbb{R}_+$. Notice that, if $w \in U$ also $-w \in U$ and so, for all $w \in U$, $F(\pm w, 0, 0) \le 0$ and $F(0, \pm w, 0) \le 0$ and so $F(\pm w, 0, 0) = 0$ and $F(0, \pm w, 0) = 0$ first for all $w \in U$ but then also for all $w \in X$. So F is the zero functional, a contradiction.

So $F(0,0,1) \ne 0$. We set

$$f_+(x) = \frac{1}{F(0,0,1)} F(x,0,0) \quad \text{and} \quad f_-(x) = \frac{1}{F(0,0,1)} F(0,x,0), \qquad x \in X$$

and obtain $f = f_+ - f_-$ with $f_\pm \in X_+^*$ from (4.11). □

REMARK 4.20. It is not sufficient for X_+^* to be generating that all points of X_+ are normal points for X_+. Let X be ℓ^1 with the variation norm and X_+ be the cone of nonnegative sequences in X. Then all points of X_+ are normal points for X_+ but X_+ is not normal (Example 2.43).

Suppose that X_+^* is generating. Then every linear functional on bv that is bounded with respect to the variation norm has a restriction to ℓ^1 that can be written as the difference of two positive linear functionals on ℓ_1 that are bounded with respect to the variation norm. Since ℓ^1 is dense in bv with respect to be variation norm, these positive linear functionals can be extended to bv with the extensions being bounded with respect to the variation norm. This would imply that bv_+^* is generating and imply by Theorem 4.17 that all points of bv_+ are normal points for bv_+. But only the points in ℓ_+^1 are normal points for bv_+ (Theorem 2.42).

4.4. Important homogeneous functionals

Let X be an ordered vector space with Archimedean order cone X_+. See Section 2.1.3. For $u \in X_+$, we consider the functionals

$$\left.\begin{array}{rl} (x/u)^\diamond = & \inf\{\alpha \in \mathbb{R}; x \le \alpha u\} \\ (x/u)_\diamond = & \sup\{\beta \in \mathbb{R}; \beta u \le x\} \end{array}\right\} x \in X, \tag{4.12}$$

with the convention that $\inf(\emptyset) = \infty$ and $\sup(\emptyset) = -\infty$. If X is a space of real-valued functions on a set Ω,

$$\left.\begin{array}{rl} (x/u)^\diamond = & \sup\left\{\frac{x(\xi)}{u(\xi)}; \xi \in \Omega, u(\xi) > 0\right\} \\ (x/u)_\diamond = & \inf\left\{\frac{x(\xi)}{u(\xi)}; \xi \in \Omega, u(\xi) > 0\right\} \end{array}\right\} x \in X.$$

Many other symbols have been used for these two functionals in the literature; see Thompson [**210**] and Bauer [**22**] for some early occurrences. For $x \in X_+$, $\|x\|_u = (x/u)^\diamond$. Since we will use this functional for $x \in X_+$ only, we will stick with the notation $\|x\|_u$. Again for $x \in X_+$ and $u \in \dot{X}_+$, $(x/u)_\diamond$ is a nonnegative real number, and we will use the leaner notation

$$[x]_u = \sup\{\beta \ge 0; \beta u \le x\}, \qquad x \in X_+, \quad u \in \dot{X}_+. \tag{4.13}$$

Since the cone X_+ is Archimedean,

$$x \ge [x]_u u, \qquad x, u \in X_+. \tag{4.14}$$

Further $[x]_u$ is the largest number for which this inequality holds. We call $[\cdot]_u$ the *infimum functional* on X_+ (with respect to u) (see (4.12)).

$(\cdot/\cdot)_\diamond : X_+ \to X_+ \to \mathbb{R}_+$ is called *order function* or *extraction grade* in [**138**] and interesting results for this function of two variables can be found there.

We notice that $[x]_u > 0$ if and only if u is x-bounded. In this and other contexts, it is convenient to introduce the following terminology:

x is *u-positive* if there exists some $\delta > 0$ such that $x \geq \delta u$.

We have the equivalences: x is u-positive, u is x-bounded, $[x]_u > 0$.

LEMMA 4.21. *Let $u \in \dot{X}_+$. Then the infimum functional $\phi = [\cdot]_u : X_+ \to \mathbb{R}_+$ is homogeneous, order-preserving and concave,*

$$\big|[y]_u - [x]_u\big| \leq \|y - x\|_u, \qquad y, x \in X_u \cap X_+. \tag{4.15}$$

If X is an ordered normed vector space and X_+ is closed, we also have the following:

The functional is bounded with respect to the original norm on X and also to the monotone companion norm,

$$[x]_u \leq \frac{\sharp x \sharp}{\sharp u \sharp} \leq \frac{\|x\|}{\sharp u \sharp}, \qquad x \in X_+,$$

and

$$\|\phi\| = \sharp \phi \sharp \leq \frac{1}{\sharp u \sharp}.$$

Finally, if X_+ is closed, ϕ is upper semicontinuous with respect to the original norm and to the monotone companion norm. More precisely, if (x_n) is a sequence and x a point in X_+ and $\psi(x_n - x) \to 0$, then $[x]_u \geq \limsup_{n\to\infty} [x_n]_u$.

Recall that $\sharp u \sharp = d(u, -X_+)$.

PROOF. We apply the monotone companion norm to (4.14),

$$\sharp x \sharp \geq [x]_u \sharp u \sharp, \qquad x \in X_+.$$

The equality $\|\phi\| = \sharp \phi \sharp$ follows from Proposition 4.12. To prove (4.15), let $x, y \in X_+ \cap X_u$. By (4.14),

$$[x]_u u \leq x \leq x - y + y \leq \|x - y\|_u u + y.$$

We reorganize,

$$y \geq ([x]_u - \|x - y\|_u) u.$$

By (4.14),

$$[y]_u \geq [x]_u - \|x - y\|_u.$$

We reorganize,

$$[x]_u - [y]_u \leq \|x - y\|_u.$$

By symmetry,

$$[y]_u - [x]_u \leq \|y - x\|_u = \|x - y\|_u.$$

This implies (4.15).

Now assume that X_+ is closed. Since the monotone companion norm is weaker than the original norm, it is sufficient to show the upper semicontinuity with respect to the companion norm. Let x be an element and (x_n) a sequence in X_+ such that $\psi(x_n - x) \to 0$. Set

$$\alpha = \limsup_{n\to\infty} [x_n]_u.$$

For every $k \in \mathbb{N}$, there exists some $n_k \geq k$ such that $[x_{n_k}]_u \geq \alpha - (1/k)$. By (4.14), $x_{n_k} \geq (\alpha - (1/k))u$. Since $\psi(x_{n_k} - x) \to 0$ and $\psi((1/k)u) \to 0$ as $k \to \infty$, by Theorem 2.47, $x \geq \alpha u$ and $[x]_u \geq \alpha$.

The other properties are readily derived from the definitions (Exercise 4.4.1).
□

Exercises.

4.4.1. Let $u \in \dot{X}_+$. Show that the infimum functional $\phi = [\cdot]_u : X_+ \to \mathbb{R}_+$ given by (4.13) is homogeneous, order-preserving and superadditive.

4.5. Mating functions

An important building block for homogeneous operators in the context of two-sex population dynamics are mating (pair formation) functions.

By themselves, *mating functions* are examples of homogeneous order-preserving functionals $\varphi : \mathbb{R}^2_+ \to \mathbb{R}_+$ where the order on $\mathbb{R}^2$ is componentwise, i.e., induced by the order cone $\mathbb{R}^2_+$ of $\mathbb{R}^2$. In other words, $\varphi(f, m)$ is an increasing function of f and m. In addition to homogeneity, one also assumes for a mating function that $\varphi(f, m) = 0$ if $fm = 0$ and $\varphi(1, 1) > 0$.

The interpretation is that $\varphi(f, m)$ is the number of mated pairs that result from f females and m males.

This section interweaves material from [**83**] and [**202**]. In [**83**], φ is the incidence function of an infectious disease, which connects the number of susceptible and infective hosts to the incidence, the rate of new infections. For pertinent literature, see [**35, 83, 106–109, 115, 117, 118, 130, 161, 202**] and the references therein.

The functional φ automatically is a bounded homogeneous functional by Theorem 4.5, but this can also be seen directly. Let $\|\cdot\|_\infty$ be the maximum-norm on $\mathbb{R}^2$. Then, for all $x \in \mathbb{R}^2_+$,

$$\varphi(x) \leq \varphi(\|x\|_\infty, \|x\|_\infty) = \|x\|_\infty \varphi(1, 1).$$

This implies $\|\varphi\| \leq \varphi(1, 1)$. Since $\|(1, 1)\|_\infty = 1$, the operator norm of φ satisfies

$$\|\varphi\| = \varphi(1, 1).$$

We introduce the *female and male per capita mating functions*

$$\varphi_1(r) = \varphi(1, r), \qquad \varphi_2(r) = \varphi(r, 1) \tag{4.16}$$

which give the chance of one female (male) to mate if r is the ratio of the opposite to the own sex. We will establish a relation between the concavity (convexity) of φ_j and φ.

LEMMA 4.22. *Let $h : (0, \infty) \to \mathbb{R}_+$. Define $\varphi : (0, \infty) \times \mathbb{R}_+ \to \mathbb{R}$ by $\varphi(f, m) = fh(m/f)$.*

Then $\varphi(\alpha f, \alpha m) = \alpha \varphi(f, m)$ for all $f, m, \alpha > 0$.

Further, if h is concave (convex) on $(0, \infty)$, then φ is concave (convex) on $(0, \infty)^2$.

PROOF. The homogeneity of φ even holds without h being concave. Assume that h is concave. The concavity of φ in m is easily seen. For $f, \tilde{f} > 0$ and $m, \tilde{m} > 0$,

$$\begin{aligned}\varphi(f,m) + \varphi(\tilde{f},\tilde{m}) &= (f+\tilde{f})\Big[\frac{f}{f+\tilde{f}}h(m/f) + \frac{\tilde{f}}{f+\tilde{f}}\tilde{h}(\tilde{m}/\tilde{f})\Big] \\ &\le (f+\tilde{f})h\Big(\frac{m+\tilde{m}}{f+\tilde{f}}\Big) = \varphi(f+\tilde{f}, m+\tilde{m}).\end{aligned}$$

Hence φ is superadditive. Since φ is also homogeneous, φ is concave.

The proof for convex h is similar. □

THEOREM 4.23. *Let $\varphi : \mathbb{R}_+^2 \to \mathbb{R}$.*

(a) *If φ is convex (concave) on $\mathbb{R}_+^2$, then φ_1 and φ_2 and $t \mapsto \varphi(t,t)$ are convex (concave) on $\mathbb{R}_+$.*
(b) *If φ is homogeneous and φ_1 or φ_2 are convex (concave) on $(0,\infty)$, then φ is convex (concave) on $(0,\infty)^2$ (on $\mathbb{R}_+^2$ if φ is also continuous).*

PROOF. (a) Let φ be convex and $\xi \in (0,1)$. For $s,t \ge 0$,

$$\varphi_1(\xi s + (1-\xi)t) = \varphi\big(\xi(1,s) + (1-\xi)(1,t)\big) \le \xi\varphi_1(s) + (1-\xi)\varphi_1(t).$$

The proof for φ_2 is similar. Let $\Delta(t) = \varphi(t,t)$, $t \ge 0$. For $s,t \ge 0$,

$$\Delta(\xi s + (1-\xi)t) = \varphi\big(\xi(s,s) + (1-\xi)(t,t)\big) \ge \xi\Delta(s) + (1-\xi)\Delta(t).$$

(b) Assume that φ is homogeneous and φ_1 is concave. Then $\varphi(f,m) = f\varphi_1(m/f)$ for $m \ge 0$, $f > 0$. By Lemma 4.22, φ is concave on $(0,\infty)^2$. If φ is continuous on $\mathbb{R}_+^2$, the concavity can be extended to $\mathbb{R}_+^2$ by an approximation argument. □

In part (b), the homogeneity is necessary. Let $\varphi(f,m) = (fm)^{2/3}$, $f,m \ge 0$. Then φ is not homogeneous and $\varphi_j(t) = t^{2/3}$ are concave functions of $t \ge 0$. But $\varphi(t,t) = t^{4/3}$ is not a concave function of $t \ge 0$ and so φ is not concave.

The proof of part (b) via Lemma 4.22 was inspired by the private communication of a different proof by Jürgen Voigt.

Recall the female and male per capita mating functions φ_i in (4.16). Notice that

$$\varphi_i(\infty) = \lim_{s\to\infty} \varphi_i(s) = \sup_{s\ge 0} \varphi_i(s)$$

exists in $[0,\infty]$ and is finite if φ_i is bounded on $\mathbb{R}_+$.

LEMMA 4.24. *Let $\varphi : \mathbb{R}_+^2 \to \mathbb{R}_+$ be a mating function.*

(a) *Then $\min\{f,m\}\varphi(1,1) \le \varphi(f,m) \le \max\{f,m\}\varphi(1,1)$ for all $f,m \ge 0$.*
(b) *If the φ_i are bounded on $\mathbb{R}_+$, then*

$$\varphi(s_1,s_2) \le s_i\varphi_i(\infty), \qquad s_1,s_2 \in \mathbb{R}_+, \quad i = 1,2,$$

and

$$\varphi(f,m) \le \min\{f,m\}\max\{\varphi_1(\infty),\varphi_2(\infty)\}.$$

PROOF. (a) Let $\alpha = \min\{f,m\}$. Since φ is order-preserving and homogeneous,

$$\varphi(f,m) \ge \varphi(\alpha,\alpha) = \alpha\varphi(1,1).$$

The other inequality is proved similarly.

(b) Let $0 < f$. Since φ is homogeneous and order-preserving,

$$\varphi(f,m) = f\varphi(1,m/f) \le f\varphi_1(\infty).$$

If $f = 0$, $\varphi(f, m) = 0$ and this inequality also holds.

Similarly, if $0 \le m$, $\varphi(f, m) \le m\varphi_2(\infty)$. □

One important class of homogeneous order-preserving functionals is

$$\varphi(s,t) = (ps^\alpha + qt^\alpha)^{1/\alpha}, \qquad s,t > 0, \tag{4.17}$$

where $\alpha \in \mathbb{R}$, $\alpha \neq 0$, and $p, q > 0$, $p + q = 1$ [**105, 109**]. If $\alpha > 0$, $\varphi(0,0) = 0$ and φ is continuous, but $\varphi(x, y) > 0$ if $x > 0$ or $y > 0$. So φ does not qualify as a mating function if $\alpha > 0$.

If $\alpha < 0$, set $\beta = -\alpha$. Then

$$\varphi(s,t) = \frac{st}{(pt^\beta + qs^\beta)^{1/\beta}}. \tag{4.18}$$

For $\beta = 1$, we obtain a generalized *harmonic mean.* Notice that, for this form of φ, $\varphi(s,t) = 0$ if either $s = 0$ or $t = 0$, and

$$\min\{s,t\} \le \varphi(s,t) \le \max\{p^{1/\alpha}, q^{1/\alpha}\} \min\{s,t\}, \qquad s,t \ge 0. \tag{4.19}$$

We conclude that $\varphi(s,t) \to 0$ as $(s,t) \to 0$ and set

$$\varphi(t,s) = 0, \quad t, s \ge 0, \quad st = 0. \tag{4.20}$$

With this definition, φ becomes continuous on $\mathbb{R}_+^2$ even if $\alpha < 0$. One readily checks that φ is homogeneous and that the φ_j are concave if $\alpha < 0$ or $0 < \alpha < 1$ and that the φ_j are convex if $\alpha \ge 1$. By Theorem 4.23, φ is concave for $0 \neq \alpha < 1$ and convex for $\alpha \ge 1$. The statements for $\alpha > 0$ also follow from Hölder's inequalities [**112**, Thm.13.4,13.6].

Another class are generalized *geometric means,*

$$\varphi(s,t) = s^p t^q, \qquad s,t \ge 0, \tag{4.21}$$

with $p, q > 0$ and $p + q = 1$. Here, it is even more obvious that φ is a concave function of each separate variable. Actually, this φ is the limit of (4.17) as $\alpha \to 0$ [**109**]. Since pointwise limits of concave functions are concave, the concavity of this φ follows from the concavity of the previous one. By the same token, the *minimum function*

$$\varphi(x,y) = \min\{x,y\} \tag{4.22}$$

is concave because it is the limit of (4.17) as $\alpha \to -\infty$ by (4.19).

Notice new homogeneous superadditive functions can be obtained from known ones by setting

$$\tilde\varphi(f,m) = \beta\varphi(\alpha f, \tilde\alpha m), \qquad f, m \ge 0,$$

where $\alpha, \tilde\alpha, \beta > 0$.

REMARK 4.25. Lemma 4.22 can be used to construct other types of mating functions. Let $h : \mathbb{R}_+ \to \mathbb{R}_+$ be concave and continuous, $h(0) = 0$ and define φ on $(0,\infty)^2$ as in Lemma 4.22 and $\varphi(s,t) = 0$ if $st = 0$. In order to make φ continuous on $\mathbb{R}_+^2$ assume that $h(t)/t \to 0$ as $t \to \infty$. See [**83**] for more.

Possible choices for h are the following well-known typical functional responses,

- the Ivlev [**122**] or Skellam [**185**] functional response $h(t) = \zeta(1 - e^{-\xi t})$,
- the logarithmic [**29**] response $h(t) = \zeta \ln(1 + \xi t)$,
- the arctangent [**19, 182**] response, $h(t) = \zeta \arctan(\xi t) = \zeta \tan^{-1}(\xi t)$,
- and the hyperbolic tangent [**124, 183**] response $h(t) = \zeta \tanh(\xi t)$ where $\tanh(s) = \frac{e^s - e^{-s}}{e^s + e^{-s}}$.

Notice that, if h is convex, the construction in Lemma 4.22 does not necessarily lead to an order-preserving φ.

Exercises.

4.5.1. Let $\varphi : \mathbb{R}_+^2 \to \mathbb{R}_+$ and $h : \mathbb{R}_+ \to \mathbb{R}_+$, $h(z) = \varphi(1, z)$ for all $z \geq 0$. Assume $\varphi(0, z) = 0 = \varphi(z, 0)$ for all $z \geq 0$. Show:

(a) φ is homogeneous if and only if $\varphi(x, y) = xh(y/x)$ for all $x, y > 0$.
(b) $\varphi(x, y)$ is increasing in $x > 0$ if and only if $h(z)/z$ is decreasing in $z > 0$.
(c) $\varphi(x, y)$ is increasing in $y \geq 0$ if and only if h is increasing.
(d) For $y \geq 0$, $\varphi(x, y) \to 0$ as $x \to 0$ if and only if $h(z)/z \to 0$ as $z \to \infty$.
(e) For $x > 0$, $\varphi(x, y) \to 0$ as $y \to 0$ if and only if $h(z) \to 0$ as $z \to 0$.
(f) φ is continuous on $\mathbb{R}_+^2$ if and only if h is continuous on $\mathbb{R}_+$ and $h(z)/z \to 0$ as $z \to \infty$.

4.5.2. Let $h : \mathbb{R}_+ \to \mathbb{R}_+$ be defined as

$$h(z) = \begin{cases} 2 - \frac{1}{z}, & z > 1 \\ z, & 0 \leq z \leq 1. \end{cases}$$

Show that h is continuous, h is increasing and concave, $h(z)/z \to 0$ as $z \to \infty$.

Notice that, if $\varphi(x, y) = xh(y/x)$, then φ is a special case of formula in [**124**, Tab.1].

4.5.3. Let $h : \mathbb{R}_+ \to \mathbb{R}_+$ be defined as

$$h(z) = \begin{cases} e^{1-z^{-1}}, & z > 1 \\ z, & 0 \leq z \leq 1. \end{cases}$$

Show that h is continuous, h is increasing and concave, $h(z)/z \to 0$ as $z \to \infty$.

Notice that, if $\varphi(x, y) = xh(y/x)$, then φ is a special case of a formula in [**124**, Tab.1].

CHAPTER 5

Spectral radii for homogeneous operators

Let X be a normed vector space.

Let $B : X_B \to X_B$ be a bounded homogeneous operator in X defined on a homogeneous subset X_B of X.

The presentation follows the one in [**198, 199**] but considers homogeneous subsets X_B of X rather than cones whenever possible.

5.1. Spectral radius and orbital spectral radius

Recall the operator norm of B defined in (4.1). Define

$$a_n = \|B^n\|.$$

By Theorem 4.7,

$$a_{n+m} \leq a_n a_m, \qquad n, m \in \mathbb{N}.$$

By Lemma A.44,

$$\inf_{n\in\mathbb{N}} \|B^n\|^{1/n} = \lim_{n\to\infty} \|B^n\|^{1/n} =: \mathbf{r}(B), \tag{5.1}$$

and $\mathbf{r}(B)$ is called the *spectral radius* of B.

Mallet-Paret and Nussbaum [**156, 157**] suggest an alternative definition of a spectral radius for homogeneous (not necessarily bounded) maps $B : X_B \to X_B$. First, define asymptotic least upper bounds for the geometric growth factors of B-orbits,

$$\gamma(x, B) := \gamma_B(x) := \gamma_x(B) := \limsup_{n\to\infty} \|B^n(x)\|^{1/n}, \qquad x \in X_B, \tag{5.2}$$

and then the *orbital spectral radius*

$$\mathbf{r}_o(B) = \sup_{x\in X_B} \gamma_B(x). \tag{5.3}$$

Here $\gamma_B(x) := \infty$ if the sequence $(\|B^n(x)\|^{1/n})$ is unbounded and $\mathbf{r}_o(B) = \infty$ if $\gamma_B(x) = \infty$ for some $x \in X_B$ or the set $\{\gamma_B(x); x \in X_B\}$ is unbounded. Since $\|B^n(x)\| \leq \|B^n\| \, \|x\|$ by Theorem 4.7,

$$\gamma(x, B) \leq \mathbf{r}(B), \quad x \in X_B, \quad \mathbf{r}_o(B) \leq \mathbf{r}(B). \tag{5.4}$$

If $X_\mathbb{C}$ is a complexification of X and $X_B = X$ and B is linear and $B_\mathbb{C}$ its linear extension to $X_\mathbb{C}$ (see Section 5.4), (5.1) is the Gelfand formula [**93**] for

$$\sup\{|\lambda|; \lambda \in \sigma(B_\mathbb{C})\} \tag{5.5}$$

where $\sigma(B_\mathbb{C}) \subseteq \mathbb{C}$ is the spectrum of $B_\mathbb{C}$.

If X_B is a closed cone in X, the number $\mathbf{r}(B)$ is called *partial spectral radius* by Bonsall [**27**], X_B spectral radius by Schaefer [**178, 180**], and *cone spectral radius* by Nussbaum [**170**]. Mallet-Paret and Nussbaum [**156, 157**] call $\mathbf{r}(B)$ the *Bonsall* *cone spectral radius* and $\mathbf{r}_o(B)$ the cone spectral radius. For $x \in X_B$, the number $\gamma_B(x)$ is called *local spectral radius* of B at x by Förster and Nagy [**87**].

Since $\mathbf{r}(B)$ makes sense as long as the domain of definition X_B is homogeneous (including the familiar special case $X_B = X$) and the domain of definition is part of the concept of an operator, we simply call $\mathbf{r}(B)$ the spectral radius of B and $\mathbf{r}_o(B)$ the orbital spectral radius. If a name were to be attached to it, *Gelfand spectral radius* might be as appropriate as Bonsall cone spectral radius.

The spectral radius and orbital spectral radius coincide for many applications (Section 5.1.4). Gripenberg [**95**], gives an example for $\mathbf{r}_o(B) < \mathbf{r}(B)$.

For a bounded positive linear map on an ordered Banach space, the spectral radius and the spectral radius of its restriction to the order cone coincide provided that the cone is generating [**157**, Thm.2.14]. See Theorem 5.15. This is not true if the cone is only total [**27**, Sec.2.8] (Section 11.6).

5.1.1. Spectral radius and eigenvalues. It may be instructive to look at the simplest possible homogeneous operator in a general ordered normed Banach space, a rank-one operator.

EXAMPLE 5.1 (Rank-one operators). Let X_B be a homogeneous subset of X, $u \in \dot{X}_B$, $\phi : X_B \to \mathbb{R}_+$ be homogeneous and bounded and $B : X_B \to X_B$ be given by $B(x) = \phi(x)u$. Then B is homogeneous and bounded with $\|B\| \le \|\phi\|\,\|u\|$. One proves by induction that

$$B^n(x) = \phi(x)[\phi(u)]^{n-1}u, \qquad x \in X_B, n \in \mathbb{N}.$$

So

$$\|B^n(x)\| = \phi(x)[\phi(u)]^{n-1}\|u\|, \qquad \|B^n\| \le \|\phi\|[\phi(u)]^{n-1}\|u\|.$$

We have $\gamma(x,B) \le \phi(u)$, $\gamma(u,B) = \phi(u)$, $\mathbf{r}(B) \le \phi(u)$. So

$$\phi(u) = \gamma(u,B) = \mathbf{r}_o(B) = \mathbf{r}(B).$$

We also see that $\phi(u)$ is an eigenvalue of B associated with the eigenvector u and the eigenfunctional ϕ,

$$B(u) = \phi(u)u, \qquad \phi \circ B = \phi(u)\phi. \qquad \square$$

In general, it is difficult to determine $\mathbf{r}(B)$ explicitly or even to derive estimates and approximations. But establishing $\mathbf{r}(B)$ as an eigenvalue will be a major program of these notes (Chapters 9 and 11). Later, we will also introduce a *Collatz–Wielandt* radius (Section 5.5). We refer to all of them as *spectral radii.*

It may be helpful to point out that (5.5) is partially preserved for bounded homogeneous B. Recall that, for linear bounded everywhere-defined B, eigenvalues are special elements of the spectrum of B.

For homogeneous $B : X_B \to X_B$, we call $\lambda \in (0,\infty)$ an *eigenvalue* of B and $v \in \dot{X}_B$ an *associated eigenvector* of B if $B(v) = \lambda v$. Recall $\dot{X}_B = X_B \setminus \{0\}$.

PROPOSITION 5.2. *Let $B : X_B \to X_B$ be homogeneous and bounded, $\lambda \in (0,\infty)$, $v \in \dot{X}_B$ and $B(v) = \lambda v$.*

Then $\lambda = \gamma(v,B) \le \mathbf{r}_o(B) \le \mathbf{r}(B)$.

PROOF. Since B is homogeneous, by induction, $\lambda^n v = B^n(v)$ for all $n \in \mathbb{N}$. So $\lambda\|v\|^{1/n} = \|B^n(v)\|^{1/n}$ for all $n \in \mathbb{N}$, and the assertion follows from (5.3) and (5.4). $\square$

We call $\lambda \in (0,\infty)$ an *approximate eigenvalue* of $B : X_B \to X_B$ if

$$\inf \{\|B(x) - \lambda x\|; x \in X_B, \|x\| = 1\} = 0. \tag{5.6}$$

PROPOSITION 5.3. *Let $\lambda \in (0,\infty)$ be an approximate eigenvalue of B and B be uniformly continuous on bounded subsets of X_B. Then $\lambda \le \mathbf{r}(B)$.*

This result is proved in [**163**, L.3.3] under the stronger assumption that B is Lipschitz continuous.

PROOF. Since B is continuous, B is bounded by Proposition 4.2, and $\mathbf{r}(B)$ is defined. By (5.6), there exists a sequence (x_k) in X_B with $\|x_k\| = 1$ for all $\in \mathbb{N}$ such that $\|B(x_k) - \lambda x_k\| \to 0$ as $k \to \infty$. By induction, for all $n \in \mathbb{N}$,

$$\|B^n(x_k) - \lambda^n x_k\| \to 0, \qquad k \to \infty. \tag{5.7}$$

Indeed, assume that $n \in \mathbb{N}$ and (5.7) holds for n. Notice that $\|B^n(x_k)\| \le \|B^n\|$ and $\|\lambda^n x_k\| = \lambda^n$ for all $k \in \mathbb{N}$.

Since, by assumption, B is uniformly continuous on bounded subsets of X_B,

$$\|B(B^n(x_k)) - B(\lambda^n x_k)\| \to 0, \qquad k \to \infty.$$

Since B and the norm are homogeneous,

$$\|B^{n+1}(x_k) - \lambda^{n+1}x_k\| \le \|B(B^n(x_k)) - B(\lambda^n x_k)\| + \lambda^n\|B(x_k) - \lambda x_k\| \stackrel{k\to\infty}{\longrightarrow} 0.$$

This shows (5.7) for $n + 1$. By the induction principle, (5.7) holds for all $n \in \mathbb{N}$.

Now, for all $n \in \mathbb{N}$, since $\|x_k\| = 1$,

$$\lambda^n = \|\lambda^n x_k\| \le \|\lambda^n x_k - B^n(x_k)\| + \|B^n(x_k)\| \le \|\lambda^n x_k - B^n(x_k)\| + \|B^n\|.$$

Since $(a + b)^{1/n} \le a^{1/n} + b^{1/n}$ for all $a, b \ge 0$ and $n \in \mathbb{N}$,

$$\lambda \le (\|\lambda^n x_k - B^n(x_k)\|)^{1/n} + \|B^n\|^{1/n}, \qquad k, n \in \mathbb{N}.$$

We take the limit as $k \to \infty$, using (5.7),

$$\lambda \le \|B^n\|^{1/n}, \qquad n \in \mathbb{N},$$

and $\lambda \le \mathbf{r}(B)$ by (5.1). □

5.1.2. Power rules for spectral radii. One readily checks for bounded homogeneous B that

$$\mathbf{r}(\alpha B) = \alpha \mathbf{r}(B), \quad \alpha \in \mathbb{R}_+, \qquad \mathbf{r}(B^m) = (\mathbf{r}(B))^m, \quad m \in \mathbb{N}. \tag{5.8}$$

See Lemma A.46. The same rules hold for $\mathbf{r}_o(B)$ though proving the second property takes some more effort [**156**, Prop.2.1]. Actually, as we show in Theorem 5.4,

$$\gamma(x, B^m) = (\gamma(x, B))^m, \qquad m \in \mathbb{N}, \quad x \in X_B, \tag{5.9}$$

which implies

$$\mathbf{r}_o(B^m) = (\mathbf{r}_o(B))^m, \qquad m \in \mathbb{N}. \tag{5.10}$$

5.1.3. Power rule for growth factors.

THEOREM 5.4. *Let $B : X_B \to X_B$ be a homogeneous operator on a homogeneous subset X_B of X and $x \in X_B$.*

Then $\gamma(x, B^m) \leq (\gamma(x, B))^m$ for all $m \in \mathbb{N}$ with equality holding if B is bounded.

PROOF. From the properties of the limit superior,

$$\gamma(x, B^m) = \limsup_{k\to\infty} \|(B^m)^k(x)\|^{1/k} = \lim_{j\to\infty} \sup_{k\geq j} \|B^{mk}(x)\|^{m/(mk)}$$

$$\leq \lim_{j\to\infty} \sup_{n\geq mj} (\|(B^n(x)\|^{1/n})^m = \Big(\limsup_{n\to\infty} \|(B^n(x)\|^{1/n}\Big)^m = (\gamma(x, B))^m.$$

To prove the opposite inequality if B is bounded, suppose that

$$\gamma(x, B^m) < (\gamma(x, B))^m.$$

Then, $\gamma(x, B) > 0$. Further, there exists some $s \in (0, 1)$ such that $\gamma(x, B^m) < s^m(\gamma(x, B))^m$. By definition of $\gamma(x, B^m)$, there exists some $N \in \mathbb{N}$ such that

$$\|B^{mn}(x)\|^{1/(mn)} < s\gamma(x, B), \qquad n \geq N. \tag{5.11}$$

Choose a sequence (k_j) in $\mathbb{N}$ such that

$$k_j \to \infty, \quad \|B^{k_j}(x)\|^{1/k_j} \to \gamma(x, B), \quad j \to \infty. \tag{5.12}$$

Then there exist sequences n_j and p_j such that $n_j \to \infty$ and $0 \leq p_j < m$ and $k_j = mn_j + p_j$. By Theorem 4.7, since B is bounded, for all $j \in \mathbb{N}$,

$$\|B^{k_j}(x)\|^{1/k_j} \leq \|B^{p_j}\|^{1/k_j} \, \|B^{mn_j}(x)\|^{1/k_j} \leq c^{1/k_j} \big(\|B^{mn_j}(x)\|^{1/(mn_j)}\big)^{(k_j-p_j)/k_j},$$

$$c = \sup_{\ell=0}^{m-1} \|B^\ell\|.$$

For $j \in \mathbb{N}$ with $n_j > N$, by (5.11),

$$\|B^{k_j}(x)\|^{1/k_j} \leq c^{1/k_j}(s\gamma(x, B))^{(k_j-p_j)/k_j} = c^{1/k_j}(s\gamma(x, B))^{1-(p_j/k_j)}.$$

By the properties of the limit superior and by (5.12),

$$\gamma(x, B) \leq \limsup_{j\to\infty} \|B^{k_j}(x)\|^{1/k_j} \leq \limsup_{j\to\infty} (s\gamma(x, B))^{1-(p_j/k_j)}.$$

Since $p_j/k_j \to 0$,

$$0 < \gamma(x, B) \leq s\gamma(x, B),$$

a contradiction. □

The companion growth bound. Let now be X be an ordered normed vector space with order cone X_+ and X_B be a homogeneous subset of X_+ and $B : X_B \to X_B$ be homogeneous.

For those $x \in X_B$ for which the sequence $\sharp B^n(x)\sharp^{1/n}$ is bounded, we define the *companion growth bound* of the B-orbit of x by

$$\gamma_x^\sharp(B) = \gamma^\sharp(x, B) := \limsup_{n\to\infty} \psi(B^n(x))^{1/n} \tag{5.13}$$

and set it equal to infinity otherwise. Here ψ is the companion half-norm (Section 2.5). We define the the *orbital companion spectral radius* of B by

$$\mathbf{r}_o^\sharp(B) := \sup_{x\in X_B} \gamma_x^\sharp(B), \tag{5.14}$$

if the set $\{\gamma_x^\sharp(B); x \in X_B\}$ is bounded; otherwise $\mathbf{r}_o^\sharp(B) = \infty$.

Since $\psi(x) \le \|x\|$ for all $x \in X$, we have

$$\gamma_x^\sharp(B) \le \gamma_x(B), \quad x \in X_B; \qquad \mathbf{r}_o^\sharp(B) \le \mathbf{r}_o(B). \tag{5.15}$$

The same proof as for Theorem 5.4 provides the following result. Recall that B is bounded with respect to the companion norm if B is bounded with respect to the original norm and order-preserving and $X_+ = X_B$ (Theorem 4.9).

PROPOSITION 5.5. *Let $B : X_B \to X_B$ be a homogeneous operator on a homogeneous subset X_B of X and $x \in X_B$. For all $n \in \mathbb{N}$ and $x \in X_B$, $\gamma^\sharp(x, B^n) \le \gamma^\sharp(x, B)^n$, with equality holding if $X_B = X_+$ and B is also bounded and order-preserving.*

5.1.4. Equality of spectral and orbital spectral radius. In the following, we explore conditions for $\mathbf{r}_o(B) = \mathbf{r}(B)$ if B is bounded.

PROPOSITION 5.6. *Let X_B be a closed homogeneous subset of the normed vector space X and $B : X_B \to X_B$ be continuous and compact. Assume that $B^n(x) \to 0$ as $n \to \infty$ for all $x \in X_B$.*

Then there exists a compact subset Q_B of X_B such that $B(Q_B) \subseteq Q_B$ and $B(x) \in Q_B$ for all $x \in X_B$ with $\|x\| \le 1$. If B is also homogeneous, $(\|B^n\|)$ is a bounded sequence and $\mathbf{r}(B) \le 1$.

The proof follows the one of [**156**, Thm.2.3]. Compactness of the operator B means that, for any bounded subset U of X_B, $B(U)$ is contained in a compact subset of X. See Definition A.43 and Exercises A.1.13 and A.1.14 for more details.

PROOF. Let $U_2 = \{x \in X_B; \|x\| < 2\}$. Let Q be the closure of $B(U_2)$. Since B is a compact operator and X_B is closed, Q is a compact subset of X_B. For every $x \in Q$, there exists some $n = n(x) \in \mathbb{N}$ such that $B^{n-1}(x) \in U_2$. Since B is continuous, there exists an open neighborhood V_x of x such that $B^{n-1}(V_x \cap X_B) \subseteq U_2$ and $B^n(V_x \cap X_B) \subseteq Q$. Since Q is compact, there exists finitely many open sets $V_1, \dots, V_k$ and $n_1, \dots, n_k \in \mathbb{N}$ such that

$$Q \subseteq X_B \cap \Big(\bigcup_{j=1}^k V_k\Big), \qquad B^{n_j}(X_B \cap V_j) \subseteq Q.$$

Choose $m = \max_{j=1}^k n_j$ and set

$$Q_B = \bigcup_{j=0}^{m-1} B^j(Q).$$

Q_B is a compact set and $B(U_2) \subseteq Q = B^0(Q) \subseteq Q_B$.

We claim that $B(Q_B) \subseteq Q_B$. It is sufficient to show that $B^m(Q) \subseteq Q_B$. Let $x \in Q$. Then $x \in V_j$ for some $j \in \{1, \dots, k\}$ and $B^{n_j}(x) \in Q$. So

$$B^m(x) = B^{m-n_j}(B^{n_j}(x)) \in B^{m-n_j}(Q) \subseteq Q_B$$

because $0 \le m - n_j \le m - 1$.

It follows that $B^n(Q_B) \subseteq Q_B$ for all $n \in \mathbb{Z}_+$ and $B^n(U_2) \subseteq Q_B$ for all $n \in \mathbb{N}$.

Now assume that X_B is a homogeneous set and B a homogeneous operator. Since Q_B is bounded as a compact set, $(\|B^n\|)$ is a bounded sequence. $\square$

THEOREM 5.7. *Let X_B be a closed homogeneous set in the normed vector space X and $B : X_B \to X_B$ be continuous, compact and homogeneous. Then $\mathbf{r}_o(B) = \mathbf{r}(B)$.*

PROOF. Suppose not. After scaling B, we can assume that $\mathbf{r}_o(B) < 1 < \mathbf{r}(B)$. Then $B^n(x) \to 0$ as $n \to \infty$ for all $x \in X_B$. By Proposition 5.6, $\mathbf{r}(B) \leq 1$, a contradiction. □

The following result is proved in the proof of [**167**, Cor.2.2].

THEOREM 5.8. *Let X be a normed vector space and B be a bounded linear map on X. Assume that there is some $m \in \mathbb{N}$ and a compact map $K : X \to X$ and a bounded linear map $L : X \to X$ such that $B^m = K + L$ and $\|L\| < \mathbf{r}(B^m)$.*

Then there is some $x \in X$ such that

$$\limsup_{n\to\infty} \frac{\|B^n x\|}{\|B^n\|} > 0. \tag{5.16}$$

For this x, $\gamma_x(B) = \mathbf{r}(B) = \mathbf{r}_o(B)$.

PROOF. For a proof by contradiction, suppose

$$b_n(x) := \frac{\|B^n x\|}{\|B^n\|} \to 0, \qquad n \to \infty, \quad x \in X. \tag{5.17}$$

For each $n \in \mathbb{N}$ choose some $x_n \in X$ with $\|x_n\| = 1$ such that $\|B^n\| \leq \|B^n x_n\| \frac{n+1}{n}$. Choose K and L according to our assumptions. For all $n \in \mathbb{N}$ and $y \in X$,

$$\begin{aligned}\|B^n\| \leq& \|B^n x_n\| \frac{n+1}{n} \leq \|B^{n-m}(y + Kx_n - y + Lx_n)\| \frac{n+1}{n} \\ \leq& \|B^{n-m} y\| \frac{n+1}{n} + \|B^{n-m}\| \left(\|Kx_n - y\| + \|L\|\right) \frac{n+1}{n}.\end{aligned}$$

For all $n \in \mathbb{N}$ and $y \in X$,

$$\|B^n\| \leq \|B^{n-m}\| \frac{n+1}{n} \Big(b_{n-m}(y) + \|Kx_n - y\| + \|L\|\Big).$$

Choose some r_2, r_1 such that $\|L\|^{1/m} < r_1 < r_2 < \mathbf{r}(B)$. Since K is compact, there exists a finite set S in X such that for each $n \in \mathbb{N}$ there exists some $y \in S$ such that $\|Kx_n - y\| < r_1^m - \|L\|$. Hence

$$\|B^n\| \leq \|B^{n-m}\| \frac{n+1}{n} \Big(\sup_{y\in S} b_{n-m}(y) + r_1^m\Big).$$

Since $\sup_{y\in S} b_n(y) \to 0$, there exists some $N \in \mathbb{N}$ such that

$$\|B^n\| \leq \|B^{n-m}\| r_2^m, \qquad n \geq N.$$

Choose some $k_0 \in \mathbb{N}$ such that $(k_0 - 1)m > N$. For $k > k_0$,

$$\|B^{(k+1)m}\| \leq \|B^{km}\| r_2^m, \qquad k \geq k_0.$$

By induction,

$$\|B^{(lm)}\| \leq \|B^{k_0 m}\| r_2^{m\ell}, \qquad \ell > k_0.$$

By (5.1), $\mathbf{r}(B) \leq r_2$, a contradiction. □

THEOREM 5.9. *Let X be an ordered normed vector space with closed order cone X_+.*

Let $B : X_B \to X_B$ be a homogeneous, bounded, and order-preserving operator and X_B a serially complete cone contained in X_+.

Assume that B satisfies the following normality condition: There exist some $m \in \mathbb{N}$ and $c > 0$ such that $\|B^m(x)\| \le c\|B^m(y)\|$ for all $x, y \in X_B$ with $x \le y$.

Then there exists some $x \in X_B$ such that

$$\mathbf{r}(B) = \gamma_B(x) = \mathbf{r}_o(B) = \lim_{k\to\infty} \|B^k(x)\|^{1/k}$$

Since B is order-preserving, the normality condition for B is weaker than condition (iv) in [**95**, Thm.2.1] (there are $m \in \mathbb{N}$ and $c > 0$ such that $\|x\| \le c\|y\|$ for all $x, y \in B^m(X_+)$ with $x \le y$). The proof remains almost the same, but requires a modification.

PROOF. We can assume that $\mathbf{r}(B) > 0$. For each $k \in \mathbb{N}$, there exists some $x_k \in X_B$ with

$$\|x_k\| = 1 \quad \text{and} \quad \|B^k\| \le \frac{k+1}{k}\|B^k(x_k)\|. \tag{5.18}$$

Since X_B is serially complete, the series $x = \sum_{k=1}^{\infty} k^{-2}x_k$ converges and $x \in X_B$. Since B is order-preserving and $x_k \le k^2 x$ because X_+ is closed,

$$B^j(x_k) \le B^j(k^2 x), \qquad j \in \mathbb{Z}_+.$$

For $k \ge m$,

$$\begin{aligned} B^k(x_k) =& B^m(y_k), \qquad B^k(k^2x) = B^m(z_k),\\ y_k :=& B^{k-m}(x_k) \le B^{k-m}(k^2 x) =: z_k. \end{aligned}$$

By the normality condition for B, $\|B^k(x_k)\| \le c\|B^k(k^2x)\|$ for $k \ge m$. Since B is homogeneous, by (5.18),

$$\|B^k\| \le \frac{k+1}{k}\|B^k(x_k)\| \le c\frac{k+1}{k}k^2\|B^k(x)\|, \qquad k > m.$$

By (5.1),

$$\mathbf{r}(B) \le \liminf_{k\to\infty} \left([c(k+1)k]^{1/k}\|B^k(x)\|^{1/k}\right) = \liminf_{k\to\infty} \|B^k(x)\|^{1/k}.$$

Since $\limsup_{k\to\infty} \|B^k(x)\|^{1/k} \le \mathbf{r}(B)$, $\mathbf{r}(B) = \lim_{k\to\infty} \|B^k(x)\|^{1/k}$. □

Unfortunately, Theorem 5.9 does not provide information about the $x \in \dot{X}_B$ for which $\gamma_B(x) = \mathbf{r}(B)$. The following terminology has been adapted from various works by Krasnosel'skii [**135**, Sec.2.1.1] and coworkers [**136**, Sec.9.4] though it has been modified.

DEFINITION 5.10. *Let X be an ordered normed vector space with closed order cone X_+.*

Let $B : X_B \to X_B$, $u \in X_B$, where X_B is a subset of X_+.

B is called pointwise u-bounded *if, for any $x \in X_B$, there exist some $n \in \mathbb{N}$ and $\zeta > 0$ such that $B^n(x) \le \zeta u$. The point u is called a* pointwise order bound *of B.*

B is called uniformly u-bounded *if there exists some $c > 0$ such that $B(x) \le c\|x\|u$ for all $x \in X_B$. The element u is called a* uniform order bound *of B.*

B is called uniformly order-bounded *if it is uniformly u-bounded for some $u \in X_B$. B is called* pointwise order-bounded *if it is pointwise u-bounded for some $u \in X_B$.*

THEOREM 5.11. *Let X be an ordered normed vector space with closed order cone X_+.*

Let $B : X_B \to X_B$ be a homogeneous, bounded, and order-preserving operator on a homogeneous subset X_B of X_+.

Let $u \in \dot{X}_B$ and B^k be uniformly u-bounded for some $k \in \mathbb{N}$.

Assume that B satisfies the following normality condition: There exist some $m \in \mathbb{N}$ and $c > 0$ such that $\|B^m(x)\| \le c\|B^m(y)\|$ for all $x, y \in X_B$ with $x \le y$.

Then $\mathbf{r}(B) = \gamma_B(u) = \mathbf{r}_o(B)$.

PROOF. We can assume that $\mathbf{r}(B) > 0$. Let $k \in \mathbb{N}$, $u \in X_B$ and B^k be uniformly u-bounded. By (5.4), we only need to show that $\mathbf{r}(B) \le \gamma_B(u)$.

By Definition 5.10, there exists some $\tilde{c} > 0$ such that $B^k(x) \le \tilde{c}\|x\|u$ for all $x \in X_B$. Since B is order-preserving and homogeneous,

$$B^{k+j}(x) \le B^j(B^k(x)) \le B^j(\tilde{c}\|x\|u) = \tilde{c}\|x\|B^j(u), \qquad j \in \mathbb{N}, \quad x \in X_B.$$

Let $m \in \mathbb{N}$ and $c > 0$ from the normality condition for B. Then

$$\|B^{k+m+j}(x)\| \le c\tilde{c}\|x\|\|B^{j+m}(u)\|, \qquad j \in \mathbb{N}, \quad x \in X_B.$$

By definition of the operator norm, (4.1),

$$\|B^n\| \le c\tilde{c}\|B^{n-k}(u)\|, \qquad n > k + j,$$

and, for $n > k + j$,

$$\begin{aligned}\|B^n\|^{1/n} =& \|B^n\|^{1/(n-k)}\|B^n\|^{(1/n)-(1/(n-k))} \\ \le& (c\tilde{c})^{1/(n-k)}\|B^{n-k}(u)\|^{1/(n-k)}(\|B^n\|^{1/n})^{-k/(n-k)}.\end{aligned}$$

Since $\mathbf{r}(B) = \inf_{n\in\mathbb{N}} \|B^n\|^n$, for $n > k + j$,

$$\|B^n\|^{1/n} \le (c\tilde{c})^{1/(n-k)}\|B^{n-k}(u)\|^{1/(n-k)}(\mathbf{r}(B))^{-k/(n-k)}.$$

We take the limit as $n \to \infty$ and obtain from (5.8) that $\mathbf{r}(B) \le \gamma_B(u)$. □

THEOREM 5.12. *Let X be an ordered normed vector space with closed order cone X_+.*

Let $B : X_B \to X_B$ be a homogeneous, bounded, and order-preserving operator on a homogeneous subset X_B of X_+.

Let $u \in \dot{X}_B$ and B be pointwise u-bounded.

Assume that B satisfies the following normality condition: There exist some $m \in \mathbb{N}$ and $c > 0$ such that $\|B^m(x)\| \le c\|B^m(y)\|$ for all $x, y \in X_B$ with $x \le y$.

Then $\gamma_B(u) = \mathbf{r}_o(B)$.

If, in addition, X_B is a serially complete cone, $\gamma_B(u) = \mathbf{r}(B)$.

PROOF. Let $u \in X_B$ and B be pointwise u-bounded.

Let $x \in \dot{X}_B$. By (5.3), we only need to show that $\gamma_B(x) \le \gamma_B(u)$.

Since B is pointwise u-bounded, there exist some $n \in \mathbb{N}$ and $\zeta > 0$ such that $B^n(x) \le \zeta u$.

By induction, since B is order-preserving and homogeneous, for all $k \in \mathbb{N}$, $B^{n+k}(x) \le B^k(\zeta u) \le \zeta B^k(u)$. Let $m \in \mathbb{N}$ and $c > 0$ from the normality condition for B. Then

$$\|B^{m+k+n}(x)\| \le \zeta \|B^{k+m}(u)\|, \qquad k \in \mathbb{N}.$$

So

$$\|B^j(x)\|^{1/j} \le \zeta^{1/j} \|B^{j-k-m}(u)\|^{1/j}, \qquad j \in \mathbb{N}, \quad j > k+m.$$

We take the limit superior as $j \to \infty$ and obtain $\gamma_B(x) \le \gamma_B(u)$.

The second assertion now follows from Theorem 5.9. □

We will revisit uniformly u-bounded operators in Chapter 6 and study them in more detail. Theorem 5.11 is related to Theorem 6.22. It is presented here because it addresses an issue raised by the other results in this section and can be proved without much preparation.

THEOREM 5.13 (cf. [**156**, Sec.2]). *Let $B : X_B \to X_B$ be a continuous additive operator on a complete wedge X_B in X. Then B is homogeneous and $\mathbf{r}(B) = \mathbf{r}_o(B)$.*

PROOF. Suppose that $r > \mathbf{r}_o(B)$. By (5.3) and (5.2),

$$r \ge \gamma_B(x) = \limsup_{n\to\infty} \|B^n(x)\|^{1/n}, \qquad x \in X_B.$$

It is sufficient to show that $\mathbf{r}(B) < r$.

For all $x \in X_B$, there exists some $m \in \mathbb{N}$ such that $\|B^n(x)\| \le r^n$ for all $n \in \mathbb{N}$, $n \ge m$. In other words,

$$X_B = \bigcup_{m\in\mathbb{N}} \bigcap_{n \ge m} C_n, \qquad C_n = \{x \in X_B; \|B^n(x)\| \le r^n\}.$$

Since B is continuous, the sets C_n are closed with respect to $\|\cdot\|$ and so are intersections of C_n. Since X_B is complete with respect to $\|\cdot\|$, by the Baire category theorem, there exists some $m \in \mathbb{N}$ such that $\bigcap_{n\ge m} C_n$ contains a subset of X_B that is relatively open with respect to $\|\cdot\|$. So there exist some $\epsilon > 0$ and $x \in X_B$ such that $\|B^n(x+\epsilon z)\| \le r^n$ whenever $z \in X$, $\|z\| < 2$, and $x + \epsilon z \in X_B$ and $n \ge m$. Let $z \in X_B$, $\|z\| \le 1$. Since X_B is a wedge, $x + \epsilon z \in X_B$. By the triangle inequality and the additivity of B,

$$\|B^n(\epsilon z)\| \le \|B^n(x+\epsilon z)\| + \|B^n(x)\| \le 2r^n.$$

Since B is homogeneous (Remark 2.11),

$$\|B^n(z)\| \le \frac{2r^n}{\epsilon}, \quad z \in X_B, \|z\| \le 1, \quad n \ge m.$$

By (5.1),

$$\mathbf{r}(B) \le \|B^n\|^{1/n} \le r \inf_{n\ge m} (2/\epsilon)^{1/n}, \qquad n \ge m.$$

We take the limit as $n \to \infty$ and obtain $\mathbf{r}(B) \le r$.

$\mathbf{r}(B) \ge \mathbf{r}_o(B)$ follows from the definition of $\mathbf{r}_o(B)$; see (5.3) and (5.2). □

5.1.5. Commutation rules. We now consider two normed vector spaces X and Y and two homogeneous subsets $\tilde{X}$ of X and $\tilde{Y}$ of Y and bounded homogeneous maps $A : \tilde{X} \to \tilde{Y}$ and $B : \tilde{Y} \to \tilde{X}$. Then the compositions BA and AB are bounded homogeneous maps on $\tilde{X}$ and $\tilde{Y}$ respectively.

THEOREM 5.14. *Let the just-mentioned assumptions be satisfied. Then* $\mathbf{r}(AB) = \mathbf{r}(BA)$ *and* $\mathbf{r}_o(AB) = \mathbf{r}_o(BA)$.

PROOF. Let $\|\cdot\|$ denote the norms on both X and Y and the respective operator norms as well.

A straightforward induction proof shows that $(AB)^{n+1} = A(BA)^n B$ for all $n \in \mathbb{N}$. Since A and B are bounded homogeneous, $\|(AB)^{n+1}\| \le \|A\| \, \|(BA)^n\| \, \|B\|$ for all $n \in \mathbb{N}$ by Theorem 4.7. We can assume that $\|A\| > 0$ and $\|B\| > 0$ because the claim is trivial otherwise. In order to show $\mathbf{r}(AB) \le \mathbf{r}(BA)$, we can also assume that $\mathbf{r}(AB) > 0$. By (5.1),

$$\mathbf{r}(AB) = \lim_{n\to\infty} \|(AB)^{n+1}\|^{1/n} \le \limsup_{n\to\infty} \|A\|^{1/n} \|(BA)^n\|^{1/n} \|B\|^{1/n} = \mathbf{r}(BA).$$

Equality holds by symmetry.

Now let $x \in X_B$. Then $(AB)^{n+1}(x) = A(BA)^n(B(x))$ for all $n \in \mathbb{N}$ and $\|(AB)^{n+1}(x)\| \le \|A\| \, \|(BA)^n(B(x))\|$. Similarly as before, by (5.2), $\gamma(x, AB) \le \gamma\big(B(x), BA\big)$. By (5.3), $\gamma(x, AB) \le \mathbf{r}_o(BA)$ and $\mathbf{r}_o(AB) \le \mathbf{r}_o(BA)$. Equality follows by symmetry. $\square$

5.2. Compatibility of the spectral radius under linear extension

If W is a wedge in a normed vector space X,

$$\tilde{X} = W - W = \{v - w; v, w \in W\}$$

is an $\mathbb{R}$-linear subspace of X. Let $B : W \to W$ be a homogeneous additive operator on W. We define an extension $A : \tilde{X} \to \tilde{X}$ by

$$Ax = Bv - Bw, \quad x = v - w, \quad v, w \in W.$$

Since B is additive, the definition does not depend on the choice of v and w. Indeed, if $v - w = \tilde{v} - \tilde{w}$ for $v, \tilde{v}, w, \tilde{w} \in W$, then $v + \tilde{w} = \tilde{v} + w \in W$ and $B(v) + B(\tilde{w}) = B(\tilde{v}) + B(w)$ and $B(v) - B(w) = B(\tilde{v}) - B(\tilde{w})$.

One readily checks that A is additive and (positively) homogeneous on $W - W$. Let $\alpha < 0$. Then

$$\begin{aligned} A(\alpha x) &= A(\alpha v - \alpha w) = A((-\alpha)w - (-\alpha)v) \\ &= B((-\alpha w) - B((-\alpha)v) = -\alpha B(w) + \alpha B(v) = \alpha A(x). \end{aligned}$$

This shows that A is linear. The linear extension A is uniquely determined by B.

5.2.1. Non-flat wedges.

THEOREM 5.15. *Let $B : W \to W$ be an additive homogeneous operator on the wedge W in X and A its unique linear extension to the linear subspace $W - W$ of X. Assume that W is non-flat and B is bounded. Then A is bounded (and thus A and B are continuous) and there is some $c > 0$ which does not depend on B and A such that $\|A\| \le c\|B\|$. Moreover, $\mathbf{r}(A) = \mathbf{r}(B)$.*

If B is a compact operator on W, A is a compact operator on $W - W$.

PROOF. Since W is non-flat, there exists some $c \geq 1$ such that for any $x \in W - W$ there exist $v, w \in W$ such that $x = v - w$ and $\|v\| + \|w\| \leq c\|x\|$. For all $n \in \mathbb{N}$,

$$\begin{aligned}\|A^n x\| &= \|B^n(v) - B^n(w)\| \leq \|B^n(v)\| + \|B^n(w)\| \\ &\leq \|B^n\|(\|v\| + \|w\|) \leq \|B^n\|\, c\, \|x\|.\end{aligned}$$

So $\|A^n\| \leq c\,|B^n\|$ for all $n \in \mathbb{N}$. This implies that $\mathbf{r}(A) \leq \mathbf{r}(B)$. The other inequality follows from $B = A$ on W.

Now let B be a compact operator on W. Let (x_n) be a bounded sequence in $W - W$. Since W is non-flat, there exist bounded sequences (v_n) and (w_n) in W such that $x_n = v_n - w_n$ for all $n \in \mathbb{N}$. Choosing subsequences twice, we find subsequence (v_{n_j}) and (w_{n_j}) such that $(B(v_{n_j})$ and $(B(w_{n_j}))$ converge in W. Then $(A(x_{n_j}))$ converges in $W - W$. □

Bonsall **[27]** presents an example of a flat cone W where $B : W \to W$ is compact, but its extension $A : W - W \to W - W$ is not compact. See Proposition 10.14. For a whole class of examples, see Corollary 10.21.

THEOREM 5.16. *Let X be complete and $B : W \to W$ be an additive homogeneous bounded operator on the wedge W in X. Assume that W is non-flat. Then there exists a unique bounded linear positive extension A to the closure of $W - W$ and there is some $c > 0$ which does not depend on B and A such that $\|A\| \leq c\|B\|$. Moreover, $\mathbf{r}(A) = \mathbf{r}(B)$.*

If B is a compact operator on W, A is a compact operator on the closure of $W - W$.

PROOF. Let $\tilde{A}$ be the unique bounded linear extension of B to $W - W$ which exists by Theorem 5.15, $\mathbf{r}(\tilde{A}) = \mathbf{r}(A)$. Let Y be the closure of $W - W$. For $y \in Y$, we define $Ay := \lim_{n\to\infty} \tilde{A}x_n$ where (x_n) is a sequence in $W - W$ with $x_n \to x$. Since $\tilde{A}$ is linear and bounded, $(\tilde{A}x_n)$ is a Cauchy sequence which converges and the limit does not depend on the choice of (x_n). One readily shows that A is linear and bounded on Y. Further, for all $k \in \mathbb{N}$, $A^k x = \lim_{n\to\infty} \tilde{A}^k x_n$. Hence, $\|A^k\| = \|\tilde{A}^k\|$ for all $k \in \mathbb{N}$ and $\mathbf{r}(A) = \mathbf{r}(\tilde{A}) = \mathbf{r}(B)$ by (5.8).

Now assume that B is compact. Let (y_n) be a bounded sequence in Y. Then there exists a sequence (x_n) in $W - W$ such that $\|y_n - x_n\| \to 0$. (y_n) is bounded. Since $\tilde{A}$ is compact on $W - W$ by Theorem 5.15, after choosing subsequences, $\tilde{A}y_n \to y$ to some $y \in Y$. Since A is bounded and linear,

$$0 = \lim_{n\to\infty} A(y_n - x_n) = \lim_{n\to\infty} (Ay_n - Ax_n) = \lim_{n\to\infty} (Ay_n - y).$$

This shows $Ay_n \to y$ and the compactness of A on the closure of $W - W$. □

5.2.2. Wedges that may be flat. Now let $B : W \to W$ be a homogeneous additive operator on the closed wedge W that is bounded with respect to $\|\cdot\|$. As before, we construct the linear extension $A : \tilde{X} \to \tilde{X}$, with $\tilde{X} = W - W$. Let $x = v - w$ with $v, w \in W$. Then $Ax = Bv - Bw$ with

$$\|Bv\| + \|Bw\| \leq \|B\|(\|v\| + \|w\|).$$

Recall the norm $\|\cdot\|^\sim$ on $\tilde{X} = W - W$,

$$\|x\|^\sim = \inf\{\|u\| + \|v\|; x = u - v, u, v \in W\}. \tag{5.19}$$

considered in Section 2.3.

THEOREM 5.17. *The extension A of B to $\tilde{X} = W - W$ with a closed wedge W and the norm $\|\cdot\|^\sim$ is bounded and $\|B\| = \|A\|^\sim$ and $\mathbf{r}(B) = \mathbf{r}^\sim(A)$.*

Here $\mathbf{r}^\sim$ denotes the spectral radius with respect to the norm $\|\cdot\|^\sim$ on $\tilde{X}$.

PROOF. Without loss of generality, $\|B\| > 0$ and so, by (5.19),

$$\|v\| + \|w\| \geq \|Ax\|^\sim / \|B\|$$

and $\|x\|^\sim \geq \|Ax\|^\sim/\|B\|$. This implies that A is bounded with respect to $\|\cdot\|^\sim$ and $\|A\|^\sim \leq \|B\|$. The oppositive inequality is obvious and we obtain

$$\|A\|^\sim = \|B\|. \tag{5.20}$$

This equality translates to the respective spectral radii. □

Let $r > \mathbf{r}(B)$. Since $\mathbf{r}(B) = \mathbf{r}(A)$, there exists some $m \in \mathbb{N}$ such that $(\|A^n\|^\sim)^{1/n} < r$ for all $n \in \mathbb{N}$ with $n > m$. We define an equivalent norm on $\tilde{X}$ by

$$\|x\|_r = \sum_{n=0}^m r^{-n} \|A^n x\|^\sim, \qquad x \in \tilde{X} = W - W.$$

More precisely,

$$\|x\| \leq \|x\|^\sim \leq \|x\|_r \leq \sum_{n=0}^m r^{-n} \|A^n\| \|x\|^\sim, \qquad x \in \tilde{X},$$

and

$$\|x\| = \|x\|^\sim \leq \|x\|_r \leq \sum_{n=0}^m r^{-n} \|A^n\| \|x\|, \qquad x \in W.$$

For $x \in \tilde{X}$,

$$\begin{aligned} \|Ax\|_r &= \sum_{n=0}^m r^{-n} \|A^{n+1}x\|^\sim = \sum_{n=1}^{m+1} r^{-n+1} \|A^n x\|^\sim \\ &= r\Big(\|x\|_r + r^{-(m+1)} \|A^{m+1}x\|^\sim - \|x\|^\sim\Big) \leq r\|x\|_r. \end{aligned}$$

THEOREM 5.18. *Let $r > \mathbf{r}(B)$. $\|\cdot\|_r$ is a norm on $W - W$ that is equivalent to $\|\cdot\|^\sim$ and satisfies $\|A\|_r < r$. In particular, there exists some $c \geq 1$ such that $\|x\| \leq \|x\|_r \leq c\|x\|$ and $\|Bx\|_r \leq r\|x\|_r$ for all $x \in W$.*

If W is the subset of a normal closed order cone X_+ and B is order-preserving on W, it can be arranged that $\|\cdot\|_B$ is order-preserving on W.

THEOREM 5.19. *Let $A : W \to W$ be an additive homogeneous continuous operator on the closed wedge W of a normed vector space X. Let $r > \mathbf{r}(A)$.*

Then there exists a subadditive homogeneous continuous functional $\theta : W \to \mathbb{R}_+$ and $c \geq 1$ such that $\|x\| \leq \theta(x) \leq c\|x\|$ and $\theta(Ax) < r\theta(x)$ for all $x \in W$.

If W is normal and A is only subadditive rather than additive, such a θ still exists and can be chosen to be order-preserving.

PROOF. Choose $m \in \mathbb{N}$ such that $r^{-(m+1)}\|A^{m+1}\| < 1$. We define

$$\theta(x) = \sum_{n=0}^m r^{-n} \|A^n x\|, \qquad x \in W.$$

Then θ is subadditive homogeneous and bounded, $\theta(x) \geq \|x\|$ for all $x \in W$. For all $x \in W$,

$$\theta(Ax) = \sum_{n=0}^{m} r^{-n}\|A^{n+1}x\| = r\Big(\theta(x) + r^{-(m+1)}\|A^{(m+1)}x\| - \|x\|\Big) \leq r\theta(x).$$

If the wedge W is normal and A is subadditive rather than additive, $\|\cdot\|$ can be chosen to be order-preserving on W by switching to an equivalent norm and θ is still subadditive and is order-preserving on W as well and $\mathbf{r}(A) < 1$ with respect to the new norm. □

5.2.3. Total cones. The next theorem only assumes that X_+ is a total cone, but that K satisfies some compactness condition.

THEOREM 5.20. *Let X be a ordered normed vector space with total cone X_+ and B be a bounded linear map on X. Assume that there is some $m \in \mathbb{N}$ and a compact map $K : X \to X$ and a bounded linear map $L : X \to X$ such that $B^m = K + L$ and $\|L\| < \mathbf{r}(B^m)$.*

Then there exists some $x \in X_+$ such that

$$\limsup_{n\to\infty} \frac{\|B^n x\|}{\|B^n\|} > 0.$$

For this $x \in X_+$, $\gamma_x(B) = \mathbf{r}(B)$.

If $B(X_+) \subseteq X_+$ and B_+ denotes the restriction of B to X_+, we have $\mathbf{r}(B_+) = \mathbf{r}(B)$.

Notice that neither X nor X_+ is assumed to be complete (cf. [**167**, Cor.2.2]).

PROOF. By Theorem 5.8, there is some $x \in X$ such that

$$\limsup_{n\to\infty} \frac{\|B^n x\|}{\|B^n\|} > 0.$$

Since X_+ is total, this inequality also holds for some $x \in X_+ - X_+$ and then for some $x \in X_+$. For this x_+, $\gamma_x(B) \leq \mathbf{r}(B)$

Let B map X_+ into itself and B_+ denote the restriction of B to X_+.

Then $\mathbf{r}(B_+) \geq \mathbf{r}(B)$. The inequality $\mathbf{r}(B_+) \leq \mathbf{r}(B)$ follows from $\|B_+\| \leq \|B\|$. □

5.3. A series characterization of the spectral radius

The condition $\mathbf{r}(B) < 1$ will later play an important role in the stability of discrete-time dynamical systems (Section 14.4). We give a characterization of Datko type [**51**]. The proof is an adaption of the one of [**111**, Prop.216]. See also [**203**, Sec.7.3].

THEOREM 5.21. *Let X be an ordered normed vectors space with normal closed order cone X_+. Consider a homogeneous order-preserving operator $B : X_B \to X_B$ on a serially complete cone $X_B \subseteq X_+$. Then the following are equivalent.*

(i) $\mathbf{r}(B) < 1$.

(ii) *For all $p > 0$, $\sum_{n=1}^{\infty} \|B^n\|^p < \infty$.*

(iii) *There is some* $p > 0$ *such that*

$$\sum_{n=1}^{\infty} \|B^n(x)\|^p < \infty, \qquad x \in X_B.$$

PROOF. (i) $\Rightarrow$ (ii): Let $\mathbf{r}(B) < 1$. Choose some $r \in (\mathbf{r}(B), 1)$. Then there exists some $c > 1$ such that $\|B^n\| \le cr^n$ for all $n \in \mathbb{N}$ and (ii) follows.

(ii) $\Rightarrow$ (iii): This is obvious.

(iii) $\Rightarrow$ (i): (iii) implies that $\|B^n(x)\| \to 0$ for each $x \in X_B$. By the Uniform Boundedness Theorem 4.8, there is some $M > 0$ such that $\|B^n\| \le M$ for all $n \in \mathbb{N}$. For all $n \in \mathbb{N}$ and $x \in X_B$,

$$n\|B^n(x)\|^p = \sum_{k=1}^{n} \|B^n(x)\|^p = \sum_{k=1}^{n} \|B^{n-k}(B^k(x))\|^p.$$

By (4.2),

$$\begin{aligned} n\|B^n(x)\|^p \le & \sum_{k=1}^{n} \|B^{n-k}\|^p \, \|B^k(x)\|^p \le \sum_{k=1}^{n} M^p \, \|B^k(x)\|^p \\ \le & M^p \sum_{k=1}^{\infty} \|B^k(x)\|^p < \infty. \end{aligned}$$

So, for each $x \in X_B$, $\{\|n^{1/p} B^n(x)\|, n \in \mathbb{N}\}$ is bounded. Again by Theorem 4.8, $\{\|n^{1/p} B^n\|, n \in \mathbb{N}\}$ is bounded. So there exists some $n \in \mathbb{N}$ such that $\|B^n\| < 1$. By (5.1), $\mathbf{r}(B) = \inf_{n \in \mathbb{N}} \|B^n\|^{1/n} < 1$. □

Recall that $\mathbf{r}(B) = \lambda$ if and only if $\mathbf{r}((1/\lambda)B) = 1$ by (5.8).

THEOREM 5.22. *Let* X *be an ordered normed vectors space with normal closed order cone* X_+. *Consider a homogeneous order-preserving operator* $B : X_B \to X_B$ *on a serially complete cone* $X_B \subseteq X_+$. *Then the following are equivalent for* $\lambda > 0$:

(i) $\mathbf{r}(B) < \lambda$.
(ii) *For all* $p > 0$, $\sum_{n=1}^{\infty} \lambda^{-pn} \|B^n\|^p < \infty$.
(iii) *There is some* $p > 0$ *such that*

$$\sum_{n=1}^{\infty} \lambda^{-pn} \|B^n(x)\|^p < \infty, \qquad x \in X_B.$$

THEOREM 5.23. *Let* X *be an ordered normed vectors space with normal closed order cone* X_+. *Consider a homogeneous order-preserving operator* $B : X_B \to X_B$ *on a serially complete cone* $X_B \subseteq X_+$. *Then, for all* $p > 0$,

$$\begin{aligned} \mathbf{r}(B) = & \inf\Big\{\lambda > 0; \sum_{n=1}^{\infty} \lambda^{-pn} \|B^n(x)\|^p < \infty, \quad x \in X_B\Big\} \\ = & \inf\Big\{\lambda > 0; \sum_{n=1}^{\infty} \lambda^{-pn} \|B^n\|^p < \infty\Big\}. \end{aligned}$$

PROOF. By Theorem 5.22, $\mathbf{r}(B)$ is a lower bound of the respective sets. Assume that $\mathbf{r}(B)$ is strictly smaller than the infimum. Then $\mathbf{r}(B)$ is not the largest lower bound and there exists a larger lower bound than $\mathbf{r}(B)$. Then, for instance, there exists some $\lambda > \mathbf{r}(B)$ such that $\sum_{n=1}^{\infty} \lambda^{-pn} \|B^n\|^p$ diverges, a contradiction. □

COROLLARY 5.24. *Let X be an ordered normed vectors space with normal closed order cone X_+. Consider a homogeneous order-preserving operator $B : X_B \to X_B$ on a serially complete cone $X_B \subseteq X_+$, $r = \mathbf{r}(B) > 0$. Then there exists some $x \in X_B$ such that $\sum_{n=0}^\infty r^{-n-1}\psi(B^n(x)) = \infty$ and*

$$\sum_{n=1}^\infty \lambda^{-n}\psi(B^n(x)) \to \infty, \qquad \lambda \to r+.$$

Here ψ is the monotone companion half-norm which is equivalent to the original norm on the normal cone X_+ (Proposition 2.45).

PROOF. We can assume that $r = \mathbf{r}(B) = 1$. By contraposition of Theorem 5.21, there exists some $x \in X_B$ such that $\sum_{n=0}^\infty \|B^n(x)\| = \infty$. Since X_+ is normal, the original norm can be replaced by the monotone companion half-norm. The last statement follows from the monotone convergence theorem with the counting measure. □

5.4. Upper estimates for growth factors

LEMMA 5.25. *Let X be an ordered normed vector space with closed order cone X_+ Let $B : X_B \to X_B$ be a homogeneous operator on a homogeneous set X_B contained in X_+ and let some power of B be compact.*

Further let $x \in X_B$ and (y_n) be a bounded sequence in X_+. Let $m \in \mathbb{N}$, and $B^n(x) \le y_n$ for all $n \in \mathbb{N}$, $n \ge m$.

Then the sequence $(B^n(x))$ is bounded.

PROOF. Assume that $(B^n(x))$ is unbounded. Set $a_n = \|B^n(x)\|$. By a lemma by Bonsall [**27**] (Lemma A.47), there exists a sequence (n_j) in $\mathbb{N}$ such that

$$a_{n_j} \to \infty, \quad j \to \infty, \qquad a_k \le a_{n_j}, \qquad k = 1, \dots, n_j, \quad j \in \mathbb{N}.$$

Set $w_j = \frac{B^{n_j}(x)}{a_{n_j}}$. Then $w_j \in X_B$ and $\|w_j\| = 1$. Choose $\ell \in \mathbb{N}$ such that B^ℓ is compact. Since B is homogeneous, $w_j = B^\ell(v_j)$ with $v_j = \frac{B^{n_j-\ell}(x)}{a_{n_j}} \in X_B$. By the properties of the (a_n), $\|v_j\| \le 1$. After choosing a subsequence, $w_j \to w$ with $w \in X_+$, $\|w\| = 1$.

By assumption, after choosing a subsequence again,

$$w_j \le \frac{y_{n_j}}{a_{n_j}}, \qquad j \in \mathbb{N}.$$

Since (y_{n_j}) is bounded and $a_{n_j} \to \infty$, $\frac{y_{n_j}}{a_{n_j}} \to 0$. Since X_+ is closed, $w = \lim_{j\to\infty} w_j \le 0$. Since $w \in X_+$ and X_+ is a cone, $w = 0$ contradicting $\|w\| = 1$. □

LEMMA 5.26. *Let X be an ordered normed vector space with closed order cone X_+. Let $B : X_B \to X_B$ be a homogeneous operator on a homogeneous set X_B contained in X_+ and let some power of B be compact. Further let (x_n) be a bounded sequence in X_+, x a point in X_B, $m, k \in \mathbb{N}$, $\lambda \ge 0$, and $B^{n+m}(x) \le \lambda^{n+k}x_n$ for all $n \in \mathbb{N}$, $n \ge m$. Then $\gamma_B(x) \le \lambda$.*

PROOF. If $\lambda = 0$, then $B^{n+m}(x) = 0$ for all $n \in \mathbb{N}$ and $\gamma^B(x) = 0$.

So we can assume that $\lambda > 0$. Set $C(x) = \lambda^{-1}B(x)$. Then C is homogeneous, and some power of C is compact. Further,

$$C^{n+m}(x) = \lambda^{-(n+m)}B^{n+m}(x) \le \lambda^{k-m}x_n, \qquad n \in \mathbb{N},$$

with the right-hand side of this inequality forming a bounded sequence. By Lemma 5.25, applied to C, $\gamma_C(x) \le 1$. By definition of the growth bounds, (5.2), and the homogeneity of B, $\gamma_B(x) \le \lambda$. □

PROPOSITION 5.27. *Let X be an ordered normed vector space with closed order cone X_+ and $B : X_B \to X_B$ be a homogeneous and order-preserving operator on a homogeneous set X_B contained in X_+. Assume that some power of B is compact. Let $u \in X_B$, $m \in \mathbb{N}$, $\lambda > 0$ such that $B^m(u) \le \lambda^m u$. Then $\gamma_B(u) \le \lambda$.*

PROOF. We can assume that $\lambda > 0$ and $u \ne 0$. For all $k \in \mathbb{N}$, $B^{km}(u) \le \lambda^{km} u$. For $q = 0, \dots, m-1$, $B^{km+q}(u) \le \lambda^{km+q}\lambda^{-q} B^q(u)$. So, for all $n \in \mathbb{N}$, $n \ge m$,

$$B^n(u) \le \lambda^n \Big(\sum_{q=0}^{m-1} \lambda^{-q} B^q(u) \Big).$$

By Lemma 5.26, $\gamma_B(u) \le \lambda$. □

Exercises.

5.4.1. Let X be a real vector space, i.e., a vector space over the field $\mathbb{R}$. Define $X_{\mathbb{C}} = X \times X = \{(x, y); x, y \in X\}$ with the addition $(u, v) + (x, y) = (u + x, v + y)$ for all $u, v, x, y \in X$ and the scalar multiplication

$$(\alpha_1 + \imath\alpha_2)(x, y) = (\alpha_1 x - \alpha_2 y, \alpha_1 y + \alpha_2 x), \qquad x, y \in X, \quad \alpha_1, \alpha_2 \in \mathbb{R}. \tag{5.21}$$

Here $\imath$ is the imaginary unit in $\mathbb{C}$ with $\imath^2 = -1$.

(a) Show that $X_{\mathbb{C}}$ with this addition and this scalar multiplication is a vector space over $\mathbb{C}$.
(b) Show that $\hat{X} = X \times \{0\} = \{(x, 0); x \in X\}$ is an $\mathbb{R}$-linear subspace of $X_{\mathbb{C}}$, i.e., $z + \tilde{z} \in \hat{X}$ for all $z, \tilde{z} \in \hat{X}$ and $\alpha z \in \hat{X}$ for all $\alpha \in \mathbb{R}$ and $z \in \hat{X}$.
(c) Let $T : X \to X$ be an $\mathbb{R}$-linear map on the real vector space X and define the extension $T_{\mathbb{C}} : X_{\mathbb{C}} \to X_{\mathbb{C}}$ of T by $T_{\mathbb{C}}(x, y) = (Tx, Ty)$ for $x, y \in X$.
Show that $T_{\mathbb{C}}$ is a linear map on the complex vector space $X_{\mathbb{C}}$.

$X_{\mathbb{C}}$ is called the *complexification* of X.

5.4.2. Let S be a nonempty set, $\mathbb{K} = \mathbb{R}$ or $\mathbb{K} = \mathbb{C}$, and $\mathbb{K}^S$ the vector space over $\mathbb{K}$ of functions $f : S \to \mathbb{K}$ with the addition $(f + g)(s) = f(s) + g(s)$ and the scalar multiplication $(\zeta f)(s) = \zeta f(s)$ for $s \in S$, $\zeta \in \mathbb{K}$ and $f, g \in \mathbb{K}^S$. Let X be the real vector space $\mathbb{R}^S$ and $X_{\mathbb{C}}$ be complexification of X. Define $J : X_{\mathbb{C}} \to \mathbb{C}^S$ by

$$[J(f, g)](s) = f(s) + \imath g(s), \qquad s \in S, \quad f, g \in X = \mathbb{R}^S. \tag{5.22}$$

Here $\imath$ denotes the imaginary unit in $\mathbb{C}$ with $\imath^2 = -1$. Show that J is a $\mathbb{C}$-linear bijective map, i.e., J is bijective,

$$J^{-1}(h) = (f, g), \qquad f(s) = \Re h(s), \quad g(s) = \Im h(s), \qquad h \in \mathbb{C}^S,$$

where $\Re$ and $\Im$ are the real and imaginary part of a complex number, and

$$J\big(\zeta(f, g) + \tilde{\zeta}(\tilde{f}, \tilde{g})\big) = \zeta J(f, g) + \tilde{\zeta} J(\tilde{f}, \tilde{g}), \qquad f, g, \tilde{f}, \tilde{g} \in X = \mathbb{R}^S, \quad \zeta, \tilde{\zeta} \in \mathbb{C}.$$

5.4.3. Let X be a normed vector space over $\mathbb{R}$ with norm $\|\cdot\|$ and $X_{\mathbb{C}}$ its complexification. We define $\|\cdot\|_{\mathbb{C}} : X_{\mathbb{C}} \to \mathbb{R}_+$ by

$$\|(x, y)\|_{\mathbb{C}} = \sup\big\{\|(\cos\theta)x + (\sin\theta)y\|; \theta \in [0, 2\pi]\big\}, \qquad x, y \in X.$$

(a) Show that $\|\cdot\|_{\mathbb{C}}$ is a norm on the vector space $X_{\mathbb{C}}$ over $\mathbb{C}$ and

$$\max\{\|x\|, \|y\|\} \le \|(x,y)\|_{\mathbb{C}} \le (\|x\|^2 + \|y\|^2)^{1/2}, \qquad x, y \in X.$$

(b) Show that $X_{\mathbb{C}}$ with norm $\|\cdot\|_{\mathbb{C}}$ is a Banach space if X with norm $\|\cdot\|$ is a Banach space.

(c) Let $T : X \to X$ be a linear map on X and $T_{\mathbb{C}}$ its linear extension to $X_{\mathbb{C}}$ as in Exercise 5.4.1.

Show that $\|T_{\mathbb{C}}\|_{\mathbb{C}} = \|T\|$ for the respective operator norms and $\mathbf{r}(T_{\mathbb{C}}) = \mathbf{r}(T)$ for the respective spectral radii given by Gelfand's formula, e.g., $\mathbf{r}(T) = \inf_{n\in\mathbb{N}} \|T^n\|^{1/n} = \lim_{n\to\infty} \|T^n\|^{1/n}$.

5.4.4. Let $\mathbb{K} = \mathbb{R}$ or $\mathbb{K} = \mathbb{C}$ and $B(S, \mathbb{K})$ be the set of of bounded functions $f : S \to \mathbb{K}$. Recall that $B(S, \mathbb{K})$ is a $\mathbb{K}$-linear subspace of $\mathbb{K}^S$ (i.e., linear with respect to the field $\mathbb{K}$) and $B(S, \mathbb{K})$ is a normed vector space over $\mathbb{K}$ with the supremum norm

$$\|f\|_\infty = \sup_{s\in S} |f(s)|$$

with $|\cdot|$ denoting the absolute value on $\mathbb{K} = \mathbb{R}, \mathbb{C}$ respectively.

(a) Show that $J(B(S, \mathbb{R})) = B(S, \mathbb{C})$ for the map J from Exercise 5.4.2.

(b) For $f \in B(S, \mathbb{R}) = X$, let

$$\|f\| = \sup_{s\in S} |f(s)| = \|f\|_\infty.$$

Show that $\|(f, g)\|_{\mathbb{C}} = \sup_{s\in S} \big((f(s))^2 + (g(s)^2)\big)^{1/2} = \|J(f, g)\|_\infty$ for $f, g \in B(S, \mathbb{R})$.

5.5. Collatz–Wielandt spectral radius

Let X be an ordered vector space with Archimedean order cone X_+, X_B be a homogeneous subset of X_+, and $B : X_B \to X_B$ be a homogeneous order-preserving operator.

5.5.1. Lower Collatz–Wielandt numbers. We define the *lower Collatz–Wielandt number* of B at $x \in \dot{X}_B$ as ([**87**] also see [**40, 41, 220**])

$$[B]_x = \sup\big\{\lambda \ge 0 : B(x) \ge \lambda x\big\} = [B(x)]_x. \tag{5.23}$$

Since B is homogeneous,

$$[B]_{\alpha x} = [B]_x, \qquad \alpha > 0. \tag{5.24}$$

By (4.14), $[B]_x \in \mathbb{R}_+$ and it is a lower eigenvalue of B,

$$B(x) \ge [B]_x x, \qquad x \in \dot{X}_B. \tag{5.25}$$

Since B is homogeneous and order-preserving,

$$[B(y)]_x \ge [B]_x [y]_x, \qquad x \in \dot{X}_+, \quad y \in X_+. \tag{5.26}$$

LEMMA 5.28. *Let $B, C : X_B \to X_B$ be homogeneous and order-preserving. Let $x \in \dot{X}_B$. Then $[CB]_x \ge [C]_x [B]_x$.*

This result, which follows from (5.25), implies

$$b_{n+m} \geq b_n b_m, \quad b_n = [B^n]_x, \quad n, m \in \mathbb{N}. \tag{5.27}$$

The *lower local Collatz–Wielandt radius* of B at $x \in \dot{X}_+$ is defined as

$$\eta_x(B) =: \sup_{n \in \mathbb{N}} [B^n]_x^{1/n}. \tag{5.28}$$

This is only a partial abuse of language. Recall that $y \in X_+$ is called x-positive if x is y-bounded.

PROPOSITION 5.29. *Let $x \in \dot{X}_B$ and assume that there is some $k \in \mathbb{N}$ such that $B^n(x)$ is x-positive for all $n \in \mathbb{N}$, $n \geq k$. Then $\eta_x(B) = \lim_{n\to\infty} [B^n]_x^{1/n}$ and $\eta_x(B^m) = (\eta_x(B))^m$ for all $m \in \mathbb{N}$.*

PROOF. Notice that $b_n > 0$ for $n \geq k$. So the first statement follows from Theorem A.45. The second statement follows from the first. □

THEOREM 5.30. *Let X be an ordered normed vector space with closed order cone X_+. Let $B : X_B \to X_B$ be a homogeneous, order-preserving operator on a homogeneous subset X_B of X_+. Then $[B]_x \leq \eta_x(B) \leq \gamma_x^\sharp(B) \leq \gamma_x(B)$ for all $x \in \dot{X}_B$.*

PROOF. Let $x \in \dot{X}_B$. The first inequality follows from (5.28). By (5.25), $B(x) \geq [B]_x x$. By induction $B^n(x) \geq [B]_x^n x$.

We apply the monotone companion half-norm ψ, which is homogeneous, and obtain $\psi(B^n(x)) \geq [B]_x^n \psi(x)$. Since $\psi(x) > 0$, $\gamma_x(B) \geq \gamma_x^\sharp(B) \geq [B]_x$. Since, for all $n \in \mathbb{N}$, B^n is homogeneous and order-preserving,

$$[B^n]_x \leq \gamma_x^\sharp(B^n) \leq \gamma_x^\sharp(B)^n.$$

The last inequality follows from Proposition 5.5. Since this holds for all $n \in \mathbb{N}$, by (5.28), $\eta_x(B) \leq \gamma_x^\sharp(B)$. The last inequality in our theorem follows from (5.15). □

5.5.2. Lower Collatz–Wielandt bound and spectral radius.

In analogy to the orbital spectral radius of B, we define the *lower Collatz–Wielandt spectral radius* as

$$\mathbf{r}_{cw}(B) = \sup_{x \in \dot{X}_B} \eta_x(B) \tag{5.29}$$

with $\eta_x(B)$ from (5.28) and $\dot{X}_B = X_B \setminus \{0\}$. We also define the *lower Collatz–Wielandt bound* as

$$cw(B) = \sup_{x \in \dot{X}_B} [B]_x = \sup\{\lambda \geq 0;\ \exists x \in \dot{X}_B : B(x) \geq \lambda x\}. \tag{5.30}$$

By (5.24),

$$cw(B) = \sup\{[B]_x; x \in X_B, \psi(x) = 1\} = \sup\{[B]_x; x \in X_B, \|x\| = 1\}. \tag{5.31}$$

By (5.28),

$$cw(B) \leq \mathbf{r}_{cw}(B). \tag{5.32}$$

The following relation is immediate from Theorem 5.30 and the respective definitions.

THEOREM 5.31. *Let X be an ordered normed vector space with closed order cone X_+. Let $B : X_B \to X_B$ be a bounded, homogeneous, order-preserving operator on a homogeneous subset X_B of X_+. Then $cw(B) \le \mathbf{r}_{cw}(B) \le \mathbf{r}_o^\sharp(B) \le \mathbf{r}_o(B) \le \mathbf{r}(B)$.*

The Collatz–Wielandt spectral radius only conditionally has the power property of a spectral radius. The following result is a consequence of Proposition 5.29.

PROPOSITION 5.32. *Assume that, for every $x \in \dot{X}_B$, there exists some $m \in \mathbb{N}$ such that $B^n(x)$ is x-positive for all $n \ge m$. Then $\mathbf{r}_{cw}(B^m) = (\mathbf{r}_{cw}(B))^m$ for all $m \in \mathbb{N}$.*

The lower Collatz–Wielandt bound has the commutative property.

THEOREM 5.33. *Let X and Y be ordered normed vector spaces with closed order cones X_+ and Y_+, respectively. Let $X_A \subseteq X_+$ and $Y_B \subseteq Y_+$ be homogeneous sets and $A : X_A \to Y_B$ and $B : Y_B \to X_A$ be homogeneous order-preserving operators. Then $cw(AB) = cw(BA)$.*

PROOF. Let $0 < \lambda < cw(BA)$. By (5.30), there exists some $x \in \dot{X}_A$ such that $BAx \ge \lambda x$. Set $y = Ax$. Then $y \in \dot{Y}_B$, otherwise $x = 0$. Since A is order-preserving and homogeneous, $ABy \ge \lambda y$ and $cw(AB) \ge \lambda$ by (5.30). Since $\lambda < cw(BA)$ has been arbitrary, $cw(AB) \ge cw(BA)$. By symmetry, we have equality. □

It may be useful to revisit Section 2.1.7 for the next result.

THEOREM 5.34. *Let $B : X_B \to X_B$ be homogeneous and order-preserving. Let B be superadditive or X_B an* sup-*semilattice. Then $cw(B) = \mathbf{r}_{cw}(B)$.*

PROOF. Let $0 \le \alpha < r_{cw}(B)$. By (5.29), there exists some $m \in \mathbb{N}$ and $x \in \dot{X}_B$ such that $B^m(x) \ge \alpha^m x$.

We first assume that X_B is an sup-semilattice. We set $y = \sup_{n=0}^{m-1} \alpha^{-n} B^n(x)$. Since B is order-preserving,

$$B(y) \ge \sup_{n=0}^{m-1} \alpha^{-n} B^{n+1}(x) = \alpha \sup_{n=1}^{m} \alpha^{-n} B^n(x) \ge \alpha y.$$

By (5.23) and (5.30), $[B]_y \ge \alpha$ and $cw(B) \ge \alpha$. Since this holds for any $0 \le \alpha < \mathbf{r}_{cw}(B)$, $cw(B) \ge \mathbf{r}_{cw}(B)$. The other inequality holds by Theorem 5.31.

If B is superadditive, we define $y = \displaystyle\sum_{n=0}^{m-1} \alpha^{-n} B^n(x)$. □

EXAMPLE 5.35. Let

$$B(x) \ge \sum_{k=1}^{m} \theta_k(x) u_k, \qquad x \in X_B,$$

with $u_k \in \dot{X}_B$ and $\theta_k : X_B \to \mathbb{R}_+$. Set $u = \sum_{j=1}^{n} u_j$. Then

$$B(u) \ge \sum_{k=1}^{m} \theta_k(u) u_k \ge \alpha u, \qquad \alpha = \inf_{k=1}^{m} \theta_k(u).$$

By (5.23) and (5.30),

$$cw(B) \ge [B]_u \ge \inf_{k=1}^{m} \theta_k\Big(\sum_{j=1}^{m} u_j\Big).$$

5.6. Estimates of the spectral radius involving functionals

Let X be a normed vector space, X_B a homogeneous subset of X and $B : X_B \to X_B$ be homogeneous.

THEOREM 5.36. *Let $\theta : X_B \to \mathbb{R}_+$ be homogeneous and bounded, $r > 0$ and $\theta(B(x)) \geq r\theta(x)$ for all $x \in X_B$. Then the following holds:*

(a) *If $u \in X_B$ and $\theta(u) > 0$, then the geometric growth factor in* (5.2) *satisfies $\gamma_u(B) \geq r$.*
(b) *If $\theta(x) > 0$ for some $x \in X_B$, then $\mathbf{r}_o(B) \geq r$.*

PROOF. By induction, for all $n \in \mathbb{N}$,

$$\theta(B^n(x)) \geq r^n\theta(x), \qquad x \in X_B.$$

Since θ is homogeneous and bounded, by (4.2),

$$\|\theta\| \, \|B^n(x)\| \geq r^n\theta(x), \qquad x \in X_B.$$

(a) Let $u \in X_B$ and $\theta(u) > 0$. Then $\|\theta\| > 0$ and

$$\|B^n(u)\|^{1/n} \geq r\big(\theta(u)/\|\theta\|\big)^{1/n}, \qquad n \in \mathbb{N}.$$

By (5.2), $\gamma_u(B) \geq r$.

(b) This follows from (a) and (5.3). □

DEFINITION 5.37. *Let $\theta : X_B \to \mathbb{R}_+$ be a homogeneous functional on the homogeneous subset X_B of X. Then θ is called* uniformly positive *if there exists some $\epsilon > 0$ such that $\theta(x) \geq \epsilon\|x\|$ for all $x \in X_B$.*

See [**135**, Sec.1.4.2].

THEOREM 5.38. *Let $\theta : X_B \to \mathbb{R}_+$ be homogeneous and uniformly positive, $r > 0$ and $\theta(B(x)) \leq r\theta(x)$ for all $x \in X_B$.*

Then $\mathbf{r}_o(B) \leq r$. If θ is also bounded, $\mathbf{r}(B) \leq r$.

PROOF. By induction, for all $n \in \mathbb{N}$,

$$\theta(B^n(x)) \leq r^n\theta(x), \qquad x \in X_B.$$

Since θ is homogeneous and uniformly positive, there exists some $\epsilon > 0$ such that by (4.2),

$$\epsilon\|B^n(x)\| \leq r^n\theta(x), \qquad x \in X_B.$$

So

$$\|B^n(x)\|^{1/n} \leq r\big(\theta(x)/\epsilon\big)^{1/n}, \qquad n \in \mathbb{N}.$$

By (5.2), $\gamma_x(B) \leq r$. By(5.3), $\mathbf{r}_o(B) \leq r$.

Let θ be also bounded. Then

$$\|B^n\| \leq r^n\|\theta\|/\epsilon.$$

By (5.1), $\mathbf{r}(B) \leq r$. □

5.7. Cartesian products

Let X be a normed vector space and X_+ a homogeneous subset of X. Then X_+^2 is a homogeneous subset of X^2. Let $B : X_+^2 \to X_+$ be homogeneous and bounded. Define $\hat{B} : X_+ \to X_+$ and $\tilde{B} : X_+^2 \to X_+^2$ by

$$\hat{B}(y) = B(y,y), \qquad \tilde{B}(y,z) = (B(y,z), B(y,z)), \qquad y,z \in X_+. \tag{5.33}$$

PROPOSITION 5.39. *Let $|\cdot|$ be any norm on $\mathbb{R}^2$ and define $\|(y,x)\|_\sim = \big|(\|x\|,\|y\|)\big|$, $x,y \in X$. Then X^2 is a normed vector space and $\hat{B}$ is a homogeneous bounded operator on X_+ and $\tilde{B}$ a homogeneous bounded operator on X_+^2 and $\mathbf{r}(\tilde{B}) = \mathbf{r}(\hat{B})$.*

PROOF. Notice that we can choose $\|(y,z)\|_\sim = \max\{\|y\|,\|z\|\}$ because the other norms on X^2 are equivalent. We claim that

$$\tilde{B}^n(y,z) = \big(\hat{B}^{n-1}(B(y,z)), \hat{B}^{n-1}(B(y,z))\big), \qquad n \in \mathbb{N}. \tag{5.34}$$

With the convention that the zeroth iterate of an operator is the identity, this claim holds for $n = 1$. Assume that $n \in \mathbb{N}$ and the claim holds for n. Then

$$\begin{aligned} \tilde{B}^{n+1}(y,z) &= \big(B(\tilde{B}^n(y,z)), B(\tilde{B}^n(y,z))\big) \\ &= \Big(B\big(\hat{B}^{n-1}(B(y,z)), \hat{B}^{n-1}(B(y,z))\big), B\big(\hat{B}^{n-1}(B(y,z)), \hat{B}^{n-1}(B(y,z))\big)\Big) \\ &= \Big(\hat{B}\big(\hat{B}^{n-1}(B(y,z))\big), \hat{B}\big(\hat{B}^{n-1}(B(y,z))\big) = \big(\hat{B}^n(B(y,z)), \hat{B}^n(B(y,z))\big). \end{aligned}$$

This shows the induction step and (5.34) follows by induction.

For all $n \in \mathbb{N}$,

$$\|\tilde{B}^n(y,z)\|_\sim \le \|\hat{B}^{n-1}(B(y,z))\| \le \|\hat{B}^{n-1}\| \, \|B\|_\sim \, \|(y,z)\|.$$

For all associated operator norms, we use $\|\cdot\|$. With this convention, for all $n \in \mathbb{N}$.

$$\|\tilde{B}^n\| \le \|\hat{B}^{n-1}\| \, \|B\|.$$

This implies that $\mathbf{r}(\tilde{B}) \le \mathbf{r}(\hat{B})$. By (5.34), for all $n \in \mathbb{N}$,

$$\tilde{B}^n(y,y) = (\hat{B}^n(y), \hat{B}^n(y)), \qquad y \in X_+.$$

So, for all $n \in \mathbb{N}$,

$$\|\tilde{B}^n(y,y)\|_\sim = \|\hat{B}^n(y)\|, \qquad y \in X_+,$$

and

$$\|\hat{B}^n(y)\| \le \|\tilde{B}^n\| \, \|(y,y)\| = \|\tilde{B}^n\| \, \|y\|.$$

This implies $\|\hat{B}^n\| \le \|\tilde{B}^n\|$ and $\mathbf{r}(\hat{B}) \le \mathbf{r}(\tilde{B})$. □

PROPOSITION 5.40. *For $j = 1,2$, let X_j be normed vector spaces and X_{j+} be homogeneous subsets of X and $B_j : X_{j+} \to X_{j+}$ be bounded homogeneous operators. Define $B : X_{1+} \times X_{2+} \to X_{1+} \times X_{2+}$ by*

$$B(x_1,x_2) = (B_1(x_1), B_2(x_2)), \qquad x_j \in X_j, \quad j = 1,2.$$

Let $|\cdot|$ be any norm on $\mathbb{R}^2$ and

$$\|(x_1,x_2)\| = \big|(\|x_1\|,\|x_2\|)\big|, \qquad x_j \in X_j.$$

Then $X_1 \times X_2$ is normed vector space.

Further, B is a bounded homogeneous operator and

$$\mathbf{r}(B) = \max\{\mathbf{r}(B_1), \mathbf{r}(B_2)\}.$$

PROOF. For all $n \in \mathbb{N}$,

$$B^n(x_1, x_2) = (B_1^n(x_1), B_2^n(x_2)), \qquad x_j \in X_{j+},$$

and

$$\|B^n(x_1, x_2)\| = \big|(\|B_1^n(x_1)\|, \|B_2^n(x_2)\|)\big|.$$

Since the spectral radius of a bounded homogeneous operator does not depend on homogeneous norms, we can choose

$$\|(x_1, x_2)\| = \max\{\|x_1\|, \|x_2\|\}.$$

Then

$$\|B^n(x_1, x_2)\| = \max\{\|B_1^n(x_1)\|, \|B_2^n(x_2)\|\}. \tag{5.35}$$

This implies

$$\|B^n(x_1, x_2)\| \le \max\{\|B_1^n\|\|x_1\|, \|B_2^n\|\|x_2\|\} \le \max\{\|B_1^n\|, \|B_2^n\|\}\|(x_1, , x_2)\|$$

and

$$\|B^n\| \le \max\{\|B_1^n\|, \|B_2^n\|\}.$$

Further,

$$\|B^n\|^{1/n} \le \max\{\|B_1^n\|^{1/n}, \|B_2^n\|^{1/n}\}.$$

By (5.1), $\mathbf{r}(B) \le \max\{\mathbf{r}(B_1), \mathbf{r}(B_2)\}$.

By (5.35),

$$\|B^n(x_1, 0)\| = \|B_1^n(x_1)\|, \qquad x_1 \in X_1,$$

and

$$\|B_1^n(x_1)\| \le \|B^n\|\|x_1\|$$

and $\|B_1^n\| \le \|B^n\|$. By (5.1), $\mathbf{r}(B_1) \le \mathbf{r}(B)$. Similarly, $\mathbf{r}(B_2) \le \mathbf{r}(B)$.

In combination, $\mathbf{r}(B) = \max\{\mathbf{r}(B_1), \mathbf{r}(B_2)\}$. □

CHAPTER 6

Order-bounded operators

In order to derive upper estimates for the spectral radius we consider a special class of order-preserving homogeneous operators in a ordered normed vector space.

The following terminology has been adopted from various works by Krasnosel'skii [**135**, Sec.2.1.1] and coworkers [**136**, Sec.9.4] though it has been modified.

Let X be an ordered normed vector space with closed order cone X_+.

DEFINITION 6.1. *Let $B : X_B \to X_B$, $u \in X_B$, where X_B is a subset of X_+.*

B is called pointwise u-bounded *if, for any $x \in X_B$, there exist some $n \in \mathbb{N}$ and $\gamma > 0$ such that $B^n(x) \le \gamma u$. The point u is called a* pointwise order bound *of B.*

B is called uniformly u-bounded *if there exists some $c > 0$ such that $B(x) \le c\|x\|u$ for all $x \in X_B$. The element u is called a* uniform order bound *of B.*

B is called uniformly order-bounded *if it is uniformly u-bounded for some $u \in X_B$. B is called* pointwise order-bounded *if it is pointwise u-bounded for some $u \in X_B$.*

If $B : X_+ \to X_+$ is homogeneous and bounded and X_+ is solid, then B is uniformly u-bounded for every interior point u of X_+.

The presentation in this chapter is similar to the one in [**198, 199**], but, whenever possible, has been generalized to $B : X_B \to X_B$ with a homogeneous subset X_B of X_+ rather than $X_B = X_+$.

6.1. Properties

Uniform order boundedness is preserved if $X_B = X_+$ and the original norm $\|\cdot\|$ is replaced by its monotone companion norm $\sharp \cdot \sharp$.

PROPOSITION 6.2. *Let $B : X_+ \to X_+$ be order-preserving. Let $u \in X_+$ and B be uniformly u-bounded. Then B is also uniformly u-bounded with respect to the monotone companion norm.*

PROOF. Let $x \in X_+$. By (2.24), for each $n \in \mathbb{N}$ there exists some $y_n \in X_+$ such that $x \le y_n$ and $\psi(x) \le \|y_n\| \le \psi(x) + (1/n)$. Since B is order-preserving and uniformly u-bounded,

$$B(x) \le B(y_n) \le c\|y_n\|u, \qquad n \in \mathbb{N}.$$

We take the limit as $n \to \infty$ and obtain $B(x) \le c\psi(x)u$. □

PROPOSITION 6.3. *Let $B : X_B \to X_B$ be continuous, order-preserving and homogeneous, X_B a closed cone contained in X_+, $u \in X_B$.*

Then the following hold.

(a) *B is uniformly u-bounded if $B(X_B) \subseteq X_u$ and X_B is serially complete.*

(b) *If B be pointwise u-bounded and X_B is complete, then some power of B is uniformly u-bounded.*

Proof. (a) This follows from Theorem 4.5 and the fact that $X_u \cap X_+$ is a normal cone under the u-norm.

(b) Define

$$M_{n,k} = \{x \in X_B; B^n(x) \le ku\}, \qquad n, k \in \mathbb{N}.$$

Since B is continuous and X_B and X_+ are closed, each set $M_{n,k}$ is closed. Since B is assumed to be pointwise u-bounded, $X_B = \bigcup_{k,n\in\mathbb{N}} M_{n,k}$. Since X_B is a complete metric space, by the Baire category theorem, there exists some $n, k \in \mathbb{N}$ such that $M_{n,k}$ contains a relatively open subset of X_B: There exists some $y \in X_B$ and $\epsilon > 0$ such that $y + \epsilon z \in M_{n,k}$ whenever $z \in X$, $\|z\| \le 1$, and $y + \epsilon z \in X_B$. Now let $z \in X_B$ and $\|z\| \le 1$. Since B is order-preserving and $y + \epsilon z \in X_B$, $B^n(\epsilon z) \le B^n(y + \epsilon z) \le ku$. Since B is homogeneous, for all $x \in \dot{X}_B$,

$$B^n(x) = \frac{\|x\|}{\epsilon} B^n\Big(\frac{\epsilon}{\|x\|} x\Big) \le \frac{k}{\epsilon} \|x\| u. \qquad \square$$

Proposition 6.4. *Let $B : X_B \to X_B$ satisfy $B(x) \le H(x) + A(x)$ for all $x \in X_B$, where $H, A : X_B \to X_B$ are homogeneous and bounded, H is uniformly u-bounded for some $u \in X_B$ and A is subadditive and order-preserving.*

Then, for each $n \in \mathbb{N}$, $B^n(x) \le B_n(x) + A^n(x)$ for all $x \in X_B$, where $B_n : X_B \to X_B$ is homogeneous and uniformly u_n-bounded with $u_n = \sum_{j=0}^n A^j(u)$.

In particular, if there is some $n \in \mathbb{N}$ and some $v \in X_B$ such that $v \ge u_n$ and A^n is uniformly v-bounded, then B^n is uniformly v-bounded.

Proof. The statement holds for $n = 1$ with $B_1 = H$. Suppose that $n \in \mathbb{N}$ is arbitrary and the statement holds for n. Since A is order-preserving and subadditive,

$$\begin{aligned} B^{n+1}(x) =& H(B^n(x)) + A(B^n(x)) \le B_{n+1}(x) + A^{n+1}(x), \\ B_{n+1}(x) =& H(B^n(x)) + A(B_n(x)), \qquad x \in X_B. \end{aligned}$$

Since H is uniformly u-bounded, A order-preserving, and B_n uniformly u_n-bounded, with appropriate $c_0 > 0$ and $c_n > 0$,

$$B_{n+1}(x) \le c_0 \|B^n(x)\| u + A(c_n \|x\| u_n).$$

Since A is homogeneous and subadditive,

$$B_{n+1}(x) \le c \|B^n\| \, \|x\| u + c_n \|x\| u_{n+1}, \qquad x \in X_B.$$

So B_{n+1} is uniformly u_{n+1}-bounded. $\square$

6.2. Upper Collatz–Wielandt numbers and radii

We continue to consider an ordered normed vector space X with closed order cone X_+ and an operator $B : X_B \to X_B \subseteq X_+$.

The *upper Collatz–Wielandt number* of B at $x \in \dot{X}_B$ is defined as **[87]**

$$\|B\|_x = \|B(x)\|_x = \inf\{r \ge 0; B(x) \le rx\}, \tag{6.1}$$

where $\|B\|_x = \infty$ if $B(x)$ is not x-bounded. See Definition 2.59. The *upper Collatz–Wielandt spectral radius* of B at $x \in \dot{X}_B$ is defined by

$$\eta^x(B) = \inf_{n\in\mathbb{N}} \|B^n\|_x^{1/n}. \tag{6.2}$$

Here the order of $[0,\infty]$ is the usual order of $[0,\infty)$ supplemented by $s \le \infty$ for all $s \in [0,\infty]$. See Appendix A.2.

LEMMA 6.5. *Let $B : X_B \to X_B, C : X_C \to X_C$ be homogeneous and order-preserving operators on homogeneous subsets X_B, X_C of X_+. $B(X_B) \subseteq X_C$. Let $x \in X_B \cap X_C$, $x \ne 0$, and $B(x)$ and $C(x)$ be x-bounded. Then $C(B(x))$ is x-bounded and $\|CB\|_x \le \|C\|_x \|B\|_x$.*

PROOF. By (2.32), $C(x) \le \|C\|_x x$ and $B(x) \le \|B\|_x x$ and

$$C(B(x)) \le \|B\|_x C(x) \le \|B\|_x \|C\|_x x;$$

so $(CB)(x)$ is x-bounded and $\|CB\|_x \le \|B\|_x \|C\|_x$ by (2.26). □

LEMMA 6.6. *Let $B : X_B \to X_B$ be a homogeneous order-preserving operator on a homogeneous subset X_B of X_+. Let $u \in \dot{X}_B$.*

(a) *Then $\eta^u(B^m) \ge (\eta^u(B))^m$ for all $m \in \mathbb{N}$.*
(b) *If there exists some $k \in \mathbb{N}$ such that $B^m(u)$ is u-bounded for all $m \ge k$, then $\eta^u(B) = \lim_{n\to\infty} \|B^n\|_u^{1/n} < \infty$ and $\eta^u(B^m) = (\eta^u(B))^m$ for all $m \in \mathbb{N}$.*

PROOF. (a) Let $u \in \dot{X}_+$, $m \in \mathbb{N}$. Then

$$\eta^u(B^m) = \inf_{n\in\mathbb{N}} \|B^{mn}\|_u^{1/n} = (\inf_{n\in\mathbb{N}} \|B^{mn}\|_u^{1/(mn)})^m \ge (\inf_{k\in\mathbb{N}} \|B^k\|_u^{1/k})^m.$$

(b) Now let $k \in \mathbb{N}$ such that $B^m(u)$ is u-bounded for all $m \ge k$. By Lemma 6.5,

$$c_{n+m} \le c_n c_m, \quad n, m \ge k, \quad c_n = \|B^n\|_u.$$

By Lemma A.44,

$$\inf_{n=k}^{\infty} c_n^{1/n} = \lim_{n\to\infty} c_n^{1/n}. \tag{6.3}$$

Let $k > 1$, $m \in \{1,\dots,k\}$ and $c_m = \|B^m\|_u < \infty$. Then $B^m(u) \le c_m u$. Since B is order-preserving, $B^{m\ell}u \le c_m^\ell u$ for all $\ell \in \mathbb{N}$ and $c_{m\ell} \le c_m^\ell$ for all $\ell \in \mathbb{N}$. Thus

$$\lim_{n\to\infty} c_n^{1/n} = \lim_{\ell\to\infty} c_{m\ell}^{1/(\ell m)} \le c_m^{1/m}.$$

If $c_m = \infty$, this inequality trivially holds.

Together with (6.3), $\lim_{n\to\infty} c_n^{1/n} \le \inf_{n=1}^{\infty} c_n$ which implies equality.

The second equality in (b) follows from the fact that every subsequence of a convergent sequence converges to the same limit. □

REMARK 6.7. If $B(u)$ is u-bounded, $\|B\|_u$ is the norm of B on $X_B \cap X_u$ with u-norm, and $\eta^u(B)$ is the spectral radius of B taken in $X_B \cap X_u$.

Recall the concepts of a normal point $u \in \dot{X}_+$ for B in Definition 4.3 (inspired by [**95**]) and of a normal point for X_B in Section 2.2.2.

REMARK 6.8.

(a) If $u \in \dot{X}_+$ is a normal point for X_B, then u is a normal point for any bounded homogeneous $B : X_B \to X_B$.
(b) If $B : X_B \to X_B$ is bounded homogeneous and order-preserving, $u \in \dot{X}_B$ and $B(u)$ is a normal point for X_B, then u is a normal point for B.

PROOF. (a) There exists some $c > 0$ such that $\|x\| \leq c$ for all $x \in X_B$ with $x \leq u$. Since B is bounded, $\|B(x)\| \leq \|B\|c$ for all $x \in X_B$ with $x \leq u$.

(b) There exists some $c > 0$ such that $\|y\| \leq c$ for all $y \in X_B$ with $y \leq B(u)$. Let $x \in X_B$ and $x \leq u$. Since B is order-preserving, $B(x) \leq B(u)$ and $\|B(x)\| \leq c$ because $B(x) \in X_B$. □

THEOREM 6.9. *Let $B : X_B \to X_B$ be homogeneous, bounded, and order-preserving on a homogeneous subset X_B of X_+. Let $u \in X_B$, $\alpha \in \mathbb{R}_+$ and $k \in \mathbb{N}$ such that $B^k(u) \leq \alpha^k u$.*

Let u be a normal point for some power of B or B be power-compact. Then $\gamma_B(u) \leq \alpha$.

PROOF. Since $\gamma(u, B^k) = (\gamma(u, B))^k$ by Theorem 5.4, we can assume that $k = 1$. Since B is homogeneous, it is enough to show that $B(u) \leq u$ implies that $\gamma_B(u) = \gamma(u, B) \leq 1$.

Let $B(u) \leq u$. Then $B^n(u) \leq u$ for all $n \in \mathbb{N}$.

We first assume that u is a normal point for some power of B. By Definition 4.3, there exist some $\tilde{c} > 0$ and $m \in \mathbb{N}$ such that $\|B^m(x)\| \leq \tilde{c}$ for all $x \in X_B$ with $x \leq u$. Then $\|B^{m+n}(u)\| \leq \tilde{c}$ for all $n \in \mathbb{N}$. By (5.2), $\gamma_B(u) \leq 1$.

Now assume that B^ℓ is compact for some $\ell \in \mathbb{N}$. Then $\gamma_u(B) \leq 1$ by Proposition 5.27. □

COROLLARY 6.10. *Let $B : X_B \to X_B$ be homogeneous, bounded, and order-preserving on a homogeneous subset X_B of X_+. Let $u \in \dot{X}_B$ and $B^k(u)$ be u-bounded for all but finitely many $k \in \mathbb{N}$. Assume that u is a normal point for some power of B or that B is power-compact. Then $\gamma_B(u) \leq \eta^u(B)$.*

PROOF. By definition, $B^k(u) \leq \|B^k\|_u u$ for all $k \in \mathbb{N}$ with $\|B^k\|_u < \infty$. By Theorem 6.9, $\gamma_B(u) \leq \|B^k\|_u^{1/k}$ for all $k \in \mathbb{N}$. So $\gamma_B(u) \leq \eta^u(B)$. □

THEOREM 6.11. *Let $B : X_B \to X_B$ be homogeneous, bounded, and order-preserving on a homogeneous subset X_B of X_+. Let $u \in X_B$, $\alpha \in \mathbb{R}_+$ and $k \in \mathbb{N}$ such that $B^k(u) \leq \alpha^k u$.*

(a) *Let u be a normal point for some power of B. Then $\mathbf{r}_o(B) \leq \alpha$ if B is pointwise u-bounded, and $\mathbf{r}(B) \leq \alpha$ if some power of B is uniformly u-bounded.*

(b) *Let B be pointwise u-bounded and power-compact. Then $\mathbf{r}_o(B) \leq \alpha$. If B is also continuous, $\mathbf{r}(B) \leq \alpha$.*

PROOF. As in the proof of Theorem 6.9, we can reduce the proof to the implication

$$B(u) \leq u \implies \mathbf{r}_o(B) \leq 1.$$

Assume that $B(u) \leq u$. Let $x \in X_B$. Since B is pointwise u-bounded, there exists some $m \in \mathbb{N}$ and $c > 0$ such that $B^m(x) \leq cu$. Since B is order-preserving and homogeneous,

$$B^{m+n}(x) \leq cB^n(u) \leq cu, \qquad n \in \mathbb{N}. \tag{6.4}$$

(a) Since u is a normal point for some power of B, by Definition 4.3, there exist some $k \in \mathbb{N}$ and some $\tilde{c} > 0$ such that $\|B^{k+m+n}(c^{-1}x)\| \leq \tilde{c}$ for all $n \in \mathbb{N}$. This implies $\gamma_B(x) \leq 1$. Since $x \in X_B$ has been arbitrary, $\mathbf{r}_o(B) \leq 1$.

If B is bounded and B^m is uniformly u-bounded, we can replace c by $c\|x\|$ and we obtain $\mathbf{r}(B) \le 1$.

(b) Now let B be power-compact. By (6.4) and Lemma 5.26, $\|B^n(x)\|$ is bounded. By (5.2), $\gamma_x(B) \le 1$. Since this holds for all $x \in X_B$, $\mathbf{r}_o(B) \le 1$ by (5.3). If B is also continuous, $\mathbf{r}(B) = \mathbf{r}_o(B) \le 1$ by Theorem 5.7. □

Similarly as for Corollary 6.10, this yields the following result.

COROLLARY 6.12. *Let $B : X_B \to X_B$ be homogeneous, bounded, and order-preserving on a homogeneous subset X_B contained in X_+.*

(a) *Let u be a normal point for some power of B. Then $\mathbf{r}_o(B) \le \eta^u(B)$ if B is pointwise u-bounded, and $\mathbf{r}(B) \le \eta^u(B)$ if B is bounded and some power of B is uniformly u-bounded.*
(b) *If B is power-compact and pointwise u-bounded, then $\mathbf{r}_o(B) \le \eta^u(B)$.*

THEOREM 6.13. *Let $B : X_B \to X_B$ be a homogeneous and order-preserving operator on a homogeneous subset $X_B \subseteq X_+$, $u \in \dot{X}_B$. Let B be pointwise u-bounded. Then*

$$\eta_x(B) \le \gamma_u^\sharp(B) \le \gamma_u(B)$$

for all $x \in \dot{X}_B$ and

$$cw(B) \le \mathbf{r}_{cw}(B) \le \gamma_u^\sharp(B) \le \gamma_u(B).$$

Assume in addition that $B^n(u)$ is u-bounded for all but finitely many $n \in \mathbb{N}$. Then

$$cw(B) \le \mathbf{r}_{cw}(B) \le \gamma_u^\sharp(B) \le \eta^u(B).$$

PROOF. Let $x \in \dot{X}_B$. We can assume $x \ne 0$. Since B is pointwise u-bounded, there exists some $k = k(x) \in \mathbb{N}$ and some $c = c(x) > 0$ such that $B^k(x) \le cu$. Since B is order-preserving and homogeneous, $B^n(x) \le cB^{n-k}(u)$ for all $n > k$. This implies

$$B^n(x) \le cB^{n-k}(u) \le c\|B^{n-k}\|_u u.$$

We apply the companion half-norm,

$$\psi(B^n(x)) \le c\psi(B^{n-k}(u)) \le c\|B^{n-k}\|_u \psi(u).$$

So

$$\psi(B^n(x))^{1/n} \le c^{1/n}\psi(B^{n-k}(u))^{1/n} \le \|B^{n-k}\|_u^{1/n} \big(c\psi(u)\big)^{1/n}.$$

We take the limit superior as $n \to \infty$,

$$\begin{aligned}\gamma_x^\sharp(B) &\le \limsup_{n\to\infty} c^{1/n}\psi(B^{n-k}(u))^{1/n}\\ &= \limsup_{n\to\infty} c^{1/n}\psi(B^n(u))^{1/n} = \gamma_u^\sharp(B) \le \gamma_u(B).\end{aligned}$$

By Theorem 5.30,

$$[B]_x \le \gamma_x^\sharp(B) \le \gamma_u(B).$$

Now, for all $n \in \mathbb{N}$, B^n is homogeneous, bounded and order-preserving, and so

$$[B^n]_x \le \gamma_x^\sharp(B^n) \le \gamma_u^\sharp(B)^n.$$

The last equality follows from Theorem 5.5. By (5.28), $\eta_x(B) \le \gamma_u^\sharp(B) \le \gamma_u(B)$.

We now additionally assume that $B^n(u)$ is u-bounded for all but finitely many $n \in \mathbb{N}$. By Lemma 6.6, $\eta^u(B) = \lim_{n\to\infty} \|B^n\|_u^{1/n}$.

If $B^n(u)$ is u-bounded, $B^n(u) \le \|B^n\|_u u$ and $\psi(B^n(u)) \le \|B^n\|_u \psi(u)$. If $B^n(u)$ is not u-bounded, this inequality holds if interpreted in $[0,\infty]$ because the right-hand side is infinity. Since $\psi(u) > 0$,

$$\gamma_u^\sharp(B) \le \limsup_{n\to\infty} \|B^n\|_u^{1/n}\psi(u)^{1/n} = \eta^u(B). \qquad \square$$

THEOREM 6.14. *Let $B : X_B \to X_B$ be a homogeneous and order-preserving on a homogeneous subset X_B of X_+, $u \in X_B$. Let some power of B be uniformly u-bounded. Then*

$$cw(B) \le \mathbf{r}_{cw}(B) \le \mathbf{r}_o^\sharp(B) \le \eta^u(B) \le \gamma_u(B).$$

If, in addition, B is bounded and u is a normal point for some power of B, then $\eta^u(B) = \mathbf{r}_o(B) = \mathbf{r}(B) = \gamma_u(B) = \lim_{n\to\infty} \|B^n(u)\|^{1/n}$.

PROOF. Let $k \in \mathbb{N}$ such that B^k is uniformly u-bounded. Then there exists some $c > 0$ such that $B^k(x) \le c\|x\|u$ for all $x \in X_B$. For all $n \in \mathbb{N}$, $B^{k+n}(x) = B^k(B^n(x)) \le c\|B^n(x)\|u$. By (6.1), with $x = u$,

$$\|B^{k+n}\|_u \le \|B^n(u)\|c, \qquad n \in \mathbb{N}.$$

By (6.2),

$$(\eta^u(B))^{(k+n)/n} \le \|B^n(u)\|^{1/n}c^{1/n}, \qquad n \in \mathbb{N}.$$

We take the limit as $n \to \infty$,

$$\eta^u(B) \le \liminf_{n\to\infty} \|B^n(u)\|^{1/n}. \tag{6.5}$$

The other inequalities and equalities follow from Theorem 6.13.

Let B be bounded and u be a normal point for B. By Corollary 6.12,

$$\eta^u(B) \ge \mathbf{r}(B) \ge \mathbf{r}_o(B) \ge \gamma_u(B) = \limsup_{n\to\infty} \|B^n(u)\|^{1/n}.$$

Together with (6.5), this implies equalities.

The other statements now follow from the previous theorems. $\square$

6.3. The upper Collatz–Wielandt bound

Let X be an ordered normed vector space with closed order cone X_+ and $B : X_B \to X_B$ be a homogeneous order-preserving operator on a homogeneous subset X_B of X_+.

Let $u \in \dot X_B$ and $B(u)$ be u-bounded. Then $B(x)$ is x-bounded for any u-comparable $x \in X_B$. Recall Definition 2.59 and (2.29).

So we define the *upper Collatz–Wielandt bound* with respect to u by

$$\begin{aligned} CW_u(B) &= \inf\{\|B\|_x; x \in X_B, x \sim u\} \\ &= \inf\Big\{\lambda > 0; \exists x \in X_B : x \sim u, B(x) \le \lambda x\Big\}. \end{aligned} \tag{6.6}$$

If $x \in X_B$ and $x \sim u$, $\eta^x(B) = \eta^u(B)$. Since $\eta^x(B) \le \|B\|_x$ by (6.2),

$$\eta^u(B) \le CW_u(B). \tag{6.7}$$

We have the following inequalities from Corollary 6.12 and Theorems 6.13 and 6.14.

THEOREM 6.15. *Let B be homogeneous and order-preserving. Let $u \in \dot{X}_B$, $B(u)$ be u-bounded and B be pointwise u-bounded.*

Then

$$cw(B) \le \mathbf{r}_{cw}(B) \le \eta^u(B) \le CW_u(B).$$

Assume in addition that B is bounded.

If u is a normal point for a power of B or B is power compact, then $cw(B) \le \mathbf{r}_o(B) \le CW_u(B)$.

If u is a normal point for a power of B and a power of B is uniformly u-bounded, then $\mathbf{r}(B) \le CW_u(B)$.

The next result may explain why for additive homogeneous operators the upper local Collatz–Wielandt spectral radius has played a lesser role than the upper Collatz–Wielandt bound.

THEOREM 6.16. *Let $B : X_B \to X_B$ be homogeneous, order-preserving and subadditive. Let $u \in \dot{X}_B$ and $B(u)$ be u-bounded.*

Then $\eta^u(B) = CW_u(B)$.

If, in addition, B is bounded and u is some normal point for a power of B and some power of B is uniformly u-bounded, then

$$CW_u(B) = \mathbf{r}_o(B) = \mathbf{r}(B) = \gamma_u(B) = \lim_{n\to\infty} \|B^n(u)\|^{1/n}.$$

PROOF. Let $\alpha > \eta^u(B)$. By (6.1), there exists some $m \in \mathbb{N}$ such that $\|B^m\|_u < \alpha^m$. Define

$$v = \sum_{n=0}^{m-1} \alpha^{-n} B^n(u).$$

Since $B(u)$ is u-bounded, v is comparable to u. Since B is homogeneous and subadditive,

$$B(v) \le \sum_{n=0}^{m-1} \alpha^{-n} B^{n+1}(u) = \alpha \sum_{n=1}^{m} \alpha^{-n} B^n(u) = \alpha(v + \alpha^{-m} B^m(u) - u) \le \alpha v.$$

Since v is u-comparable, $CW^u(B) \le \alpha$. Since this hold for any $\alpha > \eta^u(B)$, we have $CW^u(B) \le \eta^u(B)$. The opposite inequality follows from (6.7).

The last statement now follows from Theorem 6.14. □

It may be useful to revisit Section 2.1.7 for the next results.

PROPOSITION 6.17. *Let X_B be an* inf*-semilattice, B be homogeneous and order-preserving, $u \in \dot{X}_B$ and $B(u)$ be u-comparable, $n \in \mathbb{N}$, $\lambda > 0$ and $B^n(u) \le \lambda^n u$. Then*

$$v = \inf_{j=0}^{n-1} \lambda^{-j} B^j(u) \tag{6.8}$$

is u-comparable and $B(v) \le \lambda v$ and $CW_u(B) \le \lambda$.

PROOF. Since $B(u)$ is u-comparable, $B^j(u)$ is u-comparable for all $j \in \mathbb{N}$ and so v is u-comparable. For $j = 0, \dots, n-1$, since B is homogeneous and order-preserving,

$$\lambda^{-1} B(v) \le \lambda^{-1} B(\lambda^{-j} B^j(u)) = \lambda^{-(j+1)} B^{j+1}(u).$$

By the same token,

$$\lambda^{-1} B(v) \le \lambda^{-1} B(\lambda^{-(n-1)} B^{n-1}(u)) = \lambda^{-n} B^n(u) \le u = \lambda^0 B^0(u).$$

By the definition of the infimum, $\lambda^{-1}B(v) \le v$. Since v is u-comparable, $CW_u(B) \le \lambda$ by (6.6). □

THEOREM 6.18. *Let X_B be an* inf*-semilattice, B be homogeneous and order-preserving, $u \in \dot{X}_B$ and $B(u)$ be u-comparable. Then $\eta^u(B) = CW_u(B)$.*

PROOF. Since $B(u)$ is u-comparable, $\eta^u(B) \ge cw(B) > 0$. Let $n \in \mathbb{N}$. By (6.1) and (6.2), $B^n(u) \le \|B^n\|_u u$ and $\|B^n\|_u \ge \eta(B)^n > 0$. By Proposition 6.17, with $\lambda = \|B^n\|_u^{1/n}$, $CW_u(B) \le \|B^n\|_u^{1/n}$. By (6.2), $CW_u(B) \le \eta^u(B)$.

The opposite inequality holds by (6.7). □

We combine this result with Theorem 6.14.

COROLLARY 6.19. *Let X_B be an* inf*-semilattice, B be homogeneous and order-preserving, $u \in \dot{X}_B$ and $B(u)$ be u-comparable. Assume that some power of B is uniformly u-bounded.*

Then $CW_u(B) = \eta^u(B) \le \gamma_u(B)$.

If, in addition, B is bounded and u is a normal point of some power of B,

$$CW_u(B) = \mathbf{r}(B) = \gamma_u(B) = \lim_{n\to\infty} \|B^n(u)\|^{1/n}.$$

THEOREM 6.20. *Let X_B be an* inf*-semilattice, B be homogeneous and order-preserving, $u \in \dot{X}_B$ and B be uniformly u-bounded.*

Let $\theta : X_B \to \mathbb{R}_+$ be a homogeneous, order-preserving functional with $\theta(u) > 0$.

For any $\epsilon > 0$, define $B_\epsilon : X_B \to X_B$ by $B_\epsilon(x) = B(x) + \epsilon\theta(x)u$.

Then $CW_u(B_\epsilon) = \eta^u(B_\epsilon)$ for all $\epsilon > 0$.

If, in addition, B is bounded and u is a normal point for X_B,

$$CW_u(B_\epsilon) = \mathbf{r}(B_\epsilon) = \gamma_u(B_\epsilon)$$

for all $\epsilon > 0$.

PROOF. Let $\epsilon > 0$. Then $B_\epsilon(u)$ is u-comparable. Further B_ϵ is homogeneous and order-preserving and uniformly u-bounded. Since u is a normal point for X_B, it is a normal point for B_ϵ (Remark 6.8). The assertion follows from Theorem 6.18. □

EXAMPLE 6.21. Cf. Example 5.35. Let

$$B(x) \le \sum_{k=1}^m \theta_k(x)u_k, \qquad x \in X_B,$$

with $u_k \in \dot{X}_B$ and $\theta_k : X_B \to \mathbb{R}_+$ being bounded and homogeneous. Set $u = \sum_{j=1}^n u_j$. Then B is uniformly u-bounded and

$$B(u) \le \sum_{k=1}^m \theta_k(u)u_k \le \beta u, \qquad \beta = \sup_{k=1}^m \theta_k(u) = \sup_{k=1}^m \theta_k\Big(\sum_{j=1}^n u_j\Big).$$

By (6.6) and (6.1),

$$CW_u(B) \le \|B\|_u \le \beta.$$

If $B(x) = \sum_{k=1}^m \theta_k(x)u_k$ for all $x \in X_B$ and all θ_k are order-preserving in addition, u is a normal point for B by Example 4.6 and so $\mathbf{r}(B) \le \beta$ by Theorem 6.15.

6.4. Iterative calculation of the spectral radius

We extract the following result from Theorem 6.14 which may help with the numerical calculation of the spectral radius.

THEOREM 6.22. *Let $B : X_B \to X_B$ be homogeneous, bounded and order-preserving on a homogeneous subset X_B of X_+, $u \in X_B$. Let some power of B be uniformly u-bounded and u be a normal point for some power of B. Then*

$$\mathbf{r}_o(B) = \mathbf{r}(B) = \gamma_u(B) = \lim_{n\to\infty} \|B^n(u)\|^{1/n}.$$

This formula has the advantage over the definition $\mathbf{r}(B) = \lim_{n\to\infty} \|B^n\|^{1/n}$ that we only need to evaluate the powers of B at u. However, if $0 < \mathbf{r}(B) \neq 1$, $B^n(u)$ will become very big or very small, respectively, which may create numerical problems.

To address this issue, we define

$$M(x) = \frac{B(x)}{\|x\|}, \qquad x \in \dot{X}_B. \tag{6.9}$$

Induction shows that

$$M^n(x) = \frac{B^n(x)}{\|B^{n-1}(x)\|}, \qquad n \in \mathbb{N}, \tag{6.10}$$

provided that $B^n(x) \neq 0$ for all $n \in \mathbb{N}$.

If $B^n(u) = 0$ for some $n \in \mathbb{N}$, $B^k = 0$ for some $k \in \mathbb{N}$ because some power of B is uniformly u-bounded, and $\mathbf{r}(B) = 0$. So we can assume that $B^n(u) \in \dot{X}_B$ for all $n \in \mathbb{N}$. Then

$$\|B^n(u)\| = \|B^{n-1}(u)\|\|M^n(u)\| = \prod_{j=0}^{n} \|M^j(u)\|. \tag{6.11}$$

In the context of population dynamics, $\|B^n(u)\|$ measures the population size in the nth generation or year, provided that u is the initial distribution of the population. So

$$\|M^j(u)\| = \frac{\|B^j(u)\|}{\|B^{j-1}(u)\|}$$

could be interpreted as the ratio of population sizes in two successive generations (cf. [**20**, Cor.1]) and $\mathbf{r}(B)$ as the limit of geometric means of the ratio of population sizes in two successive generations. For the numerical calculation, it may be advantageous to take logarithms,

$$\ln \mathbf{r}(B) = \lim_{n\to\infty} \frac{1}{n} \sum_{j=0}^{n} \ln \|M^j(u)\|. \tag{6.12}$$

More efficiently, one may like to rewrite this formula as an iteration

$$\ln \mathbf{r}(B) = \lim_{n\to\infty} b_n \tag{6.13}$$

with

$$\begin{aligned} b_{n+1} =& \Big(1 - \frac{1}{n+1}\Big) b_n + \frac{1}{n+1} \ln \|x_{n+1}\|, \\ x_{n+1} =& M(x_n), \qquad n \in \mathbb{N}, \\ x_1 =& M(u), \qquad b_1 = \ln \|M(u)\|. \end{aligned} \tag{6.14}$$

Historically, this procedure is connected to an iterative procedure proposed by Richard von Mises and Hilda Geiringer [**217**], sometimes called *von Mises iteration or procedure* [**25**, p.61]) and sometimes called *power method* [**173**, Sec.5.3]. Under stronger assumptions than we want to make in this monograph, these iterative procedures do not only provide the spectral radius of B but also an associated eigenvector of B [**30**] [**25**, III.2.6] [**168**].

6.5. Monotonicity of spectral radii and Collatz–Wielandt bounds

Let X be an ordered normed vector space with closed order cone X_+. Let $A : X_A \to X_A$ and $B : X_B \to X_B$ be bounded homogeneous operators on homogeneous subsets X_A and X_B of X.

We write $A \preceq B$ if $X_A \subseteq X_B \subseteq X_+$ and $A(x) \le B(y)$ for all $x \in X_A$ and $y \in X_B$ with $x \le y$.

THEOREM 6.23. *Let A and B be as before and $A \preceq B$. Then the following hold:*

(a) *For all $n \in \mathbb{N}$, $A^n(x) \le B^n(x)$ for all $x \in X_A$.*
(b) *$cw(A) \le cw(B)$ and $\mathbf{r}_{cw}(A) \le \mathbf{r}_{cw}(B)$.*
(c) *For all $x \in X_A$, $\|A\|_x \le \|B\|_x$ and $\eta^x(A) \le \eta^x(B)$.*
(d) *If X_+ is a normal cone, $\mathbf{r}(A) \le \mathbf{r}(B)$ and $\mathbf{r}_o(A) \le \mathbf{r}_o(B)$.*
(e) *If $X_A = X_B$ and $u \in \dot{X}_B$ and $A(u)$ and $B(u)$ are u-bounded, then $CW^u(A) \le CW^u(B)$.*

PROOF. (a) (By induction.) Indeed, since $A \preceq B$, the statement in (a) holds for $n = 1$. For the induction step, let $n \in \mathbb{N}$ and the statement hold for n. Let $x \in X_A$. Since $A \preceq B$ and $A^n(x) \in X_A$, $B^n(x) \in X_B$, and $A^n(x) \le B^n(x)$,

$$A^{n+1}(x) = A(A^n(x)) \le B(B^n(x)) = B^{n+1}(x).$$

This proves (a).

(b) Let $x \in X_A$. By (a) and (5.25), for $n \in \mathbb{N}$, $[A^n]_x x \le A^n(x) \le B^n(x)$. By (5.23), $[A^n]_x \le [B^n]_x$. By (5.28), $\eta_x(A) \le \eta_x(B)$.

By $X_A \subseteq X_B$ and (5.30) and (5.29), $cw(A) \le cw(B)$ and $\mathbf{r}_{cw}(A) \le \mathbf{r}_{cw}(B)$.

(c) This follows in a similar way from the definitions in Section 2.6 and Section 6.2.

(d) By (a), since X_+ is a normal cone (Theorem 2.20) and $X_A \subseteq X_B \subseteq X_+$, there exists some $c > 0$ such that $\|A^n(x)\| \le c\|B^n(x)\|$ for all $n \in \mathbb{N}$ and $x \in X_A$. By (4.1), $\|A^n\| \le c\|B^n\|$ for all $n \in \mathbb{N}$. Further $\gamma_x(A) \le \gamma_x(B)$ for all $x \in X_A$ by (5.2). The assertion now follows from (5.1) and (5.3).

(e) Let $x \in X_A = A_B$, $x \sim u$. By (2.32), $A(x) \le B(x) \le \|B\|_x x$. By (6.6) and (2.26), $CW^u(A) \le \|A\|_x \le \|B\|_x$. Again by (6.6), since $X_A = X_B$, $CW^u(A) \le CW^u(B)$. □

Recall Definition 2.14.

COROLLARY 6.24. *Let X_+ be a normal cone. We assume that $X_A \subseteq X_B$, B is order-preserving, and*

$$A(x) \le B(x), \qquad x \in X_A. \tag{6.15}$$

Then $A \preceq B$ and all statements of Theorem 6.23 hold.

PROOF. Let $x \in X_A$ and $y \in X_B$ and $x \le y$. By (6.15), $A(x) \le B(x)$. Since B is order-preserving, $B(x) \le B(y)$ and so $A(x) \le B(y)$. This proves $A \preceq B$. □

In a similar way, Theorem 6.23 implies the next result.

COROLLARY 6.25. *Let X_+ be a normal cone. We assume that $X_A = X_B$, A is order-preserving, and $A(x) \le B(x)$ for all $x \in X_A$. Then $A \preceq B$ and all statements of Theorem 6.23 hold.*

We do not address the question of strict monotonicity of spectral radii; some results for positive linear operators can be found in [**96**].

CHAPTER 7

Upper semicontinuity of spectral radii

The threshold property of the spectral radius (between extinction and persistence of populations) outlined in Section 1.3 raises interest in the question whether the spectral radius of a bounded homogeneous operator B continuously depends on B.

7.1. Upper semicontinuity with respect to the operator norm

Let X be a normed vector space and $B : X_B \to X_B$ be a bounded homogeneous operator on a homogeneous subset X_B of X. Under what conditions does the following hold?

If (B_k) is a sequence of bounded homogeneous operators on X_B and $\|B_k - B\| \to 0$ as $k \to \infty$, then $\mathbf{r}(B_k) \to \mathbf{r}(B)$ as $k \to \infty$.

Here $\|\cdot\|$ is the operator norm for bounded homogeneous operators on X_B as defined in (4.1).

Conditions for continuity of the spectral radius turn out to be quite complicated and continuity does not hold in general [**146**] even if X is an ordered normed vector space and B and all B_k are order-preserving (Chapter 10).

However, upper semicontinuity of the spectral radius holds in the following sense and under the following conditions.

THEOREM 7.1. *Let X be a normed vector space and $B : X_B \to X_B$ be a bounded homogeneous operator on a homogeneous subset X_B of X that satisfies one of the following conditions:*

(a) *B is uniformly continuous on bounded sets.*
(b) *B is compact and continuous and X_B is closed.*

Then the spectral radius is upper semicontinuous at B:

If (B_k) is any sequence of bounded homogeneous operators on X_B with $\|B_k - B\| \to 0$ as $k \to \infty$, then

$$\limsup_{k\to\infty} \mathbf{r}(B_k) \le \mathbf{r}(B).$$

In condition (a), the following concept is used.

DEFINITION 7.2. *Let X and Y be normed vector spaces. A homogeneous operator $B : X_B \to Y$ on a homogeneous subset X_B of X is called* uniformly continuous on bounded sets *if B is uniformly continuous on $\{x \in X_B; \|x\| \le 1\}$.*

For the concept of the limit superior, see Appendix A.3.1 in the appendix.

Notice that Theorem 7.1 does not assume any order structure and that additional conditions are only made for B and not the approximating operators B_k (cf. [**146**, Lem.2.3]).

Theorem 7.1 is also stated and proved in [**208**], but the arguments have been completely reorganized for transparency in this monograph.

Let us first give some hint why upper semicontinuity is much easier to establish than continuity for the spectral radius. To this end, let us make the connection to the concept of upper semicontinous functions on a metric space (Appendix A.4).

Recall that the set of operators $G : X_B \to X$ is a real vector space under addition and scalar multiplication with real numbers:

$$(G_1 + G_2)(x) = G_1(x) + G_2(x), \qquad (\alpha G)(x) = \alpha G(x)$$

for $x \in X_B$, $\alpha \in \mathbb{R}$ and $G_1, G_2 : X_B \to X$.

The subset $\mathcal{H}$ of bounded homogeneous operators from $X_B \to X$ is a vector subspace and thus is a real vector space itself. Further, the operator norm defined in (4.1) is a seminorm on $\mathcal{H}$.

To see that it is a norm, indeed, let $\|G\| = 0$. Then $G(x) = 0$ for all $x \in X_B$ with $\|x\| \leq 1$. Since G is homogeneous, G is the zero operator.

The operator norm on $\mathcal{H}$ makes $\mathcal{H}$ a metric space and induces a metric on the set $\mathcal{H}_B := \{G \in \mathcal{H}; G(x) \in X_B, x \in X_B\}$. If X_B is a closed subset of X, $\mathcal{H}_B$ is a closed metric subspace of $\mathcal{H}$.

In this scenario, it is possible to discuss the continuity (and semicontinuity) of the spectral radius as a function of the operator in a standard framework (Appendix A.4). Now recall that the definition of the spectral radius involves taking an infimum (5.8) which is an operation that preserves upper semicontinuity but not continuity (Lemma A.52).

Upper semicontinuity is not only of interest because of the threshold property of the spectral radius in the dynamics of populations, but also mathematically, for instance, in order to be able to compare the spectral radius to the upper Collatz–Wielandt bound (Theorem 9.21) [**201**].

See [**146**] and Chapter 10 for the much harder question of continuity of the spectral radius.

The next result puts us on the path to conditions for upper semicontinuity of the spectral radius:

PROPOSITION 7.3. *Let B be a bounded homogeneous operator from X_B to X_B and (B_k) be a sequence of bounded homogeneous operators from X_B to X_B. Assume that, for all $n \in \mathbb{N}$, $\|B_k^n - B^n\| \to 0$ as $k \to \infty$. Then $\limsup_{k\to\infty} \mathbf{r}(B_k) \leq \mathbf{r}(B)$.*

PROOF. By assumption, for all $n \in \mathbb{N}$, $\|B_k^n\| \to \|B^n\|$ as $k \to \infty$ and so $\|B_k^n\|^{1/n} \to \|B^n\|^{1/n}$ as $k \to \infty$. Let $r > \mathbf{r}(B)$. By (5.8), there exists some $n \in \mathbb{N}$ such that $\|B^n\|^{1/n} < r$. Then there exists some $k_n \in \mathbb{N}$ such that $\|B_k^n\|^{1/n} < r$ for all $k \geq k_n$. Again by (5.8), $\mathbf{r}(B_k) < r$ for all $k \geq k_n$ and $\limsup_{k\to\infty} \mathbf{r}(B_k) < r$. Since this holds for all $r > (\mathbf{B})$, $\limsup_{k\to\infty} \mathbf{r}(B_k) < r$. □

COROLLARY 7.4. *Let $B : X_B \to X_B$ be a bounded homogeneous operator with the following property: For every sequence (B_k) of bounded homogeneous operators $B_k : X_B \to X_B$,*

$$\|B_k - B\| \overset{k\to\infty}{\longrightarrow} 0 \implies \|B_k^n - B^n\| \overset{k\to\infty}{\longrightarrow} 0, \qquad n \in \mathbb{N}.$$

Then the spectral radius is upper semicontinuous at B.

LEMMA 7.5. *Let X, Y, Z be normed vector spaces, $X_B \subseteq X$ and $Y_C \subseteq Y$ be homogeneous sets. Let $B : X_B \to Z$ and $C : Y_C \to X_B$ be bounded and homogeneous. Let (B_k) and (C_k) be sequences of bounded homogeneous maps from X_B to Z and Y_C to X_B respectively such that*

$$\|B_k - B\| \to 0, \quad \|C_k - C\| \to 0, \quad \|BC_k - BC\| \to 0, \qquad k \to \infty.$$

Then $\|B_kC_k - BC\| \to 0$ as $k \to \infty$.

PROOF. Let $x \in X_B$. By the triangle inequality and (4.1)

$$\begin{aligned} &\|B_kC_k(x) - BC(x)\| \\ &\quad \le \|B_k - B\| \, \|C_k(x)\| + \|BC_k(x) - BC(x)\| \\ &\quad \le \|B_k - B\| \, \|C_k(x) - C(x)\| + \|B_k - B\| \, \|C(x)\| + \|BC_k(x) - BC(x)\|. \end{aligned}$$

Again by (4.1),

$$\|B_kC_k - BC\| \le \|B_k - B\| \, \|C_k - C\| + \|B_k - B\| \, \|C\| + \|BC_k - BC\|.$$

The assertion now follows from the assumptions. □

Some equivalent descriptions of "uniform continuity on bounded sets" will be useful.

LEMMA 7.6. *Let $B : X_B \to Y$ be homogeneous. Then the following are equivalent:*

(a) *B is uniformly continuous on bounded sets (Definition 7.2).*
(b) *B is uniformly continuous on every bounded subset of X_B.*
(c) *For any two bounded sequences (y_n) and (z_n) in X_B with $\|y_n - z_n\| \overset{n\to\infty}{\longrightarrow} 0$ it follows that $\|B(y_n) - B(z_n)\| \overset{n\to\infty}{\longrightarrow} 0$.*

PROOF. (a) $\Rightarrow$ (b): Let Z be a bounded subset of X_B. Then there exists some $c > 0$ such that $\|x\| \le c$ for all $x \in Z$. Let $\epsilon > 0$. Then there exists some $\delta > 0$ such that $\|B(u) - B(v)\| < \epsilon/c$ for all $u, v \in X_B$, $\|u\|, \|v\| \le 1$ and $\|u - v\| < \delta$. Let $x, y \in Z$ and $\|x - y\| < c\delta$. Then x/c and y/c are elements in the homogeneous set X_B and $\|x/c\| \le 1$ and $\|y/c\| \le 1$ and $\|(x/c) - (y/c)\| = \|x - y\|/c < \delta$. Hence $\|B(x/c) - B(y/c)\| < \epsilon/c$. Since B is homogeneous, $\|B(x) - B(y)\| < \epsilon$.

(b) $\Rightarrow$ (c): Consider two bounded sequences (y_n) and (z_n) in X_B with $\|y_n - z_n\| \longrightarrow 0$ as $n \to \infty$. Then, there exists some bounded subset Z of X_B such that $y_n, z_n \in Z$ for all $n \in \mathbb{N}$. Let $\epsilon > 0$. Since B is uniformly continuous on Z, there exist some $\delta > 0$ such that $\|B(y) - B(z)\| < \epsilon$ for all $y, z \in Z$ with $\|y - z\| < \delta$. There exists some $N \in \mathcal{N}$ such that $\|y_n - z_n\| < \delta$ for all $n \ge N$. Then $\|B(y_n) - B(z_n)\| < \epsilon$ for all $n \ge N$.

(c) $\Rightarrow$ (a): We show the contrapositive. Suppose that B is not uniformly continuous on $\{y \in X_B; \|y\| \le 1\}$. Then there exists some $\epsilon > 0$ such that for any $n \in \mathbb{N}$ there are $y_n, z_n \in X_B$, $\|y_n\|, \|z_n\| \le 1$, $\|y_n - z_n\| < 1/n$ and $\|B(y_n) - B(z_n)\| \ge \epsilon$. Then the sequences (y_n) and (z_n) are bounded, $\|y_n - z_n\| \to 0$ as $n \to \infty$, but $\|B(y_n) - B(z_n)\|$ does not converge to 0 as $n \to \infty$. □

LEMMA 7.7. *Let X, Y, Z be normed vector spaces, $X_B \subseteq X$ and $Y_C \subseteq Y$ be homogeneous sets. Let $B : X_B \to Z$ and $C : Y_C \to X_B$ be bounded and homogeneous. Let (C_k) be a sequence of bounded homogeneous maps from Y_C to X_B respectively such that*

$$\|C_k - C\| \to 0, \qquad k \to \infty.$$

Then $\|BC_k - BC\| \to 0$ *as* $k \to \infty$ *if one of the following two assumptions hold:*

(a) B *is uniformly continuous on bounded sets.*

(b) C *is compact,* B *is continuous and* X_B *is closed.*

PROOF. It is sufficient to prove that both (a) and (b) imply that

$$\sup_{\|x\|\le 1} \|B(C_k(x)) - B(C(x))\| \to 0, \qquad k \to \infty.$$

Suppose that the latter does not hold. Then there exists some $\epsilon > 0$ such that, after choosing a subsequence, $\sup_{\|x\|\le 1} \|B(C_k(x)) - B(C(x))\| > \epsilon$ for all $k \in \mathbb{N}$. Then we find a sequence $(x_k) \in X_B$ with $\|x_k\| \le 1$ such that

$$\|B(C_k(x_k)) - B(C(x_k))\| > \epsilon, \qquad k \in \mathbb{N}. \tag{7.1}$$

(a) Let B be uniformly continuous on bounded sets. We have $\|C_k(x_k) - C(x_k)\| \le \|C_k - C\| \to 0$ as $k \to \infty$ and both the sequences $(C_k(x_k))$ and $(C(x_k))$ are bounded. By Lemma 7.6, $\|B(C_k(x_k)) - B(C(x_k))\| \to 0$ as $k \to \infty$, a contradiction.

(b) Let C be compact and B be continuous and X_B be closed. By (7.1), after choosing subsequences again, $C(x_k) \to y$ for some $y \in X_B$. Then, as $k \to \infty$,

$$\|C_k(x_k) - y\| \le \|C_k(x_k) - C(x_k)\| + \|C(x_k) - y\| \le \|C_k - C\| + \|C(x_k) - y\| \to 0.$$

Since B is continuous, $BC_k(x_k) \to B(y)$ and $BC_k(x_k) - BC(x_k) \to 0$, a contradiction. □

We combine Lemma 7.5 and Lemma 7.7.

LEMMA 7.8. *Let* X, Y, Z *be normed vector spaces,* $X_B \subseteq X$ *and* $Y_C \subseteq Y$ *be homogeneous sets. Let* $B : X_B \to Y_C$ *and* $C : Y_C \to Z$ *be bounded and homogeneous. Let* (B_k) *and* (C_k) *be sequences of bounded homogeneous maps from* X_B *to* Y_C *and* Y_C *to* Z *respectively such that* $\|B_k - B\| \to 0$, $\|C_k - C\| \to 0$.

Then $\|B_kC_k - BC\| \to 0$ *as* $k \to \infty$ *if one of the following two conditions holds:*

(a) B *is uniformly continuous on bounded sets.*

(b) C *is compact,* B *is continuous and* X_B *is closed.*

PROPOSITION 7.9. *Let* B *be a bounded homogeneous operator from* X_B *to* X_B *and* (B_k) *be a sequence of bounded homogeneous operators from* X_B *to* X_B *such that* $\|B_k - B\| \to 0$ *as* $k \to \infty$.

Then, for all $n \in \mathbb{N}$, $\|B_k^n - B^n\| \to 0$ *as* $k \to \infty$ *provided that one of the following two conditions hold:*

(a) B *is uniformly continuous on bounded sets.*

(b) B *is compact and continuous and* X_B *is closed.*

PROOF. The proof is by induction. It is sufficient to perform the induction step.

Let $n \in \mathbb{N}$ and $\|B_k^n - B^n\| \to 0$ as $k \to \infty$. Then $B_k^{n+1} = B_kC_k$ and $B^{n+1} = BC$ with $C_k = B_k^n$ and $C = B^n$.

(a) Apply Lemma 7.8 (a).

(b) If B is compact and continuous, so is $C = B^n$. Apply Lemma 7.8 (b).

In both (a) and (b), it follows that $\|B_k^{n+1} - B^{n+1}\| \to 0$ as $k \to \infty$. □

PROOF OF THEOREM 7.1. Combine Proposition 7.3 and Proposition 7.9. □

We now assume that X is an ordered normed vector space.

LEMMA 7.10. *Let X be an ordered normed vector space with normal order cone $X_+ \supseteq X_B$, X_B a cone as well, and let $B, C : X_B \to X_B$ be bounded, homogeneous, order-preserving, and subadditive. Let (B_k) and (C_k) be sequences of bounded homogeneous maps on X_B such that $\|B_k - B\| + \|C_k - C\| \to 0$ as $k \to \infty$.*

Then BC is bounded, homogeneous, order-preserving, and subadditive and

$$\|B_k C_k - BC\| \to 0, \qquad k \to \infty.$$

PROOF. One easily checks that BC is bounded, homogeneous, order-preserving and subadditive. For all $k \in \mathbb{N}$, set $\tilde{B}_k, \tilde{C}_k : X_B \to X_+$ by $\tilde{B}_k(x) = B_k(x) - B(x)$ and $\tilde{C}_k(x) = C_k(x) - C(x)$ for all $x \in X_B$. Then

$$B_k C_k = BC_k + \tilde{B}_k C_k = BC + D_k$$

with $D_k : X_B \to X_+$,

$$D_k = B(C + \tilde{C}_k) - BC + \tilde{B}_k C_k. \tag{7.2}$$

It is sufficient to show that $\|D_k\| \to 0$.

As $k \to \infty$, since $\|\tilde{B}_k\| + \|\tilde{C}_k\| \to 0$ as $k \to \infty$,

$$\|\tilde{B}_k C_k\| \leq \|\tilde{B}_k\| \, \|C_k\| \leq \|\tilde{B}_k\| \, (\|C\| + \|\tilde{C}_k\|) \to 0.$$

Since B is subadditive and order-preserving, for all $k \in \mathbb{N}$ and $x \in X_B$,

$$0 \leq B\big(C(x) + \tilde{C}_k(x)\big) - BC(x) \leq B(\tilde{C}_k(x)). \tag{7.3}$$

Since X_+ is normal, there is some $c > 0$ such that, for all $k \in \mathbb{N}$ and $x \in X_B$,

$$\|B(C(x) + \tilde{C}_k(x)) - BC(x)\| \leq c\|B(\tilde{C}_k(x))\| \leq c\|B\| \, \|\tilde{C}_k\| \, \|x\|.$$

By (4.1), since $\|\tilde{C}_k\| \to 0$ as $k \to \infty$,

$$\|B(C + \tilde{C}_k) - BC\| \leq c\|B\| \, \|\tilde{C}_k\| \to 0.$$

By (7.2), $\|D_k\| \to 0$ as $k \to \infty$ and the assertion follows. □

REMARK 7.11. If B is additive, instead of (7.3) we have

$$B\big(C(x) + \tilde{C}_k(x)\big) - BC(x) = B(\tilde{C}_k(x)),$$

and one does not need the order structure.

THEOREM 7.12. *Let X be an ordered normed vector space with normal order cone $X_+ \supseteq X_B$, X_B a cone as well, and $B : X_B \to X_B$ be bounded, homogeneous, order-preserving, and subadditive. Let (B_k) be a sequence of bounded homogeneous maps on X_B such that and $\|B_k - B\| \to 0$ as $k \to \infty$. Then*

$$\limsup_{k\to\infty} \mathbf{r}(B_k) \leq \mathbf{r}(B).$$

PROOF. By Lemma 7.10 and induction, for all $n \in \mathbb{N}$, B^n is bounded homogeneous, order-preserving, and subadditive and $\|B_k^n - B^n\| \to 0$ as $k \to \infty$ which implies the assertion by Proposition 7.3.

We only need to check the induction step:

Let $n \in \mathbb{N}$ and $\|B_k^n - B^n\| \to 0$ for $k \to \infty$. Then $B_k^{n+1} = B_k C_k$ and $B^{n+1} = BC$ with $C_k = B_k^n$ and $C = B^n$. Since $\|B_k - B\| \to 0$ and $\|C_k - C\| \to 0$ as $k \to \infty$, $\|B_k^{n+1} - B^{n+1}\| \to 0$ as $k \to \infty$ by Lemma 7.10. □

COROLLARY 7.13. *Let X be a normed vector space and $X \supseteq X_B$ a wedge and $B : X_B \to X_B$ be bounded, homogeneous, and additive. Let (B_k) be a sequence of bounded homogeneous maps on X_B such that and $\|B_k - B\| \to 0$ as $k \to \infty$. Then*

$$\limsup_{k\to\infty} \mathbf{r}(B_k) \le \mathbf{r}(B).$$

PROOF. This follows from Theorem 7.12 and Remark 7.11. □

Corollary 7.13 will be applied in Section 13.4.2, (13.48).

7.2. More upper semicontinuity results

We explore whether the conditions (a) or (b) of Theorem 7.1 can be replaced by order boundedness assumptions (Chapter 6).

7.2.1. Upper semicontinuity for operators with uniformly order-bounded powers. The next result is similar to [**199**, Thm.11.2] and [**201**, Thm.8.2]. Recall the concepts of pointwise and uniform order-boundedness (Chapter 6) and the definition of the upper Collatz–Wielandt spectral radius, (6.2).

THEOREM 7.14. *Let X be an ordered normed vector space with closed order cone X_+ and $B : X_B \to X_B$ be a bounded homogeneous operator on a homogeneous subset X_B of X_+, $u \in \dot{X}_B$.*

Let (B_k) be a sequence of bounded homogeneous maps $B_k : X_B \to X_B$ such that $\|B_k^n(u) - B^n(u)\| \to 0$ as $k \to \infty$ for all $n \in \mathbb{N}$.

Assume that each B_k is order-preserving and that there exist $m \in \mathbb{N}$ and $c \ge 0$ such that $B_k^m(x) \le c\|x\|u$ for all $k \in \mathbb{N}$ and all $x \in X_B$.

Then

$$\limsup_{k\to\infty} \eta^u(B_k) \le \gamma_B(u) \le \mathbf{r}(B).$$

If u is a normal point for X_B, then

$$\limsup_{k\to\infty} \mathbf{r}(B_k) \le \gamma_B(u) \le \mathbf{r}(B),$$

and, if in addition a power of B is uniformly u-bounded, $\eta^u(B) = \mathbf{r}(B)$.

PROOF. Choose $m \in \mathbb{N}$ and $c \ge 0$ as in the statement of the theorem. Then

$$B_k^{n+m}(u) \le c\|B_k^n(u)\|u, \qquad k, n \in \mathbb{N}. \tag{7.4}$$

Let $r > \gamma_B(u)$. By (5.2), there exists some $N \in \mathbb{N}$ such that

$$\|B^n(u)\| < r^n, \qquad n > N.$$

Let $n \in \mathbb{N}$, $n > N$. Since $B_k^n(u) \overset{k\to\infty}{\longrightarrow} B^n(u)$ by assumption, there exists some $k_n \in \mathbb{N}$ such that

$$\|B_k^n(u)\| < r^n, \qquad k \ge k_n.$$

We combine this inequality with inequality (7.4),

$$\|B_k^{n+m}\|_u \le cr^n, \qquad k \ge k_n.$$

Then

$$\|B_k^{n+m}\|_u^{1/(n+m)} \le c^{1/(n+m)} r^{n/(n+m)}, \qquad k \ge k_n.$$

So, for any $n > N$, since $\eta^u(B_k) = \inf_{n\in\mathbb{N}} \|B_k^n\|_u^{1/n}$ by (6.2),

$$\limsup_{k\to\infty} \eta^u(B_k) \le c^{1/(n+m)} r^{n/(n+m)}.$$

We take the limit as $n \to \infty$ and obtain

$$\limsup_{k\to\infty} \eta^u(B_k) \le r.$$

Since this holds for any $r > \gamma_B(u)$, the first assertion follows.

If u is a normal point for X_B, then u is a normal point for each B_k (Remark 6.8) and $\mathbf{r}(B_k) \le \eta^u(B_k)$ by Theorem 6.12. The same argument applies to B if some power of B is uniformly u-bounded. □

Recall that a sequence (x_k) in $X_B \subseteq X_+$ is called u-bounded, $u \in X_B$, if there is some $c > 0$ such that $x_k \le cu$ for all $k \in \mathbb{N}$. See (2.30).

LEMMA 7.15. *Let $B : X_B \to X_B$ be a bounded and homogeneous operator on a homogeneous subset X_B of X_+, $u \in \dot{X}_B$, and $B(u)$ be u-bounded. Assume that, for any u-bounded sequence (x_k) in X_B and any $x \in X_B$ with $\|x_k - x\| \to 0$, also $\|B(x_k) - B(x)\| \to 0$.*

Let (B_k) be a sequence of bounded homogeneous operators $B_k : X_B \to X_B$ such that $\|B_k - B\| \to 0$ as $k \to \infty$. Assume that there is some $c > 0$ such that $B_k(u) \le cu$ for all $k \in \mathbb{N}$.

Then, for all $n \in \mathbb{N}$, $B_k^n(u) \to B^n(u)$ as $k \to \infty$.

PROOF. We perform an induction proof over n.

By assumption, the statement is true for $n = 1$. Let $n \in \mathbb{N}$ and the statement be true for n. For $k \in \mathbb{N}$,

$$\begin{aligned}
&\|B_k^{n+1}(u) - B^{n+1}(u)\| \\
&\quad\le \|B_k(B_k^n(u)) - B(B_k^n(u))\| + \|B(B_k^n(u)) - B(B^n(u))\| \\
&\quad\le \|B_k - B\|\,\|B_k^n(u)\| + \|B(B_k^n(u)) - B(B^n(u))\| \overset{k\to\infty}{\longrightarrow} 0.
\end{aligned}$$

Indeed, since $B_k(u) \le cu$ for all $k \in \mathbb{N}$, $B_k^n(u) \le c^n u$ for all $k \in \mathbb{N}$ and $(B_k^n(u))_{k\in\mathbb{N}}$ is a u-bounded sequence. By our assumptions, the right-hand side of our inequality converges to 0 as $k \to \infty$. □

We combine Theorem 7.14 and Lemma 7.15.

COROLLARY 7.16. *Let X be an ordered normed vector space with closed order cone X_+.*

Let $B : X_B \to X_B$ be a bounded homogeneous operator on a homogeneous subset X_B of X_+, $u \in \dot{X}_B$, and $B(u)$ be u-bounded. Assume that, for any u-bounded sequence (x_k) in X_B and any $x \in X_B \cap X_u$,

$$\|x_k - x\| \overset{k\to\infty}{\longrightarrow} 0 \implies \|B(x_k) - B(x)\| \overset{k\to\infty}{\longrightarrow} 0.$$

Let (B_k) be a sequence of bounded homogeneous operators $B_k : X_B \to X_B$ such that $\|B_k - B\| \to 0$ as $k \to \infty$. Assume that there is some $c > 0$ such that $B_k(u) \le cu$ for all $k \in \mathbb{N}$.

Assume that each B_k is order-preserving and that there exist $m \in \mathbb{N}$ and $c \ge 0$ such that $B_k^m(x) \le c\|x\|u$ for all $k \in \mathbb{N}$ and all $x \in X_+$.

Then

$$\limsup_{k\to\infty} \eta^u(B_k) \le \gamma_B(u) \le \mathbf{r}(B).$$

If u is a normal point for X_B, then

$$\limsup_{k\to\infty} \mathbf{r}(B_k) \le \gamma_B(u) \le \mathbf{r}(B),$$

and, if in addition a power of B is uniformly u-bounded, $\eta^u(B) = \mathbf{r}(B)$.

Compare Corollary 7.16 to Theorem 7.1.

7.2.2. Upper semicontinuity of upper Collatz–Wielandt bounds. Let X be an ordered normed vector space with closed order cone X_+, X_B be a homogeneous subset of X_+, and $u \in \dot{X}_B$.

Let $B : X_B \to X_B$ be homogeneous and order-preserving. and let (B_k) be a sequence of homogeneous order-preserving operators $B_k : X_B \to X_B$.

THEOREM 7.17. *Let $B(u)$ be u-bounded and $B_k(u)$ be u-bounded for all $k \in \mathbb{N}$. Assume that, for all u-comparable $x \in X_B$, $\|B_k(x) - B(x)\|_u \to 0$ as $k \to \infty$. Then*

$$\limsup_{k\to\infty} CW_u(B_k) \le CW_u(B).$$

Observe that if $B(u)$ and $B_k(u)$ are in X_u, then $B(x)$ and $B_k(x)$ are in X_u for each u-comparable $x \in X_B$ and so $B_k(x) - B(x) \in X_u$. See Definition 2.59 and (2.28).

PROOF. Let $r > CW_u(B)$. By (6.6), there exists some u-comparable $w \in X_B$ such that $B(w) \le rw$. By assumption,

$$\|B_k(w) - B(w)\|_w \to 0. \tag{7.5}$$

By (2.32),

$$B_k(w) \le B(w) + \|B_k(w) - B(w)\|_w\, w \le (r + \|B_k(w) - B(w)\|_w)w.$$

By (2.26), $\|B_k\|_w \le r + \|B_k(w) - B(w)\|_w$ for all $k \in \mathbb{N}$. By (6.7),

$$CW_u(B_k) \le \|B_k\|_w \le r + \|B_k(w) - B(w)\|_w, \qquad k \in \mathbb{N}, \quad w \sim u.$$

By (7.5),

$$\limsup_{k\to\infty} CW_u(B_k) \le r.$$

Since this holds for any $r > CW_u(B)$, the assertion follows. □

REMARK 7.18. If also all B_k are uniformly u-bounded, we also get upper semicontinuity results for the spectral radius and the upper Collatz–Wielandt spectral radius:

$$\limsup_{k\to\infty} \mathbf{r}(B_k) \le \mathbf{r}(B), \qquad \limsup_{k\to\infty} \eta^u(B_k) \le \eta^u(B).$$

7.2.3. An application of upper semicontinuity of spectral radii and upper CW bounds yielding their equality. Let X be an ordered normed vector space with closed order cone X_+ and X_B a homogeneous subset of X_+ and $B : X_B \to X_B$ be a uniformly u-bounded homogeneous operator for some $u \in X_B$. We use a perturbation procedure and the upper semicontinuity of spectral radii and upper Collatz–Wieland bounds to find sufficient conditions for the equalities

$$\eta^u(B) = \gamma_u(B) = \mathbf{r}(B) = CW_u(B).$$

THEOREM 7.19. *Let $B : X_B \to X_B$ be an order-preserving bounded homogeneous operator on a homogeneous subset X_B of X_+, $u \in \dot{X}_B$, and B be uniformly u-bounded. Assume that B is continuous in the following sense:*

For any u-bounded sequence (x_k) in X_B and any $x \in X_B \cap X_u$,

$$\|x_k - x\| \overset{k\to\infty}{\longrightarrow} 0 \implies \|B(x_k) - B(x)\| \overset{k\to\infty}{\longrightarrow} 0. \tag{7.6}$$

Let $\phi : X_B \to \mathbb{R}_+$ be an order-preserving bounded homogeneous functional, $\phi(u) > 0$. (For instance, let ϕ be the monotone companion half-norm, Section 2.5.) For $\epsilon \in (0,1]$ define $B_\epsilon : X_B \to X_B$ by

$$B_\epsilon(x) = B(x) + \epsilon\phi(x)u, \qquad x \in X_B.$$

(a) *Then $CW_u(B) \le CW_u(B_\epsilon) \to CW_u(B)$ as $\epsilon \to 0+$.*

(b) *If u is a normal point for X_B and B is bounded, then*

$$\eta^u(B_\epsilon) = \gamma_u(B_\epsilon) = \mathbf{r}(B_\epsilon) \longrightarrow \eta^u(B) = \gamma_u(B) = \mathbf{r}(B), \qquad \epsilon \to 0+.$$

(c) *If X_B is an inf-semilattice, (2.7), and B is bounded and u a normal point for X_B, then $\eta^u(B) = \gamma_u(B) = \mathbf{r}(B) = CW_u(B)$.*

PROOF. For all $\epsilon \in (0,1]$, B_ϵ is an order-preserving homogeneous operator. For $\epsilon > 0$,

$$0 \le B_\epsilon(x) - B(x) = \epsilon\phi(x)u, \qquad x \in X_B.$$

Since $\phi(u) > 0$, $B_\epsilon(u)$ is u-comparable for all $\epsilon > 0$. Further,

$$\|B_\epsilon(x) - B(x)\|_u \le \epsilon\phi(x), \qquad x \in X_B.$$

and

$$\|B_\epsilon - B\| \le \epsilon\|\phi\|\,\|u\| \to 0, \qquad \epsilon \to 0.$$

By Lemma 7.15 and (7.6), $\|B_\epsilon^n(u) - B^n(u)\| \to 0$ as $\epsilon \to 0$ for all $n \in \mathbb{N}$.

Finally, for all $\epsilon \in (0,1]$

$$B_\epsilon(x) \le B(x) + \|\phi\|\|x\|u, \qquad x \in X_B.$$

Since B is uniformly u-bounded, there exists some $c > 0$ such that $B_\epsilon(x) \le c\|x\|u$ for all $\epsilon \in (0,1]$, $x \in X_B$.

(a) By Theorem 7.17, $\limsup_{\epsilon\to 0+} CW_u(B_\epsilon) \le CW_u(B)$. Since $B(x) \le B_\epsilon(x)$ for all $x \in X_E$ and B is order-preserving on X_B, $CW_u(B) \le CW_u(B_\epsilon)$ for all $\epsilon > 0$ by Corollary 6.25. This implies $CW_u(B_\epsilon) \to CW_u(B)$ as $\epsilon \to 0+$.

(b) Assume that u is a normal point for X_B. For each $\epsilon \in (0,1]$, B_ϵ is uniformly u-bounded. In fact, there is some $c > 0$ such that $B_\epsilon(x) \le c\|x\|u$ for all $\epsilon \in (0,1]$ and $x \in X_B$. By Theorem 6.14 and 6.15, for all $\epsilon \in (0,1]$,

$$\begin{aligned} \eta^u(B_\epsilon) &= \gamma_u(B_\epsilon) = r(B_\epsilon) \le CW_u(B_\epsilon), \\ \eta^u(B) &= \gamma_u(B) = \mathbf{r}(B). \end{aligned}$$

By Theorem 7.14,

$$\limsup_{\epsilon\to 0} \eta^u(B_\epsilon) \le \gamma_B(u) \le \mathbf{r}(B).$$

Since u is a normal point for X_B, then

$$\limsup_{\epsilon\to 0} \mathbf{r}(B_\epsilon) \le \gamma_B(u) \le \mathbf{r}(B).$$

By Corollary 6.25,

$$\eta^u(B) \le \eta^u(B_\epsilon), \qquad \epsilon \ge 0.$$

In combination,

$$\eta^u(B_\epsilon) = \gamma_u(B_\epsilon) = r(B_\epsilon) \longrightarrow \eta^u(B) = \gamma_u(B) = r(B), \qquad \epsilon \to 0+.$$

(c) Let B be uniformly u-bounded and u be a normal point for X_B. By Theorem 6.20, for all $\epsilon \in (0,1]$,

$$\eta^u(B_\epsilon) = \gamma_u(B_\epsilon) = r(B_\epsilon) = CW_u(B_\epsilon).$$

We take the limit as $\epsilon \to 0$ and use part (a) and (b) and the assertion follows. $\square$

CHAPTER 8

A left resolvent for homogeneous operators

There is one remnant from the usual relations between the spectral radius and the spectrum of a bounded linear operator that also holds in the homogeneous case, namely that real numbers larger than the spectral radius are in the resolvent set. However, in the homogeneous case, there only exists a left resolvent.

Let X be an ordered normed vector space with closed order cone X_+ and $B : X_B \to X_B$ be a homogeneous, bounded order-preserving operator on a serially complete cone X_B contained in X_+.

For $\lambda > \mathbf{r}_o(B)$, we introduce $R_\lambda : X_B \to X_B$,

$$R_\lambda(x) = \sum_{n=0}^{\infty} \lambda^{-n-1} B^n(x), \qquad x \in X_B. \tag{8.1}$$

The convergence of the series follows from the serial completeness of the cone X_B. See Section 2.1.2. If $\lambda > \mathbf{r}(B)$, the convergence of the series is uniform for x in bounded subsets of X_B. With this in mind, the following is easily shown.

LEMMA 8.1. *For $\lambda > \mathbf{r}_o(B)$, R_λ is defined, homogeneous, and order preserving. If $\lambda > \mathbf{r}(B)$, R_λ is bounded and, if B is continuous, R_λ is continuous.*

In order to explore to what degree R_λ acts as a resolvent to B, we introduce the following concepts.

PROPOSITION 8.2. *R_λ somewhat acts as a left resolvent,*

$$R_\lambda(B(x)) = \lambda R_\lambda(x) - x = R_\lambda(\lambda x) - x, \qquad x \in X_B, \quad \lambda > \mathbf{r}_o(B). \tag{8.2}$$

More generally, for all $m \in \mathbb{N}$ and $x \in X_B$,

$$R_\lambda(B^m(x)) = \lambda^m \Big(R_\lambda(x) - \sum_{n=0}^{m-1} \lambda^{-n-1} B^n(x) \Big). \tag{8.3}$$

If B is superadditive and continuous, $x \ge (\lambda - B)R_\lambda(x)$ for all $x \in X_B$. If B is subadditive and continuous, $x \le (\lambda - B)R_\lambda(x)$ for all $x \in X_B$.

In view of these results, we call R_λ, $\lambda > \mathbf{r}(B)$, the *(left) resolvents* of B.

PROOF. For $m \in \mathbb{N}$, $x \in X_B$, $\lambda > \mathbf{r}_o(B)$,

$$\begin{aligned} R_\lambda(B^m(x)) &= \sum_{n=0}^{\infty} \lambda^{-n-1} B^{n+m}(x) = \sum_{n=m}^{\infty} \lambda^{-n-1+m} B^n(x) \\ &= \lambda^m \Big(R_\lambda(x) - \sum_{n=0}^{m-1} \lambda^{-n-1} B^n(x) \Big). \end{aligned}$$

□

If B is superadditive and continuous, then

$$BR_\lambda(x) \geq \sum_{n=0}^{\infty} \lambda^{-n-1} B^{n+1}(x) = \sum_{n=1}^{\infty} \lambda^{-n} B^n(x) = \lambda R_\lambda(x) - x.$$

The subadditive case is proved mutandis mutatis.

PROPOSITION 8.3. *Let B be continuous and $\{p_j; j \in J\}$ be a family of homogeneous, convex, order-preserving, continuous functionals $p_j : X_B \to \mathbb{R}_+$ such that $\sup_{j\in J} \|p_j\| < \infty$. Then, for each $t > \mathbf{r}_o(B)$,*

$$\sup_{j\in J,\, \lambda \geq t} \|p_j \circ R_\lambda\| < \infty.$$

PROOF. Let $t > \mathbf{r}_o(B)$. Choose some $s \in (\mathbf{r}_o(B), t)$. For each $x \in X_B$, there exists some $c_x > 0$ such that $s^{-n}(p_j \circ B^n)(x) \leq c_x$ for all $n \in \mathbb{Z}_+$ and $j \in J$. Since $s^{-n}(p_j \circ B^n) : X_B \to \mathbb{R}_+$ are homogeneous, order-preserving and continuous on the serially complete cone X_B, by the uniform boundedness principle in Theorem 4.8, there exists some $c > 0$ such that $\|s^{-n}(p_j \circ B^n)\| \leq c$ for all $n \in \mathbb{Z}_+$ and $j \in J$. For each $x \in X_B$ with $\|x\| \leq 1$ and each $\lambda \geq t$ and each $n \in \mathbb{Z}_+, j \in J$, since the p_j are continuous, homogeneous and subadditive,

$$(p_j \circ R_\lambda)(x) \leq \sum_{n=0}^{\infty} \lambda^{-(n+1)} p_j(B^n(x)) \leq \sum_{n=0}^{\infty} t^{-(n+1)} c s^n = \frac{c}{t} \frac{1}{1-(s/t)}.$$

Thus

$$\|p_j \circ R_\lambda\| \leq \frac{c}{t-s}, \qquad \lambda \geq t, j \in J. \qquad \square$$

This result should be compared with the next one. Recall that $\mathbf{r}(B) = \mathbf{r}_o(B)$ if X_B is serially complete and X_+ is normal (Theorem 5.9).

THEOREM 8.4. *Let X_+ be a normal cone. Let B be homogeneous, bounded and superadditive. Assume that $\mathbf{r} = \mathbf{r}(B) > 0$. Then $\{\|R_\lambda\|; \lambda > \mathbf{r}(B)\}$ is unbounded.*

The proof is adapted from [**27**] and allows B to be concave rather than additive [**197**].

PROOF. R_λ inherits superadditivity from B,

$$R_\lambda(x+y) \geq R_\lambda(x) + R_\lambda(y), \qquad x, y \in X_B. \tag{8.4}$$

Suppose that the assertion is false. Then there exists some $M \geq 0$ such that $\|R_\lambda\| \leq M$ for all $\lambda > \mathbf{r}$. Let $0 < \mu < \mathbf{r} < \lambda$ and $(\lambda - \mu)M < 1$. Since X_B is serially complete.

$$E_\mu(x) = \sum_{k=1}^{\infty} (\lambda-\mu)^{k-1} R_\lambda^k(x)$$

converges for each $x \in X_B$. By (8.2),

$$E_\mu(\lambda x) = \sum_{k=1}^{\infty} (\lambda-\mu)^{k-1} R_\lambda^k(\lambda x) = \sum_{k=1}^{\infty} (\lambda-\mu)^{k-1} R_\lambda^{k-1}\big(R_\lambda(Bx) + x\big).$$

Since R_λ is superadditive,

$$\begin{aligned} E_\mu(\lambda x) &\geq \sum_{k=1}^{\infty} (\lambda-\mu)^{k-1} R_\lambda^k(Bx) + \sum_{k=1}^{\infty} (\lambda-\mu)^{k-1} R_\lambda^{k-1}(x) \\ &= E_\mu(Bx) + x + (\lambda-\mu) E_\mu(x). \end{aligned}$$

Since E_μ is homogeneous,

$$E_\mu(x) \geq \frac{1}{\mu}x + \frac{1}{\mu}E_\mu(B(x)), \qquad x \in X_B.$$

By iteration and induction,

$$E_\mu(x) \geq \sum_{k=0}^{n} \frac{1}{\mu^{k+1}} B^k(x) + \frac{1}{\mu^{n+1}} E_\mu B^{n+1}(x), \qquad x \in X_B, \quad n \in \mathbb{N}.$$

This implies that

$$E_\mu(x) \geq \frac{1}{\mu^{k+1}} B^k(x), \qquad x \in X_B, \quad k \in \mathbb{N}.$$

Since X_+ is normal, there exists some $\tilde{M} > 0$ (which depends on μ) such that

$$\|B^k(x)\| \leq \tilde{M}\mu^k \|x\|, \qquad k \in \mathbb{N}, \quad x \in X_B.$$

Thus

$$\|B^k\| \leq \tilde{M}\mu^k, \qquad k \in \mathbb{N}.$$

Since $\mu < \mathbf{r} = \mathbf{r}(B)$, this is a contradiction. □

In the next results, X_+^* denotes the wedge of bounded linear positive functionals on X.

COROLLARY 8.5. *Let X be an ordered normed vector space with a normal closed cone X_+. Let $B : X_B \to X_B$ be homogeneous, order-preserving, continuous and $B(x+y) \geq B(x) + B(y)$ for all $x, y \in X_B$. Assume that $\mathbf{r} = \mathbf{r}(B) > 0$. Then there exist some $x \in X_B$ and $x^* \in X_+^*$ such that $x^*(R_\lambda(x)) \to \infty$ as $\lambda \to \mathbf{r}+$.*

In order to reconcile this result with Proposition 8.3, recall that $\mathbf{r}_o(B) = \mathbf{r}(B)$ if X_+ is normal and X_B is serially complete (Theorem 5.9).

PROOF. By the uniform boundedness principle in Theorem 4.8, there exists some $x \in X_+$ such that $\{\|R_\lambda(x)\|; \lambda > \mathbf{r}(B)\}$ is unbounded. By the usual uniform boundedness principle, applied to the Banach space X^*, there exists some $x^* \in X^*$ such that $\{|x^* R_\lambda(x)|; \lambda > \mathbf{r}(B)\}$ is unbounded. Since X_+ is normal, the wedge X_+^* of positive bounded linear functionals is generating (Theorem 4.19) and non-flat (Theorem 2.38), and the unboundedness holds for some $x^* \in X_+^*$. Since $x^* R_\lambda(x)$ is a decreasing function of λ, $x^* R_\lambda(x) \to \infty$ as $\lambda \to \mathbf{r}(B)$. □

Recall the lower Collatz–Wielandt bound of B,

$$cw(B) = \sup\{[B]_x; x \in \dot{X}_B\} = \sup\{\lambda \geq 0;\ \exists x \in \dot{X}_B : B(x) \geq \lambda x\}. \tag{8.5}$$

Notice that $cw(B)$ is defined if X is just an ordered vector space. If it is a normed ordered vector space, the lower Collatz–Wielandt bound relates to the orbital spectral radius (Theorem 5.31) as

$$cw(B) \leq \mathbf{r}_o(B). \tag{8.6}$$

PROPOSITION 8.6. *Let $\mu \in (0,\infty)$, $x \in \dot{X}_B$ and $B(x) \geq \mu x$. Then there exists some $x_\mu^* \in X_+^*$ and $\delta_\mu \in (0,\infty)$ with $\|x_\mu^*\| = 1$ and*

$$\|x_\mu^* \circ R_\lambda\| \geq \frac{\delta_\mu}{\lambda - \mu}, \qquad \lambda > \mathbf{r}_o(B).$$

If $X_+ = X_B$ or X_+ is normal, δ_μ can be chosen independently of μ.

PROOF. Let $B(x) \geq \mu x$, $\mu \in (0, \infty)$, $x = x_\mu \in \dot{X}_B$.

For each $\lambda > \mathbf{r}_o(B)$, by Lemma 8.2, since R_λ is homogeneous and order-preserving,

$$\lambda R_\lambda(x) = R_\lambda(B(x)) + x \geq R_\lambda(\mu x) + x = \mu R_\lambda(x) + x.$$

We reorganize the inequality,

$$R_\lambda(x) \geq \frac{1}{\lambda - \mu} x. \tag{8.7}$$

Let $\psi(x) = d(x, -X_+)$, $x \in X$, be the monotone companion half-norm on X (Section 2.5).

By the Hahn–Banach theorem, there exists some $x_\mu^* \in X^*$ such that $x_\mu^* x = \psi(x)$ and $-\psi(-y) \leq x_\mu^* y \leq \psi(y)$ for all $x \in X_+$ [**223**, IV.6]. Since $\psi(-X_+) = \{0\}$, $x_\mu^* \in X_+^*$. Let $\sharp x \sharp = \max\{\psi(x), \psi(-x)\}$ be the monotone companion norm on X. Then $\sharp x_\mu^* \sharp = 1$. Since positive bounded linear functionals have the same operator norm with respect to the original norm and the monotone companion norm (Proposition 4.13), $\|x_\mu^*\| = 1$.

We apply the positive linear function x_μ^* to (8.7),

$$x_\mu^* R_\lambda(x) \geq \frac{1}{\lambda - \mu} \psi(x).$$

This implies that $\sharp x_\mu^* R_\lambda \sharp_+ \geq \frac{1}{\lambda - \mu}$. Since $\sharp x_\mu^* R_\lambda \sharp = \|x_\mu^* R_\lambda\|$ if $X_+ = X_B$ (Proposition 4.13), this implies the statement for this case. Otherwise, we obtain

$$\|x_\mu^* R_\lambda\| \, \|x\| \geq \frac{1}{\lambda - \mu} \psi(x)$$

and set $\delta_\mu = \psi(x)/\|x\|$, $x = x_\mu$. If X_+ is normal, the original norm and the monotone companion norm are equivalent, and δ can be chosen independently of x. □

THEOREM 8.7. *Let $B : X_B \to X_B$ be homogeneous, continuous, and superadditive. Then R_λ satisfies the resolvent inequality*

$$(\mu - \lambda) R_\lambda(R_\mu(x)) \geq R_\lambda(x) - R_\mu(x) \geq 0, \qquad x \in X_B, \mu > \lambda > \mathbf{r}(B).$$

PROOF. Let $x \in X_B$ and $\mu > \lambda > \mathbf{r}(B)$. Since B is continuous, homogeneous, and super-additive, so is R_λ and, by (8.1),

$$R_\lambda(R_\mu(x)) \geq \sum_{m=0}^{\infty} \mu^{-m-1} R_\lambda(B^m(x)).$$

By (8.3),

$$R_\lambda(R_\mu(x)) \geq \sum_{m=0}^{\infty} \mu^{-m-1} \lambda^m \Big(R_\lambda(x) - \sum_{n=0}^{m-1} \lambda^{n-1} B^n(x) \Big).$$

Since $\lambda < \mu$, we use the geometric series,

$$R_\lambda(R_\mu(x)) \geq \frac{1}{\mu - \lambda} R_\lambda(x) - \sum_{m=0}^{\infty} \mu^{-m-1} \lambda^m \sum_{n=0}^{m-1} \lambda^{-n-1} B^n(x).$$

We change the order of summation,

$$R_\lambda(R_\mu(x)) \geq \frac{1}{\mu-\lambda} R_\lambda(x) - \sum_{n=0}^{\infty} \Big(\sum_{m=n+1}^{\infty} (\lambda/\mu)^m \mu^{-1} \Big) \lambda^{-n-1} B^n(x).$$

We use the geometric series again,

$$\begin{aligned} R_\lambda(R_\mu(x)) \geq & \frac{1}{\mu-\lambda} R_\lambda(x) - \sum_{n=0}^{\infty} \frac{(\lambda/\mu)^{n+1}}{\mu-\lambda} \lambda^{-n-1} B^n(x) \\ = & \frac{1}{\mu-\lambda} (R_\lambda(x) - R_\mu(x)) \geq 0. \end{aligned}$$

□

If B is additive, the same arguments provide an equality. Further $R_\lambda B = B R_\lambda$ and $R_\lambda R_\mu = R_\mu R_\lambda$. So we obtain the familiar resolvent identity.

COROLLARY 8.8. *Let $B : X_B \to X_B$ be homogeneous, continuous and additive. Then R_λ satisfies the resolvent identity,*

$$(\mu - \lambda) R_\lambda R_\mu = R_\lambda - R_\mu, \qquad \lambda, \mu > \mathbf{r}(B).$$

LEMMA 8.9. *Let $B : X_B \to X_B$ be homogeneous and order-preserving, $u \in \dot{X}_B$, $B(u)$ be u-bounded and some power of B be uniformly u-bounded. Then $R_\lambda(u)$ is u-comparable for all $\lambda > \gamma_u(B)$.*

PROOF. By Theorem 6.14, $\eta^u(B) \leq \gamma_u(B)$. For all $m \in \mathbb{N}$, $\lambda > \gamma_u(B)$,

$$\sum_{n=0}^{m} \lambda^{-n-1} B^n(u) \leq \sum_{n=0}^{m} \lambda^{-n-1} \|B^n\|_u u.$$

Since there are $N \in \mathbb{N}$ and $\xi < 1$ such that $\lambda^{-1} \|B^n\|_u^{1/n} < \xi < 1$ for all $n \geq N$, $\sum_{n=0}^{\infty} \lambda^{-n-1} \| B^n\|_u$ converges in $\mathbb{R}$ and

$$\sum_{n=0}^{m} \lambda^{-n-1} B^n(u) \leq \sum_{n=0}^{\infty} \lambda^{-n-1} \|B^n\|_u u.$$

Since X_+ is closed, $R_\lambda(u) \leq \sum_{n=0}^{\infty} \lambda^{-n-1} \|B^n\|_u u$ and $R_\lambda(u)$ is u-bounded. By (8.1), $R_\lambda(u) \geq (1/\lambda) u$ and $R_\lambda(u)$ is u-positive. □

CHAPTER 9

Eigenvectors of (pseudo-)compact homogeneous operators

In the following, let X be an ordered normed vector space with closed order cone X_+.

Let $B : X_B \to X_B$ be a homogeneous, bounded and order-preserving operator on a closed cone X_B contained in X_+.

Recall the lower Collatz–Wielandt bound, $cw(B)$, lower Collatz–Wielandt radius, $\mathbf{r}_{cw}$, orbital spectral radius $\mathbf{r}_o(B)$, the spectral radius, $\mathbf{r}(B)$, and the inequality

$$cw(B) \le \mathbf{r}_{cw}(B) \le \mathbf{r}_o(B) \le \mathbf{r}(B). \tag{9.1}$$

See Theorem 5.31 and (5.4).

The simplest example of a homogeneous operator is of one-dimensional range,

$$B(x) = \phi(x)u, \qquad x \in X_B,$$

with some $u \in \dot{X}_B$ and a homogeneous bounded functional $\phi : X_B \to \mathbb{R}_+$. Notice that B is a compact operator (Appendix A.1.8).

As noted in Example 5.1, u is an eigenvector of B associated with the eigenvalue $\phi(u) = \mathbf{r}(B) = \gamma_u(B)$. Further B is uniformly u-bounded and $\eta^u(B) = \phi(u)$.

If ϕ is order-preserving, $cw(B) = \phi(u)$ and also $CW_u(B) = \phi(u)$ provided that $\phi(u) > 0$.

If $X_B = X_+$ and B is compact and continuous (Appendix A.1.7 and A.1.8), the following progressively more general results have been proved over the years.

THEOREM 9.1. *Let B be compact and continuous and r any of the numbers $cw(B), \mathbf{r}_{cw}(B), \mathbf{r}(B)$. Then, if $r > 0$, there exists some $v \in X_+ = X_B$, $v \ne 0$, with $B(v) = rv$. More specifically,*

$r = cw(B)$: Krein Rutman 1948 [**140**],

$r = \mathbf{r}_{cw}(B)$: Krasnosel'skii 1964 [**135**, Thm.2.5],

$r = \mathbf{r}(B)$: Nussbaum 1981 [**167**] *Lemmens Nussbaum* [**146**].

A proof for $r = cw(B)$ can also been found in [**28**]. [**135**, Thm.2.5] is only formulated for the case that B is defined and linear on X, but the proof also works under the assumptions made above. The case $r = \mathbf{r}_{cw}(B)$ can also been found in [**167**, Cor.2.1]. The case $r = \mathbf{r}(B)$ is basically proved in [**167**, Thm.2.1], but some finishing touches are contained in the introduction of [**146**]. See also [**15**, Chap.10]. We will give a proof of the case $r = \mathbf{r}(B)$ that does not require X or X_+ to be complete as it may be the case for the space of real Borel measures on a metric space with the flat norm.

It is well-known that the compactness of B can be substantially relaxed though not completely dropped if B is linear [**167**]. This is also possible (to a lesser degree) if B is just homogeneous using homogeneous measures of noncompactness

[**156**, **157**]. We only mention two special cases of [**156**, Thm.3.1], [**157**, Thm.4.9] [**157**, Thm.4.10], respectively. Typically, some completeness assumptions are made (Definition A.27).

THEOREM 9.2. *Let X be complete. Let $B : X_+ \to X_+$ and $B = K + H$ where $K : X_+ \to X_+$ is homogeneous, continuous, order preserving and compact and $H : X_B \to X_B$ is homogeneous, continuous, order preserving and Lipschitz continuous:*
H is Lipschitz continuous on X_+, $\|H(x) - H(y)\| \le \Lambda \|x - y\|$ for all $x, y \in X_+$, with $0 \le \Lambda < \mathbf{r}(B)$.
Then there exists some $v \in \dot{X}_+$ with $B(v) = \mathbf{r}(B)v$

Theorem 9.2 will be used in the proof of Theorem 18.21.

THEOREM 9.3. *Let X_+ be complete and normal. Let $B : X_B \to X_B$ and $B = K + H$ where $K : X_B \to X_B$ is homogeneous, continuous, order preserving and compact and $H : X_B \to X_B$ is homogeneous, continuous, order preserving, additive:*
$(H(x + y) = H(x) + H(y)$ for all $x, y \in X_B)$,
and $\mathbf{r}(H) < \mathbf{r}(B)$.
Then there exists some $v \in \dot{X}_+$ with $B(v) = \mathbf{r}(B)v$.

This theorem will be used in the proof of Corollary 9.5.
The following observation is worth mentioning [**136**, Thm.9.3] [**170**, Thm.2.2].

PROPOSITION 9.4. *Let $B : X_B \to X_B$ be additive, i.e., $B(x+y) = B(x)+B(y)$ for all $x, y \in X_B$. If $r > 0$, $w \in \dot{X}_B$, $m \in \mathbb{N}$ and $B^m(w) = r^m w$, then $B(v) = rv$ for some $v \in \dot{X}_B$.*

Simply set

$$v = \sum_{j=0}^{m-1} r^{-j} B^j(w). \tag{9.2}$$

It seems to be an open problem (with a beer barrel scent [**158**, **212**]) whether any of the results in Theorem 9.1 holds if B^2 or some higher power of B is compact rather than B itself. For additional conditions to make such a result hold, see [**8**, Thm.7.3], conditions (i) and (iii), [**197**, Sec.7] [**201**], and Section 9.2.

COROLLARY 9.5. *Let X_+ be complete and normal and $B : X_B \to X_B$. Assume that $B = A_1 + A_2$ where the A_i are homogeneous, continuous and additive on X_B for $i = 1, 2$. Assume that A_2^2 and $A_1 A_2$ are compact and $\mathbf{r}(B) > \mathbf{r}(A_1)$. Then there exists some $v \in \dot{X}_B$ with $Bv = \mathbf{r}(B)v$.*

This result will be used in the proof of Theorem 13.52 to establish the existence of an eigenfunction to a certain Feller kernel.

PROOF. By induction, for any $n \in \mathbb{N}$, $n \ge 2$, there exists a compact, continuous, additive, homogeneous $K_n : X_B \to X_B$ such that

$$B^n = A_1^n + A_2 A_1^{n-1} + K_n.$$

Indeed,

$$B^2 = A_1^2 + A_2 A_1 + K_2, \qquad K_2 = A_1 A_2 + A_2^2.$$

Suppose that $n \geq 2$ and the claim holds for n. Since A_1 and A_2 are additive,

$$B^{n+1} = A_1(A_1^n + A_2A_1^{n-1} + K_n) + A_2(A_1^n + A_2A_1^{n-1} + K_n) = A_1^{n+1} + A_2A_1^n + K_{n+1},$$

where

$$K_{n+1} = A_1A_2A_1^{n-1} + A_1K_n + A_2^2A_1^{n-1} + A_2K_n$$

is compact, additive, homogeneous, and continuous by our assumptions.

So, for all $n \in \mathbb{N}$, $n \geq 1$,

$$B^{n+1} = BA_1^n + K_{n+1}$$

and $\mathbf{r}(B^{n+1}) = \mathbf{r}(B)^{n+1}$ by (5.8). As $n \to \infty$,

$$\Big(\frac{\mathbf{r}(BA_1^n)}{\mathbf{r}(B^{n+1})}\Big)^{1/n} \leq \Big(\frac{\|B\|\,\|A_1^n\|}{\mathbf{r}(B)^{n+1}}\Big)^{1/n} = \Big(\frac{\|B\|}{\mathbf{r}(B)}\Big)^{1/n}\frac{\|A_1^n\|^{1/n}}{\mathbf{r}(B)} \longrightarrow \frac{\mathbf{r}(A_1)}{\mathbf{r}(B)} < 1.$$

So we can choose $n \in \mathbb{N}$ so large that $\mathbf{r}(BA_1^n) < \mathbf{r}(B^{n+1})$. The assertion now follows from Theorem 9.3, applied to B^{n+1} rather than B, and Proposition 9.4. □

The construction in (9.2) can be modified to yield the following result.

PROPOSITION 9.6. *Let $B : X_B \to X_B$ be homogeneous and order-preserving. Assume that X_B is a* sup-*semilattice or B is superadditive. If $r > 0$, $w \in \dot{X}_B$, $m \in \mathbb{N}$ and $B^m(w) \geq r^m w$, then $B(v) \geq rv$ for some $v \in \dot{X}_B$.*

If B is superadditive, choose v as in (9.2).

If X_B is a sup-semilattice, choose v as in the proof of [**8**, Thm.5.1],

$$v = \bigvee_{j=0}^{m-1} r^{-j}B^j(w). \tag{9.3}$$

Notice that the number r in Proposition 9.6 satisfies $r \leq [B]_v \leq cw(B)$.

This observation provides conditions for equality to hold in (9.1). In order not to burden our representation with technical language or a list of various special cases and to include possible further developments, we make the following definition.

DEFINITION 9.7. *A homogeneous bounded map $B : X_B \to X_B$ has the* KR property *(the Krein-Rutman property) if there is some $v \in \dot{X}_B$ with $B(v) = \mathbf{r}(B)v$ whenever $\mathbf{r}(B) > 0$.*

B has the lower KR property *if there is some $v \in \dot{X}_B$ with $B(v) \geq \mathbf{r}(B)v$ whenever $\mathbf{r}(B) > 0$.*

If $v \in \dot{X}_B$ with $B(v) \geq \mathbf{r}(B)v$, then $[B]_v \geq \mathbf{r}(B)$ by (4.13). If $B^m(v) \geq \mathbf{r}(B)^m v$, then $\eta_v(B) \geq \mathbf{r}(B)$ by (5.28). So, if B has the lower KR property, $cw(B) \geq \mathbf{r}(B)$ by (5.30), and, if some power of B has the lower KR property, $\mathbf{r}_{cw} \geq \mathbf{r}(B)$ by (5.29). This has the effect that the inequalities in (9.1) and in Theorem 6.14 are turned in equalities. Notice that these inequalities automatically are equalities if $\mathbf{r}(B) = 0$.

THEOREM 9.8. *Let $B : X_B \to X_B$ be a homogeneous, bounded, order-preserving operator on a homogeneous subset X_B of X_+.*

(a) *If B has the lower KR-property, then $cw(B) = \mathbf{r}_{cw}(B) = \mathbf{r}(B)$. If a power of B has the lower KR-property, then $\mathbf{r}_{cw}(B) = \mathbf{r}(B)$.*

Assume in addition that $u \in \dot{X}_B$ and some power of B is uniformly u-bounded.

(b) *If B has the lower KR property, then $cw(B) = \mathbf{r}_{cw}(B) = \eta^u(B) = \gamma_B(u) = \mathbf{r}_o(B) = \mathbf{r}(B)$.*

(c) *If some power of B has the lower KR property, then* $\mathbf{r}_{cw}(B) = \eta^u(B) = \gamma_u(B) = \mathbf{r}_o(B) = \mathbf{r}(B)$.

The maps in Theorem 9.2 and 9.3 and in Corollary 9.5 are examples that satisfy the KR property while power-compact, homogeneous, order-preserving, continuous maps on sup-semilattices and power-compact, homogeneous, superadditive, continuous maps are examples that satisfy the lower KR property (Proposition 9.6). $B : X_+ \to X_+$ is called *power-compact* if B^m is compact for some $m \in \mathbb{N}$.

We emphasize the lower KR property in addition to the KR property because of our interest in population dynamics. If B is the homogeneous first order approximation of a nonlinear map F at 0, the condition $\mathbf{r}(B) < 1$ is often enough to guarantee the local (and some sometime the global) stability of 0 for the dynamical system induced by F whether or not $\mathbf{r}(B)$ is an eigenvalue associated with a positive eigenvector. To prove persistence of the dynamical system, it is very helpful to have a positive lower eigenvector $B(v) \geq rv$ with $r > 1$ [**127, 129, 130**]. See also Theorem 15.4 and Proposition 17.10. So the lower KR property of B (more generally the equality $cw(B) = \mathbf{r}(B)$) guarantees $\mathbf{r}(B)$ to be a sharp threshold parameter that separates local stability of the extinction equilibrium 0 from persistence of the population. It should be mentioned that the KR property only refers to one part of the Krein-Rutman theorem, the existence of an eigenvector associated with the spectral radius. As for the existence of a homogeneous order-preserving eigenfunctional, see Chapter 11.

9.1. Compact operators

Recall that $B : X_B \to X_B$ is bounded, homogeneous and order-preserving on a closed cone X_B in X_+ and the large-time bound of the geometric growth factor for initial value $u \in X_B$,

$$\gamma_B(u) := \limsup_{n\to\infty} \|B^n(u)\|^{1/n}. \tag{9.4}$$

PROPOSITION 9.9. *Let $K : X_B \to X_B$ be compact and $K(x) \geq B(x)$ for all $x \in X_B$, $\|x\| = 1$. Finally let $u \in \dot{X}_B$ and K be continuous on X_B.*

Then there exists some $x \in X_B$, $\|x\| = 1$, and some $\lambda > 0$ such that $K(x)+u = \lambda x$.

If some power of B is compact or X_+ is normal, then $\lambda \geq \gamma_u(B)$.

If K maps X_B into X_u, then $B(u)$ is u-bounded and $\lambda \geq CW_u(B)$.

The idea of perturbing K and using one of the classical fixed point theorems seems as old as the theory of positive operators [**135**, Sec.2.2] [**177**, Sec.3] [**211**, Thm.3.6].

PROOF. Let $u \in \dot{X}_B$. Define

$$K_u(x) = K(x) + u, \qquad x \in X_B.$$

Then $K_u(x) \geq u$ and $\{\|K_u(x)\|; x \in X_B\}$ is bounded away from 0 because X_+ is closed. So we can define

$$\tilde{K}(x) = \frac{K_u(x)}{\|K_u(x)\|}, \qquad x \in X_B.$$

Since K is continuous on X_B, $\tilde{K}$ is continuous on X_B. $\tilde{K}$ maps the set $C = X_B \cap \bar{U}_1$, $\bar{U}_1 = \{x \in X; \|x\| \leq 1\}$, into itself. C is a closed convex subset of X, and $\tilde{K}(C)$ has compact closure because $K : X_B \to X_B$ is compact and so is $\tilde{K}$. By Tychonov's fixed point theorem [**55**, Thm.10.1], $\tilde{K}$ has a fixed point $v \in X_B$, $\|v\| = 1$,

$$K(v) + u = \lambda v, \qquad \lambda = \|K(v) + u\| > 0. \tag{9.5}$$

Since $K(v) \in X_+$, $u \le \lambda v$. By assumption, $B(v) \le K(v) \le \lambda v$. Since B is order preserving and homogeneous, $B^n(v) \le \lambda^n v$ for all $n \in \mathbb{N}$. Hence

$$B^n(u) \le \lambda^n \lambda v, \qquad n \in \mathbb{N}. \tag{9.6}$$

If a power of B is compact, $\gamma_B(u) \le \lambda$ by Lemma 5.26. If X_+ is normal, this follows as well.

Assume that K maps X_B into X_u. Then v is u-comparable and $\|B\|_v \le \lambda$. By (6.6), $CW_u(B) \le \lambda$. □

Recall that the orbital spectral radius is defined by $\mathbf{r}_o(E) = \sup_{x \in X_B} \gamma_B(x)$.

THEOREM 9.10. *Assume that $K : X_B \to X_B$ is compact and $K(x) \ge B(x)$ for all $x \in X$, $\|x\| = 1$, and that $\mathbf{r}_o(B) > 0$. Assume that K is continuous on X_B and that some power of B is compact or that X_+ is normal.*

Then there exists some $x \in X_B$, $\|x\| = 1$, and some $\lambda \ge \mathbf{r}_o(B)$ such that $K(x) = \lambda x$. If $B = K$, then $\lambda = cw(B) = \mathbf{r}_{cw}(B) = \mathbf{r}_o(B) = \mathbf{r}(B)$.

PROOF. Since $\gamma_B(\alpha u) = \gamma_B(u)$ for all $u \in X_B$, $\alpha > 0$, we can choose a sequence (u_n) in $\dot{X}_B$ such that $\|u_n\| \to 0$ and $\gamma_B(u_n) \to \mathbf{r}_o(B) > 0$.

By Proposition 9.9, for each $n \in \mathbb{N}$, there exist $x_n \in X_B$ such that $\|x_n\| = 1$ and $\lambda_n \ge \gamma_B(u_n)$ such that

$$\lambda_n x_n = K(x_n) + u_n. \tag{9.7}$$

Since K is compact, after choosing a subsequence, $K(x_n)$ converges to some $y \in X_B$. The corresponding subsequence of (λ_n) is bounded and $\lambda_n \to \lambda \ge \mathbf{r}_o(E) > 0$ after choosing another subsequence. Since $u_n \to 0$, by (9.7) $x_n \to x \in X_B$ with $\|x\| = 1$ and, since K is continuous at x, $K(x) = \lambda x$. If $B = K$,

$$\lambda \le [B]_x \le cw(B) \le \mathbf{r}_{cw}(B) \le \mathbf{r}_o(B) \le \lambda,$$

and so equality between all these numbers. The equality $\mathbf{r}(B) = \mathbf{r}_o(B)$ follows from the compactness of B (Theorem 5.7). □

COROLLARY 9.11. *Assume that B is compact and continuous on X_B and that $\mathcal{T}_0 = \mathbf{r}(B) > 0$.*

Then there exists some $x \in X_B$, $\|x\| = 1$, such that $B(x) = \mathcal{T}_0 x$.

Further, $cw(B) = \mathbf{r}_{cw}(B) = \mathbf{r}_o(B) = \mathbf{r}(B)$.

PROOF. Since B is compact, $\mathbf{r}(B) = \mathbf{r}_o(B)$ by Theorem 5.7 and the assertion follows from Theorem 9.10 with $K = B$ □

Corollary 9.11 is used in Theorem 13.39 to prove the existence of an eigenmeasure to a tight Feller kernel working with the cone $X_+ = \mathcal{M}_+(S)$ with the flat norm which is not complete if the metric space S is not complete.

REMARK 9.12. If X_+ is complete, the compactness of B can be considerably relaxed. See [**167**, Thm.2.1], the remarks in [**146**] and Theorems 9.2 and 9.3. The same holds for the next theorem. Cf. [**8**].

THEOREM 9.13. *Let $K : X_B \to X_B$ be continuous and compact and $K(x) \geq B(x)$ for all $x \in X$, $\|x\| = 1$.*

Let $u \in \dot{X}_B$ such that K maps X_B into X_u and $CW_u(B) > 0$.

Then there exists some $x \in X_B$, $\|x\| = 1$, and some $\lambda \geq CW_u(B)$ such that $K(x) = \lambda x$.

If $B = K$, then $\lambda = cw(B) = \mathbf{r}_{cw}(B) = \eta^u(B) = \gamma_u(B) = CW_u(B) = \mathbf{r}(B)$.

PROOF. Choose some sequence (ϵ_n) in $(0,1)$ such that $\epsilon_n \to 0$. Define

$$K_n(x) = K(x) + \epsilon_n u, \qquad x \in X_B, \quad n \in \mathbb{N}.$$

By Proposition 9.9, for each $n \in \mathbb{N}$, there exist $x_n \in X_B$ such that $\|x_n\| = 1$ and $\lambda_n \geq CW_{\epsilon_n u}(B)$ such that

$$\lambda_n x_n = K_n(x_n) = K(x_n) + \epsilon_n u.$$

Since $\epsilon_n u$ is equivalent to u, $CW_{\epsilon_n u}(B) = CW_u(B) > 0$. See (6.6). Since K is compact, after choosing a subsequence, $K(x_n)$ converges to some $y \in X_B$. The corresponding subsequence of (λ_n) is bounded and $\lambda_n \to \lambda \geq CW_u(B) > 0$ after choosing another subsequence. Since $\epsilon_n \to 0$, $x_n \to x \in X_B$ with $\|x\| = 1$ and $K(x) = \lambda x$. Let $B = K$.

$$\lambda \leq [B]_x \leq cw(B) \leq \mathbf{r}_{cw}(B) \leq \eta^u(B) \leq CW_u(B) \leq \lambda,$$

and so equality holds between all these numbers. In particular, $0 < cw(B) \leq \mathbf{r}(B)$. By Theorem 9.10, there exists some $v \in \dot{X}_B$ such that $B(v) = \mathbf{r}(B)v$ and $\mathbf{r}(B) \leq cw(B)$. Since $cw(B) \leq \gamma_u(B) \leq \mathbf{r}(B)$ by Theorem 6.13, equality holds for these numbers. □

REMARK 9.14. In Theorem 9.13, the assumption that K maps X_B into X_u can be replaced by any condition that makes (9.5) imply that v is u-bounded. For instance, this is the case if $K(x) \leq A(x)$ for all $x \in X_B$, where $A : X_B \to X_B$ is homogeneous, order-preserving, subadditive ($A(x + y) \leq A(x) + A(y)$ for all $x, y \in X_B$) and $A(u)$ is u-bounded and A^m maps X_B into X_u for some $m \in \mathbb{N}$.

THEOREM 9.15. *Let B be homogeneous, order-preserving, continuous, and compact. Further let $u \in \dot{X}_B$ such that B maps X_B into X_u. Then $CW_u(B) = \mathbf{r}(B) =: r$. If $r > 0$, there exists some $v \in \dot{X}_B$ such that $B(v) = rv$.*

PROOF. The claim is trivially true if $CW^u(B) = \mathbf{r}(B) = 0$. If $CW^u(B) > 0$, it follows from Theorem 9.13. So we can assume that $r = \mathbf{r}(B) > 0$. By Theorem 9.10, there exists some $v \in \dot{X}_B$ such that $B(v) = rv$. Let $\epsilon \in (0, \infty)$. Define

$$B_\epsilon(x) = B(x) + \epsilon\psi(x)u, \qquad x \in X_B. \tag{9.8}$$

Here ψ is the companion half-norm (Section 2.5). Then B_ϵ is also homogeneous, order-preserving, continuous, compact, and maps X_B into X_u. Since $B_\epsilon(v) \geq rv$, $0 < r \leq cw(B_\epsilon) \leq \mathbf{r}(B_\epsilon)$. By Theorem 9.10, there exists $v_\epsilon \in X_B$, $\|x_\epsilon\| = 1$, such that $B_\epsilon(v_\epsilon) = r_\epsilon v_\epsilon$, $r_\epsilon = \mathbf{r}(B_\epsilon)$. Since B_ϵ maps X_B into X_u and $\psi(v_\epsilon) > 0$, $v_\epsilon \sim u$. By (6.6) and (5.30), $CW^u(B_\epsilon) \leq r_\epsilon \leq cw(B_\epsilon)$. By Theorem 6.15, $r_\epsilon = CW_u(B_\epsilon) = \eta^u(B_\epsilon) \geq r > 0$. Now choose a sequence (ϵ_n) in $(0, \infty)$ with $\epsilon_n \to 0$. Set $B_n = B_{\epsilon_n}$. For all $x \in X_B$, $\|B_n(x) - B(x)\|_u = \epsilon_n \psi(x) \to 0$. By Theorem 7.17, $CW_u(B) \geq r > 0$. The assertion now follows from Theorem 9.13. □

EXAMPLE 9.16. Let X_B be serially complete and $B : X_B \to X_B$ be given by

$$B(x) = \sum_{j=1}^{\infty} \theta_j(x) y_j, \qquad x \in X_B,$$

with a sequence (y_j) in X_B and homogeneous order-preserving continuous functionals $\theta_j : X_B \to \mathbb{R}_+$. Some y_j may be zero.

Assume that $\sum_{j=1}^{\infty} \|y_j\| < \infty$ and $(\theta_j(x))$ is a bounded sequence in $\mathbb{R}$ for any $x \in X_B$. By the uniform bounded principle in Theorem 4.8, $(\|\theta\|_j)$ is a bounded sequence in $\mathbb{R}_+$. Scaling, we can assume that $\|\theta_j\| = 1$ for all $j \in \mathbb{N}$. Since X_B is serially complete, the definition of B is sound and B is order-preserving, homogeneous and continuous. To see that B is compact, let (x_n) be a bounded sequence in X_B. Then $(\theta_j(x_n))_{n\in\mathbb{N}}$ is a bounded sequence in $\mathbb{R}$ for each $j \in \mathbb{N}$. By a standard diagonalization procedure, we find a subsequence of (x_n) which we call (x_n) again such that $\theta_j(x_n) \overset{n\to\infty}{\longrightarrow} \alpha_j \in \mathbb{R}$ for each $j \in \mathbb{N}$. Since $\theta_j(x_n) \le \|x_n\|$ for all $j, n \in \mathbb{N}$, (α_j) is a bounded sequence in $\mathbb{R}_+$ and the series $z = \sum_{j=1}^{\infty} \alpha_j y_j$ converges in the serially complete cone X_B. Let $c > 0$ be an upper bound for $(\|x_n\|)$ and (α_j) and $N \in \mathbb{N}$. Then

$$\|B(x_n) - z\| \le \sum_{j=1}^{\infty} |\theta_j(x_n) - \alpha_j| \, \|y_j\| \le \sum_{j=1}^{N} |\theta_j(x_n) - \alpha_j| \, \|y_j\| + c \sum_{j=N+1}^{\infty} \|y_j\|.$$

We take the limit superior as $n \to \infty$,

$$\limsup_{n\to\infty} \|B(x_n) - z\| \le c \sum_{j=N+1}^{\infty} \|y_j\|.$$

Since this holds for any $N \in \mathbb{N}$, we can take the limit as $N \to \infty$,

$$\limsup_{n\to\infty} \|B(x_n) - z\| = 0.$$

This finishes the compactness proof for B. Actually, B is uniformly u-bounded with $u = \sum_{j=1}^{\infty} y_j$. By Theorem 9.15, $\mathbf{r}(B) = CW_u(B) =: r$. If $r > 0$, there exists some $v \in \dot{X}_B$ such that $B(v) = rv$.

To find a sufficient condition for $r > 0$ consider subsets J of $\mathbb{N}$ and bounded families $(\beta_j)_{j\in J}$ in $\mathbb{R}_+$ with $\beta_j > 0$ and $y_j \in \dot{X}_B$. Set $x_J = \sum_{k\in J} \beta_k y_k$. Then $x_J \in \dot{X}_B$ and, by (6.6),

$$B(x_J) \ge \sum_{j\in J} \theta_j\Big(\sum_{k\in J} \beta_k y_k\Big) y_j \ge \lambda_J x_J$$

with

$$\lambda_J = \inf_{j\in J} \frac{1}{\beta_j} \theta_j\Big(\sum_{k\in J} \beta_k y_k\Big) = \inf_{j\in J} \theta_j\Big(\frac{1}{\beta_j} \sum_{k\in J} \beta_k y_k\Big).$$

By (5.30), $cw(B) \ge \lambda_J$. If $\lambda_J > 0$, $r = \mathbf{r}(B) \ge cw(B) > 0$ and r is associated with an eigenvector $v \in \dot{X}_B$. $cw(B)$ is greater than or equal to the supremum of such numbers λ_J.

To obtain an upper estimate of r, consider bounded sequences $(\tilde{\beta}_j)_{j\in\mathbb{N}}$ with $\inf_{j\in\mathbb{N}} \tilde{\beta}_j > 0$. Then $\sum_{j=1}^{\infty} \tilde{\beta}_j y_j$ is comparable to u and

$$CW_u(B) \le \inf\Big\{ \sup_{j\in\mathbb{N}} \theta_j\Big(\frac{1}{\tilde{\beta}_j} \sum_{k\in J} \tilde{\beta}_k y_k\Big)\Big\}$$

where the infimum is taken over such sequences $(\tilde{\beta}_j)$.

THEOREM 9.17. *Assume that X_B is a* sup-*semilattice and $B : X_B \to X_B$ is homogeneous, continuous and order-preserving. Assume that some power of B is compact.*

Then B has the lower KR property and $cw(B) = \mathbf{r}_{cw}(B) = \mathbf{r}_o(B) = \mathbf{r}(B)$.

PROOF. Recall that the asserted equalities hold as inequalities with $\le$ and are trivially true if $\mathbf{r}(B) = 0$. Assume that $r = \mathbf{r}(B) > 0$. There exists some $m \in \mathbb{N}$ such that B^m is compact. B^m inherits from B homogeneity, continuity and order-preservation. By (5.8), $\mathbf{r}(B^m) = \mathbf{r}(B)^m > 0$. So there exists some $w \in \dot{X}_B$ such that $B^m(w) = r^m w$. By Proposition 9.6, $B(v) \ge rv$ for some $v \in \dot{X}_B$ and $cw(B) \ge r = \mathbf{r}(B)$. So we have the asserted equalities and B has the lower KR property. □

THEOREM 9.18. *Assume that X_B is a* sup-*semilattice and $B : X_B \to X_B$ is homogeneous, continuous and order-preserving. Assume that some power of B is compact. Further assume that $u \in \dot{X}_B$ and some power of B maps X_B into X_u and $B^n(u)$ is u-bounded for all but finitely many $n \in \mathbb{N}$.*

Then $cw(B) = \mathbf{r}_{cw}(B) = \mathbf{r}_o(B) = \mathbf{r}(B) = \eta^u(B)$ and, if $r = \eta^u(B) > 0$, there exists some $v \in \dot{X}_B$ such that $B(v) \ge rv$.

PROOF. By assumption, there exist $k, m \in \mathbb{N}$ such that B^m is compact and B^k maps X_B into X_u. Since B is continuous, B^n is compact for all $n \ge m$ and $B^{\ell k}$ maps X_B into X_u for all $\ell \in \mathbb{N}$. So there exists some $p \in \mathbb{N}$ such that B^p is compact and maps X_B into X_u.

Assume that $0 < \eta^u(B) = \lim_{n\to\infty} \|B^n\|_u^{1/n}$ (Lemma 6.6). Let $w \in X_B$ be u-comparable. Then $B^n(w)$ is w-bounded for all but finitely many $n \in \mathbb{N}$ and, by (6.2), $\eta^u(B) = \eta^w(B) = \inf_{n\in\mathbb{N}} \|B^n\|_w^{1/n} \le \|B^p\|_w^{1/p}$. Thus $0 < \eta^u(B) \le CW_u(B^p)^{1/p}$. By Theorem 9.13, $B^p(w) = \lambda w$ where $\lambda = CW_u(B^p) = \eta^u(B^p) = \mathbf{r}(B^p)$. By Proposition 9.6, $B(v) \ge rv$ for some $v \in \dot{X}_B$ and $r = \eta^u(B) = \mathbf{r}(B) = cw(B)$. Here we have also used (5.8) and Lemma 6.6.

If $\eta^u(B) = 0$, then $cw(B) = 0$ by Theorem 6.13 and the equalities between the spectral bounds and radii follow from Theorem 9.17. □

9.2. Pseudo-compact operators

To round off the task of finding eigenvectors of bounded homogeneous operators that are associated with the spectral radius (cf. Chapter 9), we will sketch a possibility of relaxing the assumption of compactness. Similarly as in Section 9.1, fixed point theorems will play a role, this time for generalized contractions with respect to Hilbert's metric. As a by-product, we will obtain conditions for the equality of the spectral radius and the upper Collatz–Wielandt bounds (cf. Chapter 6).

This chapter follows [**201**] which contains the detailed proofs. We assume that X is an ordered normed vector space and that the order cone X_+ is monotonically complete (equivalently, serially complete, Section 2.1.5) and contains a normal point $u \ne 0$ (Section 2.2.2) and that $B : X_+ \to X_+$ is order-preserving and bounded homogeneous.

REMARK 9.19. Monotone completeness of X_+ can be replaced by the following weaker condition:

Every u-bounded decreasing Cauchy sequence in X_+ converges (both concepts meant with respect to the original norm).

This remark may be useful if B involves integral operators and one wants to use the dominated convergence theorem.

Let us obtain some perspective of the upcoming results in comparison to those we have already obtained like the subsequent one (Theorem 6.15).

THEOREM 9.20. *Let $u \in \dot{X}_+$, $B(u)$ be u-bounded and some power of B be uniformly u-bounded. Then*

$$cw(B) \le \mathbf{r}_{cw}(B) \le \eta^u(B) \le \begin{cases} CW_u(B) \\ \gamma_u(B) \le \mathbf{r}(B). \end{cases}$$

If u is a normal point in X_+,

$$\eta^u(B) = \gamma_u(B) = \mathbf{r}(B) \le CW_u(B).$$

Actually, one more equality can be established under additional assumptions.

THEOREM 9.21. *If u is a normal point in X_+ and B is uniformly u-bounded, then*

$$\mathbf{r}(B) = \gamma_u(B) = \eta^u(B) = CW_u(B).$$

For the proof, see [**201**, Sec.9.1]. If this result is compared to [**8**, Thm.7.3], one sees that none of the conditions (i), (ii), (iii) therein is needed if one is only interested in the equality $\mathbf{r}(B) = CW_u(B)$. These conditions can be avoided by using the upper semicontinuity of the spectral radius (Chapter 7).

Theorem 9.21 is applied in Corollary 18.6.

To explore the existence of eigenvector associated with the spectral radius we introduce the concept pseudo-compactness. Later, we will display various scenarios in which it holds.

DEFINITION 9.22. *Let $r = \eta^u(B) > 0$.*

A sequence (x_n) in X_+ is called a sequence of approximate eigenvectors associated with r *if $\|x_n\| = 1$ for all $n \in \mathbb{N}$ and there exist a sequence (λ_n) in $[r, \infty)$ such that $\lambda_n \to r$ and $\|(\lambda_n - B)(x_n)\| \to 0$ and $(\lambda_n - B)(x_n) \in X_+$ for all $n \in \mathbb{N}$.*

Let $u \in \dot{X}_+$. B is called pseudo-compact *if any sequence of approximate eigenvectors (x_n) in $X_+ \cap X_u$ associated with r that is bounded with respect to the u-norm has a subsequence that converges with respect to $\|\cdot\|$.*

THEOREM 9.23. *Let u be a normal point in X_+ and B be uniformly u-bounded, pseudo-compact and continuous.*

Then there exists some $v \in \dot{X}_+$ such that $B(v) = \mathbf{r}(B)v$ and $cw(B) = \mathbf{r}_{cw}(B) = \mathbf{r}(B) = CW_u(B)$.

Compare this result to [**8**, Thm.1.1] (Generalized Collatz–Wielandt Theorem). An application of Theorem 9.23 is Proposition 18.7.

REMARK 9.24. The continuity of B can be replaced by the following weaker assumptions: B is bounded and for any u-bounded sequence (x_k) in X_+ and $x \in X_+$ with $\|x_k - x\| \to 0$ as $k \to \infty$, also $\|B(x_k) - B(x)\| \to 0$.

For the proof, we refer to [**201**]. This proof uses the same overall strategy as [**8**, Thm.7.3] by applying Hilbert's projective metric [**23, 138, 145, 211, 224**] to perturbations of the map B (like in (9.8)) and then using B being pseudo-compact for a limit process. Applying a fixed point theorem for generalized contractions [**137**, Thm.34.5] rather than the contraction mapping theorem [**8, 144**] makes it possible to shorten the proof and include certain maps that have uniformly order-bounded powers but are not uniformly order-bounded itself.

THEOREM 9.25. *Let $u \in \dot{X}_+$ be a normal point in X_+. Assume there exists a bounded subadditive homogeneous map $A : X_+ \to X_+$ such that*

$$B(x) \leq A(x), \qquad x \in X_+, \tag{9.9}$$

$A(u)$ is u-bounded and some power of A^m is uniformly u-bounded.

(a) *Then* $\mathbf{r}(B) = \gamma_u(B) = \eta^u(B) = CW_u(B)$.
(b) *If B is pseudo-compact, then there exists some $v \in \dot{X}_+$ such that $B(v) = \mathbf{r}(B)v$ and $cw(B) = \mathbf{r}_{cw}(B) = \mathbf{r}(B)$.*

Applications of these results can be found in [**201**].

Pseudo-compactness may appear like a brute force assumption, but has the advantage that it can be formulated in elementary terms, allows a succinct formulation of results, and is satisfied under a variety of conditions.

We list a couple of results that help in checking whether B is pseudo-compact. For the proofs, we refer to [**201**]. The next one is inspired by the proof of [**8**, Prop.4.2]. Notice that, compared to [**8**, Prop.4.2], the uniform continuity of B on bounded sets only needs to hold from the right.

PROPOSITION 9.26. *B is pseudo-compact if some power of B is pseudo-compact and B satisfies a righthand-sided uniform continuity condition of the following form:*

> *For any $\epsilon > 0$ and $c > 0$ there exists some $\delta > 0$ such that $\|B(x + y) - B(x)\| \leq \epsilon$ for all $x, y \in X_+$ with $\|x\| \leq c$ and $\|y\| \leq \delta$.*

REMARK 9.27. Obviously B is pseudo-compact if the following holds: If $r > 0$ and (x_n) is sequence in $X_+ \cap X_u$ such that (x_n) is bounded with respect to the u-norm and $\|(r - B)(x_n)\| \to 0$, then (x_n) has a subsequence that converges with respect to $\| \cdot \|$.

This condition is closer to the one used in [**8**, Sec.4] than the one in Definition 9.22. We choose Definition 9.22 because then we only need to assume the right-hand sided uniform continuity in Proposition 9.26 rather than assume that B is uniformly continuous on bounded subsets of X_+.

COROLLARY 9.28.

(a) *B is pseudo-compact if it is compact from $X_+ \cap X_u$ with the u-norm to X_+ with the original norm.*
(b) *B is pseudo-compact if some power of B is compact from $X_+ \cap X_u$ with the u-norm to X_+ with the original norm and B satisfies the righthand-sided uniform continuity condition of Proposition* 9.26.

More conditions for B to be pseudo-compact are formulated in [**8**] in terms of measures of noncompactness. Here, we only give some examples [**201**].

PROPOSITION 9.29. *Let* $\mathbf{r}(B) > 0$. *Assume that* $B = K + L$, *where* $K, L : X_+ \to X_+$ *are homogeneous and continuous and* K *is compact from* $X_+ \cap X_u$ *with the u-norm to* X_+ *with* $\|\cdot\|$. *Let* $r = \eta^u(B)$. *Then* B *is pseudo-compact, if* L *has the following property:*

> *If* (x_n) *is a bounded sequence in* $X_+ \cap X_u$ *with respect to the u-norm,* $\|x_n\| = 1$ *for all* $n \in \mathbb{N}$, *and* (λ_n) *a sequence in* $[r, \infty)$ *such that* $\lambda_n \to r$ *and* $(\lambda_n - L)(x_n)$ *converges with respect to* $\|\cdot\|$ *and* $(\lambda_n - L)(x_n) \in X_+$ *for all* $n \in \mathbb{N}$, *then* (x_n) *has a convergent subsequence with respect to* $\|\cdot\|$.

This property of L *is satisfied, e.g., if* X_+ *is serially complete and at least one of the following additional properties is satisfied:*

(a) L *is additive and* $\mathbf{r}(L) < r$.
(b) L *is Lipschitz continuous on* X_+ *with a Lipschitz constant* $\Lambda < r$.

CHAPTER 10

Continuity of the spectral radius

Let X be an ordered normed vector space with closed order cone X_+ and $B : X_B \to X_B$ a bounded, homogeneous, order-preserving operator on a closed cone X_B contained in X_+.

While the spectral radius of B is an upper semicontinuous function of the operator B under reasonably general assumptions (Section 7.1), it may fail to be a lower semicontinuous function even if X_B is a cone and B is additive and compact on X_B (Section 10.4). Conditions for lower semicontinuity turn out to be quite restrictive. Continuity of the spectral radius hold for bounded linear operators on a Banach space that have a compact power (Section 10.1). There are some remarkable lower semicontinuity results for the spectral radius of compact homogeneous operators [**146**]. So we concentrate later on operators that are u-bounded for some $u \in \dot{X}_B$ or have u-bounded powers (Section 10.2 and 10.2).

10.1. Continuity of the spectral radius for bounded linear maps with compact powers

To illustrate the difference (as far as continuity of the spectral radius is concerned) between bounded linear operators on a Banach space X and bounded homogeneous operators on homogeneous subsets of X, we prove the following result which seems to be well-known [**164**] (unfortunately, I did not find a more explicit reference to it) and certainly can be generalized.

This result will be applied to operators that are induced by measure kernels or Feller kernels (Section 1.7, Theorem 13.81).

THEOREM 10.1. *Let X be a Banach space. Let A be a bounded linear map on X and (A_k) a sequence of bounded linear maps on X such that $\|A_k - A\| \to 0$ as $k \to \infty$ for the operator norms. Then, if some power of A is a compact map, $\mathbf{r}(A_k) \to \mathbf{r}(A)$ as $k \to \infty$.*

PROOF. We can assume that $\mathbf{r}(A) > 0$. Otherwise, the claim follows from Corollary 7.13 with $X_B = X$. We first assume that A itself is compact.

If X is a real Banach space, we consider its complexification and extend A and A_k to the complexificaton. See Section 5.4. This does not change the spectral radii.

By [**132**, Thm.III.6.26], there exists some eigenvalue $\lambda \in \mathbb{C}$ of A with $|\lambda| = \mathbf{r}(A)$ and λ is an isolated point of the $\sigma^{\mathbb{C}}(A)$.

By [**132**, IV.5], there exists a sequence (λ_k) in $\sigma^{\mathbb{C}}(A_k)$ such that $\lambda_k \to \lambda$ as $k \to \infty$. Then $|\lambda_k| \to |\lambda| = \mathbf{r}(A)$. Since $\mathbf{r}(A_k) \geq |\lambda_k|$ for all $k \in \mathbb{N}$,

$$\liminf_{k\to\infty} \mathbf{r}(A_k) \geq \mathbf{r}(A).$$

We combine this inequality with Corollary 7.13 for the case $X_B = X$, and the assertion follows.

Now assume that there is some $m \in \mathbb{N}$ such that A^m is compact. By Lemma 7.10 and Remark 7.11, $A_k^m \to A^m$ as $k \to \infty$. By our previous consideration, $\mathbf{r}(A_k^m) \to \mathbf{r}(A^m)$ as $k \to \infty$. Since $\mathbf{r}(A^m) = (\mathbf{r}(A))^m$ and the same holds for A_k by the power rule for the spectral radius (5.8), $\mathbf{r}(A_k) \to \mathbf{r}(A)$ as $k \to \infty$. □

10.2. Lower semicontinuity of the lower Collatz–Wielandt bound

In hindsight, upper semicontinuity of the spectral radius is not so surprising because the definition of the spectral radius involves taking infima, (5.8), and taking infima preserves upper semicontinuity (Lemma A.52). Since taking suprema preserves lower semicontinuity (Lemma A.53), we consider an alternative to the spectral radius which involves taking suprema, the lower Collatz–Wielandt bound, (5.30). Recall that $\dot{X}_B = X_B \setminus \{0\}$.

THEOREM 10.2. *Let $u \in \dot{X}_B$. Assume that there is some $\lambda > 0$ and some u-comparable $v \in X_B$ such that $B(v) \geq \lambda v$. Let (B_n) be a sequence of bounded homogeneous operators $B_n : X_B \to X_B$. Assume that for all u-comparable $x \in X_B$, $B_n(x) - B(x) \in X_u$ and $\|B_n(x) - B(x)\|_u \to 0$. Then $\liminf_{n\to\infty} cw(B_n) \geq \lambda$.*

PROOF. Since v is u-comparable, $\|B_n(v) - B(v)\|_u \to 0$. By (2.31), there exists a sequence (ϵ_n) in $(0, \infty)$ such that $\epsilon_n \to 0$ as $n \to \infty$ and

$$-\epsilon_n u \leq B_n(v) - B(v) \leq \epsilon_n u, \qquad n \in \mathbb{N}.$$

Since v is u-positive, there exists some $\delta > 0$ such that $\delta u \leq v \leq cu$ and

$$-(\epsilon_n/\delta)v \leq B_n(v) - B(v), \qquad n \in \mathbb{N}.$$

For all $n \in \mathbb{N}$,

$$B_n(v) = B(v) + \big(B_n(v) - B(v)\big) \geq \lambda v - (\epsilon_n/\delta)\, v.$$

By (5.30),

$$cw(B_n) \geq [B_n]_v \geq \lambda - (\epsilon_n/\delta) \overset{n\to\infty}{\longrightarrow} \lambda.$$

This implies $\liminf_{n\to\infty} cw(B_n) \geq \lambda$. □

In order to establish the connection to $cw(B)$, we make the following definition.

DEFINITION 10.3. *B is called* conditionally pointwise u-positive *if, for any $x \in \dot{X}_B$ with $B^n(x) \neq 0$ for all $n \in \mathbb{N}$, there exists some $m \in \mathbb{N}$ such that $B^m(x)$ is u-positive, i.e., there exists some $\epsilon > 0$ such that $B^m(x) \geq \epsilon u$.*

LEMMA 10.4. *Let $u \in \dot{X}_B \subseteq X_+$ and B be order-preserving, pointwise u-bounded and conditionally pointwise u-positive.*

If $r \in (0, \infty)$ and $w \in \dot{X}_B$ with $B(r)w \geq rw$, then there exists a u-comparable $v \in X_B$ such that $B(v) \geq rv$.

PROOF. Let $r \in (0, \infty)$ and $w \in \dot{X}_B$ such that $B(w) \geq rw$. Since B is order-preserving and homogeneous, by induction, $B^n(w) \geq r^n w \neq 0$ for all $n \in \mathbb{N}$.

Since B is conditionally pointwise u-positive, there exists some $m \in \mathbb{N}$ such that $v = B^m(w)$ is u-positive. Since B is order-preserving and homogeneous,

$$B(v) = B(B^m(w)) = B^m(B(w)) \geq B^m(rw) = rv.$$

For all $k \in \mathbb{N}$, by induction, $B^k(v) \geq r^k v$ and $B^k(v) \neq 0$ for all $k \in \mathbb{N}$. Since B is pointwise u-bounded, there exists some $k \in \mathbb{N}$ such that $B^k(v)$ is u-bounded. Thus v is u-bounded and also u-comparable. □

THEOREM 10.5. *Let $u \in \dot{X}_B$, and let B be pointwise u-bounded (Definition 6.1) and conditionally pointwise u-positive. Let $X_B \subset X_+$ and B be order-preserving. Assume that, for all u-comparable $x \in X_B$ and $k \in \mathbb{N}$, $B_k(x) - B(x) \in X_u$ and $\|B_k(x) - B(x)\|_u \to 0$ as $k \to \infty$.*

Then $\liminf_{k\to\infty} cw(B_k) \ge cw(B)$.

PROOF. We can assume that $cw(B) > 0$. Let $r \in (0, cw(B))$. By (5.30), there exists some $w \in \dot{X}_B$ such that $B(w) \ge rw$. By Lemma 10.4, there exists some u-comparable $v \in X_B$ such that $B(v) \ge rv$. By Theorem 10.2,

$$\liminf_{k\to\infty} cw(B_k) \ge r.$$

Since $r \in (0, cw(B))$ has been arbitrary, the assertion follows. □

10.3. Continuity of spectral radii and CW bounds

We combine the lower semicontinuity results for the lower Collatz–Wielandt bound with the upper semicontinuity results for the spectral radius, upper Collatz–Wielandt spectral radius and upper Collatz–Wielandt bound (Chapter 7) by assuming that the first is larger or equal to either of the latter.

THEOREM 10.6. *Let X be an ordered normed vector space with a closed order cone X_+ and X_B be a homogeneous subset of X_+. Let $B : X_B \to X_B$ be a continuous homogeneous order-preserving operator and $\mathbf{r}(B) > 0$. Let (B_k) be a sequence of bounded homogeneous order-preserving operators on X_B such that $\|B_k - B\| \to 0$ as $k \to \infty$.*

Further, let $u \in \dot{X}_B$, and let B be pointwise u-bounded and conditionally pointwise u-positive. Further assume that, for all u-comparable $x \in X_B$ and $k \in \mathbb{N}$, $B_k(x) - B(x) \in X_u$ and $\|B_k(x) - B(x)\|_u \to 0$ as $k \to \infty$.

Then $\mathbf{r}(B) = cw(B)$ and

$$\mathbf{r}(B_k) \to \mathbf{r}(B), \quad cw(B_k) \to cw(B), \quad k \to \infty,$$

if one of the following two conditions holds in addition:

(a) *B is uniformly continuous on bounded sets and $cw(B) \ge \mathbf{r}(B)$.*
(b) *B is compact and X_B is a closed cone.*
(c) *$B(u)$ is u-bounded, u is a normal point for X_B, $cw(B) \ge \gamma_u(B)$, and there exist $m \in \mathbb{N}$ and $c \ge 0$ such that $B_k^m(x) \le c\|x\|u$ for all $k \in \mathbb{N}$ and all $x \in X_+$.*

PROOF. By Theorem 10.5,

$$\liminf_{k\to\infty} cw(B_k) \ge cw(B). \tag{10.1}$$

Assume (a) or (b). By Theorem 7.1,

$$\limsup_{n\to\infty} \mathbf{r}(B_k) \le \mathbf{r}(B).$$

By Theorem 5.31,

$$cw(B) \le \mathbf{r}(B), \qquad cw(B_k) \le \mathbf{r}(B_k), \quad k \in \mathbb{N}. \tag{10.2}$$

Both (a) and (b) imply $\mathbf{r}(B) = cw(B)$.

For (a), this is because we have assumed $cw(B) \ge \mathbf{r}(B)$.

Under (b), there is some $v \in \dot{X}_B$ such that $B(v) = \mathbf{r}(B)v$ (Corollary 9.11) and $cw(B) \ge \mathbf{r}(B)$ follows from (5.30).

This implies $cw(B_k) \to cw(B) = \mathbf{r}(B)$ and $\mathbf{r}(B_k) \to \mathbf{r}(B) = cw(B)$ for $k \to \infty$.

(c) Since $B(u)$ is u-bounded, $B_k(u)$ is u-bounded for all $k \in \mathbb{N}$ and $B_k(u) \leq B(u) + \|B_k(u) - B(u)\|_u u \leq cu$ for some $c > 0$ that does not depend on k.

By Corollary 7.16,

$$\limsup_{k\to\infty} r(B_k) \leq \gamma_B(u) \leq cw(B).$$

Combination with (10.1) and (10.2) implies the assertion. □

REMARK 10.7. Part (a) and (b) can already be found in [**208**, Thm.5.3] except that the assumption therein that X_+ is normal is not needed.

In part (c), continuity of B can be replaced by the following weaker assumption (Theorem 7.16):

> For any u-bounded sequence (x_k) in X_B and any $x \in X_B$ with $\|x_k - x\| \to 0$ as $k \to \infty$, also $\|B(x_k) - B(x)\| \to 0$ as $k \to \infty$.

THEOREM 10.8. *Let X be an ordered normed vector space with a normal monotonically complete closed order cone X_+.*

Let $B : X_+ \to X_+$ be a continuous homogeneous order-preserving operator and $cw(B) \geq \mathbf{r}(B) > 0$.

Further, let $u \in \dot{X}_+$ and let B be uniformly u-bounded and conditionally pointwise u-positive.

Let (B_k) be a sequence of bounded homogeneous order-preserving operators on X_+ and B_k be uniformly u-bounded for all $k \in \mathbb{N}$.

Further assume that, for all u-comparable $x \in X_B$, $\|B_k(x) - B(x)\|_u \to 0$ as $k \to \infty$.

Then $cw(B) = \mathbf{r}(B) = CW_u(B)$ and $cw(B_k) \to cw(B)$, $\mathbf{r}(B_k) \to \mathbf{r}(B)$, $CW_u(B_k) \to CW_u(B)$ as $k \to \infty$.

PROOF. By Theorem 9.20, $cw(B) = \mathbf{r}(B) = CW_u(B)$ and $cw(B_k) \leq \mathbf{r}(B_k) = CW_u(B_k)$ for all $k \in \mathbb{N}$. By Theorem 10.5,

$$cw(B) \leq \liminf_{k\to\infty} cw(B_k)$$

and, by Theorem 7.17,

$$CW_u(B) \geq \limsup_{k\to\infty} CW_u(B_k).$$

Combining all these inequalities yields the assertion. □

Let X be an ordered normed vector space with closed order cone X_+, X_B be a homogeneous subset of X_+, and $u \in \dot{X}_B$.

Let $B : X_B \to X_B$ be homogeneous and order-preserving. and let (B_k) be a sequence of homogeneous order-preserving operators $B_k : X_B \to X_B$.

THEOREM 10.9. *Let $u \in \dot{X}_B$, $B(u)$ be u-bounded and $B_k(u)$ be u-bounded for all $k \in \mathbb{N}$.*

Let some power of B be uniformly u-bounded and, for each $k \in \mathbb{N}$, some power of B_k be uniformly u-bounded.

Let B be conditionally pointwise u-positive.

Assume that, for all u-comparable $x \in X_B$, $\|B_k(x) - B(x)\|_u \to 0$ as $k \to \infty$.

Finally, assume that $cw(B) \geq CW_u(B) > 0$.

Then

$$\lim_{k\to\infty} cw(B_k) = cw(B) = CW_u(B) = \lim_{k\to\infty} CW_u(B_k).$$

If, in addition, u is a normal point for X_B,

$$cw(B) = CW_u(B) = \mathbf{r}(B) = \lim_{k\to\infty} \mathbf{r}(B_k).$$

PROOF. By Theorem 6.15 and Theorem 6.14,

$$cw(B) \le \mathbf{r}_{cw}(B) \le \eta^u(B) \le \begin{cases} CW_u(B) \\ \gamma_u(B) \le \mathbf{r}(B). \end{cases}$$

If u is a normal point in X_+,

$$cw(B) \le \eta^u(B) = \gamma_u(B) = \mathbf{r}(B) \le CW_u(B).$$

Analogous statements hold for each B_k.

By Theorem 10.5,

$$cw(B) \le \liminf_{k\to\infty} cw(B_k). \tag{10.3}$$

By Theorem 7.17,

$$\limsup_{k\to\infty} CW_u(B_k) \le CW_u(B). \tag{10.4}$$

By assumption, $CW_u(B) \le cw(B)$ and so $CW_u(B) = cw(B)$. In combination,

$$\begin{aligned}\limsup_{k\to\infty} cw(B_k) &\le \limsup_{k\to\infty} CW_u(B_k) \le CW_u(B) = cw(B) \\ &\le \liminf_{k\to\infty} cw(B_k) \le \lim\inf_{k\to\infty} CW_u(B_k).\end{aligned}$$

Since $\liminf \le \limsup$, all inequalities actually are equalities, $CW_u(B_k) \to CW_u(B)$ and $cw(B_k) \to cw(B)$ as $k \to \infty$ by Exercise A.3.1.

Now assume that u is a normal point for X_B. Then $cw(B_k) \le \mathbf{r}(B_k) \le CW_u(B_k)$ for all $k \in \mathbb{N}$, $cw(B) = \mathbf{r}(B) = CW_u(B)$ and so $\lim_{k\to\infty} \mathbf{r}(B_k) = \mathbf{r}(B)$. □

REMARK 10.10.

(a) Notice that both Theorems 10.6 (a) and 10.8 contain the assumption $cw(B) \ge \mathbf{r}(B) > 0$.

By (5.30), this is equivalent to the statement that for all $r \in (0, \mathbf{r}(B))$ there is some $w \in \dot{X}_+$ such that $B(w) \ge rw$.

The assumption $cw(B) \ge \mathbf{r}(B) > 0$ is in particular satisfied if $\mathbf{r}(B) > 0$ and there is some $v \in \dot{X}_+$ with $B(v) \ge \mathbf{r}(B)v$. This in turn is satisfied if B is compact [**203**, Thm.7.8] (Corollary 9.11) or if some power B^m is compact and the cone X_+ is a sup-semilattice. See Theorem 9.17 [**8**, Thm.5.1] [**199**, Prop.1.11]. The latter means that for two vectors x, y in X_+ the supremum $x \vee y$ exists (and is an element in X_+).

The assumption $cw(B) \ge \mathbf{r}(B)$ is also a consequence of Theorems 9.2 and 9.3.

For another example, see Section 13 and Corollary 12.8 in [**199**].

(b) Theorem 10.6 (c) contains the assumption $cw(B) \ge \gamma_u(B)$ which is satisfied if $cw(B) \ge \mathbf{r}(B)$.

(c) Theorem 10.9 contains the assumption $cw(B) \ge CW_u(B)$.

(d) All these assumption are satisfied if $X_B = X_+$ is monotonically complete, $u \in X_+$ is a normal point for X_+ and B is uniformly u-bounded, pseudo-compact and continuous (Theorems 9.21 and 9.23).

These continuity results are applied in Section 18.6.4 and in [**208**, Sec.6-10]. The applications concern spatially distributed two-sex populations. In Section 18.6.4, the continuous dependence of the spectral radius on certain parameters of the dispersal kernels are studied and in [**208**, Sec.6-10] the dependence on the basic mating/reproduction function.

10.4. Lower semicontinuity may fail

We present a generalized version of an example in [**146**] which shows that lower semicontinuity of the spectral radius may fail even if B and B_n are compact.

Remember the ordered Banach space $X = C_0(0,1]$ of continuous functions f on $[0,1]$ with $f(0) = 0$ under the supremum norm and the closed order cone X_+ of nonnegative functions.

10.4.1. The special cone. Recall the closed subcone

$$X_{con} \text{ of increasing convex functions } f \text{ on } [0,1], \qquad f(0) = 0. \tag{10.5}$$

As we have already seen (Section 2.3.2), X_{con} is total but not generating.

> We will construct compact Lipschitz continuous additive homogeneous operators B and B_n, $n \in \mathbb{N}$, on X_{con} such that $\|B_n - B\| \to 0$ as $n \to \infty$ for the operator norms associated with X_{con} and $\mathbf{r}(B) > 0 = \mathbf{r}(B_n)$ for all $n \in \mathbb{N}$ for the spectral radii with respect to X_{con}.

Every $f \in X_{con}$ can be uniformly approximated by an increasing sequence (f_n) in X_{con} where f_n is continuously differentiable. To this end, extend f to $(-\infty, 1]$ by $f(t) = 0$ for $t < 0$. The extension is continuous and convex. Set

$$f_n(t) = n \int_{t-(1/n)}^{t} f(s)ds = \int_0^1 f(t - (s/n))\, ds, \qquad t \in [0,1]. \tag{10.6}$$

The second formula shows that

$$\begin{aligned} &f_n \in X_{con}, \qquad f_n \to f \text{ uniformly on } [0,1], \\ &f_n(t) \le f_{n+1}(t) \le f(t), \qquad n \in \mathbb{N}, \quad t \in [0,1]. \end{aligned} \tag{10.7}$$

From the second formula, we also get the following:

For $0 \le s \le 1$, $1 - (s/n) = (1 - (s/n))1 + (s/n)0$. Since f is convex,

$$f(1 - (1/n)) \le (1 - (s/n))f(1) + (s/n)f(0),$$

and so

$$f_n(1) \le f(1) \int_0^1 (1 - (s/n))ds = f(1)\Big(1 - \frac{1}{2n}\Big). \tag{10.8}$$

The first formula in (10.6) shows that f_n is differentiable and

$$f_n'(t) = n\big[f(t) - f(t - (1/n))\big], \quad t \in [0,1], \qquad f_n'(0) = 0. \tag{10.9}$$

LEMMA 10.11. *Let $f \in X_{con}$. If $0 \le q \le r \le s < t \le 1$, then*

$$f(r) - f(q) \le \frac{f(t) - f(s)}{t - s}(r - q).$$

Further, $f(t)/t$ is an increasing function of $t \in (0,1]$, f is differentiable at 0, and

$$f'(0)t \le f(t) \le f(1)t, \qquad 0 \le t \le 1.$$

PROOF. Cf. [**70**, 6.3.2] The first inequality is trivially true if $q = r$. Let $q < r$. Intuitively, this is clear by drawing the appropriate secants. If f is differentiable, the first inequality follows from the mean value theorem and the fact that the derivative of any differentiable convex function is increasing. If $f \in X_{con}$ is not differentiable, it can be approximated by convex increasing differentiable functions like in (10.6).

For the second inequality, let $s \in [0,1]$. Then, for all $t \in [0,1]$,

$$f(st) = f\big((1-s)0 + st\big) \leq (1-s)f(0) + sf(t)$$

and

$$\frac{f(st)}{st} \leq \frac{f(t)}{t}. \qquad \square$$

10.4.2. The operators. Let $\xi \in X_{con}$ and $\xi(1) \leq 1$. We define a linear map $A : X \to X$ by

$$Af = f \circ \xi, \qquad f \in X. \tag{10.10}$$

A maps X_{con} into itself. Let B denote the restriction of A to X_{con},

$$Bf = f \circ \xi, \qquad f \in X_{con}. \tag{10.11}$$

LEMMA 10.12. *Let $\xi \in X_{con}$ and $\xi(1) \leq 1$. Then $\xi(t) \leq t$ for all $t \in [0,1]$.*

Either $\xi(t) = t$ for all $t \in [0,1]$ (and A and B are the identity map) or $\xi(t) < t$ for all $t \in (0,1)$ and $\xi'(0) < \xi(1) \leq 1$.

PROOF. By Lemma 10.11, $\xi(t) \leq \xi(1)t \leq t$ for all $t \in [0,1]$.

Let $\xi(s) = s$ for some $s \in (0,1)$. Since $\xi(t)/t$ is increasing and $\xi(1) \leq 1$, $\xi(t) = t$ for all $t \in [s,1]$. Suppose $f(r) < r$ some some $r \in (0,1]$. Then $0 < r < s$ and $s = \alpha r + (1-\alpha)$ with $\alpha = (1-s)/(1-r) \in (0,1)$. Since ξ is convex,

$$\xi(s) \leq \alpha\xi(r) + (1-\alpha)\xi(1) < \alpha r + 1 - \alpha = s,$$

a contradiction.

So, if ξ is not the identity map, $\xi(t)/t < 1$ for all $t \in (0,1)$ and $\xi'(0) < \xi(1) \leq 1$. $\square$

LEMMA 10.13. *Let $\xi \in X_{con}$ and $\xi(1) \leq 1$ and ξ not the identity on $[0,1]$. Then, for the iterates ξ^k,*

$$\xi^k(t) \to 0 \text{ and } (\xi^k(t))^{1/k} \to \xi'(0) < 1, \quad k \to \infty, \text{ for all } t \in [0,1).$$

If $\xi(1) < 1$, also

$$\xi^k(1) \to 0 \text{ and } (\xi^k(1))^{1/k} \to \xi'(0) < 1, \quad k \to \infty.$$

PROOF. Let $t \in [0,1)$. By Lemma 10.12, $\xi(t) < t$. By induction, $(\xi^k(t))$ is a decreasing sequence and has a limit which is a fixed point of the continuous function ξ. Since 0 is the only fixed point of ξ in [0,1), $x^k(t) \to 0$ as $t \to \infty$. If $\xi(1) < 1$, the same argument applies to $t = 1$.

Since $\xi(t) \geq \xi'(0)t$ for all $t \in [0,1]$ by Lemma 10.11, by induction $\xi^k(t) \geq (\xi'(0))^k t$ for all $t \in [0,1]$. So

$$\liminf_{k\to\infty} (\xi^k(t))^{1/k} \geq \xi'(0), \qquad t \in (0,1]. \tag{10.12}$$

Now let $t \in (0,1]$ and $\xi(t) < t$. Let $\epsilon > 0$ and $\epsilon < 1 - t^{-1}\xi(t))$. Then $\epsilon < 1 - \xi'(0)$. Since $\xi(t)/t \to \xi'(0)$ as $t \to 0$, there exists some $\delta > 0$ such that

$$\xi(t) \leq (\xi'(0) + \epsilon)t \leq \xi(t) + \epsilon t \leq t, \qquad 0 \leq t \leq \delta.$$

By induction,

$$\xi^k(t) \le (\xi'(0)+\epsilon)^k t \le t, \qquad 0 \le t \le \delta.$$

Since $\xi^k(t) \to 0$, there exists some $m \in \mathbb{N}$ such that $\xi^m(t) < \delta$. Then

$$\xi^{k+m}(t) = \xi^k(\xi^m(t)) \le (\xi'(0)+\epsilon)^k \xi^m(t), \qquad k \in \mathbb{N}.$$

So

$$\limsup_{k\to\infty} (\xi^k(t))^{1/k} \le \xi'(0) + \epsilon.$$

Since this holds for any $\epsilon \in (0, 1 - t^{-1}\xi(t))$,

$$\limsup_{k\to\infty} (\xi^k(t))^{1/k} \le \xi'(0)$$

and convergence follows in combination with (10.12). □

Recall the operator B from (10.11).

PROPOSITION 10.14 ([**27**]). *B is order-preserving with respect to the order cone X_+, homogeneous, and Lipschitz continuous. If $\xi(1) < 1$, B is compact.*

PROOF. The first statement follows from B being the restriction of the bounded positive map $A : X \to X$ to X_{con}.

Let $f \in X_{con}$ and $0 \le s \le t \le 1$. Since $\xi(s) \le \xi(t) \le \xi(1) < 1$, by Lemma 10.11,

$$0 \le (Bf)(t) - (Bf)(s) = f(\xi(t)) - f(\xi(s)) \le \frac{f(1) - f(\xi(1))}{1 - \xi(1)}(\xi(t) - \xi(s)).$$

Since ξ is continuous on the compact interval [0,1], it is uniformly continuous on [0,1]. If D is a bounded subset of X_{con}, $B(D)$ is equicontinuous and equibounded and thus has compact closure by the Arzela-Ascoli theorem.

Finally, for $f, g \in X_{con}$,

$$|(Bf)(s) - (Bg)(s)| = |f(\xi(s)) - g(\xi(s))| \le \|f - g\|_\infty$$

because $\xi(s) \in [0,1]$, and

$$\|Bf - Bg\|_\infty \le \|f - g\|_\infty. \qquad \square$$

Bonsall [**27**] gives an example where A is not compact on the linear subspace $X_{con} - X_{con}$ of X. See also Corollary 10.21.

EXAMPLE 10.15 ([**27**]). Let $\xi : [0,1] \to [0,1]$ be given by $\xi(t) = \alpha t$, $0 \le t \le 1$, with some $\alpha \in (0,1)$ and so $(Af)(t) = f(\alpha t)$, $0 \le t \le 1$.

Define sequences (g_n), (h_n) and (f_n) in X_{con} by

$$g_n(t) = (1/\alpha)^n t, \qquad 0 \le t \le 1,$$

$$h_n(t) = \left\{ \begin{array}{cc} 0, & 0 \le t \le \alpha^n \\ (1/\alpha)^n t - 1, & \alpha^n \le t \le 1 \end{array} \right\},$$

$$f_n = g_n - h_n.$$

Then

$$f_n(t) = \left\{ \begin{array}{cc} (1/\alpha)^n t, & 0 \le t \le \alpha^n \\ 1, & \alpha^n \le t \le 1 \end{array} \right\}.$$

Notice that f_n is nonnegative and increasing but concave rather than convex. Further $\|f_n\| = 1$ and, for $n \ge 2$,

$$(Af_n)(t) = \left\{ \begin{array}{cc} (1/\alpha)^n \alpha t, & 0 \le \alpha t \le \alpha^n \\ 1, & \alpha^n \le \alpha t \le 1, t \le 1 \end{array} \right\}$$

and thus $Af_n = f_{n-1}$. Finally, if $n > m \geq 1$,

$$\|f_n - f_m\| \geq |f_n(\alpha^n) - f_m(\alpha^n)| = 1 - \alpha^{k-j} \geq 1 - \alpha,$$

and (f_n) and (Af_n) have no subsequences that are Cauchy sequences.

PROPOSITION 10.16. *B is uniformly u-bounded for $u = \xi$ and for $u(t) = t$.*

Recall Definition 6.1.

PROOF. By Lemma 10.11,

$$(Bf)(t) = f(\xi(t)) \leq f(1)\xi(t) \leq f(1)\xi(1)t, \qquad 0 \leq t \leq 1. \qquad \square$$

10.4.3. Spectral radii and the failure of lower semicontinuity. Recall the lower Collatz–Wielandt bound $cw(B)$, (5.30).

THEOREM 10.17. *Let $\xi \in X_{con}$ and $\xi(1) \leq 1$. Then*

(a) $\xi'(0) \leq cw(B) \leq \mathbf{r}(B) = \lim_{k\to\infty}(\xi^k(1))^{1/k}$.
(b) *If $\xi(1) < 1$, then $\mathbf{r}(B) = \xi'(0) = cw(B)$ and, if $\xi'(0) > 0$, there exists some $f \in X_{con}$, $f(1) = 1$, such that $Bf = \xi'(0)f$.*
(c) *If $\xi(1) = 1$ and ξ is not the identity function on $[0, 1]$, then $\mathbf{r}(B) = 1 > \xi'(0) = cw(B)$.*

PROOF. (a) Let $f(t) = t$ for all $t \in [0, 1]$. By Lemma 10.11,

$$(Bf)(t) = \xi(t) \geq \xi'(0)t, \qquad t \in [0, 1].$$

So $Bf \geq \xi'(0)f$ and $cw(B) \geq \xi'(0)$ by (5.30).

By induction, $\xi^k(t) \leq \xi^k(1)t$ for all $k \in \mathbb{N}$. Let $f \in X_{con}$. For all $k \in \mathbb{N}$, $f(\xi^k(t)) \leq f(1)\xi^k(1)t$. So $\|B^k f\| \leq f(1)\xi^k(1)$. For $f(t) = t$ for all $t \in [0, 1]$, $\|f \circ \xi^k\| = \xi^k(1)$ and so $\|B^k\| = \xi^k(1)$.

(b) Now let $\xi(1) < 1$. By Lemma 10.13, $(\xi^k(1))^{1/k} \to \xi'(0)$ as $k \to \infty$ and $\mathbf{r}(B) = \xi'(0)$ by (5.1).

By Proposition 10.14, B is compact. Since B is also continuous and additive and thus homogeneous and order-preserving, existence of the eigenvector follows from Theorem 9.10.

(c) Now assume that $\xi(1) = 1$ and ξ is not the identity function on [0,1]. Then $\mathbf{r}(B) = 1$.

Suppose $cw(B) > \xi'(0)$. By (5.30), there exists some $r > \xi'(0)$ and $f \in X_{con}$, $f(1) > 0$ such that $Bf \geq rf$. Then

$$f(t) \leq (1/r)^k f(\xi^k(t)) \leq f(1)(1/r)^k \xi^k(t) = f(1)\big((\xi^k(t))^{1/k}/r\big)^k$$

for all $k \in \mathbb{N}$ and $t \in [0, 1]$. Let $t \in (0, 1)$. By Lemma 10.13, $(\xi^k(t))^{1/k} \to \xi'(0) < r$ as $k \to \infty$. So $(\xi^k(t))^{1/k}/r < c < 1$ for all but finitely many k and

$$\big((\xi^k(t))^{1/k}/r\big)^k \to 0, \qquad k \to \infty.$$

This implies $f(t) = 0$. Since this holds for any $t \in (0, 1)$ and f is continuous on $[0, 1]$, $f(t) = 0$ for all $t \in [0, 1]$, a contradiction.

This implies $cw(B) \leq \xi'(0)$. Since we already know $cw(B) \geq \xi'(0)$, equality follows. $\square$

PROPOSITION 10.18. *Let $\xi \in X_{con}$ and (ξ_n) a sequence in X_{con}. Assume that there is some $\alpha \in (0,1)$ such that $\xi(1) \le \alpha$ and $\xi_n(1) \le \alpha$ for all $n \in \mathbb{N}$. Let B be associated with ξ and B_n be associated with ξ_n and $\xi_n \to \xi$ uniformly on $[0,1]$. Then $\|B_n - B\| \to 0$ as $n \to \infty$ for the operator norms associated with the cone X_{con}.*

PROOF. By Lemma 10.11, for all $t \in [0,1]$ and $f \in X_{con}$,

$$\big|B_n(f)(t) - B(f)(t)\big| = |f(\xi_n(t)) - f(\xi(t)| \le \frac{f(1) - f(\alpha)}{1-\alpha}|\xi_n(t) - \xi(t)|.$$

We take the supremum over $t \in [0,1]$,

$$\|B_n(f) - B(f)\|_\infty \le \frac{f(1) - f(\alpha)}{1-\alpha}\|\xi_n - \xi\|_\infty.$$

We take the supremum over $f \in X_{con}$ with $\|f\|_\infty \le 1$,

$$\|B_n - B\| \le (1-\alpha)^{-1}\|\xi_n - \xi\|_\infty \to 0. \qquad \square$$

COROLLARY 10.19. *If $\xi'(0) > 0$ and $\xi(1) < 1$ or if $\xi(1) = 1$, upper semicontinuity of the spectral radius fails at B:*

Approximate ξ uniformly by (ξ_n) according to (10.6), with $\xi_n(1) \le 1-(1/2n) < 1$ by (10.8). Let B be associated with ξ and B_n be associated with ξ_n by (10.10). Then the B_n are compact on X_{con}, $\|B_n - B\| \to 0$ and $\mathbf{r}(B_n) = \xi_n'(0) = 0$ for all $n \in \mathbb{N}$ and $\mathbf{r}(B) = \xi'(0) > 0$ or $\mathbf{r}(B) = 1$.

The case $0 < \xi'(0) \le \xi(1) < 1$ is the more interesting one because B is compact as well.

The function ξ given by $\xi(t) = t/2$ [**146**] has the properties required in Corollary 10.19. In this case, $\mathbf{r}(B) = 1/2$ and ξ is an associated eigenfunction. In general, it may be difficult to find the eigenfunctions explicitly though they exist by Theorem 9.10.

10.4.4. Other cones. For completeness, we determine the spectral radius of A on $X = C_0(0,1]$ and of its restrictions A_+ to X_+ and A_0 to X_0. Recall that X_0 is the cone of increasing nonnegative functions in X.

PROPOSITION 10.20.

(a) $\mathbf{r}(A_0) \le \mathbf{r}(A_+) = \mathbf{r}(A) \le 1$.
(b) *Assume that $\xi^n(1) > 0$ for all $n \in \mathbb{N}$ (e.g., if $\xi(1) = 1$ or $\xi(t) > 0$ for all $t \in (0,1]$). Then $\mathbf{r}(A_0) = \mathbf{r}(A_+) = \mathbf{r}(A) = 1$.*
(c) *If $\xi'(0) > 0$, $cw(A_+) = 1$.*
(d) *If $\xi(1) < 1$ and $\xi(t) = 0$ for some $t \in (0,1]$, the $\mathbf{r}(A) = 0$.*

PROOF. (a) Since X_+ is a generating cone of the ordered Banach space X, X_+ is non-flat by Corollary 2.38 and $\mathbf{r}(A) = \mathbf{r}(A_+)$ by Theorem 5.15. Further $\mathbf{r}(A_+) \ge \mathbf{r}(A_0)$ by Theorem 6.23. Since $A^n f = f \circ \xi^n$, $\|A^n f\|_\infty \le \|f\|_\infty$ for all $n \in \mathbb{N}$ and so $\mathbf{r}(A) \le 1$ by (5.8).

(b) Assume that $\xi^n(1) > 0$ for all $n \in \mathbb{N}$. For $f \in X_0$ and $n \in \mathbb{N}$, $\|A_0^n f\| = f(\xi^n(1))$. Since, for all $n \in \mathbb{N}$, $\xi^n(1) > 0$, for each $n \in \mathbb{N}$, we can find some $f \in X_0$ such that $f(t) = 1$ for $t \in [\xi^n(1), 1]$, $n \in \mathbb{N}$. So $\|A_0^n\| \ge 1$ and $\mathbf{r}(A_0) = 1$.

(c) Let $\xi'(0) > 0$. Choose $f(t) = t^\alpha$ with $\alpha > 0$. By Lemma 10.11,

$$(A_+ f)(t) = (\xi(t))^\alpha \ge (\xi'(0))^\alpha f(t), \qquad t \in [0,1].$$

By (5.30), $cw(A_+) \geq (\xi'(0))^\alpha$ for all $\alpha > 0$. We take the limit as $\alpha \to 0$ and obtain $cw(A_+) = 1$.

(d) Let $\xi(1) < 1$ and $\xi(t) = 0$ for some $t \in (0, 1]$. Then $\xi = 0$ on $[0, t]$ and $\xi^k = 0$ on [0,1] for sufficiently large k. This implies $\mathbf{r}(A_+) = 0$. □

COROLLARY 10.21. *If $\xi(1) < 1$ and $\xi^n(1) > 0$ for all $n \in N$, B is compact on X_{con} but its extension to $\tilde{X} = X_{con} - S_{con}$, which is the restriction of A to $\tilde{X}$, is not compact.*

PROOF. Let $\xi(1) < 1$. Then B is compact on X_{con} by Proposition 10.14. Let $\tilde{A}$ be the restriction of A to $\tilde{X}$. Suppose that $\tilde{A}$ is compact. Since A is continuous, A is compact on the closure of $\tilde{X}$ which is X because X_{con} is a total cone of X. By Theorem 5.20, $\mathbf{r}(A) = \mathbf{r}(B)$. By Theorem 10.17, $\mathbf{r}(B) = \xi'(0) \leq \xi(1) < 1$. See Lemma 10.11. By Proposition 10.20, $\mathbf{r}(A) = 1 > \mathbf{r}(B)$, a contradiction. □

If ξ is not the identity function of $[0, 1]$, there is no lower eigenvector of A_+ associated with 1. The proof is the same as the one at the end of the proof of Theorem 10.17. So A_+ has not the lower KR property. See Definition 9.7.

COROLLARY 10.22. *The cone X_{con} is flat.*

PROOF. Suppose that X_{con} is non-flat. Since X_{con} is a total cone in the Banach space X, $\mathbf{r}(A) = \mathbf{r}(B)$ by Theorem 5.16. But for $\xi(t) = t/2$, $\mathbf{r}(A) = 1$ by Proposition 10.20 (b) and $\mathbf{r}(B) = 1/2$ by Theorem 10.17 (b), a contradiction. □

CHAPTER 11

Eigenfunctionals

For a positive linear map on an ordered Banach space, the celebrated Krein-Rutman theorem does not only prove the existence of a positive eigenvector but also shows the existence of a positive linear continuous eigenfunctional (a positive eigenvector of the dual map) provided that the map is compact and the cone is total [**140**]. An even older paper by Krein provides the existence of such a functional if the map is not necessarily compact but the cone is normal and solid [**139**]. (See also [**131**, Thm.7] [**180**, App.2.4,2.6].) We will extend these results to homogeneous operators, though the eigenfunctionals may no longer be continuous.

DEFINITION 11.1. *Let X be an ordered vector space with order cone X_+ and $B : X_B \to X_B \subseteq X_+$ be homogeneous and $\mathbf{r} > 0$. A homogeneous functional $\theta : X_B \to \mathbb{R}_+$ is called an* eigenfunctional *of B associated with $\mathbf{r}$ if $\theta \circ B = \mathbf{r}\theta$.*

In more detail, θ is an eigenfunctional if $\theta(B(x)) = \mathbf{r}\theta(x)$ for all $x \in X_B$.

Eigenfunctionals of B associated with $\mathcal{T}_0 = \mathbf{r}(B)$ will play a big role in establishing the threshold property of $\mathbf{r}(B)$ for the extinction or persistence of a population.

If $\theta : X_B \to \mathbb{R}_+$ is an eigenfunctional of B associated with $\mathcal{T}_0$, then, by induction,

$$\theta(B^n(x)) = \mathcal{T}_0^n\theta(x), \qquad x \in X_B.$$

So, if $\theta(x) > 0$, $x \in X_B$, is related to the size of a population with structural distribution x, and if the basic turnover operator B were to coincide with the turnover operator F, the population would die out if $\mathcal{T}_0 < 1$ and would persist (in the form of geometric growth) if $\mathcal{T}_0 > 1$.

If, as usually, $B \neq F$, these behaviors are somewhat inherited by F (Section 1.3).

The case $\mathcal{T}_0 = 1$, does not carry over to the case $B \neq F$.

See Chapter 14.

11.1. Preparations

We explore what still can be done if the additivity of the operator is dropped. Some ideas from the linear case are adopted [**26**] [**81**, IV.Prop.1.10]. Recall the left resolvents R_λ, $\lambda > \mathbf{r}_o(B) > 0$, in Chapter 8.

These are the overall assumptions for this section.

Let X be an ordered normed vector space with closed order cone X_+ and $B : X_B \to X_B$ be a homogeneous bounded order-preserving operator on a serially complete cone X_B contained in X_+. Recall the notation $\dot{X}_B = X_B \setminus \{0\}$.

The presentation follows [**200**] where the special case $X_B = X_+$ is considered. The situation $X_B \subseteq X_+$ better fits the examples in Section 11.6.

PROPOSITION 11.2. *Let B be uniformly order-bounded. Assume that $\mathbf{r} = \mathbf{r}_o(B) > 0$ and that there exist a sequence (λ_n) in $(\mathbf{r}, \infty)$ with $\lambda_n \to \mathbf{r}$ as $n \to \infty$ and a sequence $p_n : X_B \to \mathbb{R}_+$ of homogeneous, convex, order-preserving functionals with*

$$\|p_n\| \le 1, \quad n \in \mathbb{N}, \qquad \|p_n \circ R_{\lambda_n}\| \to \infty, \quad n \to \infty.$$

Then there exists a bounded homogeneous, order-preserving, nonzero eigenfunctional $\theta : X_B \to \mathbb{R}_+$ such that $\theta \circ B = \mathbf{r}\theta$.

If B is subadditive, so is θ. If the p_n are additive and B is additive, so is θ. If the p_n are additive and B is superadditive, θ is superadditive.

Our proof will not provide continuity of θ.

PROOF. Let (λ_n) and (p_n) be as above. Define $\psi_n : X_B \to \mathbb{R}_+$ by

$$\psi_n(x) = p_n(R_{\lambda_n}(x)), \qquad n \in \mathbb{N}, \quad x \in X_B.$$

The functionals ψ_n are homogeneous, order preserving, and bounded, $\|\psi_n\| \to \infty$ as $n \to \infty$. By (8.2),

$$p_n(R_{\lambda_n}(\lambda_n x)) = p_n\big(R_{\lambda_n}(B(x)) + x\big), \qquad x \in X_B.$$

Since the p_n are order-preserving, subadditive, and homogeneous,

$$\psi_n(B(x)) \le \psi_n(\lambda_n x) = \lambda_n \psi_n(x) \le \psi_n((B(x)) + p_n(x).$$

We set $\theta_n = \psi_n / \|\psi_n\|$. The $\theta_n : X_B \to \mathbb{R}_+$ are homogeneous and order-preserving, $\|\theta_n\| = 1$, and

$$0 \le \lambda_n \theta_n(x) - \theta_n(B(x)) \le \frac{p_n(x)}{\|\psi_n\|} \to 0, \qquad n \to \infty, \quad x \in X_B. \tag{11.1}$$

By Tychonoff's compactness theorem for topological products, there exists some

$$\theta \in \bigcap_{m \in \mathbb{N}} \overline{\Theta_m}, \qquad \Theta_m = \{\theta_n; n \ge m\},$$

where the closure is taken in the topology of pointwise convergence on $\{x \in X_B; \|x\| \le 1\}$. Notice that all θ_n are order-preserving, bounded and homogeneous. θ inherits these properties. For instance, let $x_1, x_2 \in X_+$ and $x_1 \le x_2$. Then there exist a strictly increasing sequence (n_j) of natural numbers such that $\theta_{n_j}(x_i) \to \theta(x_i)$ as $j \to \infty$, $i = 1, 2$. Since $\theta_{n_j}(x_1) \le \theta_{n_j}(x_2)$ for all $j \in \mathbb{N}$, also $\theta(x_1) \le \theta(x_2)$.

Similarly, for $x \in X_B$, there exists a strictly increasing sequence (n_j) of natural numbers such that $\theta_{n_j}(x) \to \theta(x)$ and $\theta_{n_j}(B(x)) \to \theta(B(x))$ as $j \to \infty$. By (11.1), $\theta(B(x)) = \mathbf{r}\theta(x)$.

The inheritance of various properties by θ from B and the functionals p_n follows similarly.

We need to rule out that θ is the zero functional. Since B is uniformly order-bounded, there is some $u \in X_B$, $\|u\| = 1$, such that B is uniformly u-bounded: There exist some $c \ge 0$ such that $B(x) \le c\|x\|u$ for all $x \in X_B$.

Let $x \in X_B$, $\|x\| \le 1$. Since each θ_n is order-preserving, by (11.1),

$$\lambda_n \theta_n(x) \le \theta_n(B(x)) + \frac{p_n(x)}{\|\psi_n\|} \le \theta_n(c\|x\|u) + \frac{p_n(x)}{\|\psi_n\|} \le c\theta_n(u) + \frac{\|p_n\|}{\|\psi_n\|}.$$

Since this holds for all $x \in X_B$, $\|x\| \le 1$, and since $\|p_n\| \le 1$,

$$\lambda_n = \lambda_n \|\theta_n\| \le c\theta_n(u) + \frac{1}{\|\psi_n\|}.$$

Since $\theta_{n_j}(u) \to \theta(u)$ for some strictly increasing sequence (n_j) in $\mathbb{N}$ and $\|\psi_{n_j}\| \to \infty$ and $\lambda_{n_j} \to \mathbf{r}$, we have $0 < \mathbf{r} \le c\theta(u)$. □

DEFINITION 11.3. *Let $u \in X_B$ and $B : X_B \to X_B$. Then B is called* pointwise u-positive *if for any $x \in \dot{X}_B$ there exists some $n \in \mathbb{N}$ such that $B^n(x)$ is u-positive.*

Recall Definition 2.59.

REMARK 11.4. Let $u \in \dot{X}_B$ and $\theta : X_B \to \mathbb{R}_+$ be a homogeneous, order-preserving nonzero eigenfunctional of B, $\theta \circ B = r\theta$ with some $r > 0$.

If B is pointwise u-bounded, then $\theta(u) > 0$. If B is pointwise u-bounded and pointwise u-positive, $\theta(x) > 0$ for all $x \in \dot{X}_B$.

PROOF. Suppose $\theta(u) = 0$. Let $x \in X_B$. Then there exist some $n \in \mathbb{N}$ and $c > 0$ such that $B^n(x) \le cu$. Since θ is order-preserving and homogeneous,

$$0 = c\theta(u) = \theta(cu) \ge \theta(B^n(x)) = r^n\theta(x).$$

So θ is the zero functional.

By contraposition, since θ is a nonzero functional, $\theta(u) > 0$.

Assume that B is pointwise u-positive and $x \in \dot{X}_B$. Then there exist some $n \in \mathbb{N}$ and $\epsilon > 0$ such that $B^n(x) \ge \epsilon u$. Since θ is order-preserving and homogeneous,

$$0 < \epsilon\theta(u) = \theta(\epsilon u) \le \theta(B^n x) = r^n\theta(x).$$

Since $r > 0$, $\theta(x) > 0$. □

11.2. Eigenfunctionals for maps with CW and lower KR property

As before, these are the overall assumptions for this section.

Let X be an ordered normed vector space with closed order cone X_+ and $B : X_B \to X_B$ be a homogeneous bounded order-preserving operator on a serially complete cone X_B contained in X_+.

DEFINITION 11.5. *$B : X_B \to X_B$ is said to have the* CW (Collatz–Wielandt) property *if $0 < \mathbf{r}_o(B) < \infty$ implies $cw(B) = \mathbf{r}_o(B)$, i.e., by* (8.5) *and* (8.6), *if for any $\mu < \mathbf{r}_o(B)$ there exists some $x \in \dot{X}_B$ with $B(x) \ge \mu x$.*

THEOREM 11.6. *Let $B : X_B \to X_B$ be uniformly order bounded. Assume that $\mathbf{r} = \mathbf{r}_o(B) > 0$ and B has the CW property. Further assume that the closed order cone X_+ is normal or $X_B = X_+$.*

Then there exists a homogeneous, order preserving, bounded eigenfunctional $\theta : X_B \to \mathbb{R}$ such that $\theta \circ B = \mathbf{r}\theta$ and $\theta(u) > 0$ for any uniform order bound of B. θ inherits the following properties from B: additive, subadditive, and superadditive.

PROOF. Choose sequences (λ_n) and (μ_n) with $\mu_n < \mathbf{r} < \lambda_n$ and $\lambda_n, \mu_n \to \mathbf{r}$ as $n \to \infty$. By Proposition 8.6, we have a sequence (x_n^*) in X_+^* with $\|x_n^*\| = 1$ and $\|x_n^* \circ R_{\lambda_n}\| \to \infty$ as $n \to \infty$. Apply Proposition 11.2. □

If X_B is a sup-semilattice, the following condition implies the CW property.

LEMMA 11.7. *Let X_B be a* sup-*semilattice. Then a homogeneous order-preserving map $B : X_B \to X_B$ has the CW property if for any $\mu \in (0, \mathbf{r}_o(B))$ there exist some $n \in \mathbb{N}$ and some $v \in \dot{X}_B$ such that $B^n(v) \geq \mu^n v$.*

PROOF. Let $\mu \in (0, \mathbf{r}_o(B))$, $n \in \mathbb{N}$ and $v \in \dot{X}_B$ such that $B^n(v) \geq \mu^n v$. Following [**8**], we define $w = \sup_{j=0}^{n-1} \mu^{-j} B^j v$. For $j = 0, \dots, n-1$, $B(w) \geq B(\mu^{-j} B^j(v)) = \mu^{-j} B^{j+1}(v) = \mu\mu^{-(j+1)} B^{j+1}(v)$. Since $B^n(v) \geq \mu^n v$, $B(w) \geq \mu^{-n+1} B^n(v) \geq \mu v$. Thus $B(w) \geq \mu\mu^{-j} B^j(v)$ for $j = 0, \dots, n-1$, and so $B(w) \geq \mu w$. This implies that $cw(B) \geq \mu$ for any $\mu \in (0, \mathbf{r}_o(B))$. □

A homogeneous bounded order-preserving map $B : X_+ \to X_+$ is said to have the *KR property* (Krein-Rutman property) if $r := \mathbf{r}(B) > 0$ implies that there exists some $v \in \dot{X}_+$ such that $Bv = \mathbf{r}v$. B is said to have the *lower KR property* if $r := \mathbf{r}(B) > 0$ implies that there exists some $v \in \dot{X}_+$ such that $Bv \geq \mathbf{r}v$. Obviously, the lower KR property implies the CW property. An example, where B has the CW property but not the lower KR property, is given in Example 11.24.

REMARK 11.8.

(a) B has the KR property if $B = K + A$ where $K : X_+ \to X_+$ is homogeneous, compact, continuous and order-preserving and $A : X \to X$ is linear, positive and bounded and
 (i) $\|A\| < \mathbf{r}(B)$ [**156**], or
 (ii) $\mathbf{r}(A) < \mathbf{r}(B)$ and X_+ is normal [**157**].

(b) B has the lower KR property if
 (i) some power of B has the KR property and X_+ is a sup-semilattice [**8**], or
 (ii) X_+ is an inf-semilattice, B is continuous, and there is some $u \in \dot{X}_+$ such that B is uniformly u-bounded and $B(u)$ is a regular point of X_+.

More general classes of maps with lower KR property can be found in [**198, 199**].

THEOREM 11.9. *Let $B : X_B \to X_B$ be homogeneous, be uniformly order bounded. Assume that B has the lower KR property and $\mathbf{r} = \mathbf{r}_o(B) = \mathbf{r}(B) > 0$.*

Then there exists a homogeneous, order preserving, bounded eigenfunctional $\theta : X_B \to \mathbb{R}$ such that $\theta \circ B = \mathbf{r}\theta$ and $\theta(u) > 0$ for any uniform order bound of B. θ inherits the following properties from B: additive, subadditive, and superadditive.

PROOF. Choose a sequence (λ_n) with $\mathbf{r} < \lambda_n$ and $\lambda_n \to \mathbf{r}$ as $n \to \infty$. By Proposition 8.6, with $\mu = \mathbf{r}$, we have a sequence (x_n^*) in X_+^* with $\|x_n^*\| = 1$ and $\|x_n^* \circ R_{\lambda_n}\| \to \infty$ as $n \to \infty$. Apply Proposition 11.2. □

The following result is well-known for vectors rather than functionals if B is linear [**136**, Thm.9.3][**170**, Thm.2.2].

LEMMA 11.10. *Let $B : X_B \to X_B$. Let $\mathbf{r} > 0$, $p \in \mathbb{N}$, and $\phi : X_B \to \mathbb{R}_+$ with $\phi \circ B^p = \mathbf{r}^p \phi$. Set*

$$\theta = \sum_{k=0}^{p-1} \mathbf{r}^{-k} \phi \circ B^k.$$

Then $\theta \circ B = \mathbf{r}\theta$ and $\theta(x) \geq \phi(x)$ for all $x \in X_B$.

COROLLARY 11.11. *Let X_+ be complete and $B : X_+ \to X_+$ be homogeneous, bounded, pointwise order bounded and order-preserving. Assume that $\mathbf{r} = \mathbf{r}(B) > 0$ and that some power of B has the lower KR property. Then there exists a homogeneous, order-preserving, bounded nonzero functional $\theta : X_+ \to \mathbb{R}$ such that $\theta \circ B = \mathbf{r}\theta$. θ inherits the following properties from B: additive, subadditive, and superadditive.*

PROOF. There exists some $m, \ell \in \mathbb{N}$ such that B^m has the lower KR property and B^ℓ is uniformly order-bounded. Recall that X_+ is complete and Theorem 6.3. Set $p = m\ell$. Then $B^p v \geq \mathbf{r}^p v$ for some $v \in \dot{X}_+$, and B^p is uniformly order-bounded. By Theorem 11.6, there exists some homogeneous, order preserving, bounded nonzero functional $\phi : X_B \to \mathbb{R}$ such that $\phi(B^p(x)) = \mathbf{r}^p\phi(x)$ for all $x \in X_B$. Apply the previous lemma and notice that θ inherits the desired properties from ϕ and B. □

If B has the lower KR property, order boundedness can be replaced by a growth condition for the (left) resolvent [**170, 181**]. The following result extends [**170**, Thm.2.1] to nonadditive maps.

THEOREM 11.12. *Let $B : X_B \to X_B$ be homogeneous, bounded, and order-preserving. Assume that $\mathbf{r} = \mathbf{r}(B) > 0$ and that B has the lower KR property and satisfies the following resolvent growth condition:*

For any $x^* \in X_+^*$ and $x \in X_B$, there exists some $\epsilon > 0$ such that

$$\{(\lambda - \mathbf{r})\|x^* R_\lambda(x)\|; \mathbf{r} < \lambda < \mathbf{r} + \epsilon\} \text{ is bounded.}$$

Then there exists a homogeneous, order-preserving, bounded nonzero functional $\theta : X_B \to \mathbb{R}$ such that $\theta \circ B = \mathbf{r}\theta$. θ inherits the following properties from B: additive, subadditive, and superadditive.

PROOF. Since B has the lower KR property, there exists some $v \in \dot{X}_+$ such that $B(v) \geq \mathbf{r}v$. The same proof as for Proposition 8.6 provides an $x^* \in X_+^*$ with $x^* R_\lambda(v) \geq \frac{1}{\lambda - \mathbf{r}} d(v, -X_+) > 0$ for $\lambda > r$. By the resolvent growth condition and the uniform boundedness principle (Theorem 4.8), there exists some $c > 0$ such that $\|x^* \circ R_\lambda\| \leq \frac{c}{\lambda - \mathbf{r}}$ for $\lambda \in (\mathbf{r}, \mathbf{r} + \epsilon)$. Choose a sequence (λ_n) in $(\mathbf{r}, \mathbf{r} + \epsilon)$ such that $\lambda_n \to \mathbf{r}$. Now, in the proof of Proposition 11.2, set $\psi_n = x^* \circ R_{\lambda_n}$ and $\theta_n = \psi_n / \|\psi_n\|$. The arguments now proceed as in this proof, except for showing toward the end that θ is not the zero functional. From our estimates above,

$$\theta_n(v) = \frac{x^* R_{\lambda_n}(v)}{\|x^* \circ R_{\lambda_n}\|} \geq \frac{d(v, -X_+)}{c} > 0.$$

This estimate is inherited by θ. □

Similarly as in [**170**], one notices that the resolvent growth condition is satisfied if, for any $x^* \in X_+^*$ and $x \in X_+$, there is some $c > 0$ such that $\| x^* B^n(x)\| \leq c\mathbf{r}^n \|x\|$ for all $n \in \mathbb{N}$.

REMARK 11.13. Let $v \in \dot{X}_+$ and $r > 0$ such that $B(v) \geq rv$. Define

$$\theta(x) = [x]_v, \qquad x \in X_+.$$

Here $[\cdot]_v$ is the concave bounded homogeneous functional introduced in (4.13).

Then $\theta(B(x)) \geq r\theta(x)$ for all $x \in X_+$. Moreover, $\theta(x) > 0$ for all $x \in \dot{X}_+$ if for any $x \in \dot{X}_+$ there exists some $n \in \mathbb{Z}_+$ such that $B^n(x)$ is v-positive.

In some instances this construction can even lead to an eigenfunctional (Example 11.24).

In particular, if B has the lower KR property, $r = \mathbf{r}(B)$ is associated with a lower eigenfunctional of this type. The same construction as in Lemma 11.10 shows that this also holds if some power of B has the lower KR property.

PROOF. Let $x \in X_+$. By (4.14), $x \geq [x]_v v$. Since B is order-preserving and homogeneous, $B(x) \geq [x]_v B(v) \geq [x]_v rv$. By (4.13),

$$\theta(B(x)) = [B(x)]_v \geq r[x]_v = r\theta(x).$$

The second statement is immediate. □

11.3. Eigenfunctionals for normal cones

If the order cone X_+ is closed and normal, we do not need to assume that B or some power of B has the Collatz–Wielandt or lower KR property.

As before, these are the overall assumptions for this section.

Let X be an ordered normed vector space with closed order cone X_+ and $B : X_B \to X_B$ be a homogeneous bounded order-preserving operator on a serially complete cone X_B contained in X_+.

THEOREM 11.14. *Let X_+ be a normal closed cone. Let $B : X_B \to X_B$ have a uniformly order-bounded power. Assume that $\mathbf{r} = \mathbf{r}(B) > 0$. Then there exists a homogeneous, order-preserving, bounded eigenfunctional $\theta : X_B \to \mathbb{R}$ such that $\theta(B(x)) = \mathbf{r}\theta(x)$ for all $x \in X_B$.*

The functional θ inherits the following properties from B: additive, superadditive, and subadditive.

This theorem is applied in the proof of Theorem 18.3. The normality of the cone cannot be dropped in general even if B is a bounded linear positive map on X and X_+ is solid [**27**, Sec.2(iv)].

We will first consider the special case that B is concave (superadditive) (Proposition 11.15) and then the general case (Proposition 11.16).

PROPOSITION 11.15. *Let X_+ be a normal closed cone. Let B be homogeneous, superadditive, and have a uniformly order-bounded power. Assume that $\mathbf{r} = \mathbf{r}(B) > 0$. Then there exists a homogeneous, superadditive, bounded eigenfunctional $\theta : X_B \to \mathbb{R}$ such that $\theta(B(x)) = \mathbf{r}\theta(x)$ for all $x \in X_B$. If B is additive, θ is additive.*

We emphasize that continuity of B is not required and does not necessarily imply continuity of the eigenfunctional (Section 11.5).

PROOF. By Lemma 11.10, we can assume that B itself is uniformly order bounded. By Corollary 8.5, there exists some $x^* \in X_+^*$ and some $x \in X_B$ such that $x^*(R_\lambda(x)) \to \infty$ as $\lambda \to \mathbf{r}+$. Apply Proposition 11.2. □

If we drop the superadditivity of the operator, the eigenfunctional may be no longer superadditive either. Subadditivity is inherited though.

PROPOSITION 11.16. *Let the order cone X_+ be closed and normal and $B : X_B \to X_B$ have a uniformly order-bounded power and $\mathbf{r}(B) > 0$. Then there exists a homogeneous bounded order-preserving eigenfunctional $\theta : X_B \to \mathbb{R}_+$ such that $\theta \circ B = \mathbf{r}(B)\theta$. If B is subadditive, so is θ.*

PROOF. Recall Corollary 5.24 and define

$$\psi_\lambda(x) := \sum_{n=0}^{\infty} \lambda^{-n-1}\psi(B^n(x)), \qquad x \in X_B, \quad \lambda > \mathbf{r}(B) =: r.$$

Then there exists some $x \in X_B$ such that

$$\psi_\lambda(x) \to \infty, \qquad \lambda \to r + .$$

The functionals $\psi_\lambda : X_B \to \mathbb{R}_+$ are homogeneous, order-preserving and bounded. If B is subadditive, so are the ψ_λ. For all $\lambda > r$, $x \in X_B$,

$$\psi_\lambda(B(x)) = \sum_{n=0}^{\infty} \lambda^{-n-1}\psi(B^{n+1}(x)) = \lambda\psi_\lambda(x) - \psi(x).$$

Further, $\|\psi_\lambda\| \to \infty$ as $\lambda \to r+$. We choose a sequence (λ_k) in (r,∞) with $\lambda_k \to r$. We set

$$\theta_k(x) = \frac{\psi_{\lambda_k}(x)}{\|\psi_{\lambda_k}\|}, \qquad k \in \mathbb{N},$$

and obtain order-preserving, homogeneous, bounded functionals $\theta_k : X_B \to \mathbb{R}_+$. If B is subadditive, so are the θ_k. For all $k \in \mathbb{N}$, $\|\theta_k\| = 1$ and

$$\theta_k(B(x)) = \lambda_k\theta_k(x) - \frac{\psi(x)}{\|\psi_{\lambda_k}\|}.$$

This implies

$$\theta_k(B(x)) - r\theta_k(x) \to 0, \qquad k \to \infty.$$

By Lemma 11.10, we can assume that B is uniformly order-bounded and let u be an order bound. Then there exists some $c > 0$ such that $B(x) \le c\|x\|u$ for all $x \in X_B$. Since θ_k is order-preserving and homogeneous, for all $x \in X_B$ and $k \in \mathbb{N}$,

$$c\|x\|\theta_k(u) = \theta_k(c\|x\|u) \ge \lambda_k\theta_k(x) - \frac{\psi(x)}{\|\psi_{\lambda_k}\|}.$$

Since $\|\theta_k\| = 1$,

$$c\theta_k(u) \ge \lambda_k - \frac{\|\psi\|}{\|\psi_{\lambda_k}\|}.$$

We take the limit inferior as $k \to \infty$.

$$\liminf_{k\to\infty} \theta_k(u) \ge r/c.$$

The same application of Tychonoff's compactness theorem for topological products as in the proof of Proposition 11.2 provides a homogeneous, bounded, order-preserving functional $\theta : X_B \to \mathbb{R}_+$ with $\theta(u) > 0$ and $\theta(B(x)) = r\theta(x)$ for all $x \in X_B$. If B is subadditive, θ inherits the subadditivity from the θ_k. □

The next result shows that, in Krein's theorem [**139**], completeness of X can be replaced by monotone completeness of X_+.

COROLLARY 11.17. *Let the order cone X_+ be closed, normal and solid and serially complete and $B : X \to X$ be linear and positive. Then there exists a linear bounded positive eigenfunctional $\theta : X \to \mathbb{R}$ such that $\theta \circ B = \mathbf{r}(B)\theta$.*

PROOF. Since X_+ is solid, it is generating and non-flat (Theorem 2.38). Let B_+ be the restriction of B to X_+.

Since X_+ is normal and serially complete, B_+ is bounded (Theorem 4.5). By Proposition 11.15, there is some additive homogeneous bounded $\theta_+ : X_+ \to \mathbb{R}_+$ such that $\theta_+ \circ B_+ = \mathbf{r}(B_+)\theta_+$. Since X_+ is generating and non-flat, θ_+ can be extended to a continuous linear functional θ on X and $\mathbf{r}(B) = \mathbf{r}(B_+)$ and $\theta \circ B = \mathbf{r}(B)\theta$ (Theorem 5.15). □

11.4. Eigenfunctionals if there is no norm

The next result generalizes [**26**, Thm.2] for an ordered vector space without an a priori norm.

THEOREM 11.18. *Let X be an ordered vector space with cone X_+ and $B : X_+ \to X_+$ be homogeneous and order-preserving.*

Let X have an order unit $u \in \dot{X}_+$, i.e., $X = X_u$. Further let B and u satisfy the following property:

- ◇ *For each $\epsilon > 0$ there exists some $\delta > 0$ such $B(x + \delta u) \le B(x) + \epsilon u$ for all $x \in X_+$ with $x \le u$.*

Define $\eta^u = \eta^u(B) = \inf_{n\in\mathbb{N}} \|B^n\|_u^{1/n}$ where $\|B\|_u := \|B(u)\|_u$. Then $\eta^u(B) = \lim_{n\to\infty} \|B^n\|_u^{1/n}$ and, if $\eta^u(B) > 0$, there exists an order-preserving homogeneous $\theta : X_+ \to \mathbb{R}_+$ with $\theta(u) > 0$ and $\theta \circ B = \eta^u\theta$.

The functional θ inherits the following properties from B: additive, superadditive, subadditive.

Here $\|\cdot\|_u$ is the seminorm introduced in (2.26). Notice that if $X = X_u = X_v$ for $u, v \in \dot{X}_+$, then $\eta^u = \eta^v$.

PROOF. Let

$$J = \{x \in X;\ \forall \xi > 0 : -\xi u \le x \le \xi u\} = \{x \in X : \|x\|_u = 0\}. \tag{11.2}$$

We first assume that $J = \{0\}$; equivalently, $\|\cdot\|_u$ is a norm on $X = X_u$. One readily sees that B is uniformly u-bounded and bounded and the operator norm of B is given by $\|B(u)\|_u =: \|B\|_u$. So η^u is the spectral radius of B with respect to the norm $\|\cdot\|_u$.

We claim that B is uniformly continuous on every bounded set. Since B is homogeneous, it is sufficient to show that B is continuous on the set $\{x \in X_+; x \le u\}$. Let $\epsilon > 0$. Choose $\delta > 0$ according to ◇ and $x, y \in X_+$, $x, y \le u$, $\|x - y\|_u \le \delta$. Then $-\delta u \le x - y \le \delta u$ and $x \le y + \delta u$. Since B is order-preserving, $B(x) \le B(y + \delta u) \le B(y) + \epsilon u$. Since $y \le x + \delta u$, $B(y) \le B(x) + \epsilon u$ by the same argument. So $\|B(x) - B(y)\|_u \le \epsilon$.

Now let $\tilde{X}$ be the completion of X under $\|\cdot\|_u$ and $\tilde{X}_+$ the closure of X_+ in $\tilde{X}$. We identify X and X_+ with their isometric embeddings in $\tilde{X}$. By the proof of [**26**, L.7], $\tilde{X}_+$ is a normal solid cone and u is an order-unit of $\tilde{X}$. Since B is uniformly continuous on bounded sets, it preserves Cauchy sequences in X_+. So B can be extended to a map on $\tilde{X}_+$,

$$\tilde{B}(\tilde{x}) = \lim_{n\to\infty} B(x_n), \qquad \tilde{x} \in \tilde{X}_+,$$

for any sequence (x_n) in X_+ with $x_n \to \tilde{x}$. We write $\tilde{x} \le \tilde{y}$ if $\tilde{y} - \tilde{x} \in \tilde{X}_+$. $\tilde{B}$ inherits $\diamond$ and is uniformly continuous on bounded subsets of $\tilde{X}_+$. $\tilde{B}$ is concave and homogeneous, and η^u is the spectral radius of $\tilde{B}$.

By Theorem 11.14, there exists an order-preserving and homogeneous functional $\tilde{\theta} : \tilde{X}_+ \to \mathbb{R}_+$ such that $\tilde{\theta} \circ \tilde{B} = \eta^u \tilde{\theta}$. Restricting $\tilde{\theta}$ to X_+ provides the desired functional θ. θ inherits the following properties from B: additive, superadditive, subadditive.

We now drop the assumption that $J = \{0\}$. Recall (11.2). As pointed out in the proof of [**26**, Thm.1], J is an ideal in X, $u \notin J$, and the factor space $X/J = \{x + J; x \in X\}$ is an ordered vector space with cone $X_+ + J$ and unit $u + J$. Further $\|x + J\|_{u+J} = \|x\|_u$ provides a norm on X/J. Differently from [**26**], B does not leave J invariant, but we have from $\diamond$ that

$$x, y \in X_+, x - y \in J \implies B(x) - B(y) \in J.$$

The arguments are very similar to those that provided the continuity of B before.

We can now define $B_0(x + J) = B(x) + J$, $x \in X_+$.

By our previous considerations, there exists an order-preserving homogeneous functional $\theta_0 : X_+ + J \to \mathbb{R}_+$ such that $\theta_0 \circ B_0 = \eta^u \theta_0$ where

$$\eta^u = \inf_{n\in\mathbb{N}} \|B_0^n(u + J)\|_{u+J}^{1/n} = \inf_{n\in\mathbb{N}} \|B^n(u) + J\|_{u+J}^{1/n} = \inf_{n\in\mathbb{N}} \|B^n(u)\|_u^{1/n}.$$

We define $\theta(x) = \theta_0(x + J)$ for $x \in X_+$. One easily checks that ϕ is homogeneous and order-preserving, and the same proof as in [**26**] shows that $\theta \circ B = \eta^u \theta$. θ_0 inherits the following properties from B and θ from θ_0: additive, superadditive, subadditive. □

Recall X_u, the set of u-bounded elements in X (Definition 2.59). The point $u \in \dot{X}_+$ is an order unit if $X = X_u$. If u is not an order unit but $B(u)$ is u-bounded, we introduce

$$X_u^B = \{x \in X_+;\ \exists n \in \mathbb{Z}_+ : B^n(x) \in X_u\}. \tag{11.3}$$

We note that X_u^B is a cone which is mapped into itself by B and that $X_u^B = X$ if and only if B is pointwise u-bounded (Definition 6.1).

COROLLARY 11.19. *Let X be an ordered vector space with cone X_+ and $u \in \dot{X}_+$. Assume that $B : X_+ \to X_+$ is homogeneous and superadditive and satisfies $\diamond$ in Theorem* 11.18.

Let $\eta^u = \inf_{n\in\mathbb{N}} \|B^n\|_u^{1/n}$ where $\|B\|_u := \|B(u)\|_u$. The following hold:

(a) $\eta^u = \lim_{n\to\infty} \|B^n\|_u^{1/n}$.
(b) *If $u, v \in \dot{X}_+$ and $X_u^B = X_v^B$, then $\eta^v > 0$ implies $\eta^u > 0$ and $\eta^u = \eta^v$.*
(c) *If $\eta^u > 0$, there exists a concave homogeneous $\theta : X_u^B \to \mathbb{R}_+$ with $\theta(u) > 0$ and $\theta \circ B = \eta^u \theta$ on X_u^B.*
(d) *If B is additive, θ is additive.*

PROOF. It follows from $\diamond$ with $x = 0$ and the homogeneity of B that $B(u)$ is u-bounded and B maps $X_+ \cap X_u$ into $X_+ \cap X_u$. By construction, u is an order unit of X_u. Part (a) follows from Theorem 11.18.

For Part (b), assume $X_u^B = X_v^B$ for $u, v \in \dot{X}_+$ and $\eta^v > 0$. There exist $k, m \in \mathbb{Z}_+$ such that $B^k(v) \in X_u$ and $B^m(u) \in X_v$. So $B_k(v) \le cu$ for some $c > 0$.

For all $n \in \mathbb{N}$, since B is order-preserving and homogeneous,

$$B^{n+k}(v) \leq cB^n(u) \leq c\|B^n\|_u u.$$

By the same token, for some $\tilde{c} > 0$,

$$B^{n+k+m}(v) \leq c\|B^n\|_u B^m(u) \leq c\tilde{c}\|B^n\|_u v, \qquad n \in \mathbb{N}.$$

Hence $\|B^{n+k+m}\|_v \leq c\tilde{c}\|B^n\|_u$ for all $n \in \mathbb{N}$ and

$$\left(\|B^{n+k+m}\|_v^{1/(n+k+m)}\right)^{(n+m+k)/n} \leq c\tilde{c}\|B^n\|_u^{1/n}, \qquad n \in \mathbb{N}.$$

We take the limit as $n \to \infty$ and obtain $\eta^v \leq \eta^u$. Equality follows by symmetry.

(c) We apply Theorem 11.19 to X_u and the restriction of B to $X_+ \cap X_u$. This provides an concave homogeneous $\tilde{\theta} : X_+ \cap X_u \to \mathbb{R}_+$ with $\tilde{\theta} \circ B = \eta^u \theta$. We extend $\tilde{\theta}$ to X_u^B as follows: If $B^n(x) \in X_+$ for some $n \in \mathbb{Z}_+$ we set $\theta(x) = (\eta^u)^{-n}\tilde{\theta}(B^n(x))$. This definition does not depend on $n \in \mathbb{N}$ and has the desired properties. □

REMARK 11.20. If X is an ordered normed vector space, $u \in \dot{X}_+$, $B : X_+ \to X_+$ is homogeneous, bounded, superadditive, and satisfies $\diamond$, the question arises how $\eta^u(B)$ is related to $\gamma_u(B)$ and $\mathbf{r}(B)$. If some power of B is uniformly u-bounded, $\eta^u(B) \leq \gamma_u(B)$ and, if u is a normal point of X_+ in addition, $\mathbf{r}(B) \leq \eta^u(B)$ and all three numbers are equal among themselves and to $\mathbf{r}_o(B)$ [**198, 199**]. If u is not a normal point of X_+, $\eta^u(B) < \gamma_u(B) = \mathbf{r}(B)$ may occur even if B is a positive bounded linear map on an ordered Banach space X and X_+ is a solid cone [**27**, Sec.2(iv)].

EXAMPLE 11.21. Let $X = \mathbb{R}^\mathbb{N}$ with cone $\mathbb{R}_+^\mathbb{N}$ and B the right-shift operator $B : (x_1, x_2, \dots) \mapsto (0, x_1, x_2, \dots)$. Then $cw(B) = 0$. For $\alpha > 0$, let $X_{(\alpha)}$ be the normed vector space of sequences in $\mathbb{R}^\mathbb{N}$ that are u-bounded for $u = (\alpha^n)$. Then B maps $X_{(\alpha)}$ into itself and $\eta^u = 1/\alpha$. By [**139**] (or Corollary 11.19), there exists a linear positive eigenfunctional of B on $X_{(\alpha)}$ associated with $1/\alpha$.

If $\alpha = 1$, u is the constant sequence with all terms being 1 and X_u can be identified with ℓ^∞ with the supremum norm and $\eta^u = 1$. There is an explicit superadditive eigenfunctional of B on ℓ^∞, namely, the limit inferior of the sequences. If we restrict B to c, the closed linear subspace of convergent sequences, the limits of sequences form an explicit linear positive eigenfunctional. However, no linear positive eigenfunctional that is associated with 1, the spectral radius of B, exists on c_0, the closed linear subspace of sequences that converge to 0. The same is the case for $\ell^p \subseteq c_0$ with $p > 1$ with the p-norm which shows that regularity of the cone is not suffucient for the existence of an additive eigenfunctional. On ℓ^1, with the sum-norm, $(x_n) \mapsto \sum_{n=1}^\infty x_n$ is a bounded linear positive eigenfunctional of the right-shift operator associated with one. Notice that the existence of this eigenfunctional does not follow from any of our results or the Krein-Rutman theorem and holds without uniform order-boundedness of the map.

11.5. Continuity of homogeneous order-preserving eigenfunctionals

Our existence proofs do not provide that the eigenfunctional θ is continuous and Example 11.24 will show that this cannot be expected in general even if B is additive. So some extra conditions will be needed for continuity.

PROPOSITION 11.22. *Let $\theta : X_+ \to \mathbb{R}_+$ be homogeneous and order-preserving and $u \in \dot{X}_+$. Then $\theta : X_+ \cap X_u \to \mathbb{R}_+$ is continuous with respect to the u-norm at every u-comparable point $v \in X_+$. In fact,*

$$|\theta(x) - \theta(v)| \le \|x - v\|_u \|u\|_v \theta(v), \qquad x \in X_- \cap X_u.$$

PROOF. Let $x \in X_+ \cap X_u$. Set $\epsilon = \|x - v\|_u$. Then

$$-\epsilon u \le x - v \le \epsilon u \quad \text{and} \quad v - \epsilon u \le x \le v + \epsilon u.$$

So

$$(1 - \epsilon\|u\|_v)v \le x \le (1 + \epsilon\|u\|_v)v.$$

Since θ is homogeneous and order-preserving,

$$\theta(x) \le (1 + \epsilon\|u\|_v)\theta(v).$$

If $(1 - \epsilon\|u\|_v) \ge 0$, then

$$\theta(x) \ge (1 - \epsilon\|u\|_v)\theta(v)$$

by the same token. If $(1 - \epsilon\|u\|_v) \le 0$, the previous inequality is true anyway. So

$$-\epsilon\|u\|_v\theta(v) \le \theta(x) - \theta(v) \le \epsilon\|u\|_v\theta(v)$$

and the assertion follows. □

THEOREM 11.23. *Let $B : X_+ \to X_+$ be homogeneous, continuous and order-preserving and $\theta : X_+ \to \mathbb{R}_+$ be homogeneous, bounded, and order-preserving. Moreover let $r \in (0, \infty)$ and $\theta(B(x)) = r\theta(x)$ for all $x \in X_+$*

Let $u \in \dot{X}_+$ and B be pointwise u-positive. Further assume that there exists some $k \in \mathbb{N}$ such that B^k maps X_+ with the original norm continuously into $X_+ \cap X_u$ with the u-norm.

Then θ is continuous on X_+ with respect to the original norm.

PROOF. Since B is continuous, for all $p \ge k$, B^p maps X_+ with the original norm continuously into $X_+ \cap X_u$ with the u-norm. Let $w \in \dot{X}_+$. Since B is pointwise u-positive, there exists $m \in \mathbb{N}$ and $\epsilon > 0$ such that $B^m(w) \ge \epsilon u$. Further there exists some $q \in \mathbb{N}$ and $\delta > 0$ such that $B^q(u) \ge \delta u$. Then $B^{jq+m}(w) \ge \epsilon\delta^j u$ for all $j \in \mathbb{N}$.

By choosing $n \in \mathbb{N}$ large enough, we can assume that $B^n(w)$ is u-comparable and B^n maps X_+ with the original norm continuously into $X_+ \cap X_u$ with the u-norm. Set $v = B^n(w)$. By Proposition 11.22,

$$r^n|\theta(x) - \theta(w)| = |\theta(B^n(x) - \theta(B^n(w)| \le \|B^n(x) - B^n(w)\|_u\|u\|_v\theta(v).$$

So θ is continuous at $w \in \dot{X}_+$. Since θ is homogeneous and bounded, θ is continuous at 0. □

11.6. An example by Bonsall

We give an example of a bounded linear map A on a Banach space X that maps a total, normal, regular, closed cone X_B into itself. Though the restriction B of A to X_B is compact and uniformly order-bounded, all additive homogeneous bounded eigenfunctionals of B associated with $\mathbf{r}(B) > 0$ (which exist according to Proposition 11.14) are discontinuous.

This example [**27**] has served for other counterexamples as well [**146**] (see also Section 10.4).

EXAMPLE 11.24. Let $X = C_0(0,1]$ be the Banach space of continuous functions $f : [0,1] \to \mathbb{R}$ with $f(0) = 0$ endowed with the supremum norm. Let X_{con} be the cone of nonnegative convex functions in X. X_{con} is closed and regular but not solid. As shown in [**27**, Sec.2], X_{con} is also total, but not generating. Let $u \in X_{con}$ be given by $u(t) = t$ for all $t \in [0,1]$. For any $f \in X_{con}$, $f(t)/t$ is an increasing function of t, and so $f \le f(1)u$, and u is an order unit of X_{con}. As in [**27**, Sec.4], we consider the continuous linear map A on X given by

$$A(f)(t) = f(g(t)), \qquad 0 \le t \le 1, \quad f \in X.$$

Here $g \in X_{con}$ has been chosen such that $g(1) < 1$ and $g'(0) > 0$. A maps X_{con} into itself. Then the restriction B of A to X_{con} is compact and $\mathbf{r}(B) = g'(0)$. See Lemma 10.14.

Recall that for $f \in X_{con}$, $f(t)/t$ is an increasing function of $t \in (0,1]$. So each $f \in X_{con}$ is differentiable at 0 and

$$\theta(f) := [f]_u = \lim_{t\to 0+} f(t)/t = f'(0) \le f(1), \qquad f \in X_{con}.$$

The functional θ is homogeneous, bounded, and additive, and an eigenfunctional of B satisfying $\theta(Bf) = g'(0)\theta(f)$ for all $f \in X_{con}$. The eigenfunctional θ is not continuous at u because $\theta(u) = 1$ while $\theta(f_\alpha) = 0$ with $f_\alpha(t) = t^\alpha$, $\alpha > 1$.

To see that there is no eigenfunctional θ on X_{con} that is continuous at u, we consider $h_\alpha \in X_{con}$, $\alpha \in [0,1]$, given by

$$h_\alpha(t) = \begin{cases} 0, & 0 \le t \le \alpha, \\ t-\alpha, & \alpha < t \le 1. \end{cases}$$

If $\alpha \in (0,1]$, $B^n(h_\alpha) = 0$ for sufficiently large $n \in \mathbb{N}$. So any eigenfunctional of B associated with a positive eigenvalue satisfies $\theta(h_\alpha) = 0$ if $\alpha > 0$. If θ were continuous at u, then also $\theta(u) = 0$ because $\|h_\alpha - u\|_\infty \to 0$ as $\alpha \to 0$. But this would imply that $\theta = 0$ because B is uniformly u-bounded.

Notice that $\theta = [\cdot]_u$ is continuous with respect to the u-norm $\|\cdot\|_u$ by Lemma 4.21 (an information not provided by our existence theorems).

We could also have considered the cone of $f \in X$ such that $f(t)/t$ is an increasing function of $t \in (0,1]$. It is no longer clear whether the restriction of A to that cone is compact.

The functional $[\cdot]_u$ is also defined for the cone of nonnegative functions in X, $C_{0+}(0,1]$, but it is only a lower eigenfunctional associated with $g'(0)$ of the restriction of B to $C_{0+}(0,1]$. Indeed, by (5.27),

$$[Bf]_u \ge [B]_u [f]_u$$

and

$$[B]_u = [u \circ g]_u = [g]_u = g'(0).$$

The spectral radius of B with respect to $C_{0+}(0,1]$ is one. It also shows that B has the CW property. B has not the lower KR property. Indeed, if $B(f) \ge f \in C_{0+}(0,1]$, then $f(t) \le \sup f([0, g^n(1)])$ for all $n \in \mathbb{N}$ and $t \in [0,1]$. Since f is continuous at $0 = f(0)$ and $g^n(1) \to 0$, $f \equiv 0$.

Although B is additive and has the CW property, there is no additive homogeneous eigenfunctional on $C_{0+}(0,1]$ because this cone is generating and any such eigenfunctional could be extended to a linear positive eigenfunctional on the Banach space X that would be automatically continuous. Since the continuous functions with compact support in $(0,1]$ are dense in X and $B^n f = 0$ for any such function if

n is large enough, the eigenfunctional would be zero. This shows that, in general, uniform order boundedness cannot be omitted as a condition in Theorem 11.6 or Theorem 11.15.

There is no homogeneous bounded lower eigenfunctional associated with 1. Suppose θ were such an eigenfunctional. Then $\theta(f) \le \theta(Bf)$ for all $f \in C_{0+}(0,1]$. By iteration, for all $n \in \mathbb{N}$,

$$\theta(f) \le \theta(B^n f) \le \|\theta\| \, \|B^n(f)\| \le \|\theta\| \sup f([0,2^{-n}]) \to \|\theta\| f(0) = 0.$$

So θ is the zero functional. Notice that B satisfied the resolvent growth condition in Theorem 11.12. There is no contradiction because B has not the lower KR property.

If $g(1) = 1$, then $\tilde{\theta} f = f(1)$ is a continuous eigenfunctional of A on $C_0(0,1]$ and of B on X_{con} associated with $\mathbf{r}(A) = \mathbf{r}(B) = 1$. f is linear or additive and homogeneous, respectively.

On $C[0,1]$, there is a positive linear eigenfunctional of A associated with 1, namely $f \to f(0)$.

EXAMPLE 11.25. We modify the previous example, which was suggested by Bonsall in [**27**, Sec.2]. Let $X = C_0(0,1]$ be the Banach space of continuous functions $f : [0,1] \to \mathbb{R}$ with $f(0) = 0$ endowed with the supremum norm and X_+ be cone of nonnegative functions. For $\xi \ge 0$, let X_ξ be the cone of f in X_+ such that $f(t)/t^\xi$ is an increasing function of $t \in (0,1]$. X_0) is the space of monotone nonnegative functions. X_ξ is complete and normal but not solid. Notice that $X_\zeta \subseteq X_\xi$ if $\zeta \ge \xi \ge 0$. For $\xi \in [0,1]$, X_ξ contains the cone X_{con} of convex nonnegative function and is total [**27**, Sec.2]. For no $\xi \ge 0$, X_ξ is a generating cone of X because $X_0 - X_0$ are the functions of bounded variation in X.

Let $u_\xi \in X_\xi$ be given by $u_\xi(t) = t^\xi$ for all $t \in [0,1]$. Then, for $\xi > 0$ and any $f \in X_\xi$, $f \le f(1)u^\xi$. Let $\phi : \mathbb{R}^2_+ \to \mathbb{R}_+$ be homogeneous, order-preserving, and continuous, $\phi(1,1) > 0$, and $0 < \alpha < \beta < 1$. Define $B : X_+ \to X_+$ by

$$B(f)(t) = \phi(f(\alpha t), f(\beta t)), \qquad f \in X_+, \quad t \in [0,1]. \tag{11.4}$$

B is order-preserving with respect to X_+, homogeneous, and continuous. Notice that B maps X_ξ into X_ξ and let B_ξ be the restriction to X_ξ. B_1 would not map the cone X_{con} of convex nonnegative continuous functions into itself unless we assume that ϕ is convex, an assumption we would not like to make because many important examples for ϕ are concave rather than convex. All functions f_ξ with $f_\xi(t) = t^\xi$, $\xi > 0$, are eigenvectors of B associated with the eigenvalues $\phi(\alpha^\xi, \beta^\xi)$. Since, for $\xi > 0$, B_ξ is uniformly u_ξ-bounded and X_+ is normal, $\mathbf{r}(B_\xi) = \phi(\alpha^\xi, \beta^\xi)$ by Theorem 6.14. Notice that there is an eigenfunctional associated with $\mathbf{r}(B_\xi)$, namely $\theta_\xi(f) = \lim_{t \to 0+} f(t)/t^\xi$. Surprisingly, this eigenfunctional is additive though B_ξ is not; it is not continuous though bounded, $\theta_\xi(f) \le f(1)$.

For nonnegative $f \in X_0$, $B(f)(t) \le \phi(\|f\|, \|f\|) = \phi(1,1)\|f\|$, and so $\mathbf{r}(B_0) \le \phi(1,1)$. Now the functions f_ξ with $\xi > 0$ are eigenfunctions and so $cw(B) \ge \phi(\alpha^\xi, \beta^\xi)$ for all $\xi > 0$. Taking the limit as $\xi \to 0$, $cw(B) \ge \phi(1,1)$, and so $\mathbf{r}(B_0) = \phi(1,1) = cw(B_0)$ by Theorem 6.14. In the same way, we see that $\mathbf{r}(B) = \phi(1,1) = cw(B)$.

There is no eigenvector of B_0 or B associated with $\phi(1,1)$. Since $0 < \alpha < \beta < 1$ and ϕ is order-preserving and homogeneous, for $f \in X_+$, $B(f)(t) \le \phi(\sup f([0,\beta]), \sup f([0,\beta])) = \phi(1,1)\sup f([0,\beta])$. Iterating,

$$B^n(f)(t) \le \phi(1,1)^n \sup f([0,\beta^n]), \quad n \in \mathbb{N}.$$

If f were an eigenvector of B in X_+ associated with $\phi(1,1)$, $f(t) \le \sup f([0,\beta^n])$ for all $n \in \mathbb{N}$ and $t \in [0,1]$. Since $f(0) = 0$ and f is continuous, $f \equiv 0$.

11.7. Strict monotonicity of eigenfunctionals

THEOREM 11.26. *Let X be an ordered normed vector space with closed order cone X_+ and $B : X_B \to X_B$ a homogeneous order-preserving operator on a closed homogeneous subset X_B of X_+. Let $\theta : X_B \to \mathbb{R}_+$ be an eigenfunctional of B, $\theta \circ B = r\theta$ with some $r > 0$. Assume that $u \in \dot{X}_B$ and B is pointwise u-bounded and pointwise u-positive.*

Further assume that, if $x, y \in X_B$ and $x \le y$ and $x \ne y$, there exists some $m \in \mathbb{N}$ and $\epsilon > 0$ such that $(1+\epsilon)B^m(x) \le B^m(y)$.

Then θ is strictly order-preserving: $\theta(x) < \theta(y)$ for all $x, y \in X_B$ with $x \le y$ and $x \ne y$.

PROOF. By Remark 11.4, $\theta(x) > 0$ for all $x \in \dot{X}_B$. Let $x, y \in X_B$ with $x \le y$ and $x \ne y$. If $x = 0$, $\theta(y) > 0 = \theta(x)$. So we can assume that $x \in \dot{X}_B$ and $\theta(x) > 0$. By the monotonicity properties of B, there exists some $n \in \mathbb{N}$ and $\epsilon > 0$ such that $B^n(y) \ge (1+\epsilon)B^n(x)$. Since θ is a homogeneous order-preserving eigenfunction of B,

$$r^n\theta(y) = \theta(B^n(y)) \ge (1+\epsilon)\theta(B^n(x)) = r^n\theta(x).$$

We divide by $r^n > 0$ and obtain $\theta(y) > \theta(x)$. □

PROPOSITION 11.27. *Let X_+ be a normal closed cone and X_B a monotonically complete cone contained in X_+. Let $B : X_B \to X_B$ be homogeneous, superadditive, and have a uniformly order-bounded power. Assume that $\mathbf{r} = \mathbf{r}(B) > 0$. Assume that there is some $u \in \dot{X}_B$ such that B is pointwise u-bounded and pointwise u-positive. Then there exists a strictly order-preserving, bounded homogeneous, superadditive, eigenfunctional $\theta : X_B \to \mathbb{R}$ such that $\theta(B(x)) = \mathbf{r}\theta(x)$ for all $x \in X_B$.*

PROOF. By Proposition 11.15, there exists a bounded homogeneous, superadditive eigenfunctional $\theta : X_B \to \mathbb{R}_+$ such that $\theta(B(x)) = \mathbf{r}\theta(x)$ for all $x \in X_B$. By Remark 11.4, $\theta(y) > 0$ for all $y \in \dot{X}_B$. Let $x \in X_B$ and $y \in \dot{X}_B$. Since θ is superadditive, $\theta(x+y) \ge \theta(x) + \theta(y) > \theta(x)$. This shows that θ is strictly order-preserving. □

If $\mathbf{r}(B)$ is associated with a strictly order-preserving eigenfunctional, lower eigenvectors of B associated with $\mathbf{r}(B)$ are eigenvectors.

THEOREM 11.28. *Let X_+ be a closed cone and X_B a homogeneous set contained in X_+. Let $B : X_B \to X_B$ be homogeneous and bounded. Assume that $\mathbf{r} = \mathbf{r}(B) > 0$ and that there is a strictly order-preserving homogeneous functional $\theta : X_B \to \mathbb{R}_+$ associated with $\mathbf{r}$, $\mathbf{r}\theta = \theta \circ B$.*

Then, if $v \in \dot{X}_B$, $\mathbf{r}v \le B(v)$ implies $\mathbf{r}v = B(v)$.

PROOF. Suppose that $\mathbf{r}v \leq B(v)$ and $\mathbf{r}v \neq B(v)$. Since θ is strictly order-preserving,

$$\mathbf{r}\theta(v) < \theta(B(v)) = \mathbf{r}\theta(v).$$

Since $\theta(v) > 0$, this is contradiction. $\square$

CHAPTER 12

Turnover versus reproduction number

Let X be an ordered normed vector space with closed order cone X_+ and let X_B be a closed serially complete cone contained in X_+ (Section 2.1.2).

Assume that $F : X_B \to X_B$ is an operator that maps this year's structural distribution of a population to next year's structural distribution, $F(0) = 0$, and $B : X_B \to X_B$ is its homogeneous first order approximation at the zero vector, the extinction fixed point. Then B can be seen as *the basic yearly turnover operator* of the population and is spectral radius,

$$\mathbf{r}(B) =: \mathcal{T}_0, \tag{12.1}$$

as the *basic population turnover number*. $\mathcal{T}_0$ has also been called *inherent population growth rate* [**44**] or *(population) growth factor* [**150**].

12.1. The next generation operator

Further assume that $B = Q + H$ with homogeneous, continuous, order-preserving operators $H, Q : X_B \to X_B$, Q additive, $\mathbf{r}(Q) < 1$.

The additive operator Q is associated with survival and structural development of a typical individual in the population, while the homogeneous operator H is associated with mating and reproduction. The assumption $\mathbf{r}(Q) < 1$ expresses that, at low population densities, the population dies out without mating and reproduction. If H were also additive,

$$H \sum_{n=0}^{\infty} Q^n = \sum_{n=0}^{\infty} HQ^n$$

could be interpreted as a basic *next generation operator*, from the point of view of a newborn individual. Q^n would be associated with the expectation of surviving till the nth year after its birth and HQ^n with the expected number of offspring produced in this n^{th} year.

$$\mathcal{R}_0 = \mathbf{r}\Big(H \sum_{n=0}^{\infty} Q^n\Big) = \mathbf{r}\Big(\sum_{n=0}^{\infty} Q^n H\Big) \tag{12.2}$$

would be interpreted as the *basic* [**63**] (or *inherent net* [**44, 50**]) *reproduction number*, the expected number of offspring produced by a typical newborn individual if the population density is very low. The equality in (12.2) follows from Theorem 5.14.

While the interpretations become hollow if H is only homogeneous rather than additive as well, do the familiar mathematical relations between $\mathcal{T}_0$ and $\mathcal{R}_0$ remain valid, namely that $\mathcal{T}_0$ and $\mathcal{R}_0$ are simultaneously strictly larger than 1, equal to 1 and strictly smaller than 1 [**12, 50, 128, 150**] [**196**, Thm.3.10] [**216**, Thm.5.1]? While general results for homogeneous H become too technical to be included here [**207**], we give positive answers in two very special cases, namely that H is additive or a rank-one operator. For time-periodic homogeneous evolution systems we refer to [**219**].

12.2. Both operators additive

The next result is a generalization of [**50**, Thm.3] to infinite dimensions. See also [**216**, Thm.5.1].

THEOREM 12.1 ([**196**, Thm.3.10]). *Let X be an ordered Banach space and let its closed order cone X_+ be normal and generating. Let Q and H be positive linear maps on X, $\mathbf{r}(Q) < 1$. Then $\mathbf{r}(Q + H) - 1$ and $\mathbf{r}(H(\mathbb{I} - Q)^{-1}) - 1$ have the same sign.*

12.2.1. Turnover versus reproduction number in serially complete cones.

Theorem 12.1 can be extended to serially complete cones (Section 2.1.2).

THEOREM 12.2 ([**203**]). *Let X be an ordered normed vector space and let its order cone X_+ be closed, normal and serially complete. Let Q and H be bounded homogeneous additive operators on X_+, $\mathbf{r}(Q) < 1$. Then $\mathbf{r}(Q + H) - 1$ and $\mathbf{r}\big(H \sum_{n=0}^{\infty} Q^n\big) - 1$ have the same sign.*

Recall Section 2.3.1 and 5.2.2.

PROOF OF THEOREM 12.2. By Theorem 2.35, Remark 2.34 and the preceding remarks, $\tilde{X} = X_+ - X_+$ under $\|\cdot\|^\sim$ in (2.16) is an ordered Banach space and $W = X_+$ is a normal generating closed cone of $\tilde{X}$ with $\|\cdot\|^\sim$. Let $\tilde{Q}$ be the bounded linear extension of Q to $\tilde{X}$ and A the bounded linear extension of H to $\tilde{X}$ with $\|\cdot\|^\sim$ (Section 5.2.2). Then $\tilde{Q} + A$ is the extension of $Q + H$ to $\tilde{X}$ and $A(\mathbb{I} - \tilde{Q})^{-1}$ the extension of $H \sum_{n=0}^{\infty} Q^n$ to $\tilde{X}$. By Theorem 5.17, the respective spectral radii of the extensions are equal to those of the original operators. By Theorem 12.1, $\mathbf{r}(\tilde{Q} + A) - 1 = \mathbf{r}(A(\mathbb{I} - \tilde{Q})^{-1}) - 1$ have the same sign. This implies the assertion. □

Under stronger assumptions, one obtains the more detailed results of [**216**, Thm.5.1] [**150**]. See [**56**] for limitations of the use of these results and for some historical remarks.

THEOREM 12.3 ([**207**]). *Let X be an ordered normed vector space and let its order cone X_+ be closed, normal and serially complete. Let Q and H be continuous homogeneous additive operators on X_+, $\mathbf{r}(Q) < 1$.*

Assume that at least one of the following conditions is satisfied:

(a) *Q and H are compact.*
(b) *X_+ is complete and H is compact.*
(c) *X_+ is a* sup-semilattice *and there are $\ell, m \in \mathbb{N}$ such that B^m and $(HR_\lambda)^\ell$ are compact for all $\lambda \in (\mathbf{r}(Q), \infty)$, $R_\lambda = \sum_{j=0}^{\infty} \lambda^{-j-1} Q^j$.*
(d) *X_+ has a uniform order unit.*

Then at least one (and only one) of the following three possibilities holds:

$$\text{(i) } 1 < \mathcal{T} \le \mathcal{R}. \qquad \text{(ii) } 1 = \mathcal{T} = \mathcal{R}. \qquad \text{(iii) } 1 > \mathcal{T} \ge \mathcal{R}.$$

For analogous trichotomies, see [**218**, Sec.3] [**226**, Chap.11-13].

12.2.2. Existence of eigenvectors in serially complete cones.

THEOREM 12.4 ([**203**]). *Let X be an ordered normed vector space and let its order cone X_+ be closed, normal and serially complete. Let Q and H be continuous homogeneous additive operators on X_+, $\mathbf{r}(Q) < \mathbf{r}(H+Q) =: \mathcal{T}$. Assume that H^2 and at least one of HQ or QH are compact on X_+.*

Then there exists some $w \in \dot{X}_+$ such that $(H+Q)(w) = \mathcal{T}w$.

We emphasize that the cone X_+ does not need to be complete. Notice that H^2 and at least one of HQ or QH are compact on X_+ if H is compact on X_+. We use this theorem to prove the existence of an eigenmeasure for the sum of two Feller kernels one of which is tight (Theorem 13.42) working in the cone $X_+ = \mathcal{M}_+(S)$ of nonnegative measures with the flat norm, which is not complete if the metric space S is not complete.

PROOF. We first assume that $\mathbf{r}(Q+H) =: \mathcal{T} = 1$. So $\mathbf{r}(Q) < 1$.

By Theorem 12.2, $\mathbf{r}(HR) = 1$ with

$$R = \sum_{n=0}^{\infty} Q^n = \mathbb{I} + QR = \mathbb{I} + RQ. \tag{12.3}$$

Since H is additive,

$$HR = H + HQR = H + HRQ.$$

Since all ingredients are additive and $QR = RQ$ by (12.3) ,

$$(HR)^2 = H^2R + HQR(HR) = H^2R + HR(QH)R.$$

Since by assumption, H^2 and at least one of HQ or QH are compact on X_+, $(HR)^2$ is compact (and continuous) on X_+. By Theorem 9.10, there exists some $\tilde{v} \in \dot{X}_+$ such that $\tilde{v} = (HR)^2\tilde{v}$. By Proposition 9.4, since HR is additive, there exists some $v \in \dot{X}_+$ such that $v = HRv$. Set $w = Rv$. By (12.3), $w = v + Qw$ and $w - Qw = v = Hw$, i.e., $w = (H+Q)(w)$.

We return to the general case $\mathcal{T} = \mathbf{r}(Q+H) > \mathbf{r}(Q)$ Set $H_\lambda = \frac{1}{\mathcal{T}}H$ and $Q_\mathcal{T} = \frac{1}{\mathcal{T}}Q$. Then

$$\mathbf{r}(Q_\mathcal{T} + H_\mathcal{T}) = \frac{1}{\mathcal{T}}\mathbf{r}(Q+H) = 1 > \frac{1}{\mathcal{T}}\mathbf{r}(Q) = \mathbf{r}(Q_\mathcal{T}).$$

By our previous consideration, there exists some $w \in \dot{X}_+$ such that $w = (Q_\mathcal{T} + H_\mathcal{T})w = \frac{1}{\mathcal{T}}(Q+H)w$. □

12.3. Additive perturbations of homogeneous rank-one operators

Let X be an ordered normed vector space with closed serially complete order cone X_+ and $B : X_+ \to X_+$ be given as

$$\begin{aligned} B &= Q + H, \\ H(x) &= \phi(x)u, \qquad x \in X_+, \end{aligned} \tag{12.4}$$

where $Q : X_+ \to X_+$ is a bounded additive homogeneous operator, $u \in \dot{X}_+$, and $\phi : X_+ \to \mathbb{R}_+$ a homogeneous and continuous functional.

Assume that $\mathbf{r}(Q) < 1$.

Let R_λ^Q denote the resolvent of Q,

$$R_\lambda^Q = \sum_{n=0}^{\infty} \lambda^{-(n+1)} Q^n, \qquad \lambda > \mathbf{r}(Q). \tag{12.5}$$

See Chapter 8. The operators

$$R_\lambda^Q H = (\phi \circ R_\lambda^Q)(\cdot) u$$

are of rank one and

$$\mathbf{r}(R_\lambda^Q H) = \phi(R_\lambda^Q u) =: \xi(\lambda), \quad \lambda > \mathbf{r}(Q), \quad \lambda > \mathbf{r}(Q). \tag{12.6}$$

See Example 5.1. We extend the function $\xi : (\mathbf{r}(Q), \infty) \to \mathbb{R}_+$ to $[\mathbf{r}(Q), \infty)$ by setting

$$\xi(\mathbf{r}(Q)) = \limsup_{\lambda \to \mathbf{r}(Q)+} \xi(\lambda). \tag{12.7}$$

The function ξ is somewhat analogous to the characteristic polynomial of a matrix. Though $R_1^Q H$ may not have the interpretation of a next generation operator, by abuse of notation, we set

$$\mathcal{R}_0 = \xi(1) = \phi\Big(\sum_{n=0}^{\infty} Q^n u\Big). \tag{12.8}$$

With the understanding that B is the basic turnover operator for a structured population, we again introduce the basic turnover number

$$\mathcal{T}_0 = \mathbf{r}(B).$$

PROPOSITION 12.5. *Let $\phi : X_+ \to \mathbb{R}_+$ be homogeneous and continuous.*

(a) *If $\mathcal{R}_0 > 1$, there are some $r > 1$ and $v \in \dot{X}_+$ such that $Bv = rv$ and $\mathcal{T}_0 > 1$.*
(b) *If $\mathcal{R}_0 = 1$, there is some $v \in \dot{X}_+$ such that $Bv = v$ and $\mathcal{T}_0 \geq 1$.*
(c) *If*

$$\xi(\mathbf{r}(Q)) \in (1, \infty],$$

then there exist some $r > \mathbf{r}(Q)$ and $v \in \dot{X}_+$ such that $Bv = rv$ and $\mathcal{T}_0 > \mathbf{r}(Q)$.

PROOF. Since ϕ is continuous and $\phi(0) = 0$, ξ is a continuous function on $(\mathbf{r}(Q), \infty)$ with $\xi(\lambda) \to 0$ as $\lambda \to \infty$.

(a) By the intermediate value theorem, there is some $r \in (1, \infty)$ such that

$$1 = \xi(r).$$

Set $v = R_r^Q u$. Then $v \in \dot{X}_+$, $u = (r - Q)v$ by Proposition 8.2, and

$$(Q + H)v = Qv + \phi(v)u = Qv + \xi(r)u = Qv + u = rv.$$

So r is an eigenvector of B associated with v and $\mathcal{R}_0 = \mathbf{r}(B) = \mathcal{T}_0 \geq r > 1$ by Proposition 5.2.

(b) The proof is similar, but $r = 1$.

(c) The proof is similar, but $r > \mathbf{r}(Q)$.

□

LEMMA 12.6. *Let $\phi : X_+ \to \mathbb{R}_+$ be homogeneous, continuous and order-preserving. Define $\xi : (\mathbf{r}(Q), \infty) \to [0, \infty)$ by*

$$\xi(\lambda) = \phi(R_\lambda^Q u) = \phi\big(\sum_{n=0}^{\infty} \lambda^{-n-1} Q^n u\big).$$

(a) *Then $\lambda\xi(\lambda)$ is a decreasing function of $\lambda \in (\mathbf{r}(Q), \infty)$.*
(b) *Either $\xi(1) = 0$ and ξ is identically zero on $(\mathbf{r}(Q), \infty)$*
or
$\xi(1) > 0$ and ξ is strictly positive and strictly decreasing on $(\mathbf{r}(Q), \infty)$.

PROOF. (a) Let $\mathbf{r}(Q) \leq \lambda_1 < \lambda_2 < \infty$. Since ϕ is homogeneous and order-preserving

$$\lambda_1\xi(\lambda_1) = \phi\Big(\sum_{n=0}^{\infty} \lambda_1^{-n} Q^n u\Big) \geq \phi\Big(\sum_{n=0}^{\infty} \lambda_2^{-n} Q^n u\Big) = \lambda_2 \xi(\lambda_2).$$

(b) Let $\xi(1) = 0$. By part (a), $\xi(\lambda) = 0$ for all $\lambda \geq 1$.

Let $\lambda \in (\mathbf{r}(Q), 1)$. Since ϕ is order-preserving and homogeneous, for any $m \in \mathbb{N}$,

$$\phi\Big(\sum_{n=0}^{m} \lambda^{-n-1} Q^n u\Big) \leq \phi\Big(\sum_{n=0}^{m} \lambda^{-m-1} Q^n u\Big)$$
$$= \lambda^{-m-1}\phi\Big(\sum_{n=0}^{m} Q^n u\Big) \leq \lambda^{-m-1}\xi(1) = 0.$$

Since ϕ is continuous, we can take the limit as $m \to \infty$ and obtain $\xi(\lambda) = 0$.

Now assume that $\xi(1) > 0$. By (a), $\xi(\lambda) > 0$ for all $\lambda \in (\mathbf{r}(Q), 1)$. Suppose that $\xi(\lambda) = 0$ for some $\lambda > 1$. Since ϕ is homogeneous and order-preserving, for all $m \in \mathbb{N}$,

$$0 = \phi\Big(\sum_{n=0}^{m} \lambda^{-n-1} Qu\Big) \geq \phi\Big(\sum_{n=0}^{m} \lambda^{-m-1} Qu\Big) = \lambda^{-m-1}\phi\Big(\sum_{n=0}^{m} Qu\Big).$$

So

$$\phi\Big(\sum_{n=0}^{m} Qu\Big) = 0, \qquad m \in \mathbb{N}.$$

Since ϕ is continuous, $\xi(1) = 0$ by taking the limit as $m \to \infty$, a contradiction.

This finishes the proof that $\xi(\lambda) > 0$ for all $\lambda \in (\mathbf{r}(Q), \infty)$.

The strict decrease of ξ now follows from part (a). □

THEOREM 12.7. *Let X_+ be a complete normal closed cone and $\phi : X_+ \to \mathbb{R}_+$ be homogeneous, continuous and order-preserving. Assume $\mathbf{r}(Q) < 1$.*

Then one of the following five mutually exclusive scenarios occurs:

(i) $1 < \mathcal{T}_0 \leq \mathcal{R}_0 < \xi(\mathbf{r}(Q))$.
(ii) $\mathcal{T}_0 = \mathcal{R}_0 = 1 < \xi(\mathbf{r}(Q))$.
(iii) $\mathbf{r}(Q) < \mathcal{T}_0 < 1$ *and* $0 < \mathcal{R}_0 \leq \mathcal{T}_0 < 1 < \xi(\mathbf{r}(Q))$.
(iv) $\mathbf{r}(Q) = \mathcal{T}_0 < 1$ *and* $0 < \mathcal{R}_0 < \mathcal{T}_0\xi(\mathbf{r}(Q)) \leq 1$.
(v) $\mathbf{r}(Q) = \mathcal{T}_0 < 1$ *and* $\mathcal{R}_0 = 0 = \xi(\mathbf{r}(Q))$.

In scenarios (i), (ii), (iii), *there is some $v \in \dot{X}_+$ such that $Bv = \mathbf{r}(B)v$ and $\lambda = \mathbf{r}(B)$ is the unique solution of $\xi(\lambda) = \phi(R_\lambda^Q u) = 1$.*

Notice that some of the inequalities are analogous to those in [**150**, Thm.3.3].

PROOF. Now assume that X_+ is a complete and normal closed cone. Since H maps X_+ into X_+, $\mathbf{r}(B) \geq \mathbf{r}(Q)$ by Theorem 6.23.

Assume $\lambda := \mathbf{r}(B) > \mathbf{r}(Q)$.

Since H is homogeneous, compact, continuous and order-preserving, by Theorem 9.3, there exists an eigenvector $v \in \dot{X}_+$ with

$$\lambda v = (Q + H)v.$$

We apply R_λ^Q (Proposition 8.2),

$$\lambda R_\lambda^Q v = \lambda R_\lambda^Q v - v + R_\lambda^Q H v.$$

We simplify and use the definition of H,

$$v = \phi(v) R_\lambda^Q u.$$

Since $v \in \dot{X}_+$, $\phi(v) > 0$. We apply the homogeneous functional ϕ,

$$0 < \phi(v) = \phi(v)\phi(R_\lambda^Q u).$$

We divide by $\phi(v) > 0$,

$$1 = \phi(R_\lambda^Q u) = \xi(\lambda). \tag{12.9}$$

By Lemma 12.6, ξ is strictly decreasing and strictly positive and so

$$\xi(\mathbf{r}(Q)) > 1, \qquad \mathcal{R}_0 = \xi(1) > 0,$$

and $\lambda = \mathbf{r}(B)$ is uniquely determined by (12.9).

Case 1: $\lambda > 1$.

Since $\lambda\xi(\lambda)$ is a decreasing function of λ by Lemma 12.6 and (12.9) holds,

$$1 = \xi(\lambda) \leq \lambda^{-1}\xi(1) = \lambda^{-1}\mathcal{R}_0.$$

This implies (i),

$$1 < \lambda = \mathbf{r}(B) \leq \mathcal{R}_0.$$

Case 2: $\lambda = 1$.

Then (ii), $\mathcal{R}_0 = \phi(R_1^Q)u) = 1$, by (12.9).

Case 3: $\mathbf{r}(Q) < \lambda < 1$.

Since $\lambda\xi(\lambda)$ is a decreasing function of λ by Lemma 12.6,

$$1 = \xi(\lambda) \geq \lambda^{-1}\xi(1) = \lambda^{-1}\mathcal{R}_0.$$

This implies (iii),

$$1 > \lambda \geq \mathcal{R}_0.$$

In the remaining cases we have $\mathbf{r}(B) = \mathbf{r}(Q) < 1$. Then $\mathcal{R}_0 < 1$ and $\xi(\mathbf{r}(Q)) \leq 1$ as converses of Proposition 12.5. Lemma 12.6 now implies (iv) and (v).

By Proposition 12.6,

$$\mathcal{R}_0 = \xi(1) \leq \lambda\xi(\lambda), \qquad \lambda \in (\mathcal{T}_0, 1].$$

We take the limit as $\lambda \to \mathcal{T}_0$,

$$\mathcal{R}_0 \leq \mathcal{T}_0\xi(\mathcal{T}_0) \leq \mathcal{T}_0.$$

□

Finding elegant generalizations of Theorem 12.7 turns out to be rather technical [**207**] or to require rather strong assumptions [**219**, Thm.3.4, Prop.3.8]. One reason is that, even if H is a compact additive operator on X_+, $\mathbf{r}(R_\lambda^Q H)$ may no longer be a continuous function of $\lambda > \mathbf{r}(Q)$ as it was in Proposition 12.5 in the rank-one special case. See [**146**] and Section 10.4.

In special models, it is still possible to identify rather explicit basic reproduction numbers. See Sections 17.3.1 and 16.2 as well as [**199**, Sec.14], [**130**, Sec.9] and [**207**, Sec.6].

CHAPTER 13

Linear maps on the vector space of measures

In general, the theory of positive linear operators on ordered Banach spaces is well-developed [**10, 135, 180, 181**]. Therefore, this monograph focuses on the wider class of order-preserving homogeneous operators.

However, on the ordered vector space of real Borel measures on a metric space S, $\mathcal{M}(S)$, with the flat norm (which typically is no Banach space), positive linear operators may only be bounded on the cone of nonnegative measures but not on the whole vector space. This is why we pay some attention to them here. This chapter unifies, streamlines and elaborates material from [**101**] [**203**–**206**].

Bonsall [**27**] presents an example of an ordered normed vector space X and a bounded linear map on X which is compact on a cone X_+ of X but not compact on X. In this example, X is a Banach space but the cone is only total, not generating. See Proposition 10.14, Corollary 10.21 and Section 2.3.2. For $\mathcal{M}(S)$ with the flat norm, the cone $\mathcal{M}_+(S)$ is generating but typically flat, and $\mathcal{M}(S)$ is not complete. This leads to some anomalies which are analogous to those presented in [**27**].

Let S be compact. By Theorem 3.31, $\bar{U}_1(0) \cap \mathcal{M}_+(S)$ is compact. Since $\mathcal{M}(S)$ is an infinite-dimensional normed vector space, $\bar{U}_1(0)$ is not compact. So the identity map is compact on $\mathcal{M}_+(S)$ but not on $\mathcal{M}(S)$.

In the next section, we will meet elementary linear positive maps which are bounded on $\mathcal{M}_+(S)$ but not on $\mathcal{M}^s(S)$, the space of separable Borel measures (Definition 3.17).

13.1. Multiplication maps

After the identity map, multiplication maps are the simplest ones. Let $g \in M^b(S)$, $g \geq 0$, i.e., $g : S \to \mathbb{R}$ is a bounded nonnegative Borel measurable function. Define

$$(L\mu)(T) = \int_T g d\mu, \qquad \mu \in \mathcal{M}(S), \quad T \in \mathcal{B}. \tag{13.1}$$

Then L is a linear map on $\mathcal{M}(S)$. If $g \in \mathrm{LC}^b(S)$, i.e., g is also Lipschitz continuous, L is a bounded linear map on $\mathcal{M}(S)$ with the flat norm. Let L_+ denote the restriction of L to $\mathcal{M}_+(S)$.

PROPOSITION 13.1. *If $g \in M^b(S)$ is nonnegative, then L_+ is bounded on $\mathcal{M}_+(S)$, $\|L_+\| = \|g\|_\infty = \mathbf{r}(L_+)$. L maps $\mathcal{M}^s(S)$ into itself and L_+ is continuous on $\mathcal{M}^s_+(S)$.*

PROOF. Consider $\mu \in \mathcal{M}_+(S)$. Recall the flat norm, (3.9).

Let $f : S \to [0,1]$ and $|f(x) - f(y)| \leq d(x,y)$ for all $x, y \in S$. Then

$$\Big| \int_S f\, d(L\mu) \Big| = \int_S (fg)\, d\mu \leq \|g\|_\infty \mu(S) = \|g\|_\infty \|\mu\|_\flat.$$

Thus $\|L_+\| \le \|g\|_\infty$. Moreover

$$\|L\mu\|_\flat = \int_S g d\mu.$$

Choosing $\mu = \delta_x$, $x \in S$, gives $\|L_+\|_\flat \ge \|g\|_\infty$. One readily checks that $(L^n\mu)(T) = \int_T g^n d\mu$ for all $n \in \mathbb{N}$. By our previous considerations, $\|L^n\| = \|g^n\|_\infty = \|g\|_\infty^n$. Thus $\mathbf{r}(L_+) = \|g\|_\infty$.

Obviously, L maps $\mathcal{M}^s(S)$ into itself.

Let $f \in C^b(S)$ and (μ_n) and μ a sequence and an element in $\mathcal{M}^s_+(S)$, $\|\mu_n - \mu\|_\flat \to 0$. Then $fg \in C^b(S)$ and $\int_S f d(L\mu_n - L\mu) = \int_S (fg) d(\mu_n - \mu) \to 0$ by Theorem 3.5. By Proposition 3.23, $\|L\mu_n - L\mu\|_\flat \to 0$. □

If g takes values between 0 and 1, $g(s)$ can be interpreted as the probability of an individual with structural characteristic s to survive for a year and L_+ has the interpretation of a survival map. Recall $LC^b(S)$ (Section 2.7.1).

PROPOSITION 13.2. *Let $g : S \to \mathbb{R}$ be a Lipschitz continuous bounded function and L_* on $\mathrm{LC}^b(S)$ be defined by $(L_*f)(x) = g(x)f(x)$. Then L_* is a bounded linear operator on $\mathrm{LC}^b(S)$ and $\mathbf{r}(L_*) \le \|g\|_\infty$. Further, L defined by (13.1) is a bounded linear operator on $\mathcal{M}(S)$ and $\mathbf{r}(L) \le \|g\|_\infty$. If g is nonnegative, $\mathbf{r}(L) = \|g\|_\infty$.*

PROOF. We first notice that $\|L_*f)\|_\infty \le \|g\|_\infty \|f\|_\infty$ for all $f \in C^b(S)$. Further, if $f \in \mathrm{LC}^b(S)$,

$$\begin{aligned} |(L_*f)(x) - (L_*f)(y)| \le & |f(x) - f(y)|\, |g(x)| + |f(y|\, |g(x) - g(y)| \\ \le & \Lambda(f) d(x,y) \|g\|_\infty + \Lambda(g) d(x,y) \|f\|_\infty. \end{aligned}$$

So

$$\Lambda(L_*f) \le \Lambda(f)\|g\|_\infty + \Lambda(g)\|f\|_\infty$$

where

$$\Lambda(f) = \sup\Big\{ \frac{|f(x) - f(y)|}{d(x,y)}; x, y \in S, x \ne y \Big\}.$$

Set $\|f\| = \max\{\|f\|_\infty, \Lambda(f)\}$. We introduce the following norms on $\mathrm{LC}^b(S)$ which are equivalent to $\|\cdot\|$,

$$\|f\|' = q\|f\|_\infty + (1-q)\Lambda(f), \qquad 0 < q < 1. \tag{13.2}$$

Then

$$\begin{aligned} \|L_*f\|' &\le q\|L_*f\|_\infty + (1-q)\Lambda(L_*f) \\ &\le q\|g\|_\infty \|f\|_\infty + (1-q)\big[\Lambda(f)\|g\|_\infty + \Lambda(g)\|f\|_\infty\big]. \end{aligned}$$

By (13.2),

$$\begin{aligned} \|L_*f\|' \le & \|g\|_\infty \big(q\|f\|_\infty + (1-q)\Lambda(f)\big) + (1-q)\|f\|_\infty \Lambda(g) \\ \le & \Big(\|g\|_\infty + \frac{1-q}{q}\Lambda(g)\Big)\|f\|'. \end{aligned}$$

Let $\|L_*\|'$ denote the operator norm of L_* with respect to $\|\cdot\|'$,

$$\mathbf{r}(L_*) \le \|L_*\|' \le \Big(\|g\|_\infty + \frac{1-q}{q}\Lambda(g)\Big), \qquad 0 < q < 1.$$

Since the spectral radius of L_* does not depend on equivalent norms, we take the limit as $q \to 1$ and obtain $\mathbf{r}(L_*) \le \|g\|_\infty$. Now L is dual to L_*,

$$\int_S (L_* f)\,d\mu = \int_S f\,d(L\mu), \qquad f \in C^b(s), \quad \mu \in \mathcal{M}(S).$$

By induction,

$$\int_S (L_*^n f)\,d\mu = \int_S f\,d(L^n\mu), \qquad f \in C^b(s), \quad \mu \in \mathcal{M}(S), \quad n \in \mathbb{N}.$$

By (5.8), $\mathbf{r}(L) = \mathbf{r}(L^*) \le \|g\|_\infty$. If g is nonnegative, L_+ is order-preserving and $\mathbf{r}(L) \ge \mathbf{r}(L_+) = \|g\|_\infty$ by Corollary 6.25. □

In fact, if $g \in C^b(S)$, Lipschitz continuity of g is necessary for L to be a bounded linear map on $\mathcal{M}(S)$ with respect to the flat norm.

PROPOSITION 13.3. *Let $g \in C^b(S)$ and L be given by (13.1). If L is a bounded linear map on $\mathcal{M}^s(S)$, then g is Lipschitz continuous and*

$$|g(x) - g(y)| \le (\|L\| + \|g\|_\infty) \min\{1, d(x,y)\}, \qquad x, y \in S.$$

PROOF. Let L be bounded. By the reverse triangle inequality, for all $x, y \in S$,

$$\begin{aligned}\|L\|\,\|\delta_x - \delta_y\|_\flat &\ge \|L\delta_x - L\delta_y\|_\flat = \|g(x)\delta_x - g(y)\delta_y\|_\flat \\ &\ge |g(x) - g(y)|\,\|\delta_x\|_\flat - |g(y)|\,\|\delta_x - \delta_y\|_\flat.\end{aligned}$$

We regroup and use Lemma 3.9,

$$|g(x) - g(y)| \le (\|L\| + \|g\|_\infty)\|\delta_x - \delta_y\|_\flat \le (\|L\| + \|g\|_\infty) \min\{1, d(x,y)\}.$$

□

Recall that, if $\mathcal{M}_+(S)$ were non-flat, we would have $\mathbf{r}(L_+) = \mathbf{r}(L)$ by Theorem 5.13.

PROPOSITION 13.4. *If $g \in C^b(S)$, there is a topologically equivalent metric d_g on S under which g is Lipschitz continuous and L defined by (13.1) is a bounded linear map on $\mathcal{M}(S)$ with the flat norm associated with d_g and $\mathbf{r}(L) \le \|g\|_\infty$. S is complete under the original metric if and only if S is complete under d_g.*

PROOF. We can assume that $g \ne 0$. Define d_g by

$$d_g(s,t) = d(s,t) + |g(s) - g(t)|, \qquad s, t \in S.$$

Cf. [**9**, L.3.12]. Since g is continuous and bounded, d_g is a metric which is topologically equivalent to d. S is complete under d if and only if it is complete under d_g. By construction of d_g, g is Lipschitz continuous with respect to d_g. By Proposition 13.2, $\mathbf{r}(L) \le \|g\|_\infty$. □

REMARK 13.5. Let S be a metric space with metric d and let $g \in C^b(S)$ be not Lipschitz continuous with respect to d (which exists by Proposition 2.77 if S has an accumulation point). Let d_g be a topologically equivalent metric on S such that g is Lipschitz continuous with respect to d_g (Proposition 13.4). Let $\|\cdot\|^\flat$ be the flat norm associated with d and $\|\cdot\|_g^\flat$ the flat norm associated with d_g and L be the multiplication operator on $\mathcal{M}^s(S)$ associated with g by (13.1). By Propositions 13.3 and 13.2, L is not continuous on $\mathcal{M}^s(S)$ with respect to $\|\cdot\|^\flat$, but is continuous with respect to $\|\cdot\|_g^\flat$. So these two flat norms associated with topologically equivalent metrics on S are not topologically equivalent on $\mathcal{M}^s(S)$. See the discussion in Section 3.2.5.

Multiplication operators of this kind have many eigenvectors as $L(\delta_s) = g(s)\delta_s$ for all $s \in S$.

REMARK 13.6. For $g \in M_+^b(S)$, the spectral radius of L_+, $\|g\|_\infty$, is associated with an eigenvector in $\mathcal{M}_+(S)$ if and only if g has a maximum on S.

PROOF. Let $g \in M_+^b(S)$ and L_+ be the multiplication operator associated with g on $\mathcal{M}_+(S)$.

Let g have a maximum on S at some $s \in S$, $g(s) = \|g\|_\infty$. Then

$$L_+(\delta_s)(T) = g(s)\chi_T(s) = \|g\|_\infty \delta_s(T), \qquad T \in cB,$$

and δ_s is an eigenvector associated with $\|g\|_\infty$.

Now assume that there is an eigenmeasure $\mu \in \dot{\mathcal{M}}_+(S)$ associated with $\|g\|_\infty$,

$$L_+(\mu)(T) = \|g\|_\infty \mu(T), \qquad T \in \mathcal{B}. \tag{13.3}$$

Assume that $g(s) < \|g\|_\infty$ for all $s \in S$. Since $\mu(S) > 0$ and

$$S = \{s \in S; g(s) < \|g\|_\infty\} = \bigcup_{n\in\mathbb{N}} T_n, \quad T_n = \Big\{s \in S; g(s) \le \|g\|_\infty - 1/n\Big\} \in \mathcal{B}.$$

Since μ is continuous from below, there is some $n \in \mathbb{N}$ such that $\mu(T_n) > 0$ and $\mu(T_n) > 0$. For this $n \in \mathbb{N}$,

$$(L_+\mu)(T_n) = \int_{T_n} g d\mu \le (\|g\|_\infty - 1/n)\mu(T_n),$$

contradicting (13.3). □

REMARK 13.7. Let $g \in C_+^b(S)$. Then L is compact on $\mathcal{M}_+(S)$ if g vanishes at infinity: For any $\epsilon > 0$ there exists a compact subset T of S such that $g(s) \le \epsilon$ for all $s \in S \setminus T$. If S is complete, this condition for g is also necessary.

PROOF. Use Theorems 3.30 and 3.29. For the necessity, if S is complete and L is compact, the set of measures $\{L\delta_s; s \in S\}$ is tight (Definition 3.16). □

13.2. Linear maps induced by measure kernels

To extend the class of linear maps on spaces of measures (or of bounded measurable functions) beyond multiplication maps, let $(S, \mathcal{B})$ and $(\tilde{S}, \tilde{\mathcal{B}})$ be measurable spaces with $\mathcal{B}$ and $\tilde{\mathcal{B}}$ σ-algebras of subsets of S and $\tilde{S}$, respectively.

DEFINITION 13.8. *A function $\kappa : \mathcal{B} \times \tilde{S} \to \mathbb{R}_+$ is called a* measure kernel *if*

$$\kappa(\cdot, s) \in \mathcal{M}_+(S) \textit{ for all } s \in \tilde{S}$$

and

$$\kappa(T, \cdot) \in M_b(\tilde{S}) \textit{ for all } T \in \mathcal{B}.$$

Here $M(\tilde{S})$ denotes the vector space of measurable functions from $\tilde{S}$ to $\mathbb{R}$ and $M^b(\tilde{S})$ the Banach space of bounded measurable functions from $\tilde{S}$ to $\mathbb{R}$ with the supremum norm. Cf. [**9**, Sec.19.2] [**22**, Sec.10.3]. Measure kernels are simply called measurable maps in [**13**, Sec.5.3]. See also [**114**].

In Chapter 3, we plead that measures can be meaningful representations of structural population distributions. Let us consider a plant population with size structure. Let $\tilde{S}$ be the individual state space of the feasible sizes of plants and S the individual state space of the feasible sizes of the seeds of these plants.

Then $\kappa(T, s)$ could be the number of seeds with size in set T produced by a plant of size s.

The measure kernel κ induces a linear bounded map A_* from $M^b(S)$ to $M^b(\tilde{S})$ with the supremum norm and a linear bounded map A from $\mathcal{M}(\tilde{S})$ to $\mathcal{M}(S)$ with the variation norm by

$$(A_*f)(s) = \int_S f(t)\kappa(dt, s), \qquad s \in \tilde{S}, \quad f \in M^b(S), \tag{13.4}$$

and

$$(A\mu)(T) = \int_{\tilde{S}} \kappa(T, s)\mu(ds), \qquad T \in \mathcal{B}, \quad \mu \in \mathcal{M}(\tilde{S}). \tag{13.5}$$

The following duality relation holds,

$$\int_{\tilde{S}} (A_*f) d\mu = \int_S f\, d(A\mu), \qquad f \in M^b(S), \quad \mu \in \mathcal{M}(\tilde{S}). \tag{13.6}$$

If $f = \chi_T$ is a characteristic (or indicator) function, (18.4), both sides of (13.6) equal $\int_S \kappa(T, s)\mu(ds)$. By standard integration rules, (13.6) extends to simple functions. By the monotone convergence theorem, it then extends to nonnegative bounded measurable functions and nonnegative measures because the first can be uniformly approximated by increasing sequences of nonnegative simple function. Finally (13.6) extends to bounded measurable functions and real-valued measures because they are differences of nonnegative bounded measurable functions or nonnegative measures, respectively.

There is equality of the operator norms,

$$\|A_*\| = \|A\| = \sup_{s \in \tilde{S}} \kappa(S, s). \tag{13.7}$$

The operators A and A^* will be fundamental as basic turnover operators in the nonlinear dynamics of populations that play out in the state spaces of measures (Chapter 19) or continuous functions (Chapter 18). See Section 1.7.

13.2.1. Convolution of measure kernels. Let $S = \tilde{S}$ and $\mathcal{B} = \tilde{\mathcal{B}}$.

The *convolution* of two measure kernels κ_j, $j = 1, 2$, is defined by

$$(\kappa_1 \star \kappa_2)(T, s) = \int_S \kappa_1(T, t)\kappa_2(dt, s), \qquad T \in \mathcal{B}, \quad s \in S. \tag{13.8}$$

$\kappa_1 \star \kappa_2$ is again a measure kernel. It follows from (13.8) that

$$\int_S f(z)(\kappa_1 \star \kappa_2)(dz, s) = \int_S \Big(\int_S f(z)\kappa_1(dz, t)\Big)\kappa_2(dt, s), \qquad f \in M^b(S), \quad s \in S. \tag{13.9}$$

Indeed, if $f = \chi_T$ is a characteristic (or indicator) function (18.4), (13.9) is just another way of writing (13.8). Standard integration rules extend (13.9) to simple functions. Formula (13.9) then extends to nonnegative bounded measurable functions because they can be uniformly approximated by an increasing sequence of nonnegative simple functions (monotone convergence theorem). It then holds for all bounded measurable function because these can be written as the difference of two nonnegative bounded measurable functions. (13.9) implies that this convolution is associative, i.e., for three measure kernels κ_j, $j = 1, 2, 3$,

$$\kappa_1 \star (\kappa_2 \star \kappa_3) = (\kappa_1 \star \kappa_2) \star \kappa_3. \tag{13.10}$$

LEMMA 13.9. *For $j = 1, 2$, let κ_j be measure kernels and A_{j*} and A_j be the linear bounded operators on $M^b(S)$ and $\mathcal{M}(S)$, respectively. Then $A_{2*}A_{1*}$ and A_1A_2 are induced by $\kappa_1 \star \kappa_2$.*

DEFINITION 13.10. *Let $\kappa : \mathcal{B} \times S \to \mathbb{R}_+$ be a measure kernel. We define inductively the multiple convolution kernels $\kappa^{n\star}$ by $\kappa^{1\star} = \kappa$ and $\kappa^{(n+1)\star} = \kappa^{n\star} \star \kappa$.*

13.2.2. Spectral radius of a measure kernel. By Lemma 13.9, $\kappa^{n\star}$ induces A_*^n on $M^b(S)$ and A^n on $\mathcal{M}(S)$. By (13.7) and (5.1),

$$\mathbf{r}(\kappa) := \inf_{n\to\infty} \Big(\sup_{s\in S} \kappa^{n\star}(S, s) \Big)^{1/n} = \lim_{n\to\infty} \Big(\sup_{s\in S} \kappa^{n\star}(S, s) \Big)^{1/n}, \tag{13.11}$$

and we have

$$\mathbf{r}(\kappa) = \mathbf{r}(A_*) = \mathbf{r}(A). \tag{13.12}$$

We call $\mathbf{r}(\kappa)$ the *spectral radius of the measure kernel* κ. Since the constant function with value 1, $\mathbf{1}$, is an interior point of the normal solid cone $M_+^b(S)$, we have $\mathbf{r}(\kappa) = \mathbf{r}(A_*) = \lim_{n\to\infty} \|A_*^n \mathbf{1}\|_\infty^{1/n}$ by Theorem 6.22. Since $\mathbf{1} \in M_+^b(S)$, $\mathbf{r}(\kappa) = \mathbf{r}(A_{*+}$ where A_{*+} is the restriction of A_* to $M_+^b(S)$.

Since A_* is a bounded linear map on the Banach space $M_b(S)$ with the supremum norm and A is a bounded linear operator on the Banach space $\mathcal{M}(S)$ with the variation norm, we also have the spectral formulas for the spectral radius,

$$\mathbf{r}(\kappa) = \sup \big\{ |\lambda|, \lambda \in \sigma^{\mathbb{C}}(A_*) \big\} = \sup \big\{ |\lambda|, \lambda \in \sigma^{\mathbb{C}}(A) \big\}, \tag{13.13}$$

where $\sigma^{\mathbb{C}}$ denotes the spectrum for the extensions of A_* and A to the complexifications of these Banach spaces. See (1.8).

$M_b(S)$ and $\mathcal{M}(S)$ (with variation norm) actually are Banach lattices which implies [**181**, V.4.1] that

$$\mathbf{r}(\kappa) \in \sigma(A_*), \qquad \mathbf{r}(\kappa) \in \sigma(A). \tag{13.14}$$

For the iterative calculation, Section 6.4 provides the formulas

$$\begin{aligned} \ln \mathbf{r}(\kappa) &= \lim_{n\to\infty} b_n, \\ b_{n+1} &= \Big(1 - \frac{1}{n+1}\Big) b_n + \frac{1}{n+1} \ln \sup f_{n+1}(S), \\ f_{n+1}(s) &= \frac{\int_S f_n(t)\kappa(dt, s)}{\sup f_n(S)}, \qquad s \in S, \\ f_1(s) &= \kappa(S, s), \qquad b_1 = \ln \sup f_1(S). \end{aligned} \tag{13.15}$$

Since $\mathcal{M}(S)$ with the variation norm is an ordered Banach space with the closed generating order cone $\mathcal{M}_+(S)$, $\mathcal{M}_+(S)$ is non-flat and

$$\mathbf{r}(\kappa) = \mathbf{r}(A) = \mathbf{r}(A_+), \tag{13.16}$$

where A_+ is the restriction of the bounded linear map A in (13.5) to $\mathcal{M}_+(S)$. Since $\mathcal{M}_+(S)$ is also a normal cone, we have the following result from Theorem 5.9.

PROPOSITION 13.11. *There is some* $\mu \in \mathcal{M}_+(S)$ *such that*

$$\mathbf{r}(\kappa) = \lim_{n\to\infty} \Big(\int_S \kappa^{n\star}(S,s)\mu(ds)\Big)^{1/n}.$$

This proposition implies that $\mathbf{r}(\kappa) = \mathbf{r}_o(A_+)$. Recall the definition of the lower Collatz–Wielandt bound $cw(A_+)$ by (5.30) and recall $cw(A_+) \le \mathbf{r}_o(A_+) = \mathbf{r}(A_+) = \mathbf{r}(\kappa)$. In terms of κ, these inequalities take the following form.

THEOREM 13.12. *Let* $\kappa : \mathcal{B} \times S \to \mathbb{R}_+$ *be a measure kernel. Then*

$$\mathbf{r}(\kappa) \ge \sup_{\mu \in \mathcal{M}_+(S)} \inf \Big\{ \int_S \frac{\kappa(T,s)\mu(ds)}{\mu(T)}; T \in \mathcal{B}, \mu(T) > 0 \Big\}. \tag{13.17}$$

Let $f \in M^b(S)$ be nonnegative. Then $\theta(\mu) = \int_S f d\mu$, $\mu \in \mathcal{M}_+(S)$, defines a bounded homogeneous functional with values in $\mathbb{R}_+$ on $\mathcal{M}_+(S)$ (or $\mathcal{M}_+^s(S)$ if S is a metric space). For $\mu \in \mathcal{M}_+(S)$,

$$\begin{aligned}\theta(A\mu) &= \int_S f d(A\mu) = \int_S \Big(\int_S f(s)\kappa(ds,t)\Big)\mu(dt) \\ &\ge \inf \Big\{ \int_S f(s)\kappa(ds,t)/f(t); t \in S, f(t) > 0\Big\}\theta(\mu).\end{aligned}$$

By Theorem 5.36,

$$\mathbf{r}(\kappa) = \mathbf{r}(A_+) \ge \sup_{f \in M_+^b(S)} \inf \Big\{ \frac{1}{f(t)} \int_S f(s)\kappa(ds,t); t \in S, f(t) > 0\Big\}. \tag{13.18}$$

Recall that $\breve{M}_+^b(S)$ is a solid cone and $f \in \breve{M}_+^b(S)$ if and only if $f \in M(S)$ and $0 < \inf f(S) \le \sup f(S) < \infty$. If $f \in \breve{M}_+^b(S)$, the functional $\theta(\mu) = \int_S f d\mu$ on $\mathcal{M}_+(S)$ is uniformly positive. By Theorem 5.38,

$$\mathbf{r}(\kappa) = \mathbf{r}(A_+) \le \inf_{f \in \breve{M}_+^b(S)} \sup \Big\{ \frac{1}{f(t)} \int_S f(s)\kappa(ds,t); t \in S\Big\}. \tag{13.19}$$

The right-hand side of this inequality coincides with the upper Collatz–Wielandt bound of the restriction of the map A_* to $M_+^b(S)$ with respect to any function in $\breve{M}_+^b(S)$. See (5.6) and (13.4).

Let A_{*+} be the restriction of A_* to $M_+^b(S)$. A_{*+} is subadditive, even additive, and uniformly f-bounded for any $f \in \breve{M}_+^b(S)$. Further, any two functions in $\breve{M}_+^b(S)$ are comparable to each other. By Theorem 6.16, $\mathbf{r}(A_{*+}) = CW_f(A_{*+})$ for any $f \in \breve{M}_+^b(S)$. So equality holds in (13.19). Also recall (13.18).

THEOREM 13.13.

$$\begin{aligned}&\inf_{f \in \breve{M}_+^b(S)} \sup \Big\{ \frac{1}{f(t)} \int_S f(s)\kappa(ds,t); t \in S\Big\} \\ &= \mathbf{r}(\kappa) = \mathbf{r}(A_{*+}) \\ &\ge \sup_{f \in M_+^b(S)} \inf \Big\{ \frac{1}{f(t)} \int_S f(s)\kappa(ds,t); t \in S, f(t) > 0\Big\}.\end{aligned} \tag{13.20}$$

From Theorem 5.23, we obtain the following series characterization of the spectral radius.

THEOREM 13.14. *If* $\mathbf{r}(\kappa) > 0$,

$$\mathbf{r}(B) = \inf \Big\{ \lambda > 0; \sum_{n=1}^{\infty} \lambda^{-n} \sup_{s \in S} \kappa^{n\star}(S, s) < \infty \Big\}$$
$$= \inf \Big\{ \lambda > 0; \sum_{n=1}^{\infty} \lambda^{-n} \kappa^{n\star}(S, \cdot) \textit{ bounded on } S \Big\}$$

The second equality follows from the first. The formulas are reminiscent of what has been called the Perron root of κ by Shurenkov [**184**], [**123**],

$$\mathbf{r}^{\wp}(\kappa) = \inf \Big\{ \lambda > 0;\ \forall s \in S : \sum_{n=1}^{\infty} \lambda^{-n} \kappa^{n\star}(\cdot, s) \text{ is } \sigma\text{-finite} \Big\}. \tag{13.21}$$

A measure $\mu \in \mathcal{M}_+(S)$ is called σ-finite if there is a sequence (T_n) in $\mathcal{B}$ such that $S = \bigcup_{n \in \mathbb{N}} T_n$ and $\mu(T_n) < \infty$ for all $n \in \mathbb{N}$.

Obviously, $\mathbf{r}^{\wp}(\kappa) \le \mathbf{r}(\kappa)$. Equality does not always hold (even for Feller kernels, Example 13.4.2 (13.31)). Conditions for equality appear to involve irreducibility conditions (Shurenkov) or communication conditions [**123**] that are beyond the scope of this book. In view of the fact that the equalities in Theorem 13.14 hold without any such conditions, the question whether or under which conditions $\mathbf{r}(\kappa)$ equals

$$\inf \Big\{ \lambda > 0;\ \forall s \in S : \sum_{n=1}^{\infty} \lambda^{-n} \kappa^{n\star}(S, s) < \infty \Big\}$$

may be more compelling. Corollary 5.24 implies the following for measure kernels.

REMARK 13.15. Let $r = \mathbf{r}(\kappa) > 0$. Then there exists some $\mu \in \mathcal{M}_+(S)$ such that $\displaystyle\sum_{n=1}^{\infty} r^{-n} \int_S \kappa^{n\star}(S, t) \mu(dt) = \infty$.

13.3. Feller kernels

Let S and $\tilde{S}$ be metrizable topological spaces (Section 3.2.5) and $\mathcal{B}$ and $\tilde{\mathcal{B}}$ the respective Borel σ-algebras. Then $C^b(S)$ and $C^b(\tilde{S})$ are important closed subspaces of $M^b(S)$ and $\tilde{M}^b(S)$ under the supremum norm, respectively.

DEFINITION 13.16. *A function* $\kappa : \mathcal{B} \times \tilde{S} \to \mathbb{R}_+$ *is called a* Feller kernel *if*

$\kappa(\cdot, s) \in \mathcal{M}_+(S)$ *for all* $s \in \tilde{S}$

and has the Feller property

$\int_S f(y) \kappa(dy, \cdot) \in C^b(\tilde{S})$ *for any* $f \in C^b(S)$.

A Feller kernel κ *is called a* Feller kernel of separable measures *if*

$\kappa(\cdot, s) \in \mathcal{M}_+^s(S)$ *for all* $s \in \tilde{S}$.

A Feller kernel κ *is called a* Feller kernel of tight measures *if*

$\kappa(\cdot, s) \in \mathcal{M}_+^t(S)$ *for all* $s \in \tilde{S}$.

See [**9**, Sec.19.3].

By Proposition 2.78, if κ is a Feller kernel, $\kappa(U, \cdot)$ is a Borel measurable function on $\tilde{S}$ for all open subsets U of S and thus for all Borel sets U in $\tilde{S}$. This implies that every Feller kernel is a measure kernel and induces the maps $A : \mathcal{M}(\tilde{S}) \to \mathcal{M}(S)$ and $A_* : M^b(S) \to M^b(\tilde{S})$ in (13.5) and (13.4). Since κ is a Feller kernel, A_* maps $C^b(S)$ to $C^b(\tilde{S})$.

REMARK 13.17. Consider the following statements:

(a) $\kappa : \mathcal{B} \times \tilde{S} \to \mathbb{R}_+$ is a Feller kernel.
(b) If U is an open subset of S, $\kappa(U, \cdot)$ is a lower semicontinuous function on $\tilde{S}$.
(c) If T is a closed subset of S, $\kappa(T, \cdot)$ is an upper semicontinuous function on $\tilde{S}$.

Then (a) holds if and only if (b) and (c) hold.

PROOF. Assume (a). Let U be an open subset of S. By Proposition 2.78, there exists an increasing sequence of Lipschitz continuous functions $f_n : S \to \mathbb{R}_+$ such that $f_n(s) \to \chi_U(s)$ as $n \to \infty$ pointwise in $s \in S$. Let

$$g_n(\tilde{s}) = \int_S f_n(t)\kappa(dt, \tilde{s}), \qquad \tilde{s} \in \tilde{S}.$$

Then g_n is an increasing sequence of continuous functions and $g_n(\tilde{s}) \to \kappa(U, s)$ as $n \to \infty$ pointwise in $\tilde{s} \in \tilde{S}$ by Beppo Levi's theorem of monotone convergence. So

$$\kappa(U, \tilde{s}) = \sup_{n \in \mathbb{N}} g_n(\tilde{s}), \qquad s \in \tilde{S},$$

and $\kappa(U, \cdot)$ is lower semicontinuous by Proposition A.53.

As for part (c), let T be a closed subset of S. Then $S \setminus T$ is open and, by part (a), $\kappa(S \setminus T, \cdot) = \kappa(S, \cdot) - \kappa(T, \cdot)$ is a lower semicontinuous on $\tilde{S}$. Since $\kappa(S, \cdot)$ is continuous, $\kappa(T, \cdot)$ is upper semicontinuous. See Remark A.48.

Assume (b) and (c). Then (a) follows from [**141**, Prop.A.3]. □

REMARK 13.18.

(a) A measure kernel is a Feller kernel if and only if the induced map A_* maps $C^b(S)$ to $C^b(\tilde{S})$.
(b) The convolution of two Feller kernels is a Feller kernel (by Lemma 13.9).

EXAMPLE 13.19. An example for a Feller kernel (which generalizes a multiplication map) is given by

$$\kappa(T, s) = \chi_T(\xi(s))g(s), \qquad T \in \mathcal{B}, \quad s \in \tilde{S},$$

where $\xi : \tilde{S} \to S$ and $g : \tilde{S} \to \mathbb{R}_+$. Then, for $f \in C^b(S)$,

$$\int_S f(t)\kappa(dt, s) = f(\sigma(s))g(s)$$

is a bounded continuous function of $s \in \tilde{S}$ if $g \in C^b(\tilde{S})$ and σ is continuous.

THEOREM 13.20. *Let $\kappa : \mathcal{B} \times \tilde{S} \to \mathbb{R}_+$ be a Feller kernel and $A : \mathcal{M}(\tilde{S}) \to \mathcal{M}(S)$ be induced by κ. Let $\mu \in \mathcal{M}_+(\tilde{S})$ and $A\mu \in \mathcal{M}_+(S)$ be separable.*

Then A is continuous at μ with respect to the flat norm.

PROOF. Let (μ_n) be a sequence in $\mathcal{M}_+(\tilde{S})$ with $\|\mu_n - \mu\|_\flat \to 0$. Let $f \in \mathrm{LC}^b(S)$. Since κ is a Feller kernel, $A_* f \in C^b(\tilde{S})$. By Theorem 3.5 and (13.6),

$$\int_S f\, d(A\mu_n)(dt) = \int_S (A_* f) d\mu_n \to \int_S (A_* f) d\mu = \int_S f\, d(A\mu).$$

By Proposition 3.23, $\|A\mu_n - A\mu\|_\flat \to 0$. □

PROPOSITION 13.21. *Let $A : \mathcal{M}_+^s(\tilde{S}) \to \mathcal{M}_+^s(S)$ be additive, homogeneous and continuous (with respect to the flat norm). Then A is induced by a Feller kernel of separable measures via* (13.5).

This result is somewhat analogous to results for $X = L^p(S)$ rather than $X = \mathcal{M}^s(S)$ [**16**].

PROOF. Set

$$\kappa(T, s) = (A\delta_s)(T), \qquad s \in \tilde{S}, \quad T \in \mathcal{B}.$$

Since $\delta_s \in \mathcal{M}_+^s(S)$, $\kappa(\cdot, s) \in \mathcal{M}_+^s(S)$ for any $s \in \tilde{S}$ by assumption. Let (s_n) be a sequence in $\tilde{S}$ and $s_n \to s \in \tilde{S}$. By Lemma 3.9, $\|\delta_{s_n} - \delta_s\|_\flat \to 0$ and $\|A\delta_{s_n} - A\delta_s\|_\flat \to 0$ because A is continuous by assumption. By Theorem 3.5, for any $f \in C^b(S)$,

$$\int_S f(t)\kappa(dt, s_n) = \int_S f\, dA\delta_{s_n} \to \int_S f\, dA\delta_s = \int_S f(t)\kappa(dt, s), \qquad n \to \infty.$$

So κ has the Feller property.

Let B be the additive homogeneous operator induced by κ on $\mathcal{M}_+^s(S)$. Then $B\delta_s = A\delta_s$ for all $s \in \tilde{S}$. Let $\mathcal{N}$ be the set of measures $\sum_{j=1}^k q_j \delta_{s_j}$ with $s_j \in \tilde{S}$ and $q_j \in \mathbb{Q}_+$. By Theorem 3.26, $\mathcal{N}$ is dense in $\mathcal{M}_+^s(\tilde{S})$ and $B\nu = A\nu$ for all $\nu \in \mathcal{N}$.

Let $\mu \in \mathcal{M}_+^s(\tilde{S})$. Choose a sequence (μ_n) in $\mathcal{N}$ such that $\|\mu_n - \mu\|_\flat \to 0$ as $n \to \infty$. Let $f \in C^b(S)$. Since κ has the Feller property, $g \in C^b(\tilde{S})$ where $g(s) = \int_S f(t)\kappa(dt, s)$ for $s \in \tilde{S}$. By Theorem 3.5,

$$\int_S fd(B\mu) = \int_S gd\mu = \lim_{n\to\infty} \int_S gd\mu_n = \lim_{n\to\infty} \int_S fd(B\mu_n) = \lim_{n\to\infty} \int_S fd(A\mu_n).$$

Since A is continuous by assumption, $\int_S f\, d(B\mu) = \int_S f\, d(A\mu)$. By Theorem 2.76, $(B\mu)(T) = (A\mu)(T)$ for all open subsets T of S and so for all $T \in \mathcal{B}$. □

THEOREM 13.22. *If $\kappa : \mathcal{B} \times \tilde{S} \to \mathbb{R}_+$ is a Feller kernel of separable measures and $A : \mathcal{M}(\tilde{S}) \to \mathcal{M}(S)$ is induced by κ, the following hold:*

(a) *The map $\tilde{S} \ni s \mapsto \kappa(\cdot, s)$ from $\tilde{S}$ to $\mathcal{M}_+^s(S)$ is continuous with respect to the flat norm.*
(b) *If T is a compact subset of $\tilde{S}$, the set of measures $\{\kappa(\cdot, s); s \in T\}$ in $\mathcal{M}_+^s(S)$ is pre-tight and, if S is complete, tight.*
(c) *If $\mathcal{N}$ is a tight bounded subset of $\mathcal{M}_+(\tilde{S})$, then $A(\mathcal{N})$ is a pre-tight set of measures in $\mathcal{M}(S)$ and, if S is complete, tight.*
(d) *For any separable set $\tilde{T} \in \tilde{\mathcal{B}}$, there exists a separable closed subset T of S such that $\kappa(S \setminus T, s) = 0$ for all $s \in \tilde{T}$.*
(e) *The induced map A maps $\mathcal{M}_+^s(\tilde{S})$ continuously into $\mathcal{M}_+^s(S)$.*
(f) *The induced map A maps $\mathcal{M}^s(\tilde{S})$ into $\mathcal{M}^s(S)$.*

PROOF. Let κ be a Feller kernel of separable measures. See Definition 13.16.

(a) Since κ has the Feller property, the map $\tilde{S} \ni x \mapsto \kappa(\cdot, x) \in \mathcal{M}_+^s(S)$ is continuous by Proposition 3.23.

Indeed, let (s_n) be a sequence and s a point in $\tilde{S}$ and $s_n \to s$ as $n \to \infty$. Since κ is a Feller kernel,

$$\int_S f(t)\kappa(dt, s_n) \stackrel{n\to\infty}{\longrightarrow} \int_S f(t)\kappa(dt, s), \qquad f \in C_+^b(S).$$

Since $\kappa(\cdot, s) \in \mathcal{M}_+^s(S)$,

$$\|\kappa(\cdot, s_n) - \kappa(\cdot, s)\|_\flat \to 0, \qquad n \to \infty,$$

by Proposition 3.23.

(b) Let T be a compact subset of $\tilde{S}$. By part (a), the set $\Gamma_T = \{\kappa(\cdot, s), s \in T\}$ is the continuous image of a compact set and thus a compact set in $\mathcal{M}_+^s(S)$ with the flat norm. By Proposition 3.37, it is a pre-tight set and, if S is complete, a tight set.

(c) Let $\mathcal{N}$ a tight bounded subset of $\mathcal{M}_+(\tilde{S})$ and $\epsilon > 0$. Since κ has the Feller property, there exists some $c_1 > 0$ such that $\kappa(S, x) \le c_1$ for all $x \in \tilde{S}$. Since $\mathcal{N}$ is a bounded subset of $\mathcal{M}_+(\tilde{S})$, there exists some $c_2 > 0$ such that $\mu(S) \le c_2$ for all $\mu \in \mathcal{N}$.

Since $\mathcal{N}$ is tight, by Definition 3.16 there exists a compact set V such that $\mu(\tilde{S} \setminus V) < \frac{\epsilon}{2c_1}$ for all $\mu \in \mathcal{N}$. By part (b) and Definition 3.17, there exists a closed totally bounded set T in S such that $\kappa(S \setminus T, x) < \frac{\epsilon}{2c_2}$ for all $x \in V$. Then, for all $\mu \in \mathcal{N}$,

$$\begin{aligned}(A\mu)(S \setminus T) &= \int_V \kappa(S \setminus T, x)\mu(dx) + \int_{\tilde{S}\setminus V} \kappa(S \setminus T, x)\mu(dx) \\ &\le \sup_{x\in V} \kappa(S \setminus T, x)c_2 + c_1\mu(\tilde{S} \setminus V) \le \epsilon.\end{aligned}$$

So the set of measures $A(\mathcal{N})$ is pre-tight and, if S is complete, tight.

(d) Let $\tilde{T} \in \tilde{\mathcal{B}}$ be separable. By Definition A.31, there exists a dense countable subset D of $\tilde{T}$. By Definition 3.17, for any $t \in D$, there exists a separable subset $T_t \in \mathcal{B}$ such that $\kappa(S \setminus T_t, t) = 0$. Let T be the closure of $\bigcup_{t\in D} T_t$. Since D is countable, T is a closed separable subset of S (Exercise A.1.7). Further,

$$\kappa(S \setminus T, t) \le \kappa(S \setminus T_t, t) = 0, \qquad t \in D$$

Let $s \in \tilde{T}$. Then there exists a sequence (t_j) in D such that $t_j \to s$. By (a), $\|\kappa(\cdot, t_j) - \kappa(\cdot, t)\|_\flat \to 0$ as $j \to \infty$. Since $S \setminus T$ is open,

$$\kappa(S \setminus T, s) \le \liminf_{j\to\infty} \kappa(S \setminus T, t_j) = 0$$

by Theorem 3.5 (d).

(e) Let $\mu \in \mathcal{M}_+^s(\tilde{S})$. By definition, there exists a separable set $\tilde{T} \in \tilde{\mathcal{B}}$ such that $\mu(S \setminus \tilde{T}) = 0$. We choose T according to (d). Then

$$(A\mu)(S \setminus T) = \int_{\tilde{T}} \kappa(S \setminus T, s)\mu(ds) = 0.$$

This shows that $A\mu$ is separable. By Theorem 13.20, A is continuous at μ.

(f) Let $\mu \in \mathcal{M}^s(S)$. By Definition 3.17, $|\mu| \in \mathcal{M}_+^s(S)$ and

$$\mu_\pm = \frac{1}{2}(|\mu| \pm \mu) \in \mathcal{M}_+^s(S).$$

Since A is linear, by (e), $A\mu = A\mu_+ - A\mu_- \in \mathcal{M}^s(S)$. □

REMARK 13.23. Let κ be a Feller kernel of separable measures and A^s denote the restriction of A to $\mathcal{M}^s(S)$ and A_+^s the restriction of A to $\mathcal{M}_+^s(S)$. Since the Dirac measures are separable, we still have for the operator norms that

$$\|A^s\| = \|A_+^s\| = \sup_{s\in S} \kappa(S,s),$$

(see (13.7) and, by (13.11),

$$\mathbf{r}(A^s) = \mathbf{r}(A) = \mathbf{r}(A_+^s) = \mathbf{r}(\kappa).$$

This holds both for the variation norm and the flat norm.

THEOREM 13.24. *Let d be a metric that induces the topology of S. Let $\kappa : \mathcal{B} \times \tilde{S} \to \mathbb{R}_+$ be a measure kernel of separable measures. Assume that $\int_S f(y)\kappa(dy,\cdot) \in C^b(\tilde{S})$ for any $f : S \to [0,1]$ with $|f(y) - f(z)| \le d(y,z)$ for all $y, z \in S$.*

Then κ has the Feller property. Further, the map $s \mapsto \kappa(\cdot,s)$ from $(\tilde{S},d)$ to $\mathcal{M}_+^s(S)$ endowed with the associated flat norm is continuous.

PROOF. Let (s_n) be a sequence in $\tilde{S}$ and $s \in \tilde{S}$ and $s_n \to s$. By assumption,

$$\int_S f(y)\kappa(dt,s_n) \to \int_S f(y)\kappa(dt,s)$$

for all $f : S \to [0,1]$ with $|f(y) - f(z)| \le d(y,z)$ for all $y, z \in S$.

Since $\kappa(\cdot,s)$ is separable, by Proposition 3.23, $\|\kappa(\cdot,s_n) - \kappa(\cdot,s)\|_\flat \to 0$, where $\|\cdot\|_\flat$ is the flat norm in $\mathcal{M}(S)$ associated with d. By Theorem 3.5, $\int_S g(y)\kappa(dy,s_n) \to \int_S g(y)\kappa(dy,s)$ for all $g \in C^b(S)$. This shows that κ has the Feller property. The last statement follows from Theorem 13.22 (a). □

REMARK 13.25. The map A induced by a measure kernel via (13.5) is continuous on $\mathcal{M}_+(\tilde{S})$ to $\mathcal{M}_+(S)$ with respect to the variation norms even without the Feller type property. But it seems difficult to come up with conditions for A to be compact with respect to the variation norm. See Section 13.4 for how this is different for the flat norm.

We turn to how the Feller property is inherited between kernels.

PROPOSITION 13.26. *Let S and $\tilde{S}$ be metric spaces and $\mathcal{B}$ the σ-algebra of Borel sets in S. Let $P : \mathcal{B} \times \tilde{S} \to \mathbb{R}_+$ be a Feller kernel and $g \in C_+^b(\tilde{S} \times S)$.*

Assume that, for any convergent sequence (s_n) in $\tilde{S}$ with limit $s \in \tilde{S}$,

$$\int_S \big|g(s_n,t) - g(s,t)\big| P(dt,s_n) \to 0, \qquad n \to \infty. \tag{13.22}$$

Then $\tilde{\kappa} : \mathcal{B} \times \tilde{S} \to \mathbb{R}_+$,

$$\tilde{\kappa}(T,s) = \int_T g(s,t)P(dt,s), \qquad s \in \tilde{S}, \quad T \in \mathcal{B}, \tag{13.23}$$

is a Feller kernel. In particular, $\tilde{\kappa}(S,\cdot) \in C^b(\tilde{S})$.

PROOF. Let $f \in C^b(S)$. Set $h(s) = \int_S f(t)\tilde{\kappa}(dt, s)$, $s \in \tilde{S}$. We claim that

$$h(s) = \int_S f(t)g(s,t)P(dt, s), \qquad s \in \tilde{S}.$$

Indeed, if $f = \chi_T$ with some $T \in \mathcal{B}$, this formula is just a reformulation of (13.23). The formula then holds for simple functions by the properties of the integral and for pointwise limits of increasing sequences of simple functions by the monotone convergence theorem, in particular if f is continuous and nonnegative. Finally, the formula holds for continuous f because it is the difference of two nonnegative continuous functions.

Our task is to show that $h \in C^b(\tilde{S})$. Since g and f are bounded and $P(S, \cdot)$ is bounded, h is bounded.

To demonstrate the continuity of h on $\tilde{S}$, let $s \in \tilde{S}$ and (s_n) a sequence in $\tilde{S}$ with $s_n \to s$ as $n \to \infty$.

To show that $h(s_n) \to h(s)$ as $n \to \infty$, by the triangle inequality,

$$\begin{aligned} |h(s_n) - h(s)| \le & \int_S |f(t)| \, \big|g(s_n, t) - g(s,t)\big| P(dt, s_n) \\ & + \Big| \int_S f(t)g(s,t)P(dt, s_n) - \int_S f(t)g(s,t)P(dt, s)\Big|. \end{aligned}$$

The first expression on the right-hand side converges to 0 as $n \to \infty$ by (13.22).

Since, for any $s \in \tilde{S}$, $f(t)g(s,t)$ is a continuous bounded function of $t \in S$ and P is a Feller kernel, the second expression on the right-hand side of this inequality converges to 0 as $n \to \infty$ as well. □

COROLLARY 13.27. *Let S and $\tilde{S}$ be metric spaces and $\mathcal{B}$ the σ-algebra of Borel sets in S. Let $P : \mathcal{B} \times \tilde{S} \to \mathbb{R}_+$ be a Feller kernel and $g \in C^b_+(\tilde{S} \times S)$.*

Assume that $\{g(\cdot, t); t \in S\}$ is equicontinuous on $\tilde{S}$,

Then $\tilde{\kappa} : \mathcal{B} \times \tilde{S} \to \mathbb{R}_+$,

$$\tilde{\kappa}(T, s) = \int_T g(s,t)P(dt, s), \qquad s \in \tilde{S}, \quad T \in \mathcal{B},$$

is a Feller kernel. In particular, $\tilde{\kappa}(S, \cdot) \in C^b(\tilde{S})$.

For the concept of equicontinuity see Definition 2.82.

THEOREM 13.28. *Let S and $\tilde{S}$ be metric spaces and $\mathcal{B}$ the σ-algebra of Borel sets in S. Let $P : \mathcal{B} \times \tilde{S} \to \mathbb{R}_+$ be a Feller kernel of tight measures and $g \in C^b_+(\tilde{S} \times S)$.*

Then $\tilde{\kappa} : \mathcal{B} \times \tilde{S} \to \mathbb{R}_+$,

$$\tilde{\kappa}(T, s) = \int_T g(s,t)P(dt, s), \qquad s \in \tilde{S}, \quad T \in \mathcal{B},$$

is a Feller kernel of tight measures. In particular, $\tilde{\kappa}(S, \cdot) \in C^b(\tilde{S})$.

PROOF. Since g is bounded, $\tilde{\kappa}$ inherits from P that it is a kernel of tight measures.

We check condition (13.22). Consider any convergent sequence (s_n) in $\tilde{S}$ with limit $s \in \tilde{S}$. Since P is a Feller kernel,

$$\int_S f(t)P(dt, s_n) \to \int_S f(t)P(dt, s), \qquad n \to \infty, \quad f \in C^b_+(S).$$

Since $P(\cdot, s_n), P(\cdot, s) \in \mathcal{M}_+^t(S)$, $\{P(\cdot, s_n), n \in \mathbb{N}\}$ is a tight set of measures by Theorem 3.25.

Let $\epsilon > 0$. By Definition 3.16, there exists a compact subset K of S such that $P(S \setminus K, s_n) \le \epsilon$ for all $n \in \mathbb{N}$. By the triangle inequality,

$$\limsup_{n\to\infty} \int_S \big|g(s_n,t) - g(s,t)\big| P(dt, s_n) \le 2 \sup|g| \, \epsilon + \limsup_{n\to\infty} \sup_{t\in K} \big|g(s_n,t) - g(s,t)\big| \, P(S, s_n).$$

Suppose that

$$\sup_{t\in K} \big|g(s_n,t) - g(s,t)\big|$$

does not converge to 0 as $n \to \infty$. Then there exists a sequence (t_n) in S such that $g(s_n,t_n) - g(s,t_n)$ does not converge to 0 as $n \to \infty$. After choosing subsequences, there exists some $\delta > 0$ such that $|g(s_n,t_n) - g(s,t_n) \ge \delta|$ for all $n \in \mathbb{N}$. Since K is compact, after choosing subsequences again, $t_n \to t \in K$ as $n \to \infty$ for some $t \in K$. This contradicts the continuity of g.

So

$$\limsup_{n\to\infty} \int_S \big|g(s_n,t) - g(s,t)\big| P(dt, s_n) \le 2 \sup|g| \, \epsilon.$$

Since $\epsilon > 0$ has been arbitrary,

$$\limsup_{n\to\infty} \int_S \big|g(s_n,t) - g(s,t)\big| P(dt, s_n) = 0.$$

The assertion now follows from Proposition 13.26. □

PROPOSITION 13.29. *Let S and $\tilde{S}$ be metric spaces and $\mathcal{B}$ the σ-algebra of Borel sets in S. Let $P : \mathcal{B} \times \tilde{S} \to \mathbb{R}_+$ be a Feller kernel and $\xi : S \to S$ be continuous.*

Then $\tilde{\kappa} : \mathcal{B} \times \tilde{S} \to \mathbb{R}_+$,

$$\tilde{\kappa}(T, s) = P\big(\xi^{-1}(T), \tilde{s}\big), \qquad \tilde{s} \in \tilde{S}, \quad T \in \mathcal{B}, \tag{13.24}$$

is a Feller kernel. If P is a Feller kernel of tight measures, so is κ.

PROOF. Since $\xi : S \to S$ is continuous, $\xi^{-1}(T) \in \mathcal{B}$ whenever $T \in \mathcal{B}$. This first follows for open T and then from the fact that $\{\xi^{-1}(T); T \in \mathcal{B}\}$ is a σ-algebra. Further, for any $\tilde{s} \in \tilde{S}$, $\tilde{\kappa}(\cdot, \tilde{s})$ is a measure on $\mathcal{B}$. Notice that

$$\int_S f(t)\tilde{\kappa}(dt, \tilde{s}) = \int_S f(\xi(t)) P(dt, \tilde{s}), \qquad s \in \tilde{S}, \quad f \in M^b(S). \tag{13.25}$$

This first follows for $f = \chi_T$ with $T \in \mathcal{B}$ from (13.24) because $\chi_T \circ \xi = \chi_{\xi^{-1}(T)}$. It then extends to simple functions which are linear combinations of the previous ones and then to bounded measurable functions which are uniform limits of simple functions. If $f \in C^b(S)$, then $f \circ \xi \in C^b(S)$ and the Feller property is inherited via (13.25). As for the tightness claim, let T be a compact subset of S. Since ξ is continuous, $V = \xi(T)$ is a compact subset of S. Further,

$$\xi^{-1}(S \setminus V) = S - \xi^{-1}(V) \subseteq S \setminus T.$$

By (13.24),

$$\tilde{\kappa}(S \setminus V, \tilde{s}) \le P(S \setminus T, \tilde{s}), \qquad \tilde{s} \in \tilde{S}.$$

So, if $P(\cdot, \tilde{s})$ is tight, $\kappa(\cdot, \tilde{s})$ is tight as well (Definition 3.16). □

13.4. Tight measure kernels

Let S be metric space and $\mathcal{B}$ the σ-algebra of Borel sets and $\tilde{S}$ a measurable space with σ-algebra $\tilde{B}$. A measure kernel $\kappa : \mathcal{B} \times \tilde{S} \to \mathbb{R}_+$ is called a *tight measure kernel* if the set of measures $\{\kappa(\cdot, x); x \in \tilde{S}\}$ is tight.

A measure kernel κ is called a *pre-tight measure kernel* if set of measures $\{\kappa(\cdot, x); x \in \tilde{S}\}$ is pre-tight.

See Definitions 3.16 and 3.17.

PROPOSITION 13.30. *Let S be a metric space and $\mathcal{B}$ the σ-algebra of Borel sets and let $\kappa : \mathcal{B} \times S \to \mathbb{R}_+$ be a Feller kernel of separable measures with the following property:*

For any $\epsilon > 0$, there exists a compact subset T of S such that $\kappa(S \setminus T, s) < \epsilon$ for all $s \in S \setminus T$.

Then κ is pre-tight and, if S is complete, even tight.

PROOF. Let $\epsilon > 0$. By assumption, there exists a compact subset K of S such that

$$\kappa(S \setminus K, s) < \epsilon, \quad s \in S \setminus K. \tag{13.26}$$

By Theorem 13.22 (b), the set $\{\kappa(\cdot, s); s \in K\}$ in $\mathcal{M}^s_+(S)$ is pre-tight (tight if S is complete). By Definition 3.17, there exists a closed totally bounded subset T of S (compact if S is complete) such that

$$\kappa(S \setminus T, s) < \epsilon, \qquad s \in K. \tag{13.27}$$

Then $T \cup K$ is a totally bounded subset of S (compact if S is complete) and, by (13.26) and (13.27),

$$\kappa(S \setminus (T \cup K), s) \le \min\{\kappa(S \setminus T, s), \kappa(S \setminus K, s)\} < \epsilon, \qquad s \in S. \qquad \square$$

The significance of tight Feller kernels is that the induced operators on the cone of separable measures are compact.

PROPOSITION 13.31. *Let $\kappa : \mathcal{B} \times \tilde{S} \to \mathbb{R}_+$ be a tight measure kernel. Then the induced map $A : \mathcal{M}^s_+(\tilde{S}) \to \mathcal{M}^s_+(S)$, $A\mu = \int_{\tilde{S}} \kappa(\cdot, t)\mu(dt)$, is compact and continuous with respect to the flat norms and maps $\mathcal{M}_+(\tilde{S})$ into $\mathcal{M}^t_+(S)$.*

PROOF. To show that A is compact, let (μ_n) be a bounded sequence in $\mathcal{M}_+(\tilde{S})$ with the flat norm.

Since $\{\kappa(\cdot, x); x \in \tilde{S}\}$ is tight, for any $\epsilon > 0$ there exists some compact set K in S such that $\kappa(S \setminus K, x) < \epsilon$ for all $x \in \tilde{S}$. By (13.5),

$$(A\mu_n)(S \setminus K) = \int_{\tilde{S}} \kappa(S \setminus K, t)\mu_n(dt) \le \epsilon\mu_n(\tilde{S}), \qquad n \in \mathbb{N}.$$

Since $(\mu_n(\tilde{S}))$ is a bounded sequence, the sequence $(A\mu_n)$ is tight.

Finally,

$$(A\mu_n)(S) \le \sup_{x \in \tilde{S}} \kappa(S, x)\mu_n(\tilde{S})$$

and the set $\{(A\mu_n)(S); n \in \mathbb{N}\}$ is bounded in $\mathbb{R}$. By Proposition 3.30, $(A\mu_n)$ has a convergent subsequence.

By the same arguments as above, A maps $\mathcal{M}_+(\tilde{S})$ into $\mathcal{M}^t_+(S)$ and is continuous with respect to the flat norms by Theorem 13.20. $\square$

PROPOSITION 13.32. *Let $\kappa : \mathcal{B} \times \tilde{S} \to \mathbb{R}_+$ be a measure kernel of separable measures and let the induced map $A : \mathcal{M}_+^s(\tilde{S}) \to \mathcal{M}_+^s(S)$, $A\mu = \int_S \kappa(\cdot, t)\mu(dt)$, be compact. Then κ is pre-tight and, if S is complete, tight.*

PROOF. The set $\{\delta_s; s \in \tilde{S}\}$ of point measures is a bounded subset of $\mathcal{M}_+^s(\tilde{S})$. Since the induced map A is compact and $A\delta_s = \kappa(\cdot, s)$, the set $\{\kappa(\cdot, s); s \in \tilde{S}\}$ has compact closure and is totally bounded and, by Proposition 3.37 is pre-tight and, if S is complete, tight. □

13.4.1. Preservation of tightness.

LEMMA 13.33. *Let S be a metric space and $\mathcal{B}$ the σ-algebra of its Borel sets and $\kappa_j : \mathcal{B} \times S \to \mathbb{R}_+$, $j = 1, 2$, be measure kernels and κ_1 be tight. Let A_j be the operators induced by κ_j on $\mathcal{M}(S)$. Then $\kappa_1 \star \kappa_2$ is a tight Feller kernel and its induced operator $A = A_1 A_2$ maps $\mathcal{M}_+(S)$ into $\mathcal{M}_+^t(s)$ and is continuous and compact (with respect to the flat norm).*

PROOF. Let $\epsilon > 0$. Since κ_2 is a measure kernel, $\kappa_2(S, \cdot) \in M_+^b(S)$ and there exists some $\delta > 0$ such that $\delta\kappa_2(S, s) < \epsilon$ for all $s \in S$. Since κ_1 is tight, there exists a compact subset K of S such that $\kappa_1(S \setminus K, t) < \delta$ for all $t \in S$. For all $s \in S$,

$$(\kappa_1 \star \kappa_2)(S \setminus K, s) = \int_S \kappa_1(S \setminus K, t)\kappa_2(dt, s) \leq \delta \sup_{s \in S} \kappa_2(S, s) \leq \epsilon.$$

So $\kappa_1 \star \kappa_2$ is a tight measure kernel. The other statements follow from Proposition 13.31. □

COROLLARY 13.34. *If $\kappa : \mathcal{B} \times S \to \mathbb{R}_+$ is a tight Feller kernel, so are the multiple convolutions $\kappa^{n\star}$ for all $n \in \mathbb{N}$.*

LEMMA 13.35. *Let S be a metric space and $\mathcal{B}$ the σ-algebra of its Borel sets and let $\kappa_2 : \mathcal{B} \times S \to \mathbb{R}_+$ be a tight Feller kernel and $\kappa_1 : \mathcal{B} \times S \to \mathbb{R}_+$ be a Feller kernel of separable kernels. Let A_j be the operators induced by κ_j on $\mathcal{M}(S)$. Then $\kappa_1 \star \kappa_2$ is a pre-tight (tight if S is complete) Feller kernel of separable measures and its induced operator $A = A_1 A_2$ maps $\mathcal{M}_+(S)$ into $\mathcal{M}_+^s(S)$ and is compact and continuous between these sets with respect to the flat norm.*

PROOF. By Proposition 13.31, the operator A_2 induced by κ_2 maps $\mathcal{M}_+(S)$ into $\mathcal{M}_+^t(S)$ and is compact and continuous. By Theorem 13.22 (e), the operator A_1 induced by κ_1 maps $\mathcal{M}_+^s(S)$ into itself and is continuous with respect to the flat norm. So $A = A_1 A_2$ maps $\mathcal{M}_+(S)$ into $\mathcal{M}_+^s(S)$ and is compact and continuous with respect to the flat norm. By Lemma 13.32, the convoluted kernel $\kappa_1 \star \kappa_2$ is pre-tight and, if S is complete, tight. □

COROLLARY 13.36. *Let $P : \mathcal{B} \times \tilde{S} \to \mathbb{R}_+$ be a tight Feller kernel and $g \in C_+^b(\tilde{S} \times S)$. Then $\tilde{\kappa} : \mathcal{B} \times \tilde{S} \to \mathbb{R}_+$,*

$$\tilde{\kappa}(T, s) = \int_T g(s, t)P(dt, s), \qquad s \in \tilde{S}, \quad T \in \mathcal{B},$$

is a tight Feller kernel. In particular, $\tilde{\kappa}(S, \cdot) \in C^b(\tilde{S})$.

PROOF. $\tilde{\kappa}$ inherits tightness from P via the boundedness of g. That $\tilde{\kappa}$ is a Feller kernel follows from Theorem 13.28. □

13.4.2. Survival and deterministic movement (development). We return to Example 13.19,

$$\kappa(T,s) = \chi_T\big(\xi(s)\big)\, g(s), \qquad s \in S, \quad T \in \mathcal{B}, \tag{13.28}$$

where $\xi : S \to S$ continuous, $g \in C^b_+(S)$. If the population structure is given by space, ξ describes how the individuals move: If the location is $s \in S$ at the beginning of the year, it is $\xi(s)$ at the end of the year. If g takes values between 0 and 1, $g(s)$ can be interpreted as the probability of an individual with location s at the beginning of the year to survive for a year.

Notice that

$$f \in C^b(S) \implies A_* f = \int_S f(t)\kappa(dt,\cdot) = f(\xi(\cdot))g(\cdot) \in C^b(S);$$

thus the usual Feller property is satisfied. Notice, that for each $s \in S$, $\kappa(\cdot, s) = g(s)\delta_{\xi(s)}$ is a tight measure.

The associated operator A maps $\mathcal{M}^s_+(S)$ into itself and $\mathcal{M}^t_+(S)$ into itself and is continuous on $\mathcal{M}^s_+(S)$.

PROOF. Let $\mu \in \mathcal{M}^s_+(S)$. Then there exists a separable set D such that $\mu(S \setminus D) = 0$. Since ξ is continuous, $\xi(D)$ is separable and its closure $\tilde D$ a separable closed set. Now

$$(A\mu)(S \setminus \tilde D) = \int_S \chi_{S\setminus \tilde D}(\xi(s))g(s)\mu(ds).$$

For $s \in D$, $\xi(s) \in \tilde D$ and so $\chi_{S\setminus\tilde D}(\xi(s)) = 0$. This implies $(A\mu)(S\setminus\tilde D) = 0$ and $A\mu$ is separable. By Theorem 13.20, A is continuous on $\mathcal{M}^s_+(S)$.

Let $\mu \in \mathcal{M}^t_+(S)$ and $\epsilon > 0$. Then there exists a compact subset K of S such that $\mu(S \setminus K) < \epsilon \sup g(S)$. Since ξ is continuous, $\xi(K)$ is compact. Now

$$(A\mu)(S \setminus \xi(K)) = \int_S \chi_{S\setminus\xi(K)}(\xi(s))g(s)\mu(ds)$$

Let $\chi_{S\setminus\xi(K)}(\xi(s)) = 1$. Then $\xi(s) \in S\setminus\xi(K)$ and $s \in S\setminus K$ and $\chi_{S\setminus K}(s) = 1$. This implies that $\chi_{S\setminus\xi(K)}(\xi(s)) \le \chi_{S\setminus K}(s)$. Now

$$(A\mu)(S \setminus \xi(K)) \le \int_S \chi_{S\setminus K}(s)g(s)\mu(ds) \le \sup g(S)\mu(S\setminus K) < \epsilon. \qquad \square$$

This Feller kernel κ is tight if and only if for any $\epsilon > 0$ there is a compact subset T of S such that

$$g(s) < \epsilon \quad \text{if } s \in S \text{ and } \xi(s) \notin T.$$

A^n is associated with the kernel $\chi_T\big(\xi^n(s)\big) \prod_{j=0}^{n-1} g(\xi^j(s))$,

$$\mathbf{r}(A) = \mathbf{r}(\kappa) = \inf_{n\to\infty}\Big(\sup_{s\in S}\prod_{i=0}^{n-1} g(\xi^j(s))\Big)^{1/n} = \lim_{n\to\infty}\Big(\sup_{s\in S}\prod_{i=0}^{n-1} g(\xi^j(s))\Big)^{1/n}. \tag{13.29}$$

See also (13.11).

There may or may not be an eigenvector of A associated with $\mathbf{r}(A)$.

Suppose that $s^* \in S$ and $\xi(s^*) = s^*$ and $g(s^*) \ge g(s)$ for all $s \in S$. Then $g(s^*) = \mathbf{r}(A)$ and $A\delta_{s^*} = g(s^*)\delta_{s^*}$.

Now assume that for any compact subset K of S there is some $n \in \mathbb{N}$ such that $\xi^n(s) \in S \setminus K$ for all $s \in S$. We claim that there is no $r > 0$ and $\mu \in \mathcal{M}_+^t(S)$, $\mu \neq 0$, such that $A\mu \geq r\mu$. Suppose that such $r > 0$ and μ exist. Then $\mu \leq r^{-n} A^n \mu$ for all $n \in \mathbb{N}$ and

$$\mu(T) \leq r^{-n} \int_S \chi_T(\xi^n(s)) \prod_{j=0}^{n-1} g(\xi^j(s))\mu(ds), \qquad n \in \mathbb{N}, \quad T \in \mathcal{B}. \tag{13.30}$$

By our assumption, $\mu(T) = 0$ for any compact subset T of S. Since $\mu \in \mathcal{M}_+^t(S)$, $\mu = 0$.

Alternatively, let us assume that for any compact subset T of S and any $s \in S$, $\xi^n(s) \notin T$ for all but finitely many $n \in \mathbb{N}$ and $g(s) = 1$ for all $s \in S$. Then $\chi_T(\xi^n(s)) \to 0$ as $n \to \infty$ and the right-hand side of (13.30) converges to 0 by the dominated convergence theorem and $\mu(T) = 0$. So μ is the zero measure if μ is tight.

Let us finally assume in addition that S is σ-compact, i.e., K is a countable union of compact sets. For instance, let $S = (0, 1]$ and $\xi(s) = s/2$, $s \in S$ [**27**, **146**].

Then any $\mu \in \mathcal{M}_+(S)$ is tight. So, the spectral radius is not associated with an eigenmeasure; however it is associated with all constant positive functions as eigenfunctions.

Further, for any compact set T of S and any $s \in S$, $\kappa^{n\star}(T, s) = \chi_T(\xi^n(s)) = 0$ for all but finitely many $n \in \mathbb{N}$. This implies that $\sum_{n=1}^{\infty} \lambda^{-n} \kappa^{n\star}(T, s) \in \mathbb{R}_+$ for all $\lambda \in (0, \infty)$ and $s \in S$. Since S is σ-compact, the measure $\sum_{n=1}^{\infty} \lambda^{-n} \kappa^{n\star}(\cdot, s)$ is σ-finite for all $\lambda \in (0, \infty)$ and $s \in S$ and the so-called Perron root of κ in (13.21) equals

$$\mathbf{r}^{\wp}(\kappa) = 0 \text{ while } \mathbf{r}(\kappa) = 1. \tag{13.31}$$

EXAMPLE 13.37 (Multiplication operator). Consider the special case $\xi(s) = s$ for all $s \in S$ in Example 13.4.2 with $g \in C_+^b(S)$. The induced map A on $\mathcal{M}(S)$ takes the form of a multiplication operator

$$(A\mu)(T) = \int_T g(s)\mu(ds), \qquad T \in \mathcal{B}.$$

which we already considered in Section 13.1. A maps $\mathcal{M}^s(S)$ into itself. A is a bounded linear map on $\mathcal{M}(S)$ with respect to the variation norm. A maps $\mathcal{M}_+^s(S)$ into itself continuously with respect to the flat norm.

A is continuous on $\mathcal{M}^s(S)$ with respect to the flat norm if and only if g is Lipschitz continuous.

EXAMPLE 13.38 (Deterministic movement (development)). Consider the special case $g \equiv 1$ in Example 13.4.2 such that

$$(A\mu)(T) = \int_S \chi_T(\xi(s))\mu(ds), \qquad \mu \in \mathcal{M}(S), \quad T \in \mathcal{B}.$$

Then $A\delta_s = \delta_{\xi(s)}$, $s \in S$. Let us explore which properties $\xi : S \to S$ must have to make A a bounded map on $\mathcal{M}(S)$ with respect to the flat norm, in particular,

$$\|A\|_\flat \|\delta_s - \delta_t\|_\flat \geq \|\delta_{\xi(s)} - \delta_{\xi(t)}\|_\flat, \qquad s, t \in S.$$

By Lemma 3.9,

$$\|A\|_\flat \min\{1, d(s, t)\} \geq \min\{1, d(\xi(s), \xi(t))\}. \tag{13.32}$$

This means that ξ is Lipschitz continuous with respect to the metric $d_1(s,t) = \min\{1, d(s,t)\}$ which is topologically equivalent to d. In particular, ξ is uniformly continuous on S.

13.5. Eigenmeasures of tight Feller kernels

We return to general Feller kernels κ and the maps A induced by them via (13.5). A is a bounded linear map on $\mathcal{M}(S)$ with the variation-norm. Let A_+ denote the restriction of A to $\mathcal{M}_+(S)$. Then $\|A\| = \|A_+\| = \|A_+\|_\flat$ and $\mathbf{r}(A) = \mathbf{r}(A_+)$. Since the variation norm and the flat norm coincide on $\mathcal{M}_+(S)$, this means that the spectral radius of A with respect to the variation norm coincides with the spectral radius of A_+ with respect to the flat norm and with the spectral radius of κ. Recall (13.11), (13.16), (13.18) and (13.19).

The next result gives sufficient conditions for the existence of an eigenmeasure of the kernel κ (eigenvector of the operator A_+) associated with its spectral radius.

THEOREM 13.39. *Assume that S is a metric space and κ is a tight Feller kernel and $\mathcal{T}_0 = \mathbf{r}(A) = \mathbf{r}(\kappa) > 0$.*

Then there exists some tight $\mu \in \mathcal{M}_+(S)$, $\mu \neq 0$, such that $\int_S \kappa(\cdot, s)\mu(ds) = \mathcal{T}_0\mu$.

This result does not follow from the Krein-Rutman theorem [**140**] because A is only compact and continuous on $\mathcal{M}_+(S)$ and not necessarily on $\mathcal{M}(S)$. It would follow from [**27**] if S were complete, but not if S (and then $\mathcal{M}_+(S)$, Section 3.2.4) is not complete (or cannot be made complete by transition to a topologically equivalent metric, Theorem 3.39). See Section 3.2 or [**101**] for details and further remarks.

PROOF. We apply Theorem 9.10. Since $\{\kappa(\cdot, x); x \in S\}$ is tight, the map A associated with κ maps $\mathcal{M}_+(S)$ into $\mathcal{M}_+^t(S)$. We choose as ordered normed vector space $\mathcal{M}^s(S)$ with cone $\mathcal{M}_+^s(S)$ and the flat norm. By Proposition 13.20, A_+ is continuous on $\mathcal{M}_+^s(S)$ and compact by Proposition 13.31. By Corollary 9.11, there exists some $\mu \in \mathcal{M}_+^s(S)$, $\mu \neq 0$, such that $A_+\mu = \mathcal{T}_0\mu$. Since $A_+\mu$ is tight, so is μ. □

REMARK 13.40. We do not get anything new if S is compact. Then $C(S)$, with the supremum norm, is an ordered Banach space and $C_+(S)$ is solid and A is the dual map of $f \mapsto \int_S f(t)\kappa(dt, \cdot)$. The existence of the eigenfunction then follows from [**139**].

REMARK 13.41. To see that Theorem 13.39 may fail if the kernel κ is not tight, we consider $S = (0,1)$ with the metric induced by the absolute value. Switching to a topologically equivalent metric, S becomes a complete metric space (Example 3.45). Let

$$\kappa(T, s) = \chi_T(s/2), \qquad s \in (0,1), \quad T \in \mathcal{B},$$

i.e., $\kappa(\cdot, s) = \delta_{s/2}$. κ is a Feller kernel but not tight. Further $\mathbf{r}(A) = 1$ where A is the operator induced on $\mathcal{M}_+(S)$ by κ via (13.5). Assume that there are some $r > 0$ and $\mu \in \mathcal{M}_+(S)$ such that

$$r\mu \le A\mu.$$

Then

$$r\mu(T) \le \int_{(0,1)} \chi_{2T}(s)\mu(ds), \qquad s \in S, \quad T \in \mathcal{B},$$

and $r\mu([1/2, 1)) = 0$. By induction $\mu([2^{-n}, 1) = 0$ for all $n \in \mathbb{N}$ and so $\mu((0,1)) = 0$.

The next result relaxes the assumption of tightness.

THEOREM 13.42. *Let S be a metric space. Let $\kappa_j : \mathcal{B} \times S \to \mathbb{R}_+$, $j = 1, 2$, be Feller kernels of separable measures. Assume that κ_1 is a tight measure kernel and $\mathcal{T} := \mathbf{r}(\kappa_1 + \kappa_2) > \mathbf{r}(\kappa_2)$. Then there exists some $\mu \in \mathcal{M}_+^s(S)$ with $\mu(S) = 1$ and*

$$\int_S (\kappa_1(T,s) + \kappa_2(T,s))\mu(ds) = \mathcal{T}\mu(T), \qquad T \in \mathcal{B}.$$

PROOF. We apply Theorem 12.4 with $X = \mathcal{M}^s(S)$ and $X_+ = \mathcal{M}_+^s(S)$, endowed with the flat norm $\|\cdot\|_\flat$. With respect to this norm, X_+ is a closed, normal and serially complete cone in X (Corollary 3.21).

Let $A_j : \mathcal{M}_+^s(S) \to \mathcal{M}(S)$ be given by

$$(A_j\mu)(T) = \int_S \kappa_j(T,s)\mu(ds), \qquad \mu \in \mathcal{M}_+^s(S), \quad T \in \mathcal{B}.$$

By Theorem 13.22, for $j = 1, 2$, A_j maps $\mathcal{M}_+^s(S)$ into itself and is a continuous homogeneous additive map on $\mathcal{M}_+^s(S)$. By Proposition 13.31, A_1 is compact on X_+ with respect to the flat norm.

By Theorem 12.4, there exists some $\mu \in \dot{\mathcal{M}}_+^s(S)$ such that $(A_1 + A_2)\mu = \mathcal{T}\mu$. □

COROLLARY 13.43. *Let the assumptions of Theorem 13.42 be satisfied. Then*

$$\mathbf{r}(\kappa) = \sup_{\mu \in \mathcal{M}_+(S)} \inf \Big\{ \int_S \frac{\kappa(T,s)\mu(ds)}{\mu(T)}; T \in \mathcal{B}, \mu(T) > 0 \Big\}. \tag{13.33}$$

PROOF. The inequality $\geq$ follows from Theorem 13.12. The inequality $\leq$ follows from Theorem 13.42. □

13.6. Uniform Feller kernels and their eigenfunctions

In the previous section we explored conditions under which a measure kernel κ has an eigenmeasure that is associated with its spectral radius, $\mathbf{r}(\kappa)$. We now turn to the existence of eigenfunctions f,

$$\mathbf{r}(\kappa)f(s) = \int_S f(t)\kappa(dt,s), \qquad s \in S.$$

Recall that we consider a metric space S with the σ-algebra of Borel sets, $\mathcal{B}$,

$$\kappa : \mathcal{B} \times S \to \mathbb{R}_+$$

where κ is a measure kernel or even a Feller kernel. See Definitions 13.8 and 13.16. The kernel κ induces a map A on $\mathcal{M}(S)$, the space of real measures on $\mathcal{B}$, and a map A_* on $M^b(S)$, the space of bounded measurable real-valued functions on S. See (13.5) and (13.4). An eigenfunction of κ is an eigenvector of A_* while an eigenmeasure of κ is an eigenvector of A.

We start from the observation that tight measure kernels are related to compactness in $M^b(S)$. Recall the concept of equicontinuity (Definition 2.82).

PROPOSITION 13.44. *Let κ be a tight measure kernel and Q be an equicontinuous bounded subset of $C^b(S)$. Let A_* be the map on $M^b(S)$ associated with κ. Then $A_*(Q)$ has compact closure in $M^b(S)$ (in $C^b(S)$ if κ is a tight Feller kernel).*

PROOF. Let (g_n) be a sequence in Q. Since κ is tight, there exists a sequence (K_j) of compact subsets of S such that

$$\sup_{s\in S} \kappa(S \setminus K_j, s) \to 0, \qquad j \to \infty. \tag{13.34}$$

Set $\tilde S = \bigcup_{j\in\mathbb{N}} K_j$. Then $\tilde S$ is separable. By a version of the Arzela-Ascoli theorem [**159**, Thm.8.5], there exists a subsequence (g_{n_i}) and some $g \in C^b(\tilde S)$ such that $g_{n_i} \to g$ pointwise on $\tilde S$ and uniformly on each K_j. Set

$$h_n = A_* g_n, \qquad h(s) = \int_{\tilde S} g(t)\kappa(dt,s), \quad s \in S.$$

Then $h_n \in M^b(S)$ and $h \in M^b(S)$,

$$\begin{aligned}|h_{n_i}(s) - h(s)| \le \int_{S\setminus K_j} |g_{n_i}(t)|\,\kappa(dt,s) + \int_{K_j} |g_{n_i}(t) - g(t)|\,\kappa(dt,s)\\ + \int_{\tilde S\setminus K_j} |g(t)|\,\kappa(dt,s), \qquad s\in S, \quad i,j\in\mathbb{N}.\end{aligned}$$

By our various assumptions, there is some $c > 0$ such that, for all $i, j \in \mathbb{N}$,

$$\|h_{n_i} - h\|_\infty \le 2c \sup_{s\in S} \kappa(S\setminus K_j, s) + c \sup_{t\in K_j} |g_{n_i}(t) - g(t)|.$$

For all $j \in \mathbb{N}$, since $g_{n_i} \to g$ as $i \to \infty$ uniformly on K_j,

$$\limsup_{i\to\infty} \|h_{n_i} - h\|_\infty \le 2c\kappa(S \setminus K_j).$$

By (13.34), we can take the limit as $j \to \infty$,

$$\limsup_{i\to\infty} \|h_{n_i} - h\|_\infty = 0.$$

This shows $A_*(Q)$ is a compact subset of $M^b(S)$. If κ is a tight Feller kernel, then $A_*(Q) \subseteq C^b(S)$, all h_n are continuous, and h is continuous as well. □

The preceding result motivates us to look for measure kernels that are related to equicontinuous sets of functions.

DEFINITION 13.45. *A measure kernel $\kappa : \mathcal{B} \times S \to \mathbb{R}_+$ is called a* uniform Feller kernel

$$\sup_{T\in\mathcal{B}} |\kappa(T,t) - \kappa(T,s)| \to 0, \qquad t \to s, \text{ for all } s \in S.$$

PROPOSITION 13.46. *Let $\kappa : \mathcal{B} \times S \to \mathbb{R}_+$ be a measure kernel and A the induced linear map on $\mathcal{M}(S)$ and A_* the induced linear map on $M^b(S)$ via (13.5) and (13.4), respectively.*

Then the following are equivalent:

(a) *The kernel κ is a uniform Feller kernel;*
(b) *If Q is a bounded subset of $M^b(S)$, then $A_*(Q)$ is an equicontinuous and equibounded subset of $C^b(S)$.*
(c) *If Q is a bounded subset of $C^b(S)$, then $A_*(Q)$ is an equicontinuous and equibounded subset of $C^b(S)$.*
(d) *A is continuous from $\mathcal{M}_+(S)$ with the flat norm to $\mathcal{M}_+(S)$ with the variation norm.*

PROOF. Assume (a). Let $f(t) = \sum_{i=1}^m \alpha_i \chi_{T_i}$ be a measurable function of finitely many values, $T_i \in \mathcal{B}$. Then

$$|(A_* f)(t) - A_*(f)(s)| \leq 2\|f\|_\infty \sup_{T\in\mathcal{B}} |\kappa(T,t) - \kappa(T,s)|. \tag{13.35}$$

If $f \in M^b(S)$, f is the uniform limit of a sequence of such finitely-valued measurable functions and (13.35) holds for $f \in M^b(S)$. Since κ is uniform Feller kernel, A_* maps $M^b(S)$ into $C^b(S)$. Let Q be a bounded subset of $M^b(S)$. Again, since κ is a uniform Feller kernel, $A_*(Q)$ is a bounded subset of $C^b(S)$ and an equicontinuous subset by (13.35). Thus (b) follows.

Obviously, (b) implies (c).

Assume (c). Let (μ_n) be a sequence in $\mathcal{M}_+(S)$ and $\mu \in \mathcal{M}_+(S)$ such that $\|\mu_n - \mu\|_\flat \to 0$. By (c), $\{A_* f; f \in C^b(S), 0 \leq f \leq 1\}$ is a uniformly equicontinuous and equibounded family of functions from S to $\mathbb{R}_+$. By Proposition 3.15,

$$\int_S (A_* f) d\mu_n \to \int_S (A_* f) d\mu, \qquad n \to \infty$$

uniformly for $f \in C^b(S)$ with $0 \leq f \leq 1$. Let $f \in C^b(S)$ with $\|f\|_\infty \leq 1$. Then $f = f_+ - f_-$ with $0 \leq f_\pm \leq 1$. So

$$\int_S (A_* f) d\mu_n \to \int_S (A_* f) d\mu, \qquad n \to \infty$$

uniformly for $f \in C^b(S)$ with $\|f\|_\infty \leq 1$. By the duality between A_* and A, $\int_S f d(A\mu_n) \to \int_S f d(A\mu)$ uniformly for $f \in C^b(S)$ with $\|f\|_\infty \leq 1$. Assertion (d) now follows from Theorem 3.7.

Assume (d). As $t \to s$, $\|\delta_t - \delta_s\|_\flat \to 0$ by Theorem 3.9 and, by (d), $A\delta_t \to A\delta_s$ in variation norm and $\sup_{T\in\mathcal{B}} |\kappa(T,t) - \kappa(T,s)| \to 0$ by (13.5). □

COROLLARY 13.47. *If κ is a Feller kernel and the map $A_* : C^b(S) \to C^b(S)$ associated with κ is compact, then κ is a uniform Feller kernel.*

COROLLARY 13.48. *Let κ_1 be a measure kernel on S and κ_2 a uniform Feller kernel on S. Then $\kappa_1 \star \kappa_2$ is a uniform Feller kernel on S.*

PROOF. Let A_i be the linear maps on $\mathcal{M}_+(S)$ induced by κ_i via (13.5). By Proposition 13.46, A_2 continuously maps $\mathcal{M}_+(S)$ with the flat norm into $\mathcal{M}_+(S)$ with the variation norm, while A_1 is a bounded liner map on $\mathcal{M}(S)$ with the variation norm. So $A_1 A_2$ continuously maps $\mathcal{M}_+(S)$ with the flat norm into $\mathcal{M}_+(S)$ with the variation norm. Since $A_1 A_2$ is induced by $\kappa_1 \star \kappa_2$ by Lemma 13.9, $\kappa_1 \star \kappa_2$ is a uniform Feller kernel by Proposition 13.46. □

PROPOSITION 13.49. *Let κ be a uniform Feller kernel. Let $g : S \times S \to \mathbb{R}$ be bounded and $g(s,\cdot)$ be Borel measurable and $g(\cdot,s)$ be continuous on S for every $s \in S$.*

Let $\tilde{\kappa} : \mathcal{B} \times S \to \mathbb{R}$ be given by

$$\tilde{\kappa}(T,s) = \int_T g(s,t)\kappa(dt,s), \qquad s \in S, \quad T \in \mathcal{B}.$$

Then $\tilde{\kappa}$ is a uniform Feller kernel.

PROOF. Let $s \in S$. Since $g(s, \cdot)$ is Borel measurable and bounded and $\kappa(\cdot, s)$ is a finite nonnegative measure, $\tilde{\kappa}(\cdot, s)$ is a finite nonnegative measure.

Let (s_n) be a sequence in S and $s \in S$ and $s_n \to s$ as $n \to \infty$. Then

$$\begin{aligned}
&\Big| \int_T g(s_n, t)\kappa(dt, s_n) - \int_T g(s,t)\kappa(dt, s)\Big| \\
&\quad \le \Big| \int_T g(s_n, t)\kappa(dt, s_n) - \int_T g(s_n, t)\kappa(dt, s)\Big| \\
&\qquad + \Big| \int_T g(s_n, t)\kappa(dt, s) - \int_T g(s, t)\kappa(dt, s)\Big| \\
&\quad \le 2 \sup |g| \sup_{\tilde{T}\in\mathcal{B}} \big|\kappa(\tilde{T}, s_n) - \kappa(\tilde{T}, s)\big| + \int_S |g(s_n, t) - g(s,t)|\kappa(dt, s).
\end{aligned}$$

The last integral converges to 0 as $n \to \infty$ by Lebesgue's dominated convergence theorem because $|g(s_n, t) - g(s,t)| \to 0$ as $n \to \infty$ pointwise in $t \in S$ and $|g(s_n, t) - g(s,t)| \le 2 \sup g(S \times S)$ for all $n \in \mathbb{N}$.

Notice that the last expression in the inequality converges to 0 as $n \to \infty$ uniformly for $T \in \mathcal{B}$. □

This trivially provides uniform Feller kernels.

EXAMPLE 13.50. Let $\nu \in \mathcal{M}_+(S)$ and $g : S \times S \to \mathbb{R}$ be bounded and $g(s, \cdot)$ be Borel measurable and $g(\cdot, s)$ continuous on S for every $s \in S$.

Let $\kappa : \mathcal{B} \times S \to \mathbb{R}$ be given by

$$\kappa(T, s) = \int_T g(s,t)\nu(dt), \qquad s \in S, \quad T \in \mathcal{B}.$$

Then κ is a uniform Feller kernel.

THEOREM 13.51. *Let κ_1 be a tight measure kernel and κ_2 a uniform Feller kernel. Then*

$$(A_{*1} f)(s) = \int_S f(t)\kappa_1(dt, s), \qquad s \in S, \quad f \in M^b(S),$$

defines a bounded positive linear map A_{*1} from $M^b(S)$ to $M^b(S)$ and

$$(A_{*2} f)(s) = \int_S f(t)\kappa_2(dt, s), \qquad s \in S, \quad f \in M^b(S),$$

defines a bounded positive linear map A_{*2} from $M^b(S)$ to $C^b(S)$ such that $A_{*1}A_{*2}$ is compact from $M^b(S)$ to $M^b(S)$. If κ_1 is a tight Feller kernel, A_{1*} maps $C^b(S)$ into itself and $A_{1*}A_{2*}$ is compact from $M^b(S)$ to $C^b(S)$.

PROOF. Combine Proposition 13.44 and Proposition 13.46 (b) and (c). □

The next theorem will be used in the proof of uniform weak persistence in Theorem 19.10.

THEOREM 13.52. *Let κ_2 be a uniform Feller kernel that is tight.*

Let κ_1 be a tight measure kernel and $\kappa = \kappa_1 + \kappa_2$. Assume that $\mathbf{r}(\kappa) > \mathbf{r}(\kappa_1)$. Then there exists some $f \in \dot{M}^b_+(S)$ such that $\mathbf{r}(\kappa) f(s) = \int_S f(t)\kappa(dt, s)$ for all $s \in S$. If κ_1 is a tight Feller kernel, $f \in \dot{C}^b_+(S)$.

PROOF. Let A_{*j} be the operators on $M^b(S)$ associated with κ_j. By Theorem 13.51, $A_{*1}A_{*2}$ and A_{*2}^2 are compact on $M^b(S)$. The assertion now follows from Corollary 9.5. □

COROLLARY 13.53. *Let the assumptions of Theorem* 13.52 *be satisfied. Then*

$$\begin{aligned}\mathbf{r}(\kappa) &= \inf_{f\in \dot{M}_+^b(S)} \sup\Big\{\frac{1}{f(t)}\int_S f(s)\kappa(ds,t); t\in S\Big\}\\ &= \sup_{f\in M_+^b(S)} \inf\Big\{\frac{1}{f(t)}\int_S f(s)\kappa(ds,t); t\in S, f(t)>0\Big\}.\end{aligned}$$

PROOF. The first equality follows from Theorem 13.13. The inequality $\leq$ of the second equality also follows from Theorem 13.13. The inequality $\geq$ of the second equality follows from Theorem 13.52. □

13.7. Irreducible, sustaining, and colonization kernels

Let S be a metric space and $\mathcal{B}$ the σ-algebra of Borel sets. Since a Feller kernel $\kappa : \mathcal{B}\times S \to \mathbb{R}_+$ induces a positive bounded linear map on the Banach lattice $C^b(S)$ with the supremum norm by (13.4), irreducibility of κ could be defined as irreducibility of A_* (more precisely, of its restriction to $C^b(S)$) like in [**181**, III.8]. However, the following weaker irreducibility notion seems to be better suited to a Feller kernel.

13.7.1. Topologically irreducible Feller kernels.

DEFINITION 13.54. *A Feller kernel $\kappa : \mathcal{B}\times S\to\mathbb{R}_+$ is called* topologically irreducible *if for any nonempty open strict subset U of S there exists some $s\in S\setminus U$ such that $\kappa(U,s)>0$.*

LEMMA 13.55. *Let $\kappa : \mathcal{B}\times S\to\mathbb{R}_+$ be a Feller kernel and A_* be the bounded linear map on $C^b(S)$ induced by* (13.4). *Then the following are equivalent:*

(a) *For any nonempty open subset U of S and for any $s\in S\setminus U$ there is some $n\in\mathbb{N}$ such that $\kappa^{n\star}(U,s)>0$.*
(b) *κ is topologically irreducible.*
(c) *For any $f\in \dot{C}_+^b(S)$, $S=\bigcup_{n\in\mathbb{Z}_+}\{A_*^n f>0\}=:U(f)$.*
(d) *For any Lipschitz continuous $f:S\to\mathbb{R}_+$ that is not identically equal to 0, $S=\bigcup_{n\in\mathbb{Z}_+}\{A_*^n f>0\}=:U(f)$.*

Here, $\{A_*^n f>0\}$ is a shorthand for $\{s\in S; (A_*^n f)(s)>0\}$. Recall that $\dot{C}_+^b(S)=C_+^b(S)\setminus\{0\}$.

The characterization in (a) is chosen for topological irreducibility in [**184**].

PROOF. (a) ⇒ (b): Suppose that (b) does not hold: Then there exists some nonempty open strict subset U of S such that $\kappa(U,s)=0$ for all $s\in S\setminus U$. Since $\kappa(S,\cdot)$ is bounded, there exists some $c>0$ such that $\kappa(U,s)\leq c\chi_U(s)$ for all $s\in S$. By induction, for all $n\in\mathbb{N}$,

$$\kappa^{n*}(U,s)\leq c^n\chi_U(s),\qquad s\in S, \tag{13.36}$$

and (a) does not hold.

Indeed, (13.36) holds for $n = 1$. Suppose $n \in \mathbb{N}$ and (13.36) holds for n. By definition 13.10, for all $s \in S$,

$$\begin{aligned}\kappa^{*(n+1)}(U, s) &= \int_S \kappa^{*n}(U, t)\kappa(dt, s)\\ &\le \int_S c^n \chi_U(t)\kappa(dt, s) = c^n \kappa(U, s) \le c^{n+1}\chi_U(s),\end{aligned}$$

and (13.36) holds for $n + 1$.

(b) $\Rightarrow$ (c): Since κ is a Feller kernel, the functions $A_*^n f$ in part (c) are continuous and $U(f)$ is open as union of open sets. Since f is not the zero function and $A_*^0 f = f$, $U(f)$ is nonempty. Suppose $U(f) \ne S$. By (b), there exists some $s \in S \setminus U(f)$ such that

$$0 < \kappa(U(f), s) \le \sum_{n=0}^{\infty} \kappa\big(\{A_*^n f > 0\}, s\big).$$

Here, we have used that the measure $\kappa(\cdot, s)$ is σ-subadditive. There is some $n \in \mathbb{Z}_+$ such that $\kappa\big(\{A_*^n f > 0\}, s\big) > 0$. Since $A_*^n f \ge 0$,

$$(A_*^{n+1} f)(s) = \int_S (A_*^n f)(t)\kappa(dt, s) > 0.$$

So $s \in U(f)$, a contradiction.

(c) $\Rightarrow$ (d): This is obvious.

(d) $\Rightarrow$ (a): Let U be a nonempty open subset of S. Choose some $t_0 \in U$. By Lemma 2.75, there exists some Lipschitz continuous $f : S \to [0, 1]$ such that $f(t_0) = 1$, $f(t) \le \chi_U(t)$ for all $t \in S$.

By (d), for any $s \in S$, there is some $n \in \mathbb{Z}_+$ such that $0 < (A_*^n f)(s)$. Let $s \in S \setminus U$. Then $(A_*^0 f)(s) = f(s) \le \chi_U(s) = 0$ and $0 < (A_*^n f)(s)$ for some $n \in \mathbb{N}$. Since A_*^n is induced by κ^{n*},

$$0 < (A_*^n f)(s) \le \int_S \chi_U(t)\kappa^{n*}(dt, s) \le \kappa^{n*}(U, s).$$

So (a) holds. □

PROPOSITION 13.56. *Let $\kappa : \mathcal{B} \times S \to \mathbb{R}_+$ be a topologically irreducible Feller kernel. Let $f \in \dot{C}_+^b(S)$ and $f(t) = 0$ for some $t \in S$. Then there exists some $s \in S$ such that $f(s) = 0 < (A_* f)(s)$.*

PROOF. Let $f \in \dot{C}_+^b(S)$ and there is some $t \in S$ such that $f(t) = 0$. Then $U = \{f > 0\}$ is a nonempty strict subset of S. Since κ is topologically irreducible, there is some $s \in S \setminus U$ such that $\kappa(U, s) > 0$. Since $\kappa(\cdot, s)$ is continuous from below, there is some $n \in \mathbb{N}$ such that

$$0 < \frac{1}{n}\kappa\big(\{f > 1/n\}, s\big) \le \int_S f(t)\kappa(dt, s) = (A_* f)(s).$$ □

REMARK 13.57. Assume that S is not a singleton set. If $\kappa : \mathcal{B} \times S \to \mathbb{R}_+$ is a topologically irreducible Feller kernel, then $\kappa(S \setminus \{s\}, s) > 0$ for all $s \in S$.

PROOF. Let $s \in S$ and $T = S \setminus \{s\}$. Since S is not a singleton set, T is a nonempty open strict subset of S. Since κ is topologically irreducible, by Lemma 13.55 (b), there exists some $\tilde{s} \in S \setminus T$ such that $\kappa(T, \tilde{s}) > 0$. Since $S \setminus T = \{s\}$, $\kappa(T, s) > 0$. □

All eigenfunctions of a topologically irreducible Feller kernel associated with a positive eigenvalue are strictly positive.

THEOREM 13.58. *Let κ be a topologically irreducible Feller kernel, A_* the associated linear bounded map on $C^b(S)$ via (13.4), $r > 0$. Let $f \in \dot{C}^b_+(S)$ be an eigenfunction $rf = A_* f$. Then $f(s) > 0$ for all $s \in S$.*

PROOF. For all $n \in \mathbb{N}$, $f = r^{-n} A_*^n f$ and so $f(s) > 0$ for all $s \in S$ by Lemma 13.55 (c). ☐

THEOREM 13.59. *Let κ be a topologically irreducible Feller kernel. Then, for any $\mu \in \dot{\mathcal{M}}_+(S)$ and $f \in \dot{C}^b_+(S)$, there is some $n \in \mathbb{Z}_+$ such that $\int_S f \, d(A^n \mu) = \int_S A_*^n f \, d\mu > 0$.*

PROOF. Let $\mu \in \dot{\mathcal{M}}_+(S)$ and $f \in \dot{C}^b_+(S)$. By Lemma 13.55 (c),

$$S = \bigcup_{n \in \mathbb{Z}_+} S_n(f), \qquad S_n(f) = \{A_*^n f > 0\}.$$

The last is a shorthand for $\{s \in S, (A_*^n f)(s) > 0\}$. Analogous shorthands will be used in the following.

Since the nonnegative measure μ is continuous from below and $\mu(S) > 0$, there exists some $m \in \mathbb{N}$ such that $0 < \mu\big(\bigcup_{n=0}^m S_n(f)\big)$. Since

$$\bigcup_{n=0}^m S_n(f) = \Big\{\sum_{n=0}^m A_*^n f > 0\Big\},$$

there is some $k \in \mathbb{N}$ such that

$$\mu(T_{mk}(f)) > 0, \qquad T_{mk}(f) = \Big\{\sum_{n=0}^m A_*^n f > 1/k\Big\}.$$

Now

$$\begin{aligned}\sum_{n=0}^m \int_S f \, d(A^n \mu) &= \int_S \sum_{n=0}^m (A_*^n f) \, d\mu \\ &\ge \int_{T_{mk}(f)} \Big(\sum_{n=0}^m A_*^n f\Big) d\mu \ge (1/k)\mu(T_{mk}(f)) > 0.\end{aligned}$$

So there is some $n \in \mathbb{Z}_+$ such that $\int_S f \, d(A^n \mu) = \int_S A_*^n f \, d\mu > 0$. ☐

COROLLARY 13.60. *Let κ be a topologically irreducible Feller kernel, A the associated linear map on $\mathcal{M}_+(S)$, $r > 0$. Let $\mu \in \dot{\mathcal{M}}_+(S)$ be an eigenmeasure $r\mu = A\mu$. Then $\int_S f d\mu > 0$ for any $f \in \dot{C}^b_+(S)$.*

PROOF. For all $n \in \mathbb{N}$, $\mu = r^{-n} A^n \mu$ and the assertion follows from Theorem 13.59. ☐

PROPOSITION 13.61. *Let κ be a topologically irreducible Feller kernel and let $\mathcal{N}$ be a tight subset of $\mathcal{M}_+(S)$ with $\inf_{\mu \in \mathcal{N}} \mu(S) > 0$. Then, for any $f \in \dot{C}^b_+(S)$, there exist some $m \in \mathbb{N}$ and $\delta > 0$ such that*

$$\sum_{n=0}^m \int_S A_*^n f \, d\mu \ge \delta, \qquad \mu \in \mathcal{N}.$$

PROOF. Let $\eta = (1/2)\inf_{\mu\in\mathcal{N}} \mu(S)$. Then $\eta > 0$. Since $\mathcal{N}$ is tight, there exists some compact subset K of S such that $\mu(S \setminus K) \leq \eta$ for all $\mu \in \mathcal{N}$ and so

$$\mu(K) \geq \eta, \qquad \mu \in \mathcal{N}. \tag{13.37}$$

Let $f \in C_+^b(S)$, $f \neq 0$. Since κ is topologically irreducible, $S = \bigcup_{n\in\mathbb{Z}_+} S_n(f)$ with open sets $S_n(f) = \{A_*^n f > 0\}$ by Lemma 13.55 (c). Since K is compact, there exists some $m \in \mathbb{N}$ such that $K \subseteq \bigcup_{n=0}^m S_n(f)$. Since A_*^n is continuous, there exists some $\tilde{\delta} > 0$ such that

$$\sum_{n=0}^m (A_*^n f)(s) \geq \tilde{\delta}, \qquad s \in K.$$

For all $\mu \in \mathcal{N}$, by (13.37),

$$\sum_{n=0}^m \int_S A_*^n f \, d\mu \geq \int_K \Big(\sum_{n=0}^m A_*^n f\Big) \, d\mu \geq \tilde{\delta}\mu(K) \geq \tilde{\delta}\eta > 0. \qquad \square$$

PROPOSITION 13.62. *Let $P : \mathcal{B} \times S \to \mathbb{R}_+$ be a Feller kernel, $g \in C_+^b(S \times S)$ and $\tilde{\kappa} : \mathcal{B} \times S \to \mathbb{R}_+$ be defined by*

$$\tilde{\kappa}(T, s) = \int_T g(s,t)P(dt, s), \qquad s \in S, \quad T \in \mathcal{B}. \tag{13.38}$$

If P is topologically irreducible and $g(s,t) > 0$ for all $s, t \in S$, then $\tilde{\kappa}$ is topologically irreducible.

PROOF. Consider a nonempty open strict subset U of S. Since P is topologically irreducible, there is some $s \in S \setminus U$ such that $P(U, s) > 0$. Since $g(s,t) > 0$ for all $t \in S$,

$$U = \bigcup_{n\in\mathbb{N}} U_n, \qquad U_n = \{t \in U; g(s,t) > 1/n\}.$$

(U_n) is an increasing sequence of open subsets of S. Since $\tilde{\kappa}(\cdot, s)$ is increasing, for all $n \in \mathbb{N}$,

$$\tilde{\kappa}(U, s) \geq \tilde{\kappa}(U_n, s) \geq \frac{1}{n} P(U_n, s).$$

Since $P(\cdot, s)$ is continuous from below, there is some $n \in \mathbb{N}$ such that $P(U_n, s) > 0$. This implies $\tilde{\kappa}(U, s) > 0$. $\square$

13.7.2. Sustaining Feller kernels.

DEFINITION 13.63. *Let $\kappa : \mathcal{B} \times S \to \mathbb{R}_+$ be a Feller kernel. The kernel κ is called* sustaining *if $\kappa(U, s) > 0$ for any open subset U of S and any $s \in U$.*

PROPOSITION 13.64. *Let $\kappa : \mathcal{B} \times S \to \mathbb{R}_+$ be a Feller kernel and A_* be operator on $C^b(S)$ induced by κ via*

$$(A_* f)(s) = \int_S f(t)\kappa(dt, s), \qquad s \in S.$$

Then the following are equivalent:

(a) *The kernel κ is sustaining.*
(b) *For any $f \in C_+^b(S)$, $\{f > 0\} \subseteq \{A_* f > 0\}$.*
(c) *For any Lipschitz continuous $f : S \to \mathbb{R}_+$, $\{f > 0\} \subseteq \{A_* f > 0\}$.*

PROOF. (a) $\Rightarrow$ (b): Let $f \in C_+^b(S)$. Let $s \in S$ and $f(s) > 0$. Set $U = \{t \in S; f(t) > f(s)/2\}$. Then $s \in U$. Since f is continuous, U is open. Since κ is sustaining, $\kappa(U, s) > 0$ and

$$(A_* f)(s) \geq \frac{f(s)}{2}\kappa(U, s) > 0.$$

This implies $\{f > 0\} \subseteq \{A_* f > 0\}$.

(b) $\Rightarrow$ (c): This is obvious.

(c) $\Rightarrow$ (a): Let U be an open subset of S and $s \in U$. Then $\{s\}$ is compact, $S \setminus U$ is closed and $\{s\}$ and $S \setminus U$ are disjoint. By Lemma 2.75, there exists a Lipschitz continuous function $f : S \to [0, 1]$ such that $f(s) = 1$ and $f(t) = 0$ for $t \in S \setminus U$. Since (c) is assumed, $\{f > 0\} \subseteq \{A_* f > 0\}$ and

$$0 < (A_* f)(s) \leq \int_U f(t)\kappa(dt, s) \leq \kappa(U, s). \qquad \square$$

COROLLARY 13.65. *Let $\kappa_1, \kappa_2 : \mathcal{B}\times \to \mathbb{R}_+$ be sustaining Feller kernels. Then $\kappa_1 \star \kappa_2$ is a sustaining Feller kernel.*

PROOF. Let A_{j*} be the operators induced by κ_j on $C^b(S)$. By Lemma 13.9, $\kappa_1 \star \kappa_2$ induces $A_{2*}A_{1*}$. Since κ_1 and κ_2 are sustaining, by Proposition 13.64,

$$\{f > 0\} \subseteq \{A_{1*} f > 0\} \subseteq \{A_{2*}A_{1*} f > 0\}.$$

Again by Proposition 13.64, $\kappa_1 \star \kappa_2$ is sustaining. $\square$

THEOREM 13.66. *Let κ_1, κ_2 be sustaining Feller kernels and κ_1 be topologically irreducible. Then $\kappa_1 \star \kappa_2$ is sustaining and topologically irreducible.*

PROOF. By Corollary 13.65, $\kappa_1 \star \kappa_2$ is sustaining. Let U be a nonempty open strict subset of S. Since κ_1 is topologically irreducible, there is some $s \in S \setminus U$ such that $\kappa_1(U, s) > 0$. By Lemma 2.75, there exists a Lipschitz continuous function $f : S \to [0, 1]$ such that $f(s) = 1$ and $f(t) = 0$ for all $t \in S \setminus U$. So

$$\kappa_1(U, t) \geq \int_S f(t)\kappa_1(dt, s) = (A_{1*} f)(t), \qquad t \in S,$$

and

$$(\kappa_1 \star \kappa_2)(U, s) \geq \int_S (A_{1*} f)(t)\kappa_2(dt, s) = (A_{2*}(A_{1*} f))(s).$$

Since $f(s) > 0$ and κ_1 is sustaining, $(A_{1*} f)(s) > 0$ by Proposition 13.64. By the same token, since κ_2 is sustaining, $(A_{2*}(A_{1*} f))(s) > 0$ and so $(\kappa_1 \star \kappa_2)(U, s) > 0$. $\square$

THEOREM 13.67. *Let $\kappa : \mathcal{B} \times S \to \mathbb{R}_+$ be a Feller kernel. Then the following are equivalent.*

(a) *κ is topologically irreducible and sustaining.*
(b) *For any $f \in \dot{C}_+^b(S)$, $S = \bigcup_{n\in\mathbb{Z}_+} U_n(f)$ where $U_n(f) = \{A_*^n f > 0\}$, $n \in \mathbb{Z}_+$, forms an increasing sequence of sets.*

PROOF. (a) $\Rightarrow$ (b): Since κ is topologically irreducible, for any $f \in \dot{C}_+^b(S)$, $S = \bigcup_{n\in\mathbb{Z}_+} U_n(f)$ by Lemma 13.55. The $U_n(f)$ form an increasing sequence of sets by Proposition 13.64.

(b) $\Rightarrow$ (a): Let (b) hold. Then κ is topologically irreducible by Lemma 13.55. Further, with $n = 0$, $\{f > 0\} \subseteq \{A_* f > 0\}$ and κ is sustaining by Proposition 13.64. $\square$

DEFINITION 13.68. *Feller kernels $\kappa_1, \dots, \kappa_m$, $m \in \mathbb{N}$, from $\mathcal{B} \times S \to \mathbb{R}_+$ are called* jointly topologically irreducible *if for any nonempty open strict subset U of S there is some $s \in S \setminus U$ such that $\kappa_j(U,s) > 0$ for $j = 1, \dots, m$.*

THEOREM 13.69. *Let $m \in \mathbb{N}$ and $\kappa_1, \dots, \kappa_m$ be sustaining Feller kernels that are jointly topologically irreducible. Let $b : S \times \mathbb{R}_+^m \to \mathbb{R}_+$ be continuous and*

$$b(s,x) > 0, \qquad s \in S, \quad x \in (0,\infty)^m. \tag{13.39}$$

Define $K_j : M_+^b(S) \to M_+^b(S)$, $j = 1, \dots, m$, by

$$(K_j h)(s) = \int_S h(t)\kappa_j(dt,s), \qquad h \in M_+^b(S), \quad s \in S, \tag{13.40}$$

and $F : M_+^b(S) \to M_+^b(S)$ by

$$F(h)(s) = b\big(s, (K_1 h)(s), \dots, (K_m h)(s)\big), \qquad h \in M_+^b(S), \quad s \in S. \tag{13.41}$$

Then F maps $C_+^b(S)$ into itself. Further, for any $h \in \dot{C}_+^b(S)$,

$$S = \bigcup_{n \in \mathbb{N}} U_n(h) \text{ where } U_n(h) = \big\{s \in S; F^n(h)(s) > 0\big\}$$

forms an increasing sequence of open sets in S.

PROOF. Let $h \in \dot{C}_+^b(S)$. Then $K_j h$ in $C_+^b(S)$ because κ_j is a Feller kernel and $F(h) \in C_+^b(S)$ as the composition of continuous functions.

For the increase of the sequence $U_n(h)$, it is sufficient to show that $\{h > 0\} \subseteq U_1(h)$ because $U_{n+1}(h) = U_1(U_n(h))$.

Let $s \in S$ and $h(s) > 0$. Since κ_j is a sustaining kernel, $(K_j h)(s) > 0$ for $j = 1, \dots, m$ by Proposition 13.64 and so $F(h)(s) > 0$ by (13.39).

Set $U(h) = \bigcup_{n\in\mathbb{N}} U_n(h)$. $U(h)$ is an nonempty open subset of S. Suppose that $S \neq U(h)$. Since the κ_j are jointly topologically irreducible, there is some $s \in S \setminus U(h)$ such that $\kappa_j(U(h), s) > 0$ for $j = 1, \dots, m$. Since the measures $\kappa_j(\cdot, s)$ are continuous from below,

$$0 < \kappa_j(U(h), s) = \lim_{n\to\infty} \kappa_j(U_n(h), s).$$

So there exists some $n \in \mathbb{N}$ such that $\kappa_j(U_n(h), s) > 0$ for $j = 1, \dots, m$. Now $U_n(h) = \bigcup_{\ell\in\mathbb{N}} T_{n,\ell}$ with $T_{nj} = \{t \in S; F^n(h)(t) > 1/\ell\}$. By the same argument as before, there exists some $\ell \in \mathbb{N}$ such that $\kappa_j(T_{n,\ell}, s) > 0$ for $j = 1, \dots, m$. So

$$(K_j F^n(h))(s) \ge \int_{T_{m,\ell}} F^n(h)(t)\kappa_j(dt,s) \ge \frac{1}{\ell}\kappa_j(T_{n\,\ell}), s) > 0$$

and

$$F^{n+1}(h)(s) = b\Big(s, \big(K_1 F^n(h)\big)(s), \dots, \big(K_m F^n(h)\big)(s)\Big) > 0$$

by (13.39). By definition, $s \in U_{n+1}(h)$ and so $s \in U(h)$, a contradiction. $\square$

13.7.3. Strongly topologically irreducible Feller kernels. We strengthen the concept of topological irreducibility as follows.

DEFINITION 13.70. *A Feller kernel $\kappa : \mathcal{B} \times S \to \mathbb{R}_+$ is called* strongly topologically irreducible *if for any nonempty open subset U of S and any nonempty compact subset K of S there exists some $n \in \mathbb{N}$ such that $\kappa^{n\star}(U,s) > 0$ for all $s \in K$.*

LEMMA 13.71. *Let $\kappa : \mathcal{B} \times S \to \mathbb{R}_+$ be a Feller kernel and A_* be the associated bounded linear map on $C^b(S)$. Then the following are equivalent:*

(a) *κ is strongly topologically irreducible.*

(b) *For any $f \in \dot{C}^b_+(S)$ and any nonempty compact subset K of S there exists some $n \in \mathbb{N}$ such that $(A_*^n f)(s) > 0$ for all $s \in K$.*

(c) *For any Lipschitz continuous $f : S \to \mathbb{R}_+$ that is not identically equal to 0 and any nonempty compact subset K of S, there exists some $n \in \mathbb{N}$ such that $(A_*^n f)(s) > 0$ for all $s \in K$.*

PROOF. (a) $\Rightarrow$ (b): Let $f \in \dot{C}^b_+(S)$ and K be a compact subset of S. Since f is continuous,

$$U = \{t \in S; f(t) > \|f\|_\infty/2\}$$

is a nonempty open subset of S. Since κ is strongly topologically irreducible, there exists some $n \in \mathbb{N}$ such that, for all $s \in K$.

$$0 < \kappa^{n\star}(U, s) \le \int_U \frac{2f(t)}{\|f\|_\infty} \kappa^{n\star}(dt, s) \le \frac{2}{\|f\|_\infty} (A_*^n f)(s).$$

(b) $\Rightarrow$ (c): This is obvious.

(c) $\Rightarrow$ (a): Let U be a nonempty open subset of S. Choose some $t_0 \in U$. By Lemma 2.75, there exists some Lipschitz continuous $f : S \to [0, 1]$ such that

$$f(t_0) = 1, \qquad f(t) \le \chi_U(t), \quad t \in S.$$

Let K be a nonempty compact subset of S. By (c), there is some $n \in \mathbb{N}$ such that $0 < (A_*^n f)(s)$ for all $s \in K$. Since A_*^n is induced by κ^{n*}, for all $s \in K$,

$$0 < (A_*^n f)(s) \le \int_S \chi_U(t) \kappa^{n*}(dt, s) \le \kappa^{n*}(U, s).$$

So (a) holds. □

REMARK 13.72. It follows from Lemmata 13.55 and 13.71 that every strongly topologically irreducible Feller kernel is topologically irreducible.

We give a sufficient condition for strong topological irreducibility.

PROPOSITION 13.73. *Let $\kappa : \mathcal{B} \times S \to \mathbb{R}_+$ be a Feller kernel with the following property for any $f \in \dot{C}^b_+(S)$:*

For all $s \in S$ there exists some neighborhood $U_s \subseteq S$ of s and some $n_s \in \mathbb{N}$ such that $\int_S f(t)\kappa^{n\star}(dt, \tilde{s}) > 0$ for all $n \in \mathbb{N}$, $n \ge n_s$, and all $\tilde{s} \in U_s$.

Then κ is a strongly topologically irreducible Feller kernel.

PROOF. The neighborhoods U_s can be chosen as open sets. Let K be a nonempty compact subset of S. Then $K \subseteq \bigcup_{s \in S} U_s$ and there exists a finite subset $\tilde{S}$ of S such that $K \subseteq \bigcup_{s \in \tilde{S}} U_s$. Set $m = \max_{s \in \tilde{S}} n_s$. Then $m \in \mathbb{N}$ and

$$\int_S f(t)\kappa^{m\star}(dt, \tilde{s}) > 0, \qquad \tilde{s} \in K.$$

By Lemma 13.9, $(A_*^m f)(\tilde{s}) > 0$ for all $\tilde{s} \in K$, and κ is strongly topologically irreducible by Definition 13.70. □

THEOREM 13.74. *Let κ be a Feller kernel. Then κ is strongly topologically irreducible if it is topologically irreducible and sustaining.*

PROOF. Let $f \in \dot{C}_+^b(S)$. By Theorem 13.67 $S = \bigcup_{n\in\mathbb{Z}_+} U_n(f)$ where $U_n(f) = \{A_*^n f > 0\}$, $n \in \mathbb{Z}_+$, forms an increasing sequence of sets. Let K be a compact subset of S. Then there exists some $m \in \mathbb{N}$ such that $K \subseteq \bigcup_{n=0}^m U_n(f)$. Since the $U_n(f)$ form an increasing sequence of sets, $K \subseteq U_m(f)$ and $(A_*^m f)(s) > 0$ for all $s \in K$. So κ is strongly topologically irreducible by Theorem 13.71. □

A similar proof as for Proposition 13.61 yields the following result.

PROPOSITION 13.75. *Let κ be a strongly topologically irreducible Feller kernel and let $\mathcal{N}$ be a tight subset of $\mathcal{M}_+(S)$ with $\inf_{\mu\in\mathcal{N}} \mu(S) > 0$. Then, for any $f \in \dot{C}_+^b(S)$, there exist some $n \in \mathbb{N}$ and $\delta > 0$ such that*

$$\int_S A_*^n f \, d\mu \geq \delta, \qquad \mu \in \mathcal{N}.$$

PROOF. Let $\eta = (1/2)\inf_{\mu\in\mathcal{N}} \mu(S)$. Then $\eta > 0$. Since $\mathcal{N}$ is tight, there exists some compact subset K of S such that $\mu(S \setminus K) \leq \eta$ for all $\mu \in \mathcal{N}$ and so

$$\mu(K) \geq \eta, \qquad \mu \in \mathcal{N}. \tag{13.42}$$

Let $f \in \dot{C}_+^b(S)$. Then $U = \{t \in S; f(t) > \|f\|_\infty/2\}$ is a nonempty open subset of S. Since κ is strongly topologically irreducible, there exists some $n \in \mathbb{N}$ such that $\kappa^\star(U, s) > 0$ for all $s \in K$. For all $s \in K$,

$$(A_*^n f)(s) \geq \frac{\|f\|_\infty}{2} \kappa^\star(U, s) > 0.$$

Since K is compact and $A_*^n f$ is continuous, there is some $\tilde{\delta} > 0$ such that $(A_*^n f)(s) \geq \tilde{\delta}$ for all $s \in K$. For all $\mu \in \mathcal{N}$, by (13.42),

$$\int_S A_*^n f \, d\mu \geq \tilde{\delta}\mu(K) = \tilde{\delta}\eta. \qquad \square$$

13.7.4. Colonization kernels. The next concept seems quite suited for spatially structured populations, but less for populations with other structures.

DEFINITION 13.76. *Let $\kappa : \mathcal{B} \times S \to \mathbb{R}_+$ be a Feller kernel. κ is called a* colonization kernel *if for any $s \in S$ there is an open subset $U \ni s$ of S such that $\kappa(V, s) > 0$ for all nonempty open subsets V of U.*

LEMMA 13.77. *Any colonization kernel is sustaining.*

PROOF. Let κ be a colonization kernel. Let U be an open subset of S and $s \in U$. Then there exists an open set W such that $s \in W$ and $\kappa(V, s) > 0$ for any nonempty open set V in W. Set $V = U \cap W$. Then, $s \in V$ and V is an open subset of W and of U. Thus, $0 < \kappa(V, s) \leq \kappa(U, s)$. □

LEMMA 13.78. *Let $\kappa : \mathcal{B} \times S \to \mathbb{R}_+$ be a Feller kernel of tight measures and $g : S \times S \to (0, \infty)$ be continuous and bounded. Then $\tilde{\kappa} : \mathcal{B} \times S \to \mathbb{R}_+$ defined by*

$$\tilde{\kappa}(T, s) = \int_T g(s, t)\kappa(dt, s), \qquad T \in \mathcal{B}, \quad s \in S,$$

is also a Feller kernel of tight measures, and it is a colonization kernel if and only if κ is a colonization kernel.

PROOF. By Corollary 13.36 and Theorem 13.28, $\tilde{\kappa}$ is a Feller kernel of tight measures.

Assume that κ is a colonization kernel. Let $s \in S$. By Definition 13.76, there is some open subset $U \ni s$ of S such that $\kappa(V,s) > 0$ for all nonempty open subsets V of U. Since g is strictly positive and continuous, $V = \bigcup_{n\in\mathbb{N}} V_n$ with open subsets $V_n = \{t \in V; g(s,t) > 1/n\}$ of V. For all $n \in \mathbb{N}$,

$$\tilde{\kappa}(V,s) \geq \int_{V_n} g(s,t)\kappa(dt,s) \geq (1/n)\kappa(V_n,s).$$

Since $\kappa(\cdot,s)$ is continuous from below, $\kappa(V_n,s) \to \kappa(V,s)$ as $n \to \infty$ and $\tilde{\kappa}(V,s) > 0$.

Assume that $\tilde{\kappa}$ is a colonization kernel. Since $\tilde{\kappa}(T,s) \leq \sup g \; \kappa(T,s)$ for all $T \in \mathcal{B}$, $s \in S$, κ is a colonization kernel. □

LEMMA 13.79. *Let $\kappa_j : \mathcal{B} \times S \to \mathbb{R}_+$, $j = 1,\dots,m$, be colonization kernels and S be connected. Then $\kappa_1,\dots,\kappa_m$ are jointly topologically irreducible.*

PROOF. Let U be a nonempty open strict subset of S. Since S is connected, U is a strict subset of $\bar{U}$. Choose some $s \in \bar{U} \setminus U$.

Let $j \in \{1,\dots,m\}$. Since the κ_j are colonization kernels, there exist open sets U_j with $s \in U_j \subseteq S$ and $\kappa_j(V,s) > 0$ for any nonempty subset V of U_j.

Set $V = \bigcap_{j=1}^m U_j$. Then V is a nonempty open set and $s \in V$. Since s is a limit point of U, $W = V \cap U$ is a nonempty open set contained in U_j for $j = 1,\dots,m$. This implies that $0 < \kappa_j(V,s) \leq \kappa_j(U,s)$ for $j = 1,\dots,m$. □

13.7.5. Preservation of irreducibility properties.

PROPOSITION 13.80. *Let $P : \mathcal{B} \times \tilde{S} \to \mathbb{R}_+$ be a Feller kernel, $g \in C_+^b(S \times S)$, and $\kappa : \mathcal{B} \times S \to \mathbb{R}_+$ be defined by*

$$\kappa(T,s) = \int_T g(s,t)P(dt,s), \qquad s \in S, \quad T \in \mathcal{B}. \tag{13.43}$$

Assume that κ is also a Feller kernel and that $g(s,t) > 0$ for all $s,t \in S$.

(a) *P is top-irreducible if and only if κ is top-irreducible.*
(b) *P is strongly top-irreducible if and only if κ is strongly top-irreducible.*
(c) *P is sustaining if and only if κ is sustaining.*

Assumptions for κ to inherit the Feller property from P can be found in Proposition 13.26, Corollary 13.27, and Theorem 13.28.

PROOF. Let $f_0 \in \dot{C}_+^b(S)$ and $h_0 = f_0$. For $n \in \mathbb{Z}_+$, set $U_n = \{f_n > 0\}$ and $V_n = \{h_n > 0\}$, where

$$f_{n+1} = \int_S f_n(t)P(dt,\cdot), \qquad h_{n+1} = \int_S h_n(t)\kappa(dt,\cdot).$$

Let $U(f_0)$ and $V(f_0)$ be the respective unions over $n \in \mathbb{N}$.

For any $f \in \dot{C}_+^b(S)$ and $s \in S$, we have the equivalence of the following two statements:

(i) $\displaystyle\int_S f(t)P(dt,s) > 0$,
(ii) $P(\{f > 0\},s) > 0$.

An analogous equivalence holds for κ replacing P.

Since g is strictly positive on S^2, statement (ii) for F is equivalent to the statement (ii) for κ replacing P.

With this observation, it follows by induction that $U_n = V_n$ for all $n \in \mathbb{N}$ such that $U(f) = V(f)$. So $S = U(f)$ if and only if $S = V(f)$.

The equivalence in (a) follows from Lemma 13.55 (c).

The equivalence in (b) follows from Lemma 13.71 (b).

The equivalence in (c) follows from Proposition 13.64 (b).

In these lemmata, $U_n = \{A_*^n f > 0\}$ if A_* is induced by P and $V_n = \{A_*^n f > 0\}$ if A_* is induced by κ. $\square$

13.8. Continuous dependence of the spectral radius on kernels

In Chapter 19, the spectral radius of a measure (or Feller) kernel will be the threshold parameter between extinction and survival of a population the structural distributions of which are modeled by measures. This motivates us to study under which conditions the spectral radius continuously depends on the kernel.

Let S be a measurable space with a σ-algebra $\mathcal{B}$. Consider a measure kernel $\kappa : \mathcal{B} \times S \to \mathbb{R}_+$ and a sequence of measure kernels (κ_j) from $\mathcal{B} \times S$ to $\mathbb{R}_+$. Notice that $\kappa_j, \kappa \in B(\mathcal{B} \times S)$, the Banach space of bounded real-valued functions on $\mathcal{B} \times S$ with the supremum norm. See Example 2.18.

THEOREM 13.81. *Assume that*

$$\|\kappa_j - \kappa\|_\infty := \sup_{T\in\mathcal{B}, s\in S} \big|\kappa_j(T,s) - \kappa(T,s)\big| \to 0, \qquad j \to \infty. \tag{13.44}$$

Then

$$\limsup_{j\to\infty} \mathbf{r}(\kappa_j) \le \mathbf{r}(\kappa). \tag{13.45}$$

If S is a metric space and $\mathcal{B}$ the Borel σ-algebra and κ is a tight uniform Feller kernel, then $\mathbf{r}(\kappa_j) \to \mathbf{r}(\kappa)$ as $j \to \infty$.

PROOF. Let A_* and A_{j*} be the bounded linear operators induced by κ and κ_j on $M_b(S)$ with the supremum norm. For all $s \in S$ and $f \in M_b(S)$,

$$\big|(A_{j*}f)(s) - (A_*f)(s)\big| \le 2 \sup_{T\in\mathcal{B}} \big|\kappa_j(T,s) - \kappa(T,s)\big| \, \|f\|_\infty.$$

For all $f \in M^t(S)$,

$$\|A_{j*}f - A_*f\|_\infty \le 2\|\kappa_j - \kappa\|_\infty \, \|f\|_\infty.$$

For the operator norms,

$$\|A_{j*} - A_*\| \le 2\|\kappa_j - \kappa\|_\infty \to 0, \qquad j \to \infty.$$

By Theorem 7.1,

$$\limsup_{j\to\infty} \mathbf{r}(A_{j*}) \le \mathbf{r}(A_*). \tag{13.46}$$

Since $\mathbf{r}(\kappa) = \mathbf{r}(A_*)$ and a similar statement holds for the sequences, (13.45) follows.

If S is a metric space and κ is a tight uniform Feller kernel, A_*^2 is compact on $C^b(S)$ and $M^t(S)$ by Theorem 13.51. The convergence of the spectral radii now follows from Theorem 10.1. $\square$

13.8.1. Upper semicontinuity. Let S be a metric space with metric d. We restrict our discussion to the spectral radius of Feller kernels.

We explore what conditions we obtain when we use the flat norm. Recall

$$\mathcal{F} = \Big\{ f \in [0,1]^S; \qquad |f(s) - f(t)| \le d(s,t), \quad s,t \in S \Big\}.$$

THEOREM 13.82. *Let $\kappa : \mathcal{B} \times S \to \mathbb{R}_+$ be a Feller kernel and (κ_j) be a sequence of Feller kernels. Let A_* and A_{j*} be the operators on $C^b(S)$ induced by κ and κ_j respectively by* (13.4). *Assume*

$$\sup_{f \in \mathcal{F}} \|A_{j*} f - A_* f\|_\infty = \sup_{s \in S} \|\kappa_j(\cdot, s) - \kappa(\cdot, s)\|_\flat \to 0, \qquad j \to \infty. \tag{13.47}$$

Here $\|\cdot\|_\flat$ is the flat norm on $\mathcal{M}(S)$. See Section 3.2.

Then

$$\limsup_{j \to \infty} \mathbf{r}(\kappa_j) \le \mathbf{r}(\kappa).$$

Recall that $\mathbf{r}(\kappa) = \mathbf{r}(A_*)$ is also given by (13.11).

PROOF. Recall that $\mathbf{r}(\kappa) = \mathbf{r}(A)$ as well, where A is the operator induced by κ on $\mathcal{M}_+(S)$ by (13.5).

By Theorem 7.12, it is sufficient to show that $\|A_j - A\| \to 0$ as $j \to \infty$, where $\|\cdot\|$ is the operator norm associated with the flat norm on $\mathcal{M}_+(S)$. By the duality formula (13.6),

$$\begin{aligned} \|A_j - A\| &= \sup_{f \in \mathcal{F}, \mu \in \mathcal{M}_+(S), \mu(S) \le 1} \left| \int_S f\, d(A_j \mu) - \int_S f\, d(A\mu) \right| \\ &= \sup_{f \in \mathcal{F}, \mu \in \mathcal{M}_+(S), \mu(S) \le 1} \left| \int_S (A_{j*} f - A_* f)\, d\mu \right|. \end{aligned}$$

Since $\{\mu \in \mathcal{M}_+(S), \mu(S) \le 1\}$ contains all the Dirac measures δ_s with $\delta_s(T) = \chi_T(s)$ on the one hand and since

$$\left| \int_S (A_{j*} f - A_* f) d\mu \right| \le \|A_{j*} f - A_* f\|_\infty\, \mu(S)$$

on the other hand, we have

$$\begin{aligned} \|A_j - A\| &= \sup_{f \in \mathcal{F}} \|A_{j*f} - A_* f\|_\infty \\ &= \sup_{f \in \mathcal{F}, s \in S} \left| \int_S f(t) \kappa_j(dt, s) - \int_S f(t) \kappa(dt, s) \right| \\ &= \sup_{s \in S} \|\kappa_j(\cdot, s) - \kappa(\cdot, s)\|_\flat. \end{aligned}$$

Here, in the last equality, we have used (3.9). □

While (13.47) is weaker than (13.44), in many applications one may not be able to take advantage of it differently from the next example.

EXAMPLE 13.83 (Deterministic movement and survival). Let us now consider a sequence of Feller kernels $\kappa_j : \mathcal{B} \times S \to \mathbb{R}_+$

$$\kappa_j(T, s) = \chi_T\big(\xi_j(s)\big)\, g_j(s), \qquad s \in S, \quad T \in \mathcal{B},$$

where $\xi_j : S \to S$ continuous, $g_j \in C_+^b(S)$, $j \in \mathbb{N}$. Assume that $g_j \to g$ and $\xi_j \to \xi$ as $j \to \infty$ uniformly on S. Then g and ξ are continuous. Let κ be given by (13.28).

We claim that

$$\limsup_{j\to\infty} \mathbf{r}(\kappa_j) \le \mathbf{r}(\kappa). \tag{13.48}$$

While it may be possible to derive this from formula (13.29), we will use Theorem 13.82. It is enough to show that

$$\sup_{f\in\mathcal{F}} \left\|A_{j*}f - A_*f\right\|_\infty \to 0, \qquad j\to\infty,$$

where $\mathcal{F}$ is the set of functions

$$f: S\to[0,1], \quad |f(s)-f(t)| \le d(s,t) \text{ for all } t,s\in S, \tag{13.49}$$

See (3.9). For all $s\in S$, $f\in\mathcal{F}$,

$$\begin{aligned}|(A_{j*}f)(s)-(A_*f)(s)| &= \left|\int_S f(t)\kappa_j(dt,s) - \int_S f(t)\kappa(dt,s)\right| \\ &= \left|f(\xi_j(s))g_j(s) - f(\xi(s))g(s)\right|.\end{aligned}$$

By the triangle inequality and (13.49),

$$\begin{aligned}&\left|(A_{j*}f)(s)-(A_*f)(s)\right| \\ &\quad\le f(\xi_j(s))|g_j(s)-g(s)| + |f(\xi_j(s))-f(\xi(s))|g(s) \\ &\quad\le |g_j(s)-g(s)| + d\big(\xi_j(s),\xi(s)\big)\,\|g\|_\infty.\end{aligned}$$

Since $g_j \rightrightarrows g$ and $\xi_j\to\xi$ as $j\to\infty$ uniformly on S,

$$\sup_{f\in\mathcal{F}} \|A_{j*}f) - A_*f\|_\infty \le \|g_j - g\|_\infty + \sup_{s\in S} d\big(\xi_j(s),\xi(s)\big)\,\|g\|_\infty \overset{j\to\infty}{\longrightarrow} 0.$$

As mentioned before, the upper semicontinuity may also follow from the formula (13.29). Consider a Feller kernel κ that is a sum of kernels of the form (13.28) and let κ_j be a sequence of kernels of analogous sums. By the triangle inequality, we still have $\|A_j - A\|\to 0$ as $j\to\infty$ for the induced operators and (13.48) holds. The formula (13.29), however, is no longer valid.

We can take this example to a more general level.

EXAMPLE 13.84. Consider a Feller kernel $P:\mathcal{B}\times S\to\mathbb{R}_+$ and a continuous function $\xi: S\to S$ and sequences of Feller kernels $P_j:\mathcal{B}\times S\to\mathbb{R}_+$ and of continuous functions $\xi_j: S\to S$, $j\in\mathbb{N}$, and define

$$\begin{aligned}\kappa_j(T,s) =& P_j(\xi_j^{-1}(T),s), \qquad j\in\mathbb{N},\\ \kappa(T,s) =& P(\xi^{-1}(T),s), \qquad T\in\mathcal{B},\quad s\in S.\end{aligned}$$

By Proposition 13.29,

$$\int_S f(t)\kappa_j(dt,s) = \int_S f(\xi_j(t))P_j(dt,s), \qquad s\in S, \tag{13.50}$$

an analogous formula holds for κ, ξ and P, and the κ_j,κ are Feller kernels.

Assume that $\kappa_j\to\kappa$ uniformly on $\mathcal{B}\times S$ and $\xi_j\to\xi$ uniformly on S as $j\to\infty$. Then

$$\limsup_{j\to\infty} \mathbf{r}(\kappa_j) \ge \mathbf{r}(\kappa).$$

Proof. By Theorem 13.82, it is enough to show that $\left\|\kappa_j(\cdot,s)-\kappa(\cdot,s)\right\|_\flat \to 0$ as $j\to\infty$ uniformly for $s\in S$. Here $\|\cdot\|_\flat$ is the flat norm on $\mathcal{M}(S)$. See Section 3.2.

Recall the set $\mathcal{F}$ of functions

$$f : S \to [0,1], \quad |f(s)-f(t)| \le d(s,t) \text{ for all } t,s\in S, \tag{13.51}$$

Then, for $f\in\mathcal{F}$ and $s\in S$, by the triangle inequality and (13.50),

$$\begin{aligned}&\left|\int_S f(t)\kappa_j(dt,s) - \int_S f(t)\kappa(dt,s)\right|\\ &\quad\le \left|\int_S f(\xi_j(t))P_j(dt,s) - \int_S f(\xi_j(t))P(dt,s)\right|\\ &\quad\quad+\left|\int_S (f(\xi_j(t)) - f(\xi(t))P(dt,s)\right|.\end{aligned}$$

By (13.51), this can be estimated by

$$2\|P_j-P\|_\infty + \|\xi_j-\xi\|_\infty\,\|P\|_\infty. \tag{13.52}$$

Recall that $P_j, P\in B(\mathcal{B}\times S)$, the space of bounded real-valued functions on $\mathcal{B}\times S$ with the supremum norm. Since (13.52) does not depend on $s\in S$ or on $f\in\mathcal{F}$, by the definition of the flat norm, (3.9)

$$\sup_{s\in S}\|\kappa_j(\cdot,s)-\kappa(\cdot,s)\|_\flat \le 2\|P_j-P\|_\infty + \|\xi_j-\xi\|_\infty\,\|P\|_\infty \to 0, \qquad j\to\infty.$$

The assertion now follows from Theorem 13.82. □

CHAPTER 14

Nonlinear dynamics

In this central chapter, the theory of homogeneous operators is connected to the question of extinction/persistence of structured populations in general discrete-time models. See Section 1.1.

14.1. Homogeneous maps as derivatives

Let us recall the following mathematical motivation to consider homogeneous operators (Section 1.1).

Let X and Y be normed vector spaces, X_F be a homogeneous subset of X and $F : X_F \to Y$, $F(0) = 0$. Let X_B be the set of those $x \in X_F$ such that the directional derivative at 0 in the direction of x exists,

$$\lim_{0<b\to 0} \frac{1}{b} F(bx) = \partial F(0,x) =: B(x). \tag{14.1}$$

Since $0 \in X_F$ and $F(0) = 0$, $0 \in X_B$ and $B(0) = 0$. If $X_B = X_F$, $\partial F(0, \cdot)$ is called the *Gateaux derivative* of F at 0.

THEOREM 14.1. *$X_B \subseteq X_F$ is a homogeneous subset of X and the operator $B : X_B \to Y$ is (positively)* homogeneous *(of degree one).*

PROOF. Let $x \in X_B$. Let $\alpha \in \mathbb{R}_+$. Obviously, if $\alpha = 0$, $\alpha x = 0 \in X_B$ and $B(\alpha x) = 0 = \alpha B(x)$.

So we assume $\alpha \in (0, \infty)$. Then

$$\frac{F(t[\alpha x])}{t} = \alpha \frac{F([t\alpha]x)}{t\alpha}, \qquad t \in (0,\infty).$$

As $t \to 0$, also $\alpha t \to 0$. Since $x \in X_B$, the directional derivative in direction αx exists and equals $\alpha \partial F(0,x)$. So $\alpha x \in X_B$ and

$$\partial F(0, \alpha x) = \alpha \partial F(0,x). \qquad \square$$

In turn, we want to explore in what sense homogeneous operators can serve as derivatives. Cf. [**130**].

14.2. (Lower and upper) order derivatives

Let X_+ be the closed order cone of an ordered normed vector space X.

A homogeneous $B : X_+ \to X_+$ is called an *order-derivative* of $F : X_+ \to X_+$ at $0 = F(0)$ if for any $\epsilon \in (0,1)$ there is some $\delta > 0$ such that

$$(1-\epsilon)B(x) \le F(x) \le (1+\epsilon)B(x), \qquad x \in X_+, \quad \|x\| \le \delta. \tag{14.2}$$

A homogeneous $B : X_+ \to X_+$ is called a *lower order-derivative* of $F : X_+ \to X_+$ at $0 = F(0)$ if one part of (14.2) holds:

$$\text{For any } \epsilon \in (0,1) \text{ there is some } \delta > 0 \text{ such that}\\ F(x) \ge (1-\epsilon)B(x) \text{ for all } x \in X_+ \text{ with } \|x\| \le \delta. \tag{14.3}$$

B is called an *upper order-derivative* of F at 0 if the other part of (14.2) holds:

$$\text{For any } \epsilon \in (0,1) \text{ there is some } \delta > 0 \text{ such that}\\ F(x) \le (1+\epsilon)B(x) \text{ for all } x \in X_+ \text{ with } \|x\| \le \delta. \tag{14.4}$$

PROPOSITION 14.2. *Let $F : X_+ \to X_+$, $F(0) = 0$, and let $B_1 : X_+ \to X_+$ be a lower order-derivative and $B_2 : X_+ \to X_+$ an upper order-derivative of F at 0. Then $B_1(x) \le B_2(x)$ for all $x \in X_+$.*

PROOF. Let $\epsilon \in (0,1)$. Then there exist some $\delta > 0$ such that

$$(1-\epsilon)B_1(y) \le F(y) \le (1+\epsilon)B_2(y), \qquad y \in X_+, \quad \|y\| \le \delta.$$

Let $x \in X_+$. Then there exists some $t > 0$ such that $\|tx\| < \delta$ and

$$(1-\epsilon)B_1(tx) \le (1+\epsilon)B_2(tx).$$

Since B_1 and B_2 are homogeneous,

$$(1-\epsilon)B_1(x) \le (1+\epsilon)B_2(x).$$

Since $\epsilon \in (0,1)$ has been arbitrary and X_+ is closed, $B_1(x) \le B_2(x)$. □

COROLLARY 14.3. *Let $F : X_+ \to X_+$, $F(0) = 0$, and $B : X_+ \to X_+$ be an order-derivative of F at 0. Then B is uniquely determined.*

Chain rules hold for upper and lower order-derivatives under some extra assumptions.

PROPOSITION 14.4. *Let $F_i : X_+ \to X_+$ and $F_i(0) = 0$, $j = 1,2$. Let B_1 be a lower order-derivative of F_1 at 0 and B_2 be a lower order-derivative of F_2 at 0. Assume that F_1 is continuous at 0 and B_2 is order-preserving. Then $B_2 \circ B_1$ is a lower order-derivative of $F_2 \circ F_1$ at 0.*

PROOF. Let $\tilde\epsilon \in (0,1)$. Choose some $\epsilon \in (0,1)$ such that

$$(1-\epsilon)^2 \ge 1 - \tilde\epsilon. \tag{14.5}$$

Then there exist $\delta_i > 0$ such that

$$F_i(x) \ge (1-\epsilon)B_i(x), \qquad x \in X_+, \quad \|x\| \le \delta_i, \quad i = 1,2. \tag{14.6}$$

Since F_1 is continuous at 0 and $F_1(0) = 0$, we can arrange that

$$\|F_1(x)\| \le \delta_2, \qquad x \in X_+, \quad \|x\| \le \delta_1 \tag{14.7}$$

(by making δ_1 smaller).

Let $x \in X_+$, $\|x\| \le \delta_1$. By (14.6) and (14.7),

$$F_2(F_1(x)) \ge (1-\epsilon)B_2(F_1(x)).$$

Since B_2 is order-preserving and homogeneous, by (14.6) and (14.5),

$$\begin{aligned}(F_2 \circ F_1)(x) &\ge (1-\epsilon)B_2\big((1-\epsilon)B_1(x)\big)\\ &= (1-\epsilon)^2 B_2(B_1(x)) \ge (1-\tilde\epsilon)(B_2 \circ B_1)(x).\end{aligned}$$ □

REMARK 14.5. An analogous chain rule holds for upper order-derivatives.

LEMMA 14.6. *Let $F : X_+ \to X_+$ and $B : X_+ \to X_+$ be the order-derivative of F at $0 = F(0)$. Assume that X_+ is normal and B is bounded. Then*

$$\frac{\|F(x) - B(x)\|}{\|x\|} \to 0, \qquad \dot{X}_+ \ni x \to 0.$$

This means that B is kind of a Frechet derivative of F at 0, with the exception that it is not linear.

PROOF. Let $\epsilon > 0$. Choose $\delta > 0$ according to the definition of an order-derivative, (14.2). Let $x \in X_+$ and $\|x\| \le \delta$. Then

$$-\epsilon B(x) \le F(x) - B(x) \le \epsilon B(x).$$

Since X_+ is normal and B is homogeneous and bounded, by Corollary 2.52, there exists some $c > 0$ such that, for all $x \in X_+$, $\|x\| \le \delta$,

$$\|F(x) - B(x)\| \le c\|\epsilon B(x)\| \le c\epsilon\|B\|\|x\|.$$

This implies the assertion. □

A similar proof shows the following result.

LEMMA 14.7. *Let $F : X_+ \to X_+$ and $B : X_+ \to X_+$ be the order-derivative of F at $0 = F(0)$. Assume that X_+ is normal. Then*

$$B(x) = \lim_{t \to 0+} \frac{1}{t} F(tx), \qquad x \in X_+$$

This means that B is the Gateaux derivative of F at 0.

REMARK 14.8. If B is the order-derivative of F at $0 = F(0)$ and B is not additive on X_+, then F is not Frechet-differentiable at 0. The Frechet-derivative at 0, by definition, would be linear on X and equal the nonadditive Gateaux-derivative B on X_+.

14.3. Partial stability and persistence results

Let X be an ordered normed vector space with closed order cone X_+. Recall the definition of a normal point in Section 2.2.2 and the notation $\dot{X}_- = X_+ \setminus \{0\}$.

14.3.1. A conditional stability and extinction result.

Our first result gives conditions for the extinction of a population that has come too close to the extinction state in a specific order sense.

THEOREM 14.9. *Let X_+ be the closed cone of an ordered normed vector space. Let $F, B : X_+ \to X_+$ and let B be homogeneous and order-preserving.*

Let $v \in \dot{X}_+$ and $r \in (0, 1)$ such that $B(v) \le rv$.

Assume that B is an upper order-derivative of F at 0 in a v-modified way: For any $\eta > 0$, there exists some $\tilde{\delta} > 0$ such that

$$F(x) \le (1 + \eta)B(x) \qquad x \in X_+, \quad x \le \tilde{\delta}v. \tag{14.8}$$

Then, for each $\alpha \in (r, 1)$, there exists some $\delta_\alpha > 0$ such that the following holds for all $x \in X_-$ with $x \le \delta_\alpha v$:

(a) *$F^n(x) \le \alpha^n \delta_\alpha v$ for all $n \in \mathbb{N}$.*

(b) *$\|F^n(x)\| \le c_v \alpha^n \delta_\alpha \to 0$ as $n \to \infty$, for some $c_v \ge 1$ independent of n, if v is a normal point for X_+.*

(c) $\|F^n(x)\| \le \alpha^n \delta_\alpha \|v\| \to 0$ *as* $n \to \infty$ *provided that the norm is order-preserving on* X_+.

In particular, the semiflow induced by F *is not uniformly weakly persistent (in the sense of Theorem* 1.6*).*

PROOF. Let $v \in \dot{X}_+$ and $r \in (0,1)$ such that $B(v) \le rv$. Let $\alpha \in (r,1)$.

Choose $\eta > 0$ such that $(1+\eta)r = \alpha < 1$. Choose $\delta = \delta_\alpha > 0$ such that (14.8) holds.

Let $x \in X_+$, $x \le \delta_\alpha v$.

By induction,

$$F^n(x) \le \alpha^n \delta_\alpha v, \qquad n \in \mathbb{Z}_+. \tag{14.9}$$

Indeed, interpreting $F^0(x) = x$, this holds for $n = 0$. Assume that $n \in \mathbb{Z}_+$ and (14.9) holds for n.

Since $\alpha \le 1$, by (14.8),

$$F^{n+1}(x) = F(F^n(x)) \le (1+\eta)B(F^n(x)).$$

Since B is order-preserving,

$$F^{n+1}(x) \le (1+\eta)B(\alpha^n \delta_\alpha v).$$

Since B is homogeneous, $\alpha \in (r,1)$ and $Bv \le rv$,

$$F^{n+1}(x) \le (1+\eta)\alpha^n \delta_\alpha B(v) \le (1+\eta)\alpha^n \delta_\alpha r v \le \alpha^{n+1}\delta_\alpha v.$$

By the induction principle, (14.9) holds and part (a) has been proved.

Part (b) and (c) follow from (a). □

We obtain the following result as a corollary.

THEOREM 14.10 ([**205**, Thm.4.1]). *Let* X_+ *be the closed cone of an ordered normed vector space. Let* $F, B : X_+ \to X_+$ *and let* B *be homogeneous and order-preserving.*

Assume that B *is an upper order-derivative of* F *at 0,* (14.4).

Let $v \in \dot{X}_+$ *be a normal point for* X_+ *and* $r \in (0,1)$ *such that* $B(v) \le rv$.

Then, for each $\alpha \in (r,1)$*, there exists some* $\delta_\alpha > 0$ *such that the following holds for all* $x \in X_+$ *with* $x \le \delta_\alpha v$*:*

(a) $F^n(x) \le \alpha^n \delta_\alpha v$ *for all* $n \in \mathbb{N}$.

(b) $\|F^n(x)\| \le c_v \alpha^n \delta_\alpha \to 0$ *as* $n \to \infty$*, for some* $c_v \ge 1$ *independent of* n.

(c) $\|F^n(x)\| \le \alpha^n \delta_\alpha \|v\| \to 0$ *as* $n \to \infty$ *provided that the norm is order-preserving on* X_+.

In particular, the semiflow induced by F *is not uniformly weakly persistent (in the sense of Theorem* 1.6*).*

PROOF. Let v be a normal point for X_+ (Section 2.2.2). Then there exists some $c \ge 1$ such that $\|x\| \le c$ for all $x \in X_+$ with $x \le v$. Let B be an upper order-derivative of F at 0. Let $\epsilon > 0$. Then there exists some $\delta > 0$ such that

$$F(x) \le (1+\epsilon)B(x), \qquad x \in X_+, \quad \|x\| \le \delta.$$

Let $\tilde\delta = \delta/c$. Let $0 \le x \le \tilde\delta v$. Then $\|x\| \le \tilde\delta c = \delta$ and $F(x) \le (1+\epsilon)B(x)$.

This shows that B is an upper order-derivative of F in the sense of (14.8), and the assertion follows from Theorem 14.9. □

14.3.2. A conditional instability result. The second class of results gives conditions under which populations that are close to extinction can avoid extinction for at least for some time though it is not excluded that extinction finally occurs. These results can also been found in [**203**, Sec.8.1] but their presentation has been thoroughly edited in order to fit the larger framework of this book.

THEOREM 14.11. *Let $F, B : X_+ \to X_+$ and let B be homogeneous and order-preserving and a lower order-derivative of F at 0 as in* (14.3).

Let $r > 1$ and $\theta : X_+ \to \mathbb{R}_+$ be a homogeneous bounded order-preserving functional with

$$\theta(B(x)) \geq r\theta(x), \qquad x \in X_+. \tag{14.10}$$

Then there exists some $\delta_0 > 0$ such that

$$\sup_{n \in \mathbb{Z}_+} \|F^n(x)\| \geq \delta_0, \qquad x \in X_+, \quad \theta(x) > 0.$$

In particular, 0 is an unstable equilibrium of F.

PROOF. Choose some $\epsilon \in (0, 1)$ such that $1 < (1 - \epsilon)r =: s$. Since B is the lower order-derivative of F at 0, we can choose some $\delta > 0$ such that

$$F(x) \geq (1 - \epsilon)B(x), \qquad x \in X_+, \quad \|x\| \leq \delta.$$

Suppose that the statement of the theorem is false. By contraposition, there exists some $x \in X_+$ such that

$$\theta(x) > 0, \qquad \sup_{n \in \mathbb{Z}_+} \|F^n(x)\| < \delta. \tag{14.11}$$

Since θ is bounded, this implies that $(\theta(F^n(x)))$ is a bounded sequence.

We will show by induction that

$$\theta(F^n(x)) \geq s^n \theta(x), \quad n \in \mathbb{Z}_+, \tag{14.12}$$

which is a contradiction because $s > 1$.

Statement (14.12) holds for $n = 0$. Let $n \in \mathbb{Z}_+$ and (14.12) hold for n. Since $\|F^n(x)\| \leq \delta$ by (14.11),

$$F^{n+1}(x) = F(F^n(x)) \geq (1 - \epsilon)B(F^n(x)).$$

Since B and θ are order-preserving and homogeneous and θ satisfies (14.10),

$$\begin{aligned}\theta(F^{n+1}(x)) \geq& (1 - \epsilon)\theta(B(F^n(x))) \geq (1 - \epsilon)r\theta(F^n(x)) \\ \geq& ss^n\theta(x) = s^{n+1}\theta(x).\end{aligned}$$

□

Recall that, if $x, v \in X_+$, x is called v-positive provided that $x \geq \delta v$ for some $\delta > 0$. See Section 4.4.

COROLLARY 14.12. *Let $F, B : X_+ \to X_+$ and let B be an order-preserving lower order-derivative of F at 0. Let $r > 1$ and $v \in \dot{X}_+$ such that $B(v) \geq rv$.*

Then there exists some $\delta_0 > 0$ such that

$$\sup_{n \in \mathbb{Z}_+} \|F^n(x)\| \geq \delta_0$$

for all v-positive $x \in X_+$.

PROOF. We apply Theorem 14.11 with $\theta(x) = [x]_v$, $x \in X_+$. See Section 4.4. Notice that $\theta(x) > 0$ if and only if x is v-positive and use Remark 11.13. □

14.4. Local asymptotic stability of the extinction state in the subthreshold case

We establish one half of the threshold property of the basic turnover number $\mathcal{T}_0$ which is the spectral radius of the basic turnover operator, B, a homogeneous first order approximation of the turnover operator, F. See Sections 1.1 and 1.2. In the subthreshold case $\mathcal{T}_0 < 1$, populations the size of which falls under a certain level go extinct geometrically fast.

THEOREM 14.13. *Let X_+ be the normal closed cone of an ordered normed vector space. Let $F, B : X_+ \to X_+$ and let B be homogeneous, bounded and order-preserving, $\mathcal{T}_0 = \mathbf{r}(B) < 1$, and let B an upper order-derivative of F at 0, (14.4).*

Then F is locally asymptotically stable in the following sense:

For each $\alpha \in (\mathcal{T}_0, 1)$, there exist some $\delta_0 > 0$ and $M \geq 1$ such that $\|F^n(x)\| \leq M\alpha^n\|x\|$ for all $n \in \mathbb{N}$ and all $x \in X_+$ with $\|x\| \leq \delta_0$.

The proof only relies on the definition of the spectral radius of the basic turnover operator B; existence of associated eigenvectors or eigenfunctionals is not needed.

PROOF. Since X_+ is normal, we can replace the original norm by an equivalent monotone norm (like the monotone companion norm, see Theorem 2.48). This does not affect $\mathcal{T}_0 = \mathbf{r}(B)$ given by (5.1).

Let $\alpha \in (\mathcal{T}_0, 1)$ Choose some s such that $\mathcal{T}_0 < s < \alpha$.

Then there exists some $m \in \mathbb{N}$ such that

$$\|B^n\| < s^n, \qquad n \in \mathbb{N}, \quad n \geq m. \tag{14.13}$$

Further there exists some $c \geq 1$ such that

$$\|B^n\| \leq c, \qquad n = 1, \dots, m. \tag{14.14}$$

Choose $\eta > 0$ such that

$$\alpha = (1 + \eta)s \tag{14.15}$$

and then, by (14.4), choose $\delta > 0$ such that $F(x) \leq (1+\eta)B(x)$ for all $x \in X_+$ with $\|x\| \leq \delta$. (Since we switched to an equivalent norm, the δ may not be the same as in the assumption of the theorem.) Choose $\delta_0 \in (0, \delta)$ such that

$$(1 + \eta)^m c\delta_0 \leq \delta. \tag{14.16}$$

Let $x \in X_+$ and $\|x\| \leq \delta_0$. Then $F(x) \leq (1+\eta)B(x)$. Since the norm is monotone, by (4.2), (14.14), and (14.16),

$$\|F(x)\| \leq (1 + \eta)\|B\| \; \|x\| \leq (1 + \eta)c\delta_0 \leq \delta.$$

Since B is order-preserving and homogeneous,

$$F^2(x) \leq (1 + \eta)B(F(x)) \leq (1 + \eta)^2 B^2(x).$$

Since the norm is monotone, by (4.2), (14.14), and (14.16),

$$\|F^2(x)\| \leq (1 + \eta)^2\|B^2\|\|x\| \leq (1 + \eta)^2 c\delta_0 \leq \delta.$$

Proceeding this way we obtain that

$$F^n(x) \leq (1 + \eta)^n B^n(x), \qquad n = 1, \dots, m,$$

and

$$\|F^n(x)\| \leq (1 + \eta)^n c\|x\| \leq (1 + \eta)^n c\delta_0 \leq \delta, \qquad n = 1, \dots, m. \tag{14.17}$$

For $n = m + 1$,

$$F^n(x) \le (1+\eta)B(F^m(x)) \le (1+\eta)^n B^n(x).$$

Since the norm is monotone,

$$\|F^n(x)\| \le (1+\eta)^n \|B^n\| \|x\| \le (1+\eta)^n s^n \|x\| = \alpha^n \|x\| < \delta.$$

Now, by induction,

$$F^n(x) \le (1+\eta)^n B^n(x), \qquad n \ge m.$$

By (14.15) and (14.13),

$$\|F^n(x)\| \le \alpha^n \|x\|, \qquad n \ge m.$$

We combine this estimate with the one in (14.17) using (14.15) again and obtain the assertion. □

14.5. Local stability of the extinction state in the threshold case

In the threshold case $\mathcal{T}_0 = \mathbf{r}(B) = 1$, under appropriate assumptions (see Chapter 9), there exists some $u \in \dot{X}_+ = X_+ \setminus \{0\}$ such that $B(u) \le u$.

THEOREM 14.14. *Let X_+ be the normal closed cone of an ordered normed vector space X. Let $F, B : X_+ \to X_+$ and let B be homogeneous, bounded and order-preserving and $F(x) \le B(x)$ for all $x \in X_+$. Let $u \in \dot{X}_+$ such that $B(u) \le u$ and a power of B is uniformly u-bounded (Definition 6.1).*

Then there exists some $c > 0$ such that $\|F^n(x)\| \le c\|x\|$ for all $x \in X_+$ and $n \in \mathbb{N}$.

Instead of assuming that X_+ is normal, it is sufficient that u is a normal point for X_+. See Section 2.2.2. As it is well-known, $\mathbf{r}(B) = 1$ is not sufficient for this result even if B can be represented by a matrix.

PROOF. Since B is order-preserving, by induction, $F^n(x) \le B^n(x)$ for all $x \in X_+$. Since some power of B is uniformly u-bounded, there is some $m \in \mathbb{N}$ and $\tilde{c} > 0$ such that $B^m(x) \le \tilde{c}\|x\|u$ for all $x \in X_+$. Since B is order-preserving and homogeneous, for all $n \in \mathbb{Z}_+$,

$$F^{n+m}(x) \le B^n(B^m(x)) \le \tilde{c}\|x\|B^n(u) \le \tilde{c}\|x\|u, \qquad x \in X_+.$$

Since X_+ is normal (or u is a normal point for X_+), there exists some $\check{c} > 0$ such that

$$\|F^{n+m}(x)\| \le \check{c}\tilde{c}\|x\| \, \|u\|, \qquad n \in \mathbb{Z}_+, \quad x \in X_+.$$

For $n = 1, \dots, m$, $F^n(x) \le B^n(x)$ and

$$\|F^n(x)\| \le \check{c}\|B^n(x)\| \le \check{c}\|B^n\| \, \|x\|.$$ □

14.6. Point-dissipativity and compact attractors

For many aspects of nonlinear dynamics, it is useful if the semiflow induced by F (or otherwise) has a compact attractor of points [**187**, Sec.2.2].

$F : X_+ \to X_+$ is called *point dissipative* if there exists a bounded subset B of X such that $d(F^n(x), B) \to 0$ as $n \to \infty$ for all $x \in X_+$. Here, $d(y, B)$ denotes the distance from y to B in X.

Since X is a normed vector space, equivalently, F is point-dissipative if there exists some constant $c > 0$ such that

$$\limsup_{n\to\infty} \|F^n(x)\| \le c, \qquad x \in X_+.$$

F is said to be *asymptotically smooth* if for every closed, bounded subset U of X_+ with $F(U) \subseteq U$ and for every sequence (n_i) in $\mathbb{Z}_+$, $n_i \to \infty$ as $i \to \infty$, and every sequence (u_i) in U, $(F^{n_i}(u_i))$ has a convergent subsequence.

Recall that a homogeneous functional $\theta : X_B \to \mathbb{R}_+$ on a homogeneous subset X_B of X is called *uniformly positive* if there exists some $\epsilon > 0$ such that $\theta(x) \ge \epsilon\|x\|$ for all $x \in X_B$. See Definition 5.37.

THEOREM 14.15. *Let X_+ be the closed cone of an ordered normed vector space X. Let $F : X_+ \to X_+$ map bounded subsets of X_+ into bounded subsets of X_+. Let $\theta : X_+ \to \mathbb{R}_+$ be homogeneous, subadditive, continuous and uniformly positive. Assume that*

$$\limsup_{\|x\|\to\infty} \frac{\theta(F(x))}{\theta(x)} < 1. \tag{14.18}$$

Then, for any bounded subset B of X_+, there exists a bounded convex subset $\tilde{B}$ of X_+ such that $F^n(B) \subseteq \tilde{B}$ for all $n \in \mathbb{N}$.

Further, there exists a bounded convex subset $\hat{B}$ of X_+ which absorbs all points in X_+: For each $x \in X_+$ there exists some $m \in \mathbb{N}$ such that $F^n(x) \in \hat{B}$ for all $n \ge m$. In particular, F is point-dissipative.

If, in addition, F is continuous and asymptotically smooth, the semiflow induced by F has a compact attractor of bounded sets [**187**, Sec.2.2.3].

PROOF. By (14.18) and the other properties of θ, there exists some $\xi \in (0,1)$ and $R_1 > 0$ such that

$$\theta(F(x)) \le \xi\theta(x), \qquad x \in X_+, \quad \theta(x) \ge R_1. \tag{14.19}$$

We claim that there exists some $R_2 > 0$ such that, for all $x \in X_+$,

$$\theta(x) \le R_2 \implies \theta(F(x)) \le R_2. \tag{14.20}$$

If not, for any $n \in \mathbb{N}$, there exists some $x_n \in X_+$ such that

$$\theta(x_n) \le n < \theta(F(x_n)). \tag{14.21}$$

Since F maps bounded sets in X_+ into bounded sets of X_+ and θ is bounded and uniformly positive, $\theta(x_n) \to \infty$ as $n \to \infty$. By (14.19) and (14.21), this leads to a contradiction for n large enough such that $\theta(x_n) \ge R_1$:

$$n < \theta(F(x_n)) \le \xi\theta(x_n) \le n.$$

So (14.20) holds. Let $R_3 = \max\{R_1, R_2\}$. Let $R \ge R_3$ and $B_R^+ = \{x \in X_+; \theta(x) \le R\}$. Since θ is convex and continuous, B_R^+ is convex and closed. Since θ is uniformly positive, B_R^+ is bounded. By (14.20), $F(B_R^+) \subseteq B_R^+$.

Let B be a bounded subset of X_+. Then there exists some $R > R_3$ such that $B \subseteq B_R^+$ and $F^n(B) \subseteq B_R^+$ for all $n \in \mathbb{N}$.

Let $x \in X_+$. If $\|x\| \le R_3$, $\theta(F^n(x)) \le R_3$ for all $n \in \mathbb{N}$.

If $\theta(x) > R_3$, by (14.19), $\theta(F^{n+1}(x)) \le \xi\theta(F^n(x))$ as long as $\theta(F^n(x)) \ge R_3$. So $\theta(F^m(x)) \le R_3$ for some $m = m_x \in \mathbb{N}$ and $\theta(F^n(x)) \le R_3$ for all $n \ge m_x$.

In either case, $\limsup_{n\to\infty} \theta(F^n(x)) \le R_3$.

Set $\hat{B} = B_{R_3}$ and $F^n(x) \in \hat{B}$ for all $n \geq m_x$.
Since θ is uniformly positive, there exists some $c > 0$ such that

$$\limsup_{n\to\infty} \|F^n(x)\| \leq c \qquad x \in X_+. \qquad \square$$

14.6.1. Normal cones.

THEOREM 14.16. *Let X_+ be the normal closed cone of an ordered normed vector space X. Let $F : X_+ \to X_+$ satisfy*

$$F(x) \leq Ax + G(x), \qquad x \in X_+, \tag{14.22}$$

with a continuous, subadditive, homogeneous map $A : X_+ \to X_+$ and continuous $G : X_+ \to X_+$ such that

$$\mathbf{r}(A) < 1, \qquad \frac{\|G(x)\|}{\|x\|} \to 0, \quad \|x\| \to \infty. \tag{14.23}$$

Then, for any bounded subset B of X_+ there exists a bounded convex subset $\tilde{B}$ of X_+ such that $F^n(B) \subseteq \tilde{B}$ for all $n \in \mathbb{N}$. Further there exists a bounded convex subset B_0 of X_+ such that for each $x \in X_+$ there exists some $m \in \mathbb{N}$ such that $F^n(x) \in B_0$ for all $n \geq m$.

If, in addition, F is asymptotically smooth, the semiflow induced by F has a compact attractor of bounded sets [**187**, Sec.2.2.3].

This theorem follows by combining Theorem 14.15 with the subsequent result.

LEMMA 14.17. *Let X_+ be the normal closed cone of an ordered normed vector space X. Let $F : X_+ \to X_+$ satisfy*

$$F(x) \leq Ax + G(x), \qquad x \in X_+, \tag{14.24}$$

with a continuous, subadditive, homogeneous map $A : X_+ \to X_+$ and continuous $G : X_+ \to X_+$ such that

$$\mathbf{r}(A) < 1, \qquad \frac{\|G(x)\|}{\|x\|} \to 0, \quad \|x\| \to \infty. \tag{14.25}$$

Then there is some homogeneous, continuous, subadditive, uniformly positive, order-preserving functional $\theta : X_+ \to \mathbb{R}_+$ such that

$$\limsup_{\|x\|\to\infty} \frac{\theta(F(x))}{\theta(x)} < 1.$$

PROOF. By Theorem 5.19, since X_+ is normal, there is some homogeneous, continuous, subadditive, order-preserving $\theta : X_+ \to X_+$ and some $r \in (\mathbf{r}(A), 1)$, $c > 0$ such that $\theta(Ax) \leq r\theta(x)$ and $\|x\| \leq \theta(x) \leq c\|x\|$ for all $x \in X_+$. By (14.25), $\theta(G(x))/\theta(x) \to 0$ as $\|x\| \to \infty$. Since θ is order-preserving and subadditive,

$$\frac{\theta(F(x))}{\theta(x)} \leq \frac{\theta(Ax + G(x))}{\theta(x)} \leq \frac{\theta(Ax)}{\theta(x)} + \frac{\theta(G(x))}{\theta(x)}.$$

So

$$\limsup_{\|x\|\to\infty} \frac{\theta(F(x))}{\theta(x)} \leq r < 1. \qquad \square$$

14.6.2. Variation of constants formula. Since some problems are better studied in ordered Banach spaces X with nonnormal cones X_+ [**167**], we add a point-dissipativity result [**128**, Thm.2.3] in which normality of the cone is replaced by G mapping bounded sets into bounded sets. More severely, we assume the equality $F(x) = Ax + G(x)$, $x \in X_+$, rather than an inequality (Theorem 14.16). By induction and the additivity of A, the sequence of iterates $(x_n = F(x_{n-1}))$ of $x_0 \in X_+$ satisfies

$$x_n = A^n x_0 + \sum_{k=1}^{n} A^{k-1} G(x_{n-k}), \qquad n \in \mathbb{N}. \tag{14.26}$$

THEOREM 14.18. *Let $F = A + G$ on X_+ with a bounded additive homogeneous operator A on X_+, $\mathbf{r}(A) < 1$. Assume that $G : X_+ \to X_+$ maps bounded subsets of X_+ into bounded subsets of X_+ and*

$$\frac{\|G(x)\|}{\|x\|} \to 0, \qquad \|x\| \to \infty. \tag{14.27}$$

Then the semiflow induced by F is point-dissipative and is bounded on bounded sets.

PROOF. By (14.26) and the triangle inequality,

$$\|x_n\| \le \|A^n\| \, \|x_0\| + \sum_{k=1}^{n} \|A^{k-1}\| \, \|G(x_{n-k})\|, \qquad n \in \mathbb{N}.$$

Let $\epsilon > 0$, to be chosen later. Since G maps bounded sets into bounded sets, by (14.27), there exists some $c_\epsilon > 0$ such that

$$\|G(x)\| \le c_\epsilon + \epsilon \|x\|, \qquad x \in X_+.$$

We substitute this inequality into the previous one,

$$\|x_n\| \le \|A^n\| \, \|x_0\| + \sum_{k=1}^{n} \|A^{k-1}\| \, c_\epsilon + \sum_{k=1}^{n} \|A^{k-1}\| \, \epsilon \, \|x_{n-k}\|, \qquad n \in \mathbb{N}. \tag{14.28}$$

For $m \in \mathbb{N}$, we define

$$\alpha_m = \sup_{n=0}^{m} \|x_n\|.$$

Since $\mathbf{r}(A) < 1$,

$$\sum_{k=1}^{\infty} \|A^{k-1}\| =: \zeta \tag{14.29}$$

converges. For all $m \in \mathbb{N}$, by (14.28),

$$\alpha_m \le \max_{n=0}^{m} \|A^n\| \, \|x_0\| + \zeta \, c_\epsilon + \epsilon \zeta \alpha_m.$$

Choose $\epsilon > 0$ such that

$$\epsilon \zeta \le 1/2. \tag{14.30}$$

Then, for all $m \in \mathbb{N}$,

$$\|x_m\| \le \alpha_m \le 2 \sup_{n=0}^{\infty} \|A^n\| \, \|x_0\| + 2\zeta \, c_\epsilon.$$

This shows that the semiflow induced by F is bounded on bounded sets.

We apply Fatou's lemma to (14.28) with the counting measure on $\mathbb{N}$. Then, for $\alpha^\infty = \limsup_{n\to\infty} \|x_n\|$,

$$\alpha^\infty \le \sum_{k=1}^\infty \|A^{k-1}\|(c_\epsilon + \epsilon\alpha^\infty).$$

We choose $\epsilon > 0$ as in (14.30),

$$\alpha^\infty \le 2\zeta\|c_\epsilon.$$

This shows that the semiflow induced by F is point-dissipative. □

14.6.3. Asymptotic smoothness. The previous results motivate us to look for conditions that make the semiflow induced by F asymptotically smooth.

THEOREM 14.19. *Let X_+ be the closed cone of an ordered Banach space X and $F : X_+ \to X_+$ and $F = A + G$ with a bounded linear positive map A on X, $\mathbf{r}(A) < 1$, and a compact continuous map $G : X_+ \to X_+$. Then the semiflow induced by F is asymptotically smooth.*

The semiflow has a compact attractor of bounded sets if, in addition,

$$\|G(x)\|/\|x\| \to 0 \text{ as } \|x\| \to \infty$$

and if X_+ is normal or G maps bounded sets into bounded sets.

PROOF. By induction, $F^n = A^n + K_n$ with compact continuous maps $K_n : X_+ \to X_+$. We apply [**187**, Thm.2.46]. Let C be a bounded subset of X_+. Then diam $A^n(C) \le \|A^n\|$ diam $(C) \to 0$, $n \to \infty$. Further, $K_n(C)$ has compact closure. So the semiflow induced by F is asymptotically smooth.

The last statement follows from Theorem 14.16 or from Theorem 14.18. □

14.6.4. Weak point-dissipativity. A possible road to point-dissipativity consists in showing weak point-dissipatitivity first.

DEFINITION 14.20. *Let X be a normed vector space and $F : X_F \to X_F$ be an operator on a subset X_F of X.*

Recall that F is point-dissipative *if there exists some $c > 0$ such that*

$$\limsup_{n\to\infty} \|F^n(x)\| < c \quad \text{for all } x \in X_F.$$

F is called weakly point-dissipative *if there exists some $c > 0$ such that*

$$\liminf_{n\to\infty} \|F^n(x)\| < c \quad \text{for all } x \in X_F.$$

THEOREM 14.21. *Let X be a normed vector space and $F : X_F \to X_F$ be a compact operator on a closed subset X_F of X such that $\|F(x)\|$ is a continuous function of $x \in X_F$.*

Assume that F is weakly point-dissipative. Then F is point-dissipative and, if F is also continuous, the semiflow induced by F has a compact attractor of points.

PROOF. Suppose that F is weakly point-dissipative but not dissipative. Then there exists some $c > 0$ such that

$$\liminf_{n\to\infty} \|F^n(x)\| < c \qquad x \in X_F, \tag{14.31}$$

and sequences (x_k) and (c_k) in X_F and $(0,\infty)$ such that

$$\limsup_{n\to\infty} \|F^n(x_k)\| > c_k, \quad k \in \mathbb{N}, \quad c_k \overset{k\to\infty}{\longrightarrow} \infty. \tag{14.32}$$

Then there exist sequences (r_k), (s_k), (t_k) in $\mathbb{N}$ such that $r_k \to \infty$ and

$$
\begin{aligned}
&\|F^{r_k}(x_k)\| \le c+1, \\
&\|F^{r_k+s_k}(x_k)\| \ge c_k, \\
&\|F^{r_k+s}(x_k)\| \ge c+1, \quad s=1,\dots,s_k+t_k, \\
&\|F^{r_k+s_k+t_k+1}(x_k)\| < c+1.
\end{aligned} \tag{14.33}
$$

We set $y_k = F^{r_k}(x_k)$. Then

$$
\begin{aligned}
&\|y_k\| \le c+1, \\
&\|F^{s_k}(y_k)\| \ge c_k, \\
&\|F^{s}(y_k)\| \ge c+1, \quad s=1,\dots,s_k+t_k, \\
&\|F^{s_k+t_k+1}(y_k)\| \le c+1.
\end{aligned} \tag{14.34}
$$

Since F is compact, after choosing subsequences, $F(y_k) \to x$ for $k \to \infty$ with some $x \in X_F$. Since $\|F(z)\|$ is a continuous function of $z \in X_F$, $\|F^n(y_k)\| \to \|F^{n-1}(x)\|$ for all $n \in \mathbb{N}$.

We claim that $s_k \to \infty$ as $k \to \infty$.

If not, there exists a constant subsequence (s_{k_j}) with $s_{k_j} = m \in \mathbb{N}$ for all $j \in \mathbb{N}$ and $c_{k_j} \le \|F^{s_{k_j}}(y_{k_j})\| \to \|F^m(x)\|$, a contradiction.

Let $s \in \mathbb{N}$. Since $s_k \to \infty$, $s_k > s$ for all but finitely many $k \in \mathbb{N}$. So $\|F^s(y_k)\| \ge c+1$ for all but finitely many $k \in \mathbb{N}$ and thus $\|F^{s-1}(x)\| \ge c+1$. This contradicts (14.31).

This proves that F is dissipative. Since F is compact, F is asymptotically smooth and the induced semiflow has a compact attractor of point by [**187**, Thm.2.28]. □

14.7. Population extinction

We derive conditions that imply that $F^n(x) \to 0$ as $n \to \infty$ for all $x \in X_+$. The first result, for the subthreshold case $\mathcal{T}_0 = \mathbf{r}(B) < 1$, is proved by using the definition of the spectral radius, (1.7) .

THEOREM 14.22. *Let X_+ be the normal closed cone of an ordered normed vector space. Let $F, B : X_+ \to X_+$ and let B be homogeneous, bounded and order-preserving , $\mathbf{r}(B) < 1$, and $F(x) \le B(x)$ for all $x \in X_+$.*

Then $F^n(x) \to 0$ as $n \to \infty$ uniformly for x in bounded subsets of X_+.

PROOF. By induction, since B is order-preserving, $F^n(x) \le B^n(x)$ for all $n \in \mathbb{N}$ and $x \in X_+$. Since X_+ is a normal closed cone, there is some $c > 0$ such that $\|F^n(x)\| \le c\|B^n(x)\| \le c\|B^n\|\,\|x\|$ for all $x \in X_+$ and $n \in \mathbb{N}$. Since $\mathbf{r}(B) < 1$, $\|B^n\| \to 0$ as $n \to \infty$ by (1.7) and the assertion follows. □

The next extinction results deal with the threshold case that the basic turnover number equals 1, $\mathcal{T}_0 = \mathbf{r}(B) = 1$. It is useful if $\mathcal{T}_0 = 1$ is an eigenvalue of B associated with a homogeneous upper eigenfunctional; see Theorem 14.23(b) and see Chapter 11 for scenarios where this is the case.

THEOREM 14.23. *Let X_+ be the normal closed cone of an ordered normed vector space. Let $F, B : X_+ \to X_+$ and let B be homogeneous, bounded and order-preserving, $\mathcal{T}_0 = \mathbf{r}(B) = 1$, and $F(x) \le B(x)$ for all $x \in X_+$.*

Further, by assumption, for any $\delta > 0$ there is some $\epsilon \in (0,1)$ such that $F(x) \leq \epsilon B(x)$ for all $x \in X_+$ with $\|x\| \geq \delta$ and $\|F(x)\| \geq \delta$.

(a) *Then $\liminf_{n\to\infty} \|F^n(x)\| \to 0$ for all $x \in X_+$.*

(b) *If there exists a bounded homogeneous order-preserving eigenfunctional $\theta : X_+ \to \mathbb{R}_+$ of B, $\theta(B(x)) \leq \theta(x)$ for all $x \in X_+$, then $\theta(F_n(x)) \to 0$ as $n \to \infty$.*

If, in addition, $\theta(x) > 0$ for all $x \in \dot{X}_+$ and θ is continuous and the semiflow induced by F is asymptotically smooth and point-dissipative, then $\|F_n(x)\| \to 0$ as $n \to \infty$, $x \in X_+$.

PROOF. (a) Suppose that $x \in X_+$ and $\liminf_{n\to\infty} \|F^n(x)\| > \delta > 0$.

Then there is some $m \in \mathbb{N}$ such that $\|F^n(x)\| > \delta$ for all $n \geq m$. Choose $\epsilon \in (0,1)$ according to our assumption.

We claim that $F^{n+m}(x) \leq \epsilon^n B^n(F^m(x))$ for all $n \in \mathbb{Z}_+$.

This holds for $n = 0$. Suppose that $n \in \mathbb{Z}_+$ and the claim holds for n.

$$F^{n+1+m}(x) = F(F^{n+m}(x)) \leq \epsilon B(F^{n+m}(x))$$

because both $\|F(F^{n+m}(x))\| \geq \delta$ and $\|(F^{n+m}(x))\| \geq \delta$. Since B is order-preserving,

$$F^{n+1+m}(x) \leq \epsilon B \epsilon^n B^n(F^m(x)).$$

This concludes our induction proof.

Since X_+ is normal, there exists some $c > 0$ such that

$$\|F^{n+m}(x)\| \leq c\|\epsilon B\|^n \|F^m(x)\|.$$

Since $\mathbf{r}(\epsilon B) = \epsilon \mathbf{r}(B) < 1$, $\|F^{n+m}(x)\| \to 0$ as $n \to \infty$, a contradiction.

This finishes the proof of part (a).

(b) Let $\theta : X_+ \to \mathbb{R}_+$ be a bounded homogeneous order-preserving eigenfunctional of B, $\theta \circ B(x) \leq \theta(x)$ for all $x \in X$, and let $x \in X_+$. Then $(\theta(F^n(x))$ is a decreasing sequence,

$$\theta(F^{n+1}(x)) \leq \theta(B(F^n(x))) = \theta(F^n(x)), \qquad n \in \mathbb{N}.$$

Since θ is bounded, $\liminf_{n\to\infty} \theta(F^n(x)) = 0$ and thus $\theta(F^n(x)) \to 0$ as $n \to \infty$.

Let, in addition, F be point-dissipative and asymptotically smooth. Then, for each $x \in X_+$, there exists a nonempty compact invariant ω-limit set $\omega(x)$ which attracts $(F^n(x))$ [**187**, Thm.2.28]. For each $y \in \omega(x)$, since θ is continuous, $\theta(y) = 0$ and so $y = 0$. This implies $F^n(x) \to 0$ as $n \to \infty$. □

Alternatively, it is useful if $\mathcal{T}_0 = 1$ is an eigenvalue associated with an (upper) eigenvector of B, see Theorem 14.24 and Chapter 9 for scenarios where this is the case.

THEOREM 14.24. *Let X_+ be the normal closed cone of an ordered normed vector space. Let $F, B : X_+ \to X_+$ and let B be homogeneous, bounded and order-preserving, $\mathbf{r}(B) = 1$, and $F(x) \leq B(x)$ for all $x \in X_+$.*

Further, by assumption, for any $\delta > 0$ there is some $\epsilon \in (0,1)$ such that $F(x) \leq \epsilon B(x)$ for all $x \in X_+$ with $\|x\| \geq \delta$.

Finally, assume that there is some $v \in \dot{X}_+$ such that $B(v) \leq v$ and that some power of B is uniformly v-bounded (Definition 6.1).

Then $\|F^n(x)\| \to 0$ as $n \to \infty$.

PROOF. Let $m \in \mathbb{N}$ such that B^m is uniformly v-bounded (Definition 6.1). We define the functional $\theta : X_+ \to \mathbb{R}_+$ by

$$\theta(x) = \|B^m(x)\|_v, \qquad x \in X_+. \tag{14.35}$$

Then θ is homogeneous, order-preserving and bounded. By (2.32), for $x \in X_+$, $B^m(x) \le \theta(x)v$. Since B is homogeneous and order-preserving,

$$B(B^m(x)) \le \theta(x)B(v) \le \theta(x)v.$$

By (2.26), $\|B^m(B(x))\|_v \le \theta(x)$ and, by (14.35), $\theta(B(x)) \le \theta(x)$. By Theorem 14.23, $\|B^m(F^n(x))\|_v \to 0$ as $n \to \infty$. Since $F(y) \le B(y)$ for all $y \in X_+$, $\|F^{m+n}(x)\|_v \to 0$ as $n \to \infty$. Since X_+ is normal, $\|F^n(x)\| \to 0$ as $n \to \infty$ by Lemma 2.69 (b). □

ALTERNATIVE PROOF. By Theorem 14.23,

$$\liminf_{n\to\infty} \|F^n(x)\| = 0, \qquad x \in X_+.$$

By Theorem 14.14, there exists some $c > 0$ such that $\|F^n(y)\| \le c\|y\|$ for all $y \in X_+$, $n \in \mathbb{Z}_+$.

Let $x \in X_+$ and $\epsilon > 0$. Then there exists some $m \in \mathbb{N}$ such that $\|F^m(x)\| \le \epsilon/c$. For all $n \ge m$,

$$\|F^n(x)\| = \|F^{n-m}F^m(x)\| \le c\|F^m(x)\| \le \epsilon. \qquad \square$$

14.8. Nonzero fixed points

One way of showing that populations the dynamics of which are governed by the turnover operator F do not always go extinct consists in establishing the existence of a nonzero fixed point $F(x) = x$ (persistence at equilibrium). For a more condensed presentation see [**203**, Sec.8.3].

THEOREM 14.25. *Let X_+ be the closed cone of an ordered normed vector space X and $F : \dot{X}_+ \to X_+$ be continuous and compact. Assume that there are some homogeneous subadditive functional $\theta : X_+ \to \mathbb{R}_+$ and some order-preserving lower order-derivative $B : X_+ \to X_+$ of F at 0 with the following properties:*

(i) *θ is uniformly positive and continuous.*

(ii) $\displaystyle\limsup_{\|x\|\to\infty} \frac{\theta(F(x))}{\theta(x)} < 1.$

(iii) *There is some $v \in \dot{X}_+$ and $r > 1$ such that $B(v) \ge rv$.*

Then there exists some $x \in \dot{X}_+$ such that $F(x) = x$.

We emphasize that X_+ does not need to be complete. Recall that θ is uniformly positive if there is some $\zeta \in (0,1)$ such that $\theta(x) \ge \zeta\|x\|$ for all $x \in X_+$. The concept of a lower order-derivative can be found in Section 14.2. Recall the notation $\dot{X}_+ = X_+ \setminus \{0\}$.

We use the well-established idea [**135**, Sec.2.2] [**177**, Sec.3] [**211**, Thm.3.6]. of considering perturbed operators F_λ, $\lambda > 0$, with $F_\lambda(x) \to F(x)$ for $\lambda \to 0$, $x \in X_+$. The F_λ inherit continuity and compactness from F and have fixed points x_λ by a fixed point theorem by Tychonoff which does not require completeness [**55**, Thm.10.1]. Because of the perturbation, $0 \ne x_\lambda \in X_+$. Because of the compactness of F, there exists a sequence (λ_n) in $(0,1)$ with $\lambda_n \to 0$ and $x_{\lambda_n} \to x$ for some fixed point $F(x) = x \in X_+$. Condition (iii) is used to show that $x \ne 0$.

PROOF. For $\lambda \in (0,1)$, we define $F_\lambda : X_+ \to X_+$ by

$$F_\lambda(x) = F(x + \lambda v) + \lambda v, \qquad x \in X_+, \tag{14.36}$$

where v is chosen according to assumption (iii). We claim:

($\diamondsuit$) There exists some $R > 0$ such that $\theta(F_\lambda(x)) \le R$ for all $\lambda \in (0,1]$ and all $x \in X_+$ with $\theta(x) \le R$.

Suppose that ($\diamondsuit$) does not hold. Then, for any $n \in \mathbb{N}$, there exist some $x_n \in X_+$ and $\lambda_n \in (0,1]$ such that $\theta(x_n) \le n$ and $\theta(F_{\lambda_n}(x_n)) \ge n$.

So, for all $n \in \mathbb{N}$, since θ is subadditive and homogeneous,

$$n \le \theta\big(F(x_n + \lambda_n v) + \lambda_n v\big) \le \theta(F(x_n + \lambda_n v)) + \theta(v).$$

Since F is compact on $\dot{X}_+$ and θ bounded, (x_n) is unbounded. After choosing subsequences, $\|x_n\| \to \infty$ as $n \to \infty$. By assumption (ii), there exists some $\alpha \in (0,1)$ and $\beta > 0$ such that

$$\theta(F(x)) \le \alpha\theta(x), \qquad x \in X_+, \quad \|x\| \ge \beta.$$

So, for large enough n,

$$n \le \alpha\theta(x_n + \lambda_n v) + \theta(v) \le \alpha n + 2\theta(v),$$

a contradiction because $\alpha \in (0,1)$.

So, we have shown that ($\diamondsuit$) holds which can be rephrased as follows:

There is some $R > 0$ such that, for any $\lambda \in (0,1]$, F_λ maps the set $\{x \in X_+; \theta(x) \le R\}$ into itself. This set is bounded because θ is uniformly positive by assumption (i). It is convex and closed because θ is subadditive, homogeneous and continuous. Since F_λ is compact and continuous on X_+, by Tychonov's fixed point theorem [**55**, Thm.10.1], there exists some $x_\lambda \in X_+$ with

$$F_\lambda(x_\lambda) = x_\lambda, \qquad \theta(x_\lambda) \le R, \quad \lambda \in (0,1]. \tag{14.37}$$

Now, choose a sequence (λ_n) in $(0,1]$ such that $\lambda_n \to 0$ for $n \to \infty$. Since θ is uniformly positive by assumption (i) and since (14.36) holds, there exists a bounded sequence (x_n) in X_+such that

$$F(x_n + \lambda_n v) + \lambda_n v = x_n, \qquad n \in \mathbb{N}. \tag{14.38}$$

We choose $\epsilon \in (0,1)$ such that $(1-\epsilon)r > 1$ with the number r from assumption (iii). Since B is a lower order-derivative of F, by (14.3) we choose $\delta > 0$ such that $F(x) \ge (1-\epsilon)B(x)$ for all $x \in \dot{X}_+$ with $\|x\| \le 2\delta$.

We claim that

$$\liminf_{n\to\infty} \|x_n\| \ge \delta. \tag{14.39}$$

Suppose not. Then, for some large $n \in \mathbb{N}$, $\|x_n\| \le \delta$ and $\lambda_n\|v\| \le \delta$ and so $\|x_n + \lambda_n v\| \le 2\delta$ and by (14.38),

$$x_n \ge (1-\epsilon)B(x_n + \lambda_n v) + \lambda_n v.$$

Since $v \in X_+$ and B is order-preserving, $x_n \ge (1-\epsilon)B(x_n)$. Since B is homogeneous and order-preserving, by induction,

$$x_n \ge (1-\epsilon)^k B^k(x_n), \qquad k \in \mathbb{N}.$$

We also have that $x_n \ge \lambda_n v$ and $B(v) \ge rv$. Since B is order-preserving and homogeneous,

$$x_n \ge (1-\epsilon)^k B^k(\lambda_n v) \ge [(1-\epsilon)r]^k \lambda_n v, \qquad k \in \mathbb{N}.$$

Since X_+ is a closed cone and $(1-\epsilon)r > 1$, this implies that $v = 0$, a contradiction.

This proves (14.39). Since F is compact on $\dot{X}_+$ and the sequence $(x_n + \lambda_n v)$ in $\dot{X}_+$ is bounded by (14.38) and $\lambda_n \to 0$, $F(x_n + \lambda_n v) \to y \in X_+$ after choosing a subsequence. By (14.38) and (14.39), $x_n \to y$ as $n \to \infty$ with $y \in X_+$ and $\delta \le \|y\|$. Since F is continuous on $\dot{X}_+$, $F(y) = y$ by (14.38). □

By Corollary 9.11, Theorem 14.25 has the following consequence for the superthreshold case $\mathcal{T}_0 > 1$. Since we want to apply this result to the cone of non-negative Borel measures with the flat norm (Theorem 19.18), we emphasize again that X_+ does not need to be complete.

COROLLARY 14.26. *Let X_+ be the closed cone of an ordered normed vector space X and $F : \dot{X}_+ \to X_+$ be continuous and compact,*

$$\limsup_{\|x\|\to\infty} \frac{\|F(x)\|}{\|x\|} < 1.$$

Further assume that there is some lower order-derivative $B : X_+ \to X_+$ of F at 0 that is order-preserving, compact and continuous and $\mathcal{T}_0 = \mathbf{r}(B) > 1$.

Then there exists some $x \in \dot{X}_+$ such that $F(x) = x$.

14.9. Population persistence

The results in this section are related to the other half of the threshold property of the basic turnover number $\mathcal{T}_0$ which is the spectral radius of the basic turnover operator, B, a homogeneous first order approximation of the turnover operator, F. See Sections 1.1 and 1.2. If $\mathcal{T}_0 > 1$, populations persist in a certain sense under suitable extra conditions. See Section 1.2.

14.9.1. Uniform weak persistence. As before, let X_+ be the closed order cone of an ordered normed vector space X and $\dot{X}_+ = X_+ \setminus \{0\}$. Recall the definitions of a normal cone (Section 2.2.1) and of a serially complete cone (Section 2.1.2).

THEOREM 14.27. *Let X_+ be a serially complete normal closed cone and $u \in \dot{X}_+$. Let $F, B : X_+ \to X_+$ and B be bounded homogeneous and order-preserving and B a lower order-derivative of F at 0 (Section 14.2). Assume $\mathcal{T}_0 = \mathbf{r}(B) > 1$.*

Let the following properties be satisfied.

(a) *$F(\dot{X}_+) \subseteq \dot{X}_+$.*
(b) *A power of B is uniformly u-bounded (Definition 6.1).*
(c) *B is pointwise u-positive (Definition 11.3).*

Then the population is uniformly weakly persistent: There exists some $\delta > 0$ such that $\limsup_{n\to\infty} \|F^n(x)\| \ge \delta$ for all $x \in \dot{X}_+$.

The subsequent results, which are of their own interest, prepare the proof of this theorem, which is at the end of this section. Eventually, they need to be combined with Theorem 11.14, which provides a homogeneous eigenfunctional of B associated with $\mathcal{T}_0 = \mathbf{r}(B)$.

The presentation loosely follows [**130**].

PROPOSITION 14.28. *Let $F, B : X_+ \to X_+$ and B be homogeneous and order-preserving and a lower order-derivative of F at 0 (Section 14.2). Let $\theta : X_+ \to \mathbb{R}_+$ be homogeneous, bounded, and order-preserving, and $\rho : X_+ \to \mathbb{R}_+$.*

Let the following properties be satisfied.

(a) *If $x \in X_+$ and $\rho(x) > 0$, then $\theta(F^n(x)) > 0$ for infinitely many $n \in \mathbb{N}$.*

(b) *There exists some $r > 1$ such that $\theta(B(x)) \geq r\theta(x)$ for all $x \in X_+$.*

Then the extinction state $0 \in X_+$ is uniformly weakly ρ-repelling for the semiflow induced by F: There exists some $\delta > 0$ such that

$$\limsup_{n\to\infty} \|F^n(x)\| \geq \delta \text{ for all } x \in X_+ \text{ with } \rho(x) > 0.$$

Notice that assumption (b) in Proposition 14.28 and Theorem 14.29 implies that $\mathcal{T}_0 = \mathbf{r}(B) > 1$ (Section 5.6).

In turn, additional general assumptions under which $\mathcal{T}_0 = \mathbf{r}(B) > 1$ implies assumption (b) can be found in Chapter 11 and in [**130**, Sec.7].

Further assumptions tailored to specific ordered normed vector spaces are presented in Section 13.5 and 13.6.

PROOF. Choose some $r > 1$ according to (b) and some $\epsilon \in (0,1)$ such that

$$1 < r(1-\epsilon) =: s. \tag{14.40}$$

Since B is a lower order-derivative of F at 0, we can choose $\delta > 0$ according to (14.3):

$$F(x) \geq (1-\delta)B(x), \qquad x \in X_+, \quad \|x\| \leq \delta. \tag{14.41}$$

Suppose that the extinction state is not uniformly weakly ρ-repelling: There exists some $x \in X_+$ such that

$$\rho(x) > 0 \text{ and } \limsup_{n\to\infty} \|F^n(x)\| < \delta.$$

By (a), after a shift in time, we can assume that $x \in X_+$ and

$$\theta(x) > 0, \qquad \|F^n(x)\| < \delta, \quad n \in \mathbb{Z}_+. \tag{14.42}$$

Set $x_n = F^n(x)$ for $n \in \mathbb{Z}_+$. Then

$$x_{n+1} = F(x_n), \quad n \in \mathbb{Z}_+, \qquad x_0 = x. \tag{14.43}$$

By (14.42), $\|x_n\| \leq \delta$ for all $n \in \mathbb{N}$, and the sequence $(\theta(x_n))$ is bounded because θ is bounded. We claim that

$$x_n \geq (1-\epsilon)^n B^n(x), \qquad n \in \mathbb{Z}_+. \tag{14.44}$$

This trivially holds for $n = 0$. Assume it holds for n. By (14.43) and (14.41), since B is order-preserving,

$$x_{n+1} = F(x_n) \geq (1-\epsilon)B(x_n) \geq (1-\epsilon)B((1-\epsilon)^n B^n(x_0)).$$

Since B is homogeneous, $x_{n+1} \geq (1-\epsilon)^{n+1}B^{n+1}(x_0)$. So, (14.44) holds by the induction principle.

By (14.44), since θ is order-preserving and homogeneous,

$$\theta(x_n) \geq (1-\epsilon)^n \theta(B^n(x)). \tag{14.45}$$

We claim that, for all $n \in \mathbb{Z}_+$,

$$\theta(B^n(x)) \geq r^n \theta(x). \tag{14.46}$$

This is trivially true for $n = 0$. Assume it holds for some arbitrary $n \in \mathbb{Z}_+$. Then, by (b), since θ is homogeneous and order-preserving,

$$\theta(B^{n+1}(x)) = \theta(B(B^n(x))) \geq r\theta(B^n(x)) \geq r\theta(r^n x) = r^{n+1}\theta(x).$$

We combine (14.44) and (14.46) and obtain that $\theta(x_n) \ge (1-\epsilon)^n r^n \theta(x) = s^n\theta(x)$.

Since $s > 1$ by (14.40) and $\theta(x) > 0$ by (14.42), this implies that $\theta(x_n) \to \infty$. Since we have established above that $(\theta(x_n))$ is bounded, we have arrived at a contradiction. □

The following result implies that the origin is unstable.

THEOREM 14.29. *Let $F, B : X_+ \to X_+$ and B be homogeneous and order-preserving and B a lower order-derivative of F at 0 (Section 14.2),*

(a) $F(\dot X_+) \subseteq \dot X_+$.

Further, assume that there is some bounded homogeneous $\theta : X_+ \to \mathbb{R}_+$ that is order-preserving and satisfies the following properties:

(b) *There exists some $r > 1$ such that $\theta(B(x)) \ge r\theta(x)$ for all $x \in X_+$.*

(c) *For any $x \in \dot X_+$, there exists some $n \in \mathbb{Z}_+$ such that $\theta(F^n(x)) > 0$.*

Then the semiflow induced by F is uniformly weakly norm-persistent: There exists some $\delta > 0$ such that $\limsup_{n\to\infty} \|F^n(x)\| \ge \delta$ for all $x \in \dot X_+$.

Since $F(0) = 0$, assumption (a) is necessary for uniform weak persistence. This theorem is used in the proof of Theorem 18.14 and Theorem 19.10.

PROOF. We apply Proposition 14.28 with $\rho(x) = \|x\|$.

It is sufficient to show assumption (a) of Proposition 14.28 is satisfied.

Suppose that there is $x \in X_+$ with $\rho(x) > 0$ and that the set $\{n \in \mathbb{N}; \theta(F^n(x)) > 0\}$ is finite.

Then there exists some $m \in \mathbb{N}$ such that $\theta(F^n(x)) = 0$ for all $n \in \mathbb{N}$, $n \ge m$. By assumption (a), $F^m(x) \in \dot X_+$. By assumption (c), there exists some $k \in \mathbb{N}$ such that $\theta(F^k(F^m(x))) > 0$. So $\theta(F^{m+k}(x)) > 0$, a contradiction.

All assumptions of Proposition 14.28 are satisfied and the assertion follows. □

The following result is different from Proposition 14.28 in so far as the norm is replaced by ρ.

THEOREM 14.30. *Let $F, B : X_+ \to X_+$ and B be homogeneous and order-preserving and a lower order-derivative of F at 0. Let $\theta : X_+ \to \mathbb{R}_+$ be homogeneous, bounded, and order-preserving, and $\rho : X_+ \to \mathbb{R}_+$. Let the following properties be satisfied.*

(a) *If $x \in X_+$ and $\rho(x) > 0$, then $\theta(F^n(x)) > 0$ for infinitely many $n \in \mathbb{N}$.*

(b) *There exists some $r > 1$ such that $\theta(B(x)) \ge r\theta(x)$ for all $x \in X_+$.*

(c) *For any $\epsilon > 0$ there exists some $\delta > 0$ such that*

$$\sup_{n\in\mathbb{N}} \rho(F^n(x)) \le \delta \implies \limsup_{n\to\infty} \|F^n(x)\| \le \epsilon.$$

Then the semiflow induced by F is uniformly weakly ρ-persistent: There is some $\delta > 0$ such that $\limsup_{n\to\infty} \rho(F^n(x)) \ge \delta$ for all $x \in X_+$ with $\rho(x) > 0$.

PROOF. By assumptions (a) and (b), all assumptions of Proposition 14.28 are satisfied. So, there exists some $\epsilon > 0$ such that $\limsup_{n\to\infty} \|F^n(x)\| \ge \epsilon$ for all $x \in X_+$ with $\rho(x) > 0$.

According to (c), choose some $\delta > 0$ such that $\limsup_{n\to\infty} \|F^n(x)\| \le \epsilon/2$ if $x \in X_+$ and $\sup_{n\in\mathbb{N}} \rho(F^n(x)) \le \delta$.

Suppose that $x \in X_+$, $\rho(x) > 0$ and $\limsup_{n\to\infty} \rho(F^n(x)) < \delta$. Then there exists some $m \in \mathbb{N}$ such that

$$\delta > \sup_{n \ge m} \rho(F^n(x)) = \sup_{n\in\mathbb{Z}_+} \rho(F^n(F^m(x)).$$

So

$$\epsilon/2 \ge \limsup_{n\to\infty} \|F^n(F^m(x))\| = \limsup_{n\to\infty} \|F^n(x)\|,$$

a contradiction. □

The following is an example for condition (c) in Theorem 14.30 to hold.

REMARK 14.31. Let X_+ be normal and $w \in X_+$ such that $F(x) \le \rho(x)w + A(x)$ for all $x \in X_+$ with $A : X_+ \to X_+$ homogeneous, subadditive and $\sum_{j=0}^\infty |A^j(w)\| < \infty$ and $A^n(x) \to 0$ as $n \to \infty$ for all $x \in X_+$.

Then there exists some $c > 0$ such that

$$\limsup_{n\to\infty} \|F^n(x)\| \le c \sum_{j=0}^\infty \|A^j(w)\| \limsup_{n\to\infty} \rho(F^n(x)).$$

PROOF. Let $x \in X_+$ and $x_n = F^n(x)$ for all $n \in \mathbb{Z}_+$. By induction, since A is subadditive and homogeneous,

$$x_n \le \sum_{j=0}^{n-1} \rho(x_{n-1-j}) A^j(w) + A^n(x), \quad n \in \mathbb{N}.$$

Since X_+ is normal, there exists some $c > 0$ such that

$$\|x_n\| \le c \sum_{j=0}^{n-1} \rho(x_{n-1-j}) \|A^j(w)\| + \|A^n(x)\|.$$

Now apply Fatou's lemma with the counting measure. □

PROOF OF THEOREM 14.27. We apply Theorem 14.29. By Theorem 11.14, there exists some homogeneous, order-preserving bounded eigenfunctional $\theta : X_+ \to \mathbb{R}_+$ such that $\theta(B(x)) = \mathcal{T}_0\, \theta(x)$ for all $x \in X_+$. So assumption (b) of Theorem 14.29 is satisfied.

Since a power of B is uniformly u-bounded, $\theta(u) > 0$. Since B is pointwise u-positive, $\theta(x) > 0$ for all $x \in \dot{X}_+$. See Remark 11.4. By assumption (a), $\theta(B^n(x)) > 0$ for all $n \in \mathbb{N}$ and $x \in \dot{X}_+$. So assumption (c) of Theorem 14.29 is satisfied.

All assumptions of Theorem 14.29 are satisfied, and it implies our assertion. □

14.9.2. Uniform persistence. There are many assumptions that lead from uniform weak to uniform population persistence. They can be obtained by combining the previous results in various ways guided by [**187**, Sec.4.1,4.2]. We choose one set of assumptions that may be stronger than needed but can be stated as succinctly as possible.

THEOREM 14.32. *Let X_+ be a serially complete normal closed cone and $u \in \dot{X}_+$. Let $F, B : X_+ \to X_+$ and let B be bounded homogeneous and order-preserving and let B a lower order-derivative of F at 0 (Section 14.2). Assume $\mathcal{T}_0 = \mathbf{r}(B) > 1$.*

Let the following properties be satisfied.

(a) *$F(0) = 0$ and $F(\dot{X}_+) \subseteq \dot{X}_+$.*
(b) *A power of B is uniformly u-bounded (Definition 6.1).*

(c) *B is pointwise u-positive (Definition* 11.3*).*

(d) *F is point-dissipative and asymptotically smooth (Section* 14.6*).*

Then the population is uniformly persistent: There exists some $\epsilon > 0$ such that $\liminf_{n\to\infty} \|F^n(x)\| \ge \epsilon$ for all $x \in \dot{X}_+$.

The proof combines Theorem 14.29 with [**187**, Thm.4.5].

PROOF. By Theorem 14.29, the population is uniformly weak persistence; in the language of [**187**, Thm.4.5] uniformly ρ-persistent with $\rho = \|\cdot\|$.

We apply [**187**, Thm.4.5].

Since F (more precisely, the semiflow induced by F) is asymptotically smooth and point-dissipative, there exists a compact attractor of points [**187**, Thm.2.28]: There exists a nonempty compact subset Y of X_+ such that $d(F^n(x), Y) \to 0$ as $n \to \infty$ for all $x \in X_+$. Here $d(y, Y)$ is the distance from y to Y. This is condition $\tilde{\clubsuit}_2$ in [**187**, Thm.2.28]. Condition $\tilde{\clubsuit}_1$ follows from $F(0) = 0$ and $F(\dot{X}_+) \subseteq \dot{X}_+$.

By [**187**, Thm.4.5], the semiflow induced by F is uniformly ρ-persistent with $\rho = \|\cdot\|$, and the assertion has been proved. □

CHAPTER 15

Unstructured population models

Many populations have two sexes, female and male, and mating between females and males is required for reproduction. The two sexes may have different survival rates, and the sex-ratio of neonates (newborn individuals) may be not one-to-one. Our first model will have no other structure in addition to the sexual one. Intraspecific competition affects birth rates and mortality rates.

15.1. A very basic two-sex population model

In discrete-time, a simple two-sex population model can be formulated as follows [**202**],

$$x_n = F(x_{n-1}), \quad x_n = (f_n, m_n), \qquad n \in \mathbb{N}, \tag{15.1}$$

with the yearly population turnover operator $F : \mathbb{R}^2_+ \to \mathbb{R}^2_+ = [0,\infty)^2$,

$$F(x) = (F_1(x), F_2(x)), \qquad x \in \mathbb{R}^2_+. \tag{15.2}$$

Here, f_n and m_n are the numbers of female and male individuals in the nth year, and

$$\left.\begin{aligned} F_1(x) =&\ p_1(x)\, f + \beta_1(x)\,\varphi(x) \\ F_2(x) =&\ p_2(x)\, m + \beta_2(x)\,\varphi(x) \end{aligned}\right\} \quad x = (f,m) \in \mathbb{R}^2_+. \tag{15.3}$$

The numbers $p_1(x), p_2(x) \in [0,1]$ are the one-year survival probabilities for females and males if $x \in \mathbb{R}^2_+$ is the distribution of sexes in that year, and $\beta_1(x), \beta_2(x) > 0$ are the amounts of surviving female and male offspring per pair and per year, respectively. The functions p_j and β_j are continuous from $\mathbb{R}^2_+ \to \mathbb{R}_+$.

We also assume the following for $j = 1,2$:

$$p_j(0,0) = 0 \implies p_j(x) = 0 \quad \text{for all } x \in \mathbb{R}^2_+. \tag{15.4}$$

This means that if females and males do not survive without competition, then they also do not survive with competition. We allow $p_j(0,0) = 0$ in order to include the case that reproduction occurs only once in a lifetime and is soon followed by death (annual plants, semelparous animals). If $p_j(0,0) > 0$ for $j = 1,2$, we have an iteroparous population where individuals reproduce several times during their lifetime.

$\varphi : \mathbb{R}^2_+ \to \mathbb{R}_+$ is a *mating* or *pair formation* function as introduced in Section 4.5: φ is homogeneous, order-preserving, and continuous,

$$\varphi(0,1) = 0 = \varphi(1,0), \qquad \varphi(1,1) > 0. \tag{15.5}$$

(15.5), together with homogeneity, rules out that φ is convex.

Two-sex models with spatial or rank structure are considered in [**127, 129, 130, 207**] and in Chapters 17 and 18. But the gist of the mathematical approach can be illustrated by handling this simple model without having to cope with many of the technicalities.

For certain forms of φ, (15.1), (15.3) can be written as a matrix population model [**36**, Ch.10] [**37**, Ch.17], but for general homogeneous φ we prefer not to do that.

The presentation is a more detailed version of the one in [**202**].

15.1.1. The first order approximation at origin. Since the homogeneous function φ is not differentiable at $(0,0)$ (unless it is linear), the first order approximation of the nonlinear population turnover operator F at the origin is not a linear positive operator but the homogeneous order-preserving operator $B : \mathbb{R}^2_+ \to \mathbb{R}^2_+$ given by

$$\begin{aligned} B(x) &= (B_1(x), B_2(x)), \qquad x = (f, m) \in \mathbb{R}^2_+, \\ B_1(f,m) &= q_1\, f + \alpha_1\, \varphi(f,m), \\ B_2(f,m) &= q_2\, m + \alpha_2\, \varphi(f,m), \end{aligned} \tag{15.6}$$

with

$$q_j = p_j(0,0), \quad \alpha_j = \beta_j(0,0), \quad j = 1,2. \tag{15.7}$$

The numbers q_j are the probabilities of surviving at least for one year without resource competition and α_j are the amounts of surviving female and male offspring per pair and per year without resource competition (there is still the competition for sexual partners built into the mating function φ).

It follows from the continuity of β_j and p_j from (15.4) that B is the *order-derivative* of F in the sense of Section 14.2.

The operator B can be interpreted as the *basic (next year population) turnover operator* and its spectral radius $\mathcal{T}_0 = \mathbf{r}(B)$ as the *basic turnover number.*

Since $\varphi(1,0) = 0 = \varphi(0,1)$, B has the eigenvectors $(1,0)$ and $(0,1)$ associated with the eigenvalues q_1 and q_2, respectively.

We notice that B is an order-preserving compact, continuous, homogeneous operator on the order cone $\mathbb{R}^2_+$ of the ordered vector space $\mathbb{R}^2_+$ which we endow with any of the usual norms. The order induced by $\mathbb{R}^2_+$ is the coordinate-wise order, $(f_1, m_1) \le (f_2, m_2)$ if and only if $f_1 \le f_2$ and $m_1 \le m_2$.

With respect to this order,

$$\begin{aligned} B(\alpha_1, \alpha_2) &= \big(q_1\alpha_1 + \alpha_1\varphi(\alpha_1,\alpha_2), q_2\alpha_2 + \alpha_2 q_1 + \varphi(\alpha_1,\alpha_2)\big) \\ &\ge \big[\min\{q_1,q_2\} + \varphi(\alpha_1,\alpha_2)\big](\alpha_1,\alpha_2). \end{aligned}$$

By (5.30) and Theorem 5.31,

$$\mathcal{T}_0 = \mathbf{r}(B) \ge cw(B) \ge \max\big\{q_1, q_2, \min\{q_1,q_2\} + \varphi(\alpha_1,\alpha_2)\big\}.$$

Further,

$$B(\alpha_1,\alpha_2) \le \big[\max\{q_1,q_2\} + \phi(\alpha_1,\alpha_2)\big](\alpha_1,\alpha_2).$$

Since we have assumed that $\alpha_j > 0$, $j = 1,2$, (α_1,α_2) is in the interior of the cone $\mathbb{R}^2_+$ and a uniform order unit such that B is uniformly (α_1,α_2)-bounded. By Theorem 6.11,

$$\mathbf{r}(B) \le \max\{q_1,q_2\} + \varphi(\alpha_1,\alpha_2).$$

If $q_1 = q_2$, we can combine the two inequalities for $\mathbf{r}(B)$ into an equality.

THEOREM 15.1. *If* $q_1 = q_2$, $\mathcal{T}_0 = \mathbf{r}(B) = q_2 + \phi(\alpha_1, \alpha_2)$.

Anyway, since $\phi(\alpha_1, \alpha_2) > 0$, $\mathcal{T}_0 = \mathbf{r}(B) > 0$. By Theorem 9.10, with $B = K$, there is an eigenvector of B associated with $\mathcal{T}_0$.

15.1.2. The homogeneous eigenvalue problem. The eigenvalue problem $\lambda x = B(x)$ leads to the eigensystem

$$\lambda f = q_1 f + \alpha_1 \varphi(f, m), \qquad \lambda m = q_2 m + \alpha_2 \varphi(f, m). \tag{15.8}$$

This eigensystem has some similarity to the one in [**109**, Sec.5]. We will proceed in a similar way, except that we are rather interested in $\lambda > 0$ than $\lambda \in \mathbb{R}$. After all, our eigensystem comes from a discrete-time rather than a continuous-time model.

As mentioned before, since $\varphi(0,1) = 0 = \varphi(1,0)$, (15.8) always has the eigenvectors $(1,0)$ and $(0,1)$ associated with the eigenvalues $\lambda = q_1$ and $\lambda = q_2$.

We are now looking for a third (!!!) eigenvalue (with strictly positive eigenvector). Since φ is homogeneous, we can scale the eigenvector (f, m) in (15.8) such that

$$\varphi(f, m) = 1, \quad \frac{\alpha_1}{\lambda - q_1} = f, \quad \frac{\alpha_2}{\lambda - q_2} = m. \tag{15.9}$$

Since one may find the existence of a third eigenvalue to a two by two system surprising at the first glance, we mention that homogeneous operators on finite-dimensional cones can have infinitely many eigenvalues [**145**, Thm.5.2.4, Example 5.2.5].

The characteristic equation. We substitute the second and the third equation in (15.9) into the first equation and obtain the equivalent *characteristic equation*

$$\varphi\Big(\frac{\alpha_1}{\lambda - q_1}, \frac{\alpha_2}{\lambda - q_2}\Big) = 1, \qquad \lambda > \tilde{q} = \max\{q_1, q_2\}. \tag{15.10}$$

This equation has a solution $\lambda > \tilde{q}$ if and only if

$$\mathcal{R}^\natural := \lim_{\lambda \to \tilde{q}} \varphi\Big(\frac{\alpha_1}{\lambda - q_1}, \frac{\alpha_2}{\lambda - q_2}\Big) > 1; \tag{15.11}$$

the existence follows from the intermediate value theorem. The solution is unique because $\varphi(x_1, x_2) < \varphi(y_1, y_2)$ if $0 \le x_1 < y_1$ and $0 \le x_2 < y_2$.

If $q_1 = q_2$, the existence condition is satisfied anyway, $\mathcal{R}^\natural = \infty$. In this case $\lambda = q_1 + \varphi(\alpha_1, \alpha_2)$. See Theorem 15.1.

Since φ is homogeneous, the existence condition takes the form,

$$\frac{\alpha_2}{q_1 - q_2} \varphi(\infty, 1) > 1 \quad \text{if } q_1 > q_2; \qquad \frac{\alpha_1}{q_2 - q_1} \varphi(1, \infty) > 1 \quad \text{if } q_2 > q_1.$$

This can be rewritten as

$$\alpha_2 \varphi(\infty, 1) > q_1 - q_2 \quad \text{and} \quad \alpha_1 \varphi(1, \infty) > q_2 - q_1. \tag{15.12}$$

Once λ has been determined (if it exists), we obtain the eigenvector from (15.9).

THEOREM 15.2. *If* (15.11) *holds, equivalently* (15.12), *then* $\mathcal{T}_0 = \mathbf{r}(B)$ *is the unique* $\lambda > \max\{q_1, q_2\}$ *that satisfies the characteristic equation* (15.10), *and there exists some* $v \in (0, \infty)^2$ *such that* $B(v) = \mathcal{T}_0 v$.

If (15.11) *does not hold, such a* λ *does not exist and* $\mathcal{T}_0 = \mathbf{r}(B) = \max\{q_1, q_2\}$.

PROOF. Since $B(v_1) = q_1 v_1$ and $B(v_2) = q_2 v_2$ for $v_1 = (1,0)$ and $v_2 = (0,1)$, $\mathbf{r}(B) \geq \max\{q_1, q_2\} > 0$ by Proposition 5.2. Since B is compact, continuous, homogeneous and order-preserving, if $\mathbf{r}(B) > \max\{q_1, q_2\}$, there exists some $v \in \dot{\mathbb{R}}^2_+$ such that $B(v) = \mathbf{r}(B)v$ by Theorem 9.10. The statement now follows from the preceding considerations. □

To compare (15.11) to analogous conditions found in [**109**, Sec.5], we observe that the homogeneity of φ implies for any $x > 0$ that

$$\begin{aligned}\varphi(1,\infty) &= \lim_{t\to\infty} \varphi(1, tx) = \lim_{t\to\infty} t\varphi(1/t, x)\\ &= \lim_{s\to 0+} \frac{\varphi(s,x) - \varphi(0,x)}{s} = \partial_1\varphi(0,x)\end{aligned}$$

and $\varphi(\infty, 1) = \partial_2\varphi(x, 0)$, where $\partial_j\varphi$ denotes the partial derivative of φ with respect to the jth coordinate.

We conjecture (and will prove in Theorem 15.4) that the extinction state $(0,0)$ is locally asymptotically stable if $\mathbf{r}(B) < 1$ and that the population persists if $\mathbf{r}(B) > 1$. To distinguish these two cases, we define the *basic sexual reproduction number*

$$\mathcal{R}_0 = \varphi\Big(\frac{\alpha_1}{1-q_1}, \frac{\alpha_2}{1-q_2}\Big). \tag{15.13}$$

Recall that $q_1 = p_1(0,0)$ is the probability that a female survives for at least one year if there is no competition for resources. Then the power q_1^n is the probability that a female survives for at least n years without resource competition. So, the fraction

$$\frac{1}{1-q_1} = \sum_{n=0}^{\infty} q_1^n$$

is the life expectation of females without resource competition. $\mathcal{R}_1 = \alpha_1/(1-q_1)$ would be the basic reproduction number of the female part of the population if females were to clone themselves and not reproduce sexually at all and α_1 were the amount of surviving daughters per year and per female. An analogous interpretation holds for males.

THEOREM 15.3. *Let* $\tilde{q} = \max\{q_1, q_2\} < 1$ *and* $\mathcal{T}_0 = \mathbf{r}(B)$. *Recall* (15.11),

$$\mathcal{R}^\natural = \lim_{\lambda\to\tilde{q}} \varphi\Big(\frac{\alpha_1}{\lambda - q_1}, \frac{\alpha_2}{\lambda - q_2}\Big).$$

Then $\mathcal{R}^\natural > \mathcal{R}_0$ *and exactly one of the following four possibilities occurs.*

(i) $\mathcal{R}_0 < \mathcal{R}^\natural \leq 1$ *and* $\mathcal{R}_0 \leq \mathcal{T}_0 = \tilde{q} < 1$,
(ii) $\mathcal{R}_0 < \mathcal{T}_0 < 1 < \mathcal{R}^\natural$ *and* $\mathcal{T}_0 > \tilde{q}$,
(iii) $\tilde{q} < \mathcal{R}_0 = \mathcal{T}_0 = 1 < \mathcal{R}^\natural$,
(iv) $\tilde{q} < 1 < \mathcal{T}_0 < \mathcal{R}_0 < \mathcal{R}^\natural$.

In the cases (ii), (iii), *and* (iv), *there exists some* $v \in \dot{\mathbb{R}}^2_+$ *such that* $Bv = \mathcal{T}_0 v$.

PROOF. (i) Assume $\mathcal{R}^\natural \leq 1$. Then $q_1 \neq q_2$. By symmetry, it is sufficient to consider $0 \leq q_2 < q_1 < 1$. Then $\mathcal{T}_0 = q_1$. Since ϕ is homogeneous,

$$1 \geq \lim_{\lambda\to q_1} \varphi\Big(\frac{\alpha_1}{\lambda - q_1}, \frac{\alpha_2}{\lambda - q_2}\Big) = \phi(\infty, 1)\frac{\alpha_2}{q_1 - q_2}.$$

Further,

$$\mathcal{R}_0 = \varphi\Big(\frac{\alpha_1}{1-q_1}, \frac{\alpha_2}{1-q_2}\Big) \le \phi(\infty, 1)\frac{\alpha_2}{1-q_2}.$$

Since $\phi(\infty, 1)\alpha_2 \le q_1 - q_2$, $\mathcal{R}_0 \le \frac{q_1-q_2}{1-q_2} \le q_1 = \mathcal{T}_0 \le 1$.

(ii)–(iv) If $\mathcal{R}^\natural > 1$, $\mathcal{T}_0 = \mathbf{r}(B)$ is the unique solution of (15.10), and since φ is homogeneous, by (15.13),

$$\varphi\Big(\frac{\alpha_1}{\mathcal{T}_0 - q_1}, \frac{\alpha_2}{\mathcal{T}_0 - q_2}\Big) = 1 = \varphi\Big(\frac{\alpha_1}{\mathcal{R}_0 - \mathcal{R}_0 q_1}, \frac{\alpha_2}{\mathcal{R}_0 - \mathcal{R}_0 q_2}\Big).$$

The claims now follow from the fact that $\varphi(x_1, x_2) < \varphi(y_1, y_2)$ if $0 \le x_1 < y_1$ and $0 \le x_2 < y_2$.

Existence of the eigenvector follows from Theorem 15.2. □

Alternatively, we could have applied Theorem 12.7.

15.1.3. Extinction and persistence for the fully nonlinear model. Recall the model (15.1), (15.3) and the assumptions that we have made at the beginning of this section.

THEOREM 15.4. *Let $q_1, q_2 \in [0, 1)$.*

(a) *If $\mathcal{R}_0 < 1$, the origin is locally asymptotically stable.*

(b) *If $\mathcal{R}_0 > 1$, the population persists in the following sense: There exists some $\epsilon > 0$ such that*

$$\limsup_{n\to\infty} \|(f_n, m_n)\|_\infty \ge \epsilon$$

for all solutions with $\varphi(f_0, m_0) > 0$.

Here $\|\cdot\|_\infty$ is the sum or the maximum norm on $\mathbb{R}^2$.

PROOF. If $\mathcal{R}_0 < 1$, it follows from Theorem 15.3 and 15.2 that $\mathcal{T}_0 = \mathbf{r}(B) < 1$. The local stability of the origin now follows from the general local stability result Theorem 14.13.

To show the persistence result, we apply Proposition 14.28.

Let $X_+ = \mathbb{R}^2_+$, the nonnegative quadrant. Since $\mathcal{R}_0 > 1$, by Theorem 15.3 and 15.2 there exists some $r > 1$ and $v \in (0, \infty)^2$ such that $B(v) = rv$. (Actually, $r = \mathcal{T}_0 = \mathbf{r}(B)$.) To match the notation in Proposition 14.28, let $\rho = \varphi$ and

$$\theta(x) = \min\{\frac{x_1}{v_1}, x_2 v_2\}, \qquad x = (x_1, x_2).$$

Then $\theta(B(x)) \ge r\theta(x)$ for all $x \in X_+$, which implies assumption (b) of Proposition 14.28.

Notice that $\theta(x) > 0$ if and only if $x_1 > 0$ and $x_2 > 0$ and that $\rho(x) > 0$ if and only if $x_1, x_2 > 0$. By our assumptions, $\rho(x) > 0$ implies $\theta(F(x)) > 0$ and then $\theta(F^n(x)) > 0$ for all $n \in \mathbb{N}$. So assumption (a) of Proposition 14.28 is satisfied. □

COROLLARY 15.5. *Also assume that the β_i are bounded on $\mathbb{R}^2_+$, $\sup p_i(\mathbb{R}^2_+) < 1$ for $i = 1, 2$, and $\mathcal{R}_0 > 1$.*

(a) *Then there exists some $\epsilon > 0$ such that*

$$\limsup_{n\to\infty} \varphi(f_n, m_n) \ge \epsilon$$

for all solutions with $\varphi(f_0, m_0) > 0$.

(b) *If also* $\beta_j(x)\varphi(x)/\|x\| \to 0$ *as* $\|x\| \to \infty$, *then there exists some* $\delta > 0$ *such that*

$$\liminf_{n\to\infty} \min\{f_n, m_n\} \geq \delta$$

for all solutions with $\varphi(f_0, m_0) > 0$.

PROOF. (a) Suppose the statement is false. Then, for all $\epsilon > 0$, there exists some solution (x_n) with $\varphi(x_0) > 0$ such that

$$\limsup_{n\to\infty} \varphi(x_n) < \epsilon.$$

After a shift in time, we can assume that $0 < \varphi(x_n) < \epsilon$ for all $n \in \mathbb{Z}_+$. Let $\|\cdot\|$ be the maximum norm on $\mathbb{R}^2$. Our assumptions imply that there is some $q \in (0,1)$ and some $c > 0$ such that

$$\|x_{n+1}\| \leq q\|x_n\| + c\varphi(x_n) \leq q\|x_n\| + c\epsilon, \qquad n \in \mathbb{N}.$$

By induction, $\|x_n\| \leq \frac{c\epsilon}{1-q}$. Since we find such a solution (x_n) for any $\epsilon > 0$, this contradicts Theorem 15.4 (b).

(b) We first show that the semiflow induced by F has a bounded attractor of bounded sets (which has compact closure in $\mathbb{R}^2_+$). To this end, we apply Theorem 14.16.

If $X = \mathbb{R}^2$ is endowed with the maximum norm, $X_+ = \mathbb{R}^2_+$ is a normal cone. Set $a_j = \sup p_j(\mathbb{R}^2_+)$. Then $F(x) \leq Ax + G(x)$ with

$$Ax = (a_1x_1, a_2x_2), \quad G(x) = \varphi(x)\big(\beta_1(x), \beta_2(x)\big), \qquad x = (x_1, x_2).$$

By our assumptions, $\|G(x)\|/\|x\| \to 0$ as $\|x\| \to \infty$ and $\mathbf{r}(A) < 1$. By Theorem 14.16, F has a compact attractor of bounded sets.

We now apply [**187**, Thm.4.21]. In the language of that theorem, the semiflow induced by F is uniformly weakly ρ-persistent with $\rho = \varphi$ by part (a). Set $\tilde{\rho}(x) = \min\{x_1, x_2\}$. Then $\tilde{\rho}$ is lower semicontinuous. Further, if $x \in X_+$ and $\varphi(x) > 0$, then $\tilde{\rho}(F(x)) > 0$. So the assumptions of [**187**, Thm.4.21] are satisfied and the statement in part (b) follows. □

For more general models, it may not be possible to derive a characteristic equation which only depends on the eigenvalue λ and not on the eigenvector [**36**, 10.3.3] [**37**, 17.3.4]. Rather, the analog of the eigensystem (15.8) may need to be solved for λ and f and m simultaneously.

15.2. A simple two-sex population model in hybrid time

We assume that mating and reproduction only occur once a year during a very short season. We let this season mark the transition from one year to the next so that the reproductive season it at the beginning of the year. During the year, female and male individuals are subject to mortality that depends on the densities of both sexes. Our aim is to construct a map $F : \mathbb{R}^2_+ \to \mathbb{R}^2_+$ on the pairs of nonnegative numbers that maps this year's numbers of females and males to next year's numbers of females and males and gives rise to a discrete semiflow $(F^n)_{n\in\mathbb{Z}_+}$ and a difference equation

$$x_n = F(x_{n-1}), \qquad n \in \mathbb{N}. \tag{15.14}$$

15.2.1. The construction of the yearly turnover operator. Let $u_1(t)$ and $u_2(t)$ represent the number of females and males at time $t \in [0,1]$ during a given year where 0 and 1 mark the beginning and the end of the year,

$$\begin{aligned} u_i' &= -\mu_i(t,u)u_i, \qquad 0 < t \le 1, \quad i = 1,2, \\ u &= (u_1, u_2), \\ u_i(0) &= \sigma_i y_i + \beta_i \varphi(y), \qquad y = (y_1, y_2). \end{aligned} \tag{15.15}$$

Here $\mu_i(t,u)$ are the per capita mortality rates of females/males at time t if $u = (u_1, u_2)$ gives the numbers of female and male competitors for resources.

σ_i is the probability of a female/male to survive the mating/reproduction season. Further $\varphi : \mathbb{R}_+^2 \to \mathbb{R}$ is the mating function, and β_i is the number of female/male offspring produced per mating. y_i is the number of females/males at the beginning of the year right before the mating season, and $u_i(0)$ is the number of females/males at the beginning of the year right after the mating/reproductive season.

Assume that $\mu : [0,1] \times \mathbb{R}_+^2 \to \mathbb{R}_+$ is continuous and satisfies a local Lipschitz condition: For any $c > 0$ there is some Lipschitz constant $\Lambda > 0$ such that, for $i = 1,2$ and all $t \in [0,1]$,

$$|\mu_i(t,u) - \mu_i(t,\tilde{u})| \le \Lambda \|u - \tilde{u}\|_\infty, \qquad u, \tilde{u} \in \mathbb{R}_+^2, \quad \|u\|_\infty + \|\tilde{u}\|_\infty \le c. \tag{15.16}$$

Here $\|\cdot\|_\infty$ is the maximum norm on $\mathbb{R}^2$.

It follows from standard ODE theory that, for any $y \in \mathbb{R}_+^2$, there is a unique solution u of (15.15) on $[0,1]$.

The yearly population turnover operator F is defined by

$$F(y) = u(1), \tag{15.17}$$

where $u(1) = (u_1(1), u_2(1))$ gives the numbers of females and males at the end of the year and u is the solution of (15.15).

15.2.2. The homogeneous first order approximation. We consider the solutions $\tilde{u}$ of the system (15.15) with $\mu_i(t,0,0)$ replacing $\mu_i(t,u(t))$,

$$\begin{aligned} \tilde{u}_i' &= -\mu_i(t,0,0)\tilde{u}_i, \qquad 0 < t \le 1, \quad i = 1,2, \\ \tilde{u}_i(0) &= \sigma_i y_i + \beta_1 \varphi(y), \qquad y = (y_1, y_2), \end{aligned} \tag{15.18}$$

and set

$$B(y) = \tilde{u}(1), \qquad \tilde{u} = (\tilde{u}_1, \tilde{u}_2). \tag{15.19}$$

This decoupled system can be explicitly solved,

$$\begin{aligned} &\left.\begin{aligned} B(y) &= (B_1(y), B_2(y)) \\ B_i(y) &= \tilde{q}_i \sigma_i y_i + \tilde{q}_i \beta_i \varphi(y) \end{aligned}\right\} \qquad y = (y_1, y_2), \\ &\tilde{q}_i = \exp\Big(- \int_0^1 \mu_i(t,0,0)dt \Big). \end{aligned} \tag{15.20}$$

Notice that B is of the same form as in (15.6) and that its spectral radius can be analyzed along the same lines as in Section 15.1.2 using the basic sexual reproduction number $\mathcal{R}_0$.

Because of the Lipschitz condition (15.16), the solution u of (15.15) depends continuously on its initial data and so on y. Even more, the definition $\Phi(t,y) = u(t)$, $t \in [0,1]$, $y \in \mathbb{R}_+^2$, defines a continuous map from $[0,1] \times \mathbb{R}_+^2$ to $\mathbb{R}_+^2$. Since $\varphi(0,0) = 0$,

$u \equiv (0,0)$ is a solution of (15.15) for $y = 0$ and so $\Phi(t,0,0) = (0,0)$ for all $t \in [0,1]$. The continuity of Φ implies that

$$\Phi(t,y) \to (0,0 \text{ uniformly for } t \in [0,1] \text{ as } y \to (0,0). \tag{15.21}$$

By (15.15),

$$u_i(t) = u_i(0) \exp\Big(- \int_0^t \mu_i(s,u(s))ds\Big).$$

For $i = 1,2$, by (15.17),

$$F_i(y) = B_i(y) \exp\Big(\int_0^1 [\mu_i(s,\Phi(s,y)) - \mu_i(s,0,0)]ds\Big).$$

Since $\int_0^1 \mu_i(s,\Phi(s,y))ds \to \int_0^1 \mu_i(s,0,0)ds$ as $y \to (0,0)$ by (15.21) and (15.16), B is the order-derivative of F.

The same proof as the one for Theorem 15.4 provides the following result.

THEOREM 15.6. *Let $\sigma_1, \sigma_2 \in [0,1]$ and $q_i = \sigma_i \tilde{q}_i < 1$ and $\alpha_i = \tilde{q}_i \beta_i(0,0)$. Set*

$$\mathcal{R}_0 = \varphi\Big(\frac{\alpha_1}{1-q_1}, \frac{\alpha_1}{1-q_1}\Big).$$

(a) *If $\mathcal{R}_0 < 1$, the origin is locally asymptotically stable.*

(b) *If $\mathcal{R}_0 > 1$, the population persists in the following sense: There exists some $\epsilon > 0$ such that*

$$\limsup_{n\to\infty} \|(f_n, m_n)\|_\infty \geq \epsilon$$

for all solutions $x_n = (f_n, m_n)$ of $x_n = F(x_{n-1})$ with $\varphi(f_0, m_0) > 0$.

Here $\|\cdot\|_\infty$ is the sum or the maximum norm on $\mathbb{R}^2$.

15.2.3. Point-dissipativity. Assume that there are continuous functions ν_1, $\nu_2 : \mathbb{R}^2_+ \to \mathbb{R}_+$ such that

$$\mu_i(t,y) \geq \nu_i(t,y_i), \qquad t \in [0,1], \quad y = (y_1,y_2) \in \mathbb{R}^2_+, \tag{15.22}$$

where

$$\int_0^1 \nu_i(t,\xi)dt \to \infty, \qquad \xi \to \infty. \tag{15.23}$$

Assume that the function ν_i satisfy a local Lipschitz condition: For any $c > 0$, there exists some Lipschitz constant $\Lambda > 0$ such that

$$|\nu_i(t,z) - \nu_i(t,\tilde{z})| \leq \Lambda |z - \tilde{z}|, \qquad z, \tilde{z} \in \mathbb{R}_+, \quad z + \tilde{z} \leq c. \tag{15.24}$$

Further assume that there exists some constant $c_\varphi > 0$ such that

$$\varphi(1,s) + \varphi(s,1) \leq c_\varphi, \qquad s \geq 0. \tag{15.25}$$

This is a plausible assumption: If there is only one female or one male, then the number of pairs being formed is bounded however many members of the other sex are present.

Assumption (15.25) and the other properties of φ imply that

$$\varphi(y) \leq c_\varphi \min\{y_1, y_2\}, \qquad y = (y_1,y_2) \in \mathbb{R}^2_+. \tag{15.26}$$

Recall that $F(y) = (F_1(y), F_2(y))$ where $F_i(y) = u_i(1)$ and u_1 and u_2 solve (15.15).

It follows from (15.22) and from standard comparison arguments for ODEs that

$$F_i(y_1, y_2) \le G_i(y_i), \qquad y_i \in \mathbb{R}_+, \quad i = 1, 2, \tag{15.27}$$

where $G_i(y_i) = U_i(1)$ and U_i is the solution of

$$\begin{aligned} U_i' &= -\nu_i(t, U_i)U_i, \\ U_i(0) &= y_i(\sigma_i + \beta_i c_\varphi). \end{aligned} \tag{15.28}$$

LEMMA 15.7. *G_i is point-dissipative.*

See Definition 14.20.

PROOF. We first show that G_i is weakly point-dissipative and then apply Theorem 14.21.

According to (15.23), choose some $c > 0$ such that

$$\exp\Big(\int_0^1 \nu_i(t, \xi)dt\Big) \ge p_i + \beta_i c_\varphi + 1, \qquad \xi \ge c.$$

Suppose there is some $y_i \ge 0$ such that $\liminf_{n\to\infty} G_i^n(y_i) > c$. After a shift in time, $G_i^n(y_i) > c$ for all $n \in \mathbb{N}$.

As an intermezzo, consider $y_i \in \mathbb{R}_+$ with $G_i(y_i) > c$. Notice that $U_i(t) \ge G_i(y_i) > c$ for all $t \in [0, 1]$ where U_i solves (15.28). So

$$U_i(1) = y_i(p_i + \beta_i c_\varphi) \exp\Big(-\int_0^1 \nu_i(t, U_i(t))dt\Big) \le y_i \frac{p_i + \beta_i c_\varphi}{p_i + \beta_i c_\varphi + 1}.$$

This implies $G_i(y_i) \le \zeta_i y_i$ with some $\zeta_i \in (0, 1)$.

By our intermezzo, $G_i^n(y_i) \le \zeta_i G_i^{n-1}(y_i)$ for all $n \in \mathbb{N}$ with ζ_i being independent of n. Thus $G_i^n(y_i) \to 0$ as $n \to \infty$, a contradiction.

This contradiction argument shows that $\liminf_{n\to\infty} G_i^n(y_i) \le c$ for all $y_i \in \mathbb{R}_+$.

In the language of Definition 14.20, G_i is weakly point-dissipative. Obviously, G_i maps bounded subset of $\mathbb{R}_+$ into bounded subsets of $\mathbb{R}_+$ and is thus compact. G_i is also continuous, because solutions of ODEs with locally Lipschitz continuous vector fields depend continuously on their initial data. By Theorem 14.21, G_i is point-dissipative. □

Since (15.28) involves a scalar ODE, $G_i : \mathbb{R}_+ \to \mathbb{R}_+$ is increasing. By induction, $G_i^n : \mathbb{R}_+ \to \mathbb{R}_+$ is increasing for all $n \in \mathbb{N}$ and $i = 1, 2$. Again by induction, for all $n \in \mathbb{N}$,

$$F^n(y_1, y_2) \le (G_1^n(y_1), G_2^n(y_2)), \qquad y_1, y_2 \in \mathbb{R}. \tag{15.29}$$

Indeed, if $n \in \mathbb{N}$ and (15.29) holds for n, then

$$F^{n+1}(y_1, y_2) = F^n(F(y_1, y_2)) \le \big(G_1^n(F_1(y_1, y_2)), G_2^n(F_2(y_1, y_2))\big).$$

Since G_1^n and G_2^n are increasing, by (15.27),

$$F^{n+1}(y_1, y_2) \le \big(G_1^n(G_1(y_1)), G_2^n(G_2(y_2))\big) = (G_1^{n+1}(y_1), G_2^{n+1}(y_2)),$$

and (15.29) holds for $n + 1$.

The inequality (15.29) implies that F inherits point-dissipativity from G_1 and G_2.

THEOREM 15.8. *Assume that* $\sigma_i \in [0,1]$ *and* $\mu_i(t,y) \ge \tilde{\mu}_i(t)$, $0 \le t \le 1$, $y \in \mathbb{R}^2_+$, *and* $\int_0^1 \tilde{\mu}_i(t)dt > 0$, $i = 1,2$, *and that* $\mathcal{R}_0 > 1$.

(a) *Then there exists some* $\epsilon > 0$ *such that*

$$\limsup_{n\to\infty} \varphi(f_n, m_n) \ge \epsilon$$

for all solutions (x_n), $x_n = (f_n, m_n)$, *of* $x_n = F(x_{n-1}$, $n \in \mathbb{N}$, *with* $\varphi(f_0, m_0) > 0$.

(b) *Assume* (15.22), (15.23), (15.23), *and* (15.25) *in addition.*
Then there exists some $\delta > 0$ *such that*

$$\liminf_{n\to\infty} \min\{f_n, m_n\} \ge \delta$$

for all solutions (x_n), $x_n = (f_n, m_n)$, *of* $x_n = F(x_{n-1}$, $n \in \mathbb{N}$, *with* $f_0 > 0$, $m_0 > 0$.

PROOF. (a) Suppose the statement is false. Then, for all $\epsilon > 0$, there exists some solution (x_n) with $\varphi(x_0) > 0$ such that

$$\limsup_{n\to\infty} \varphi(x_n) < \epsilon.$$

After a shift in time, we can assume that $0 < \varphi(x_n) < \epsilon$ for all $n \in \mathbb{Z}_+$. Let $\|\cdot\|$ be the maximum norm on $\mathbb{R}^2$. Our assumptions imply that there is some $q \in (0,1)$ and some $c > 0$ such that

$$\|x_{n+1}\| \le q\|x_n\| + c\varphi(x_n) \le q\|x_n\| + c\epsilon, \qquad n \in \mathbb{N}.$$

In fact, q can be chosen as the maximum of $\sigma_i \exp\Big(-\int_0^1 \tilde{\mu}_i(s)ds\Big)$, $i = 1, 2$.

By induction, $\|x_n\| \le \frac{c\epsilon}{1-q}$. Since we find such a solution (x_n) for any $\epsilon > 0$, this contradicts Theorem 15.4 (b).

(b) Recall that $\varphi(f,m) > 0$ if and only if $f > 0$ and $m > 0$.

We apply [**187**, Thm.4.21]. In the language of that theorem, the semiflow induced by F is uniformly weakly ρ-persistent with $\rho = \varphi$ by part (a). Set $\tilde{\rho}(x) = \min\{x_1, x_2\}$. Then $\tilde{\rho}$ is lower semicontinuous.

Since F is point-dissipative and all bounded closed subsets in $\mathbb{R}^2_+$ are compact, assumption (i) of [**187**, Thm.4.21] is satisfied with a compact subset C of $\mathbb{R}^2_+$.

Further, if $x \in X_+$ and $\varphi(x) > 0$, then $\tilde{\rho}(F(x)) > 0$. So assumption (ii) of [**187**, Thm.4.21] is satisfied as well, and the statement in part (b) follows. □

15.2.4. An alternative turnover operator. We now place the short reproductive season at the end rather than be beginning of the year.

Let $u_1(t)$ and $u_2(t)$ represent the number of females and males at time $t \in [0,1]$ during a given year where 0 and 1 mark the beginning and the end of the year,

$$\begin{aligned} u_i' &= -\,\mu_i(t,u)u_i, \qquad 0 < t \le 1, \quad i = 1,2, \\ u &= (u_1, u_2), \\ u(0) &= y \in \mathbb{R}^2_+. \end{aligned} \tag{15.30}$$

Here $\mu_i(t,u)$ are the per capita mortality rates of females/males at time t if $u = (u_1, u_2)$ gives the numbers of female and male competitors for resources. y_i is the number of females/males at the beginning of the year.

σ_i is the probability of a female/male to survive the mating/reproduction season. Further $\varphi : \mathbb{R}^2_+ \to \mathbb{R}$ is the mating function, and β_i is the number of female/male offspring produced per mating.

The yearly population turnover operator $\tilde{F}$ is defined by

$$
\begin{aligned}
\tilde{F}(y) &= (\tilde{F}_1(y), \tilde{F}_2(y)), \qquad && y \in \mathbb{R}^2_+, \\
\tilde{F}_i(y) &= \sigma_i u_i(1) + \beta_i \varphi(u(1)), \qquad && i = 1, 2,
\end{aligned}
\tag{15.31}
$$

where $u(1) = (u_1(1), u_2(1))$ gives the numbers of females and male towards the end of the year, right before the reproductive season, and u is the solution of (15.30). $\tilde{F}(y)$ gives the number of female and male individuals at the very end of the year, after the reproductive season.

Define $\Psi : \mathbb{R}^2_+ \to \mathbb{R}^2_+$ by

$$
\begin{aligned}
\Psi(y) &= (\Psi_1(y), \Psi_2(y)), \qquad && y = (y_1, y_2) \in \mathbb{R}^2_+, \\
\Psi_i(y) &= \sigma_i y_1 + \beta_i \varphi(y), \qquad && i = 1, 2.
\end{aligned}
$$

Then $\Psi \circ F = \tilde{F} \circ \Psi$ where F is the turnover operator constructed in Section 15.2.1. By induction, $\Psi \circ F^n = \tilde{F}^n \circ \Psi$ for all $n \in N$ such that the dynamics are the same as far as extinction and persistence are concerned. We have worked with F rather than $\tilde{F}$ because then the work in Section 15.1 can be used. We present the construction of $\tilde{F}$ because, when we add spatial spread to the model in Chapter 17, we will extend this construction rather than the one of F.

CHAPTER 16

A rank-structured population with mating

We consider a population in which each individual has a rank $j \in \mathbb{N}$. Within a year, an individual (provided that it survives) either moves to the next higher rank or stays at its rank.

Reproduction is assumed to be of Leslie type: All neonates (newborn individuals) have the lowest rank 1 [**36, 37, 50, 150**]. Individuals at rank $j > 1$ are at least $j-1$ years old. Reproduction is also assumed to involve mating which is influenced by the ranks of the mates. The population is iteroparous: The individuals can reproduced several times in their lives.

Similar models have been considered in [**199**, Sec.14], [**130**, Sec.9] and [**207**, Sec.6].

16.1. The model equations and well-posedness

A natural ordered normed vector space as population state space is the sequence space $X = \ell^1$ with the norm

$$\|x\| = \sum_{j=1}^{\infty} |x_j| < \infty, \qquad x = (x_j) \in \ell^1, \tag{16.1}$$

and the cone ℓ^1_+ of nonnegative sequences.

If x is a nonnegative sequence, x_j is the number of individuals at rank j and $\|x\|$ is the total number of individuals and x is the rank distribution of the population.

Given a rank distribution $x \in \ell^1_+$, let $p_j(x)$ denote the probability that an individual of rank j survives the year and moves to rank $j+1$ and $q_j(x)$ be the probability that it survives and stays at rank j,

$$q_j(x), p_j(x) \ge 0, \quad p_j(x) + q_j(x) < 1, \quad j \in \mathbb{N}. \tag{16.2}$$

If $q_j = 0$ for all $j \in \mathbb{N}$, the model becomes an age-structured population model. The assumption $p_j(x) + q_j(x) < 1$ means that at any rank some individuals die during the year.

Let $b(x)$ denote the population birth rate. In preparation for a turnover operator define a map $F : X_+ \to \mathbb{R}^{\mathbb{N}}_+$, $F(x) = (F_j(x))$, by

$$\left.\begin{aligned} F_1(x) = {} & q_1(x)x_1 + b(x) \\ F_j(x) = {} & p_{j-1}(x)x_{j-1} + q_j(x)x_j, \qquad j \ge 2 \end{aligned}\right\} \qquad x = (x_j) \in X_+. \tag{16.3}$$

Assume that $q_j, p_j : \ell^1_+ \to \mathbb{R}_+$ are continuous.

For all $x \in \ell^1_+$,

$$\sum_{j=1}^{\infty} F_j(x) = \sum_{j=1}^{\infty} (p_j(x) + q_j(x))x_n + b(x). \tag{16.4}$$

By (16.2).

$$\|F(x)\| \le \|x\| + b(x), \qquad x \in \ell^1_+,$$

and $F : \ell^1_+ \to \ell^1_+$.

LEMMA 16.1. *If $b : \ell^1_+ \to \mathbb{R}_+$ is continuous, then $F : \ell^1_+ \to \ell^1_+$ is continuous.*

PROOF. Let (x^n) be a sequence in ℓ^1_+, $x^n = (y^n_j)$, and $x = (x_j) \in \ell^1_+$, $\|x^n - x\| \to 0$ as $n \to \infty$. By the triangle inequality and (16.2), for all $n \in \mathbb{N}$,

$$\begin{aligned}\|F(x^n) - F(x)\| \le{}& \|x^n - x\| + |b(x^n) - b(x)| \\ &+ \sum_{j=1}^{\infty} \big(|q_j(x^n) - q_j(x)| + |p_j(x^n) - p_j(x)|\big)x_j.\end{aligned}$$

Since, for all $n \in \mathbb{N}$, $|q_j(x^n) - q_j(x)| + |p_j(x^n) - p_j(x)| \to 0$ as $j \to \infty$ and $|q_j(x^n) - q_j(x)| + |p_j(x^n) - p_j(x)| \le 2$ for all $j \in \mathbb{N}$, the last term converges to 0 as $n \to \infty$ by the dominated convergence theorem. □

We turn to the birth rate b. We assume that reproduction involves mating,

$$b(x) = \sum_{j,k=1}^{\infty} \beta_{jk}(x)\varphi(x_j, x_k), \qquad x \in \ell^1_+, \tag{16.5}$$

with $\beta_{jk}(x) \ge 0$ for all $j,k \in \mathbb{N}$. $\sum_{j,k=1}^{\infty}$ is to be understood as $\lim_{n\to\infty} \sum_{j,k=1}^{n}$.

This form of a mating function means that mating is assumed to be rank-selective. The numbers $\beta_{jk}(x)$ combine the probabilities that, at rank distribution x, individuals of rank j and k mate and the fertility of such a pair. The expression $\varphi(x_j, x_k)$ represents the number of pairs that result from x_j individuals at rank j and x_k individuals of rank k. For a two-sex population, one may like to assume a 1:1 sex ratio at each rank.

The mating function $\varphi : \mathbb{R}^2_+ \to \mathbb{R}_+$ is homogeneous and order-preserving, $\varphi(x, 0) = 0 = \varphi(0, x)$ for all $x \in \mathbb{R}_+$ and $\varphi(x, y) > 0$ if $x, y \in (0, \infty)$. See Section 4.5.

We assume that there is a Lipschitz constant $\Lambda > 0$ such that

$$\left.\begin{aligned}|\varphi(z_1, z_3) - \varphi(z_2, z_3)| \le{}& \Lambda|z_1 - z_2| \\ |\varphi(z_1, z_2) - \varphi(z_1, z_3)| \le{}& \Lambda|z_2 - z_3|\end{aligned}\right\} \qquad z_1, z_2, z_3 \ge 0. \tag{16.6}$$

By the triangle inequality, for all $x_j, x_k, y_j, y_k \in \mathbb{R}_+$,

$$|\varphi(x_j, x_k) - \varphi(y_j, y_k)| \le \Lambda(|x_j - y_j| + |x_k - y_k|). \tag{16.7}$$

ASSUMPTION 16.2. *We make the following assumptions:*

(a) *For all $j \in \mathbb{N}$ and $x \in \ell^1_+$, $0 \le p_j(x) \le 1$ and $0 \le q_j(x) \le 1$.*

(b) $\dfrac{\beta_{jk}(x)}{\beta_{jk}(0)} \to 1$ *as $x \to 0$ uniformly for those $j, k \in \mathbb{N}$ for which $\beta_{jk}(0) > 0$.*

(c) $\dfrac{p_j(x)}{p_j(0)} \to 1$ *as $x \to 0$ uniformly for those $j \in \mathbb{N}$ for which $p_j(0) > 0$.*

(d) $\dfrac{q_j(x)}{q_j(0)} \to 1$ *as $x \to 0$ uniformly for those $j \in \mathbb{N}$ for which $q_j(0) > 0$.*

(e) *$p_j, q_j : X_+ \to \mathbb{R}_+$, $\beta_{jk} : X_+ \to \mathbb{R}_+$ are continuous for all $j, k \in \mathbb{N}$.*

(f) *There exist numbers* $M_{jk} \in \mathbb{R}_+$, $j,k \in \mathbb{N}$, *and* $M \in \mathbb{R}_+$ *such that*

$$\beta_{jk}(x) \le M_{jk}, \qquad j,k \in \mathbb{N}, \quad x \in \ell^1_+,$$
$$\sum_{k=1}^{\infty}(M_{jk} + M_{kj}) \le M, \qquad j \in \mathbb{N}.$$

(g) *For any* $c > 0$ *there exists* $u \in \ell^1$ *such that* $p_{j-1}(x) + q_j(x) \le u_j$ *for* $j \ge 2$ *and all* $x \in X_+$ *with* $\|x\| \le c$.

(h) $p_j(x) + q_j(x) \to 0$ *uniformly for* $j \in \mathbb{N}$ *as* $\|x\| \to \infty$, *and*

$$\sup_{j\in\mathbb{N}} \sum_{k=1}^{\infty}(\beta_{jk}(x) + \beta_{kj}(x)) \to 0, \qquad \|x\| \to \infty.$$

To see that $b(x)$ is well-defined, for $n \in \mathbb{N}$, set

$$b_n(x) = \sum_{j=1}^{n}\sum_{k=1}^{n} \beta_{jk}(x)\phi(x_j, x_k), \qquad x \in \ell^1_+.$$

For each $x \in \ell^1_+$, $(b_n(x))$ is an increasing sequence. By Assumption 16.2 (e),

$$\begin{aligned} 0 \le b_n(x) &\le \sum_{j=1}^{n}\sum_{k=1}^{n} \beta_{jk}(x)(x_j + x_k) = \sum_{j=1}^{n}\sum_{k=1}^{n} \big(\beta_{jk}(x) + \beta_{kj}(x)\big)x_j \\ &\le \sum_{j=1}^{n}\sum_{k=1}^{\infty}(M_{jk} + M_{kj})x_j \le M\|x\|. \end{aligned} \tag{16.8}$$

Then $b_n(x) \le M\|x\|$ for all $n \in \mathbb{N}$ and $b(x) = \lim_{n\to\infty} b_n(x)$ exists, $b(x) \le M\|x\|$.

We next show that $b : \ell^1_+ \to \mathbb{R}$ is continuous.

Let $x \in \ell^1_+$ and (x^m) be a sequence in ℓ^1_+ with $\|x^m - x\| \to 0$ as $m \to \infty$.

Then, for all $m \in \mathbb{N}$,

$$\begin{aligned} |b(x^m) - b(x)| \le & \sum_{j,k=1}^{\infty} \big|\beta_{jk}(x^m) - \beta_{jk}(x)\big|\, \phi(x_j, x_k) \\ & + \sum_{j,k=1}^{\infty} \beta_{jk}(x^m)\, \big|\phi(x^m_j, x^m_k) - \phi(x_j, x_k)\big|. \end{aligned}$$

The first term after the inequality can be estimated by

$$\sum_{j,k=1}^{\infty} |\beta_{jk}(x^m) - \beta_{jk}(x)| \Lambda(x_j + x_k).$$

We will use the dominated convergence theorem to show that this expression converges to 0 as $m \to \infty$.

For each $j, k \in \mathbb{N}$, $|\beta_{jk}(x^m) - \beta_{jk}(x)| \to 0$ as $m \to \infty$ because β_{jk} is continuous.

For each $m \in \mathbb{N}$, $|\beta_{jk}(x^m) - \beta_{jk}(x)| \le 2M_{jk}$.

By Tonelli's theorem and Assumption 16.2 (e),

$$\sum_{j,k=1}^{\infty} 2M_{jk}\Lambda(x_j+x_k) = 2\sum_{j=1}^{\infty}\sum_{k=1}^{\infty}\Lambda x_j(M_{jk}+M_{kj})$$
$$= 2\sum_{j=1}^{\infty}|x_j|\sum_{k=1}^{\infty}\Lambda x_j(M_{jk}+M_{kj}) \leq 2\Lambda\|x\|M < \infty.$$

So the first term after the inequality converges to 0 by the dominated convergence theorem.

As for the second term after the inequality sign ,

$$\sum_{j,k=1}^{\infty} \beta_{jk}(x^m)\big|\phi(x_j^m,x_k^m)-\phi(x_j,x_k)\big| \leq \sum_{j,k=1}^{\infty}\beta_{jk}(x^m)\Lambda\big(|x_j^m-x_j|+|x_k^m-x_k|\big).$$

By Tonelli's theorem, this equals

$$\sum_{j=1}^{\infty}\Lambda|x_j^m-x_j|\Big(\sum_{k=1}^{\infty}(\beta_{jk}(x^m)+\beta_{kj}(x^m))\Big)$$

and, because of Assumption 16.2 (e) can be estimated by

$$\sum_{j=1}^{\infty}\Lambda|x_j^m-x_j|\Big(\sum_{k=1}^{\infty}(M_{jk}+M_{kj})\Big) \leq \Lambda\|x^m-x\|M,$$

which converges to 0 as $m\to\infty$.

This finishes the proof that $b:\ell_+^1\to\mathbb{R}_+$ is continuous and shows that $F:\ell_+^1\to\ell_+^1$ is continuous (Lemma 16.1).

Let $B:\ell_1^+\to\ell_+^1$ be the operator obtained from F by substituting $p_j(x)$, $q_j(x)$, and $\beta_{jk}(x)$ by $p_j(0)$, $q_j(0)$, and $\beta_{jk}(0)$ where 0 stands for the sequences consisting of zeroes. B is homogeneous and order-preserving. The similar (easier) proof as for F shows that B is continuous.

Assumption 16.2 (a)–(d) imply that for each $\epsilon\in(0,1)$ there exists some $\delta>0$ such that $(1-\epsilon)B(x)\leq F(x)\leq(1+\epsilon)B(x)$ for all $x\in X_+$ with $\|x\|\leq\delta$. In other words, B is the (upper and lower) order derivative of F (Section 14.2).

Assumptions 16.2 (f) makes $F,B:\ell_+^1\to\ell_+^1$ compact operators. See Definition A.43. In particular, F is asymptotically smooth. Assumption 16.2 (g) makes F point-dissipative (Section 14.6).

Indeed, by (16.4), for $x\in\dot{\ell}_+^1$,

$$\frac{\|F(x)\|}{\|x\|}\leq\sup_{j\in\mathbb{N}}(p_j(x)+q_j(x))+\frac{|b(x)|}{\|x\|}. \tag{16.9}$$

By (16.8),

$$\frac{|b_n(x)|}{\|x\|}\leq\sup_{j\in\mathbb{N}}\sum_{k=1}^{\infty}\big(\beta_{jk}(x)+\beta_{kj}(x)\big).$$

We take the limit as $n\to\infty$,

$$\frac{|b(x)|}{\|x\|}\leq\sup_{j\in\mathbb{N}}\sum_{k=1}^{\infty}\big(\beta_{jk}(x)+\beta_{kj}(x)\big)\to 0,\qquad \|x\|\to\infty,$$

with the convergence following from Assumption 16.2 (g) which now also applies that $\frac{\|F(x)\|}{\|x\|}\to 0$ as $\|x\|\to\infty$ via (16.9). Apply Theorem 14.16 with $\theta=\|\cdot\|$.

PROPOSITION 16.3. *Let* $\mathcal{T}_0 = \mathbf{r}(B) > 0$.

(a) *Then there exists some eigenvector* $v \in \ell^1_+$ *such that* $B(v) = \mathcal{T}_0 v$, $v_j > 0$ *for all* $j \in \mathbb{N}$.

(b) *For any* $r \in (0, \mathcal{T}_0)$, *there exists some* $m \in \mathbb{N}$ *and* $w \in \ell^1_+$ *such that* $B(w) \geq rw$, $w_j > 0$ *for* $j = 1, \dots, m$, $w_j = 0$ *for* $j > m$.

PROOF. (a) This part follows from Corollary 9.11.

(b) Let v be the eigenvector (eigensequence) from part (a).

$$\begin{aligned} \mathcal{T}_0 v_1 = &\quad B_1(v) = q_1 v_1 + \textstyle\sum_{j,k=1}^{\infty} \beta_{jk}(0)\varphi(v_j, v_k) \\ \mathcal{T}_0 v_j = &\quad B_j(v) = p_{j-1} v_{j-1} + q_j v_j, \qquad j \geq 2. \end{aligned}$$

Let $m \in \mathbb{N}$ and set $w_j = v_j$ for $j = 1, \dots, m$ and $w_j = 0$ for $j > m$. Then

$$\begin{aligned} B_1(w) = &\quad q_1 v_1 + \textstyle\sum_{j,k=1}^{m} \beta_{jk}(0)\varphi(v_j, v_k) \\ B_j(w) = &\quad p_{j-1} w_{j-1} + q_j w_j, \qquad j = 2, \dots, m, \\ B_j(w) \geq 0 = &\quad \mathcal{T}_0 w_j, \qquad j > m. \end{aligned} \tag{16.10}$$

By the monotone convergence theorem, as $m \to \infty$,

$$\begin{aligned} B_1(w) &= q_1 v_1 + \sum_{j,k=1}^{m} \beta_{jk}(0)\varphi(v_j, v_k) \\ &\to q_1 v_1 + \sum_{j,k=1}^{\infty} \beta_{jk}(0)\varphi(v_j, v_k) \\ &= B_1(v) = \mathcal{T}_0 v_1 > r w_1. \end{aligned}$$

By (16.10), if m is chosen sufficiently large, $B_1(w) \geq r w_1$ and $B(w) \geq rw$. □

16.2. A basic reproduction number

In view of our previous considerations, we define the basic turnover operator $B : \ell^1 \to \mathbb{R}^{\mathbb{N}}$, $B(x) = (B_j(x))$, by

$$\left.\begin{aligned} B_1(x) = &\quad b_0(x) + q_1 x_1 \\ B_j(x) = &\quad p_{j-1} x_{j-1} + q_j x_j, \qquad j \geq 2 \end{aligned}\right\} \qquad x = (x_j) \in X_+. \tag{16.11}$$

Here $b_0 : \ell^1_+ \to \mathbb{R}_+$ is homogeneous, order-preserving and continuous and describes the mating process between individuals of different ranks,

$$b_0(x) = \sum_{j,k=1}^{\infty} \beta_{jk} \varphi(x_j, x_k), \qquad x = (x_j) \in \ell^1_+, \tag{16.12}$$

with $\beta_{jk} \geq 0$ for all $j, k \in \mathbb{N}$. $\displaystyle\sum_{j,k=1}^{\infty}$ is to be understood as $\displaystyle\lim_{n \to \infty} \sum_{j,k=1}^{n}$.

For shorter notation, we have written β_{jk} for $\beta_{jk}(0)$, p_j for $p_j(0)$ and q_j for $q_j(0)$.

The mating function $\varphi : \mathbb{R}^2_+ \to \mathbb{R}_+$ is homogeneous and order-preserving, $\varphi(x, 0) = 0 = \varphi(0, x)$ for all $x \in \mathbb{R}_+$ and $\varphi(x, y) > 0$ if $x, y \in (0, \infty)$. We assume that there is a Lipschitz constant $\Lambda > 0$ such that

$$\left.\begin{aligned} |\varphi(z_1, z_3) - \varphi(z_2, z_3)| \leq &\quad \Lambda |z_1 - z_2| \\ |\varphi(z_1, z_2) - \varphi(z_1, z_3)| \leq &\quad \Lambda |z_2 - z_3| \end{aligned}\right\} \qquad z_1, z_2, z_3 \geq 0. \tag{16.13}$$

We impose the Assumptions 16.2. The same considerations as in the previous section provide that $b_0 : \ell^1_+ \to \mathbb{R}$ is Lipschitz continuous,

$$|b_0(x) - b_0(y)| \le 2\Lambda M \|x - y\|.$$

B is the additive perturbation of a rank one-operator, $B(x) = Ax + b_0(x)u$ with $A(x) = (A_j(x))$,

$$\text{(16.14)} \qquad \left.\begin{aligned} A_1(x) &= \quad q_1 x_1 \\ A_j(x) &= \quad p_{j-1}x_{j-1} + q_j x_j, \qquad j \ge 2 \end{aligned}\right\} \qquad x = (x_j) \in \ell^1_+,$$

the sequence $u = (u_j) \in \dot{\ell}^1_+$ with $u_1 = 1$ and $u_j = 0$ for $j \ge 2$ and

$$\text{(16.15)} \qquad b_0(x) = \sum_{j,k=1}^{\infty} \beta_{jk}\varphi(x_j, x_k), \qquad x = (x_j) \in \ell^1.$$

Since $\|A(x)\| = |q_1 x_1| + \sum_{j=2}^{\infty} |p_{j-1}x_{j-1} + q_j x_j| \le \sum_{j=1}^{\infty} |q_j + p_j|\, |x_j|$, A maps ℓ^1 into ℓ^1 and so does B. It is easy to see that A is a linear map. The previous inequality shows that A is a bounded linear map and that the operator norm of A satisfies $\|A\| \le \sup_{j\in\mathbb{N}}(q_j + p_j)$. So $\|A\| < 1$ if

$$\text{(16.16)} \qquad \sup_{j\in\mathbb{N}}(q_j + p_j) < 1,$$

which we assume from now on. This may be a brute-force assumption, but makes biological sense because $q_j + p_j$ is the probability at which an individual at rank j survives the year.

Recall the (left) resolvent of A (Chapter 8).

Since A is a bounded linear map on ℓ^1, the resolvent of A satisfies $R_\lambda = (\lambda - A)^{-1}$ and is the resolvent in the traditional sense, the inverse of $(\lambda - A)$.

To calculate $(\lambda - A)^{-1}u$, we solve the equation $\lambda x - Ax = u$, that is, the system

$$\lambda x_1 - q_1 x_1 = 1, \qquad \lambda x_j - q_j x_j = p_{j-1}x_{j-1}, \quad j \ge 2.$$

This gives the recursion

$$x_1 = \frac{1}{\lambda - q_1}, \qquad x_j = \frac{p_{j-1}}{\lambda - q_j}x_{j-1}, \quad j \ge 2.$$

It is solved by

$$(R_\lambda u)_j = x_j = \frac{1}{\lambda - q_j}\prod_{i=1}^{j-1}\frac{p_i}{\lambda - q_i}, \qquad j \in \mathbb{N}, \quad \lambda > \sup_{k\in\mathbb{N}}(p_k + q_k),$$

where $\prod_{i=1}^{0} := 1$. We conclude that the turnover/reproduction trichotomy of Theorem 12.7 holds with $\xi(\lambda) = b_0(R_\lambda u)$ and $Q = A$,

$$\text{(16.17)} \qquad \begin{aligned} \mathcal{R}_0 =& \xi(1) = \sum_{j,k=1}^{\infty} \beta_{jk}\varphi(P_j, P_k), \\ P_1 =& \frac{1}{1 + q_1}, \qquad P_j = \frac{1}{1 - q_j}\prod_{i=1}^{j-1}\frac{p_i}{1 - q_i}, \quad j \ge 2. \end{aligned}$$

THEOREM 16.4. *Let Assumption* 16.2 *hold and*

$$\text{(16.18)} \qquad \sup_{j\in\mathbb{N}}(q_j + p_j) < 1.$$

Then the following trichotomy holds for $\mathcal{T}_0 = \mathbf{r}(B)$ and $\mathcal{R}_0$ from (16.17).

(a) $1 < \mathcal{T}_0 \le \mathcal{R}_0$.
(b) $\mathcal{T}_0 = \mathcal{R}_0 = 1$.
(c) $0 < \mathcal{R}_0 \le \mathcal{T}_0 < 1$.

These results generalize reproduction/turnover trichotomies for Leslie age-structured models in several ways [**50, 150**]. Age-structure is replaced by rank-structure where yearly promotion to the next higher rank is not obligatory. There is no a priori maximum rank (or age). Mating is accounted for under the assumption of one-to-one sex ratios and same promotion probabilities for the two sexes but possibly sex-specific mating preferences as far as the rank of the mate is concerned.

16.3. The basic reproduction number as threshold parameter

Differently from the turnover number $\mathcal{T}_0 = \mathbf{r}(B)$, the basic reproduction number $\mathcal{R}_0$ in (16.17) gives us an explicit threshold number for population extinction and persistence in the rank-structured population model.

Recall that $\varphi : \mathbb{R}_+^2 \to \mathbb{R}_+$ is a Lipschitz continuous mating function, (16.6). In Section 16.1, we have established the following.

PROPOSITION 16.5. *Let Assumption 16.2 be satisfied. Then the operators $F, B : \ell_+^1 \to \ell_+^1$ are continuous and compact, F is point-dissipative, B is homogeneous and order-preserving and the order derivative of F.*

16.3.1. $\mathcal{R}_0 < 1$.

We have the following result from Theorems 14.13 and 16.4.

THEOREM 16.6. *Let Assumption 16.2 be satisfied. If $\mathcal{R}_0 < 1$ and $\sup_{i\in\mathbb{N}}(p_i(0)+q_i(0)) < 1$, then the origin is locally asymptotically stable in the sense of Theorem 14.13.*

THEOREM 16.7. *Let Assumption 16.2 be satisfied. Let $p_j(x) \le p_j(0), q_j(x) \le p_j(0)$ and $\beta_{jk}(x) \le \beta_{jk}(0)$ for all $j, k \in \mathbb{N}$ and all $x \in \ell_+^1$.*

If $\mathcal{R}_0 < 1$ and $\sup_{i\in\mathbb{N}}(p_i(0) + q_i(0)) < 1$, then the origin is globally asymptotically stable: $F^n(x) \to 0$ as $n \to \infty$ uniformly for $x \in \ell_+^1$.

PROOF. By our assumptions and (16.11) and (16.12), $F(x) \le B(x)$ for all $x \in \ell_+^1$. Apply Theorems 14.22 and 16.4. □

16.3.2. $\mathcal{R}_0 = 1$.

THEOREM 16.8. *Let Assumption 16.2 be satisfied. Let $p_j(x) \le p_j(0), q_j(x) \le p_j(0)$ and $\beta_{jk}(x) \le \beta_{jk}(0)$ for all $j, k \in \mathbb{N}$ and all $x \in \ell_+^1$.*

Further assume that, for any $\delta > 0$ there is some $\epsilon \in (0, 1)$ such that $p_j(x) \le \epsilon p_j(0), q_j(x) \le \epsilon p_j(0)$ and $\beta_{jk}(x) \le \epsilon\beta_{jk}(0)$ for all $j, k \in \mathbb{N}$ and all $x \in \ell_+^1$ with $\|x\| \ge \delta$.

Finally assume that $\mathcal{R}_0 = 1$ and $\sup_{i\in\mathbb{N}}(p_i(0) + q_i(0)) < 1$.

Then $\liminf_{n\to\infty} \|F^n(x)\| \to 0$ as $n \to \infty$ for all $x \in \ell_+^1$.

PROOF. By our assumptions and (16.11) and (16.12), $F(x) \le B(x)$ for all $x \in \ell_+^1$. Further, for any $\delta > 0$ there is some $\epsilon \in (0, 1)$ such that $F(x) \le \epsilon B(x)$ for all $x \in X_+$ with $\|x\| \ge \delta$.

Apply Theorems 16.4 and 14.23 (a). □

16.3.3. $\mathcal{R}_0 > 1$. The existence of a persistence equilibrium of the population can be shown without further assumptions.

THEOREM 16.9. *Let Assumption* 16.2 *be satisfied,* $\sup_{i\in\mathbb{N}}(p_i(0)+q_i(0)) < 1$*, and* $\mathcal{R}_0 > 1$ *with* $\mathcal{R}_0$ *from* (16.17)*. Then there exists a nonzero equilibrium population rank-distribution* $F(x) = x \in \dot{\ell}^1_+$*.*

PROOF. Apply Corollary 14.26. Since $\mathcal{R}_0 > 1$, $\mathcal{T}_0 = \mathbf{r}(B) > 1$ by Theorem 16.4. All other assumptions of Corollary 14.26 have been verified in Section 16.1. See Proposition 16.5. □

Establishing uniform weak persistence faces a problem known from age-structured populations, namely that the a population cannot survive if its members are beyond reproductive age. So assumptions must be imposed on the initial rank distribution that reproduction will occur if not right away so in the future.

ASSUMPTION 16.10. *There is some rank* $j \in \mathbb{N}$ *such that mating can occur between individuals that both have rank* j *or both have rank* $j+1$*,*

$$\beta_{jj}(x) > 0 \quad \text{and} \quad \beta_{j+1,j+1}(x) > 0, \qquad x \in \ell^1_+. \tag{16.19}$$

Further, we assume that, at any rank, some individuals move to the next one:

$$0 < p_i(x) \le 1, \qquad i \in \mathbb{N}, \quad x \in \ell^1_+. \tag{16.20}$$

THEOREM 16.11. *Let the Assumptions* 16.2 *and* 16.10 *be satisfied and* j *be the rank from* (16.19)*. Then, if* $\mathcal{R}_0 > 1$ *and* $\sup_{i\in\mathbb{N}}(p_i(0) + q_i(0)) < 1$*, there exists some* $\delta > 0$ *such that*

$$\limsup_{n\to\infty} \|F^n(x)\| \ge \delta$$

for all $x \in \ell^1_+$ *with* $\sum_{i=1}^j x_i > 0$*.*

PROOF. Since $\mathcal{R}_0 > 1$, $\mathcal{T}_0 = \mathbf{r}(B) > 1$ by Theorem 16.4.
We apply Proposition 14.28. We choose $\rho, \theta : X_+ \to \mathbb{R}_+$ as

$$\begin{aligned} \theta(x) =& [x]_w = \min\Big\{\frac{x_i}{w_i};\ i \in \mathbb{N},\ w_i > 0\Big\}, \\ \rho(x) =& \sum_{i=1}^j x_i, \qquad x = (x_i) \in X_+, \end{aligned} \tag{16.21}$$

with w from Proposition 16.3 (b). Then $\rho(0) = 0$ and ρ is continuous. Further, θ is bounded homogeneous and order-preserving. For any $x \in \ell^1_+$, by (16.21),

$$x \ge \theta(x) w.$$

Since B is order-preserving and homogeneous, by Proposition 16.3 (b),

$$B(x) \ge \theta(x) B(w) \ge r\theta(x) w, \qquad x \in \ell^1_+.$$

By (16.21),

$$\theta(B(x)) \ge r\theta(x).$$

and part (b) of Proposition 14.28 is satisfied.

Let $\rho(x) > 0$. Then $x_i > 0$ for some $i \in \{1, \dots, j\}$. Recall (16.3). If $i < j$, then $F(x)_{i+1} > 0$ because $p_i(x) > 0$ by (16.20). If $i = j$, then $F(x)_1 > 0$ because $\beta_{jj}(x) > 0$ by (16.19). In any case, $\rho(F(x)) > 0$.

By induction,

$$\rho(x) > 0 \implies \rho(F^m(x)) > 0, \quad m \in \mathbb{Z}_+. \tag{16.22}$$

To check part (a) of Proposition 14.30, let $z \in \ell^1_+$ and $\rho(z) > 0$.

It is sufficient to show that $\theta(F^n(z)) > 0$ for infinitely many $n \in \mathbb{N}$.

Suppose not. Then there exists some $N \in \mathbb{N}$ such that

$$\theta(F^n(z)) = 0, \qquad n \in \mathbb{N}, \quad n \geq N. \tag{16.23}$$

Set $x = F^N(z)$. By (16.22), $\rho(x) > 0$. By (16.21), $x_i > 0$ for some $i \in \{1, \dots, j\}$. By Assumption 16.10, for some sufficiently large n we have that $F^{n-2}(x)_j > 0$. Then $F^{n-1}(x)_1 > 0$ by (16.19) and $F^{n-1}(x)_{j+1} > 0$ by (16.20). So $F^n(x)_2 > 0$ by (16.20) and $F^n(x)_1 > 0$ by (16.19). For some even larger $n \in \mathbb{N}$, we have that $F^n(x)_k > 0$ for $k = 1, 2, 3$. Finally, for some sufficiently large $n \in \mathbb{N}$, we have $F^n(x)_k > 0$ for $k = 1, \dots j$. For this n, we then also have the $F^{n+i}(x)_k > 0$ for $k = 1, \dots, j + i$, $i \in \mathbb{N}$. Since the sequence w from Proposition 16.3 (b) has all but finitely many terms zero, we find some $i \in \mathbb{N}$ such that $\theta(F^{n+i}(x)) > 0$, with θ from (16.21). Since $x = F^n(z)$, $\theta(F^{N+n+i}(z)) > 0$. This contradicts (16.23) and part (a) of Proposition 14.28 has been verified.

Proposition 14.28 now implies the assertion. $\square$

CHAPTER 17

Two diffusing sexes and short reproductive season

We consider a model for a spatially distributed population of male and female individuals that mate and reproduce during a very short reproductive season. There is only one reproductive season in a year, and individuals start reproducing during the reproductive season that follows the one during which they have been born. Between the reproductive seasons, females and males move by diffusion on a bounded domain Ω under Dirichlet boundary conditions and are subject to density-dependent per capita mortality rates. Mating and reproduction are described by a (positively) homogeneous function (of degree one). For simplicity, we assume that per capita reproduction rates are not density-dependent. The model is a special case of a periodically impulsive reaction diffusion system, but similarly as in [**127**, **147**] we treat it as an abstract difference equation or a discrete semiflow on an infinite dimensional state space [**187**]. Differently from [**127**], individuals may reproduce several times during their life-times so that generations overlap (iteroparity). This requires a system of two difference equations, one for females and one for males, rather than one difference equation for offspring.

17.1. The model

We assume that mating and reproduction only occur once a year during a very short season. We let this season mark the transition from one year to the next and count the years in such a way that the reproductive season is a the end of the year. During the year, females and males move by diffusion on a bounded domain Ω in some Euclidean space $\mathbb{R}^N$, subject to Dirichlet boundary conditions.

Our aim is to construct a population turnover operator $F : X_+ \to X_+$, $X_+ = C_+(\bar{\Omega}) \times C_+(\bar{\Omega})$, on the pairs of nonnegative continuous functions that maps this year's spatial densities of females and males to next year's spatial densities of females and males and thus gives rise to a discrete semiflow $(F^n)_{n\in\mathbb{Z}_+}$ and a difference equation

$$(f_n, m_n) = F(f_{n-1}, m_{n-1}), \qquad n \in \mathbb{N}. \tag{17.1}$$

Here f_n and m_n are the spatial densities of females and males at the beginning of the nth year, right after the reproduction season.

$C(\bar{\Omega})$ denotes the Banach space of continuous functions on $\bar{\Omega}$ and $C_+(\bar{\Omega})$ the cone of nonnegative continuous functions on $\bar{\Omega}$. $X_+ = C_+(\bar{\Omega}) \times C_+(\bar{\Omega})$ is the cone of $X = C(\bar{\Omega}) \times C(\bar{\Omega})$.

17.1.1. The model for within-year dispersion. The dispersion within one year is modeled by a system of partial differential equations.

The density of females and males at location $x \in \Omega$ at the end of the reproductive season is denoted by $g_1(x) = f(x)$ and $g_2(x) = m(x)$, respectively, with continuous functions $f, m : \bar{\Omega} \to \mathbb{R}_+$.

Let $u_1(t,x)$ and $u_2(t,x)$ represent the densities of females and males at location $x \in \Omega$ at time $t \in [0,1]$, where the number 1 marks the end of the year,

$$
\begin{aligned}
(\partial_t - d_i \Delta) u_i &= -\mu_i\big(t, x, u(t,x)\big) u_i, \qquad 0 < t \le 1, \quad x \in \Omega, \quad i = 1,2, \\
u &= (u_1, u_2),
\end{aligned} \tag{17.2}
$$

with Dirichlet boundary conditions

$$
u_i(t,x) = 0, \qquad x \in \partial\Omega, \tag{17.3}
$$

and initial conditions

$$
u_i(0,x) = g_i(x), \qquad i = 1,2. \tag{17.4}
$$

Here Δ is the Laplace operator in x, d_i are diffusion constants of females and males respectively, μ_i are density dependent per capita mortality rates, $g_i(x)$ are the numbers of female/male individuals at x at the beginning of the year and $u_i(0,x)$ the respective numbers towards the end of the year right before the mating/reproductive season.

We also consider the linear partial differential equations (17.2), (17.3), and (17.4) with $\mu_i(t,x,u)$ being replaced by $\mu_i(t,x,0)$.

ASSUMPTION 17.1. *We assume that $\mu_i : [0,\infty) \times \Omega \times [0,\infty)^2 \to \mathbb{R}$ are nonnegative and differentiable and that the derivatives and μ_i themselves are bounded on all sets $[0,a] \times \Omega \times [0,a]^2$, $a > 0$. We also assume that the boundary $\partial\Omega$ is C^4.*

Under these assumptions, unique nonnegative mild solutions u_1, u_2 exist on $[0,1] \times \bar{\Omega}$ as can be seen by a standard fixed point argument, u_i continuous on $(0,1] \times \bar{\Omega}$ and $[0,1] \times \Omega$. These solutions are classical in the sense that they have partial derivatives $\partial_t u_i$ and first and second partial derivatives with respect to x on $(0,1] \times \Omega$, $u_i(x,t) = 0$ for $t \in (0,1]$, $x \in \partial\Omega$. They depend continuously on the initial data. This can be seen by a bootstrap argument similar to the one in the proof of [**186**, Thm.3.1].

17.1.2. The model for the mating/reproductiove season. If $h_1, h_2 \in C_+(\bar{\Omega})$ are the spatial densities of females and males immediately before the short mating and reproduction season, the densities of females and males immediately after that season are given by

$$
H_i(h_1,h_2)(x) = p_i(x) h_i(x) + \beta_i(x)\, \varphi\big(x, h_1(x), h_2(x)\big), \qquad i = 1,2, \quad x \in \bar{\Omega}. \tag{17.5}
$$

For $i = 1,2$, $p_i(x)$ is the probability of a female/male at x to survive the mating/reproductive season. $\beta_i(x)$ are the numbers of female/male offspring at x per mating.

ASSUMPTION 17.2. *The functions $p_i, \beta_i : \bar{\Omega} \to [0,1]$ are continuous. The β_i are not zero everywhere on $\bar{\Omega}$.*

Further, $\varphi : \Omega \times \mathbb{R}_+^2 \to \mathbb{R}_+$ is the mating and reproduction function. If there are u_1 females and u_2 males at location x in Ω, $\varphi(x, u_1, u_2)$ is the amount of matings occurring at x.

ASSUMPTION 17.3 (Properties of the mating and reproduction function).

(a) *φ is continuous on $\bar{\Omega} \times \mathbb{R}^2_+$,*

(b) *$\varphi(x, \cdot)$ is order preserving on $\mathbb{R}^2_+$ for each $x \in \Omega$,*

(c) *$\varphi(x, \cdot)$ is (positively) homogeneous for each $x \in \Omega$,*

$$\varphi(x, \alpha u) = \alpha \varphi(x, u), \qquad \alpha \geq 0, \quad x \in \Omega, \quad u \in \mathbb{R}^2_+,$$

(d) *For $i = 1, 2$, $\beta_i(x)\varphi(x, 1, 1) > 0$ for some $x \in \Omega$.*

For examples, see Section 4.5.

These assumptions make $H : X_+ \to X_+$ a continuous homogeneous order-preserving map.

17.1.3. The yearly population turnover operator. The yearly population turnover operator $F : X_+ \to X_+$, $X_+ = C_+(\bar{\Omega}) \times C_+(\bar{\Omega})$, is given by

$$F(g) = H(u(1, \cdot)), \qquad g \in X_+, \quad x \in \bar{\Omega}, \tag{17.6}$$

with H from (17.5) and $u = (u_1, u_2)$ being the solution of (17.2), (17.3), and (17.4). In other words,

$$F(g_1, g_2)(x) = \big(F_1(g_1, g_2)(x), F_2(g_1, g_2)(x)\big), \qquad g_i \in C_+(\bar{\Omega}), \quad x \in \bar{\Omega} \tag{17.7}$$

where, for $i = 1, 2$, $F_i(g_1, g_2)(x)$ gives the number of females and males at the very end of the year, right after the reproductive season,

$$F_i(g_1, g_2)(x) = p_i(x)u_i(1, x) + \beta_i(x)\varphi(x, u(1, x)), \tag{17.8}$$

where $u(t, x) = (u_1(t, x), u_2(t, x))$ and u_1 and u_2 are the solutions of (17.2), (17.3) and (17.4).

Since u depends continuously on the initial data and H is continuous, $F : X_+ \to X_+$ is continuous.

17.1.4. Main result. We introduce the map $B : X_+ \to X_+$,

$$X_+ = C_+(\bar{\Omega}) \times C_+(\bar{\Omega}),$$

as an approximation of F at zero, which means B is F with $\mu_i(t, x, u)$ being replaced by $\mu_i(t, x, 0)$ in (17.2),

$$B(g) = H(\tilde{u}(1, \cdot)), \qquad g = (g_1, g_2) \in X_+, \tag{17.9}$$

with H in (17.5) and $\tilde{u}(t, x) = (\tilde{u}_1(t, x), \tilde{u}_2(t, x))$ and $\tilde{u}_1$ and $\tilde{u}_2$ the solutions of the partial differential equations

$$(\partial_t - d_i\Delta)\tilde{u}_i = -\mu_i(t, x, 0)\tilde{u}_i, \qquad 0 \leq t \leq 1, \quad x \in \Omega,$$

with Dirichlet boundary conditions $\tilde{u}_i(t, x) = 0$, $x \in \partial\Omega$, and initial conditions

$$\tilde{u}_i(0, x) = g_i(x), \qquad x \in \bar{\Omega}.$$

B is the yearly population turnover operator under optimal conditions (when the per capita death rate is not increased by overcrowding effects) and is called the *basic turnover operator*.

We will show that the map B is homogeneous, order-preserving, continuous and compact. We can therefore define the spectral radius $\mathcal{T}_0 = \mathbf{r}(B)$ of B (see Section 5.1) which is the threshold parameter separating extinction from persistence. If no distinction were made between males and females, the map B would be linear and bounded. $\mathcal{T}_0 = \mathbf{r}(B)$ can be interpreted as the basic turnover number of the population.

In the following theorem, $\|\cdot\|_\infty$ denotes the supremum norm on $C(\bar{\Omega})$, $\|(f,m)\| = \max\{\|f\|_\infty, \|m\|_\infty\}$ for $f, m \in C(\bar{\Omega})$, and $w \in C^2(\bar{\Omega})$ denotes the unique solution of

$$-\Delta w = 1 \text{ on } \Omega, \qquad w = 0 \text{ on } \partial\Omega.$$

THEOREM 17.4. *There exists a threshold parameter $\mathcal{T}_0 = \mathbf{r}(B)$ that separates population extinction and persistence as follows:*

- *If $\mathcal{T}_0 < 1$, the extinction state 0 is locally asymptotically stable in the following sense: For all $\alpha \in (\mathcal{T}_0, 1)$ there exist $\delta_0 > 0$ and $M \geq 1$ such that $\|F^n(g)\| \leq M\alpha^n \|g\|$ for all $n \in \mathbb{N}$ and all $g \in X_+$ with $\|g\| \leq \delta_0$.*
- *If $\mathcal{T}_0 < 1$ and $\mu_i(t,x,u) \geq \mu_i(t,x,0)$ for all $t \geq 0, x \in \Omega, u \in \mathbb{R}^2_+$, then the population always dies out: $F^n(f) \to 0$ as $n \to \infty$, uniformly on Ω and uniformly for f in bounded subsets of X_+.*
- *If $\mathcal{T}_0 > 1$, the population persists uniformly weakly: There exists some $\epsilon > 0$ such that, for any $g = (g_1, g_2) \in X_+$, $g_1 \not\equiv 0$, $g_2 \not\equiv 0$,*

$$\limsup_{n\to\infty} \max_{i=1,2} \int_\Omega w(x) g_{n,i}(x)dx \geq \epsilon,$$

where $(g_{n,1}(x), g_{n,2}(x)) = F^n(g)(x)$ are the densities of females and males at the end of the nth year.

Weak persistence implies instability of the origin. It would be desirable to prove uniform persistence in which the limit superior is replaced by the limit inferior, but this will require additional assumptions on the per capita mortality rates μ (Section 17.5). Actually, under these additional assumptions, order permanence will be shown (Section 17.6).

As in the linear case, no closed formula is available in general for the important threshold parameter $\mathcal{T}_0$, but estimates and approximations can be found in a similar way. For instance, the following convergence theorem holds, which is proved at the end of Section 17.2.

THEOREM 17.5. *For all $f, m \in C_+(\bar{\Omega})$, $f \not\equiv 0$ and $m \not\equiv 0$,*

$$\mathcal{T}_0 = \mathbf{r}(B) = \lim_{n\to\infty} \|B^n(f,m)\|^{1/n} = \lim_{n\to\infty} \Big(\int_\Omega [f_n(x) + m_n(x)]dx \Big)^{1/n}$$

where $(f_n, m_n) = B^n(f,g)$.

Semiexplicit formulas can be found in terms of the principal eigenvalue of the Laplacian if the per capita mortality rates and the mating and reproduction function do not depend on the location $x \in \bar{\Omega}$ (Section 17.3). For alternative characterizations of $\mathcal{T}_0$ and its approximation by upper and lower Collatz–Wielandt numbers and bounds see [**8**, **198**, **199**] and Sections 5.5 and 6.2.

17.2. Proof of main result

Let $Z = C(\bar{\Omega})$ be the Banach space of continuous functions on $\bar{\Omega}$ with supremum norm $\|f\|_\infty = \sup_{x\in\bar{\Omega}} |f(x)|$ and $Z_+ = C_+(\bar{\Omega})$ be the cone of nonnegative functions. Z_+ is normal, generating and solid; actually, the supremum norm is monotone itself. We consider $X = Z^2$ which is an ordered Banach space with order cone $X_+ = Z^2_+$ and the norm $\|(f,m)\| = \max\{\|f\|_\infty, \|m\|_\infty\}$. X_+ inherits the properties of being normal, generating, solid and closed from Z_+.

The maximum principle implies the following.

LEMMA 17.6. *$\|u_i(t,\cdot)\|_\infty \le \|u_i(0,\cdot)\|_\infty$ for all $t \in [0,1]$, $i = 1,2$, for the solutions of (17.2).*

Part of the proof of the main result will be that the homogeneous map B is an order approximation of the next-year-offspring map F, given by (17.8), at the origin.

PROPOSITION 17.7. *B is compact, homogeneous, continuous, and order-preserving on X_+. For any $\eta \in (0,1)$, there exists some $\delta > 0$ such that*

$$(1-\eta)B(g) \le F(g) \le (1+\eta)B(g), \qquad g \in X_+, \quad \|g\| \le \delta.$$

If $\mu_i(t,x,0) \le \mu_i(t,x,u)$ for $i = 1,2$, $t \in [0,1]$, $x \in \Omega$, and $u \in \mathbb{R}^2_+$, then $F(g) \le B(g)$ for all $g \in X_+$.

PROOF. Recall that

$$B(g)(x) = (B_1(g)(x), B_2(g)(x)), \qquad g = (g_1, g_2) \in X_+.$$

Here, for $i = 1,2$,

$$B_i(g)(x) = p_i(x)\tilde{u}_i(1,x) + \beta_i(x)\varphi(x, \tilde{u}(1,x)),$$

where $\tilde{u} = (\tilde{u}_1, \tilde{u}_2)$ and $\tilde{u}_1$ and $\tilde{u}_2$ are the solutions of the partial differential equations

$$(\partial_t - d_i\Delta)\tilde{u}_i = -\mu_i(t,x,0)\tilde{u}_i, \qquad 0 \le t \le 1, \quad x \in \Omega, \tag{17.10}$$

with Dirichlet boundary conditions $\tilde{u}_i(t,x) = 0$, $x \in \partial\Omega$, and initial conditions

$$\tilde{u}_i(0,x) = g_i(x), \qquad x \in \bar{\Omega}. \tag{17.11}$$

It follows from the smoothing properties of parabolic differential equations that the map $g \mapsto \tilde{u}(1,\cdot)$ is compact from X to X. Since the solutions of (17.10) depend continuously on their initial data, this map is also continuous. It is linear because the equations are linear. By a standard comparison principle (see [**186**, Sec.7.3], e.g.), this map is also order-preserving. It follows from the properties of the mating and reproduction function φ (Section 17.1.3) that B is compact, homogeneous, continuous, and order-preserving on X_+.

We first prove the second assertion: We assume $\mu_i(t,x,0) \le \mu_i(t,x,u)$ for $u \in \mathbb{R}^2_+$. By a standard comparison principle (see [**186**, Sec.7.3], e.g.), we have $u_i \le \tilde{u}_i$ for the solutions u_i of (17.2) and (17.3) and so $F(g) \le B(g)$ for all $g \in X_+$.

We now check this: For any $\eta \in (0,1)$ there exists some $\delta > 0$ such that $F(g) \ge (1-\eta)B(g)$ for all $g \in X_+$ with $\|g\| \le \delta$.

Recall that $F(g)(x) = (F_1(g), F_2(g))$ for $g = (g_1, g_2) \in X_+$, where

$$F_i(g)(x) = p_i(x)u_i(1,x) + \beta_i(x)\varphi(x, u(1,x))$$

and $u = (u_1, u_2)$ and u_1 and u_2 are the solutions of

$$\begin{aligned}(\partial_t - d_i\Delta)u_i &= -\mu_i(t,x,u)u_i, \qquad 0 \le t \le 1, \quad x \in \Omega,\\ u &= (u_1, u_2),\end{aligned} \tag{17.12}$$

with Dirichlet boundary conditions $u_i(t,x) = 0$, $x \in \partial\Omega$, and initial conditions

$$u_i(0,x) = g_i(x), \qquad x \in \bar{\Omega}, \tag{17.13}$$

while the operator B is associated with the solutions of (17.10).

We also consider solutions of

$$(17.14)\qquad \begin{aligned} (\partial_t - d_i\Delta)\bar{u}_i &= 0, && 0 \le t \le 1, \quad x \in \Omega,\\ \bar{u}_i(t,x) &= 0, && x \in \partial\Omega,\\ \bar{u}_i(0,x) &= g_i(x), && x \in \bar{\Omega}. \end{aligned}$$

Since $\mu_i \ge 0$, it follows from a comparison principle [**186**, Sec.7.3] that

$$(17.15)\qquad \tilde{u}_i \le \bar{u}_i, \qquad u_i \le \bar{u}_i,$$

and, for any $\tilde{\delta} > 0$ and female/male density $g_i \in C_+(\bar{\Omega})$,

$$(17.16)\qquad g_i \le \tilde{\delta} \implies \bar{u}_i \le \tilde{\delta}.$$

Let $\eta \in (0,1)$ and $\epsilon > 0$ such that $e^{-\epsilon} = 1 - \eta$. Since μ_i is locally Lipschitz in u, by choosing $\tilde{\delta} \in (0,1)$ small enough,

$$\mu_i(t,x,u) \le \mu_i(t,x,0) + \epsilon, \qquad t \ge 0, \quad x \in \Omega, \quad u = (u_1,u_2), \quad 0 \le u_i \le \tilde{\delta}.$$

Set $u_i^\epsilon(t,x) = e^{\epsilon t}u(t,x)$. Since $u_i \le \bar{u}_i \le \tilde{\delta}$,

$$(\partial_t - d_i\Delta)u_i^\epsilon(t,x) = -(\mu_i(t,x,u) - \epsilon)u_i^\epsilon(t,x) \ge -\mu_i(t,x,0)u_i^\epsilon.$$

We recall (17.10), use a comparison principle again [**186**] and obtain,

$$u_i^\epsilon(t,x) \ge \tilde{u}_i(t,x), \qquad i = 1,2,$$

which implies that

$$u(t,x) \ge e^{-\epsilon t}\tilde{u}(t,x) \ge (1-\eta)\tilde{u}(t,x), \qquad 0 \le t \le 1, \quad x \in \Omega.$$

Since $\varphi(x,\cdot)$ is order-preserving on $\mathbb{R}_+^2$,

$$\begin{aligned} F_i(g)(x) &= p_i(x)u_i(1,x) + \beta_i(x)\varphi(x,u(1,x))\\ &\ge p_i(x)(1-\eta)\tilde{u}_i(1,x) + \beta_i(x)\varphi(x,(1-\eta)\tilde{u}(1,x)). \end{aligned}$$

Since $\varphi(x,\cdot)$ is also homogeneous,

$$F_i(g)(x) \ge (1-\eta)p_i(x)\tilde{u}_i(1,x) + (1-\eta)\beta_i(x)\varphi(x,\tilde{u}(1,x)) = (1-\eta)B_i(g)(x).$$

Similarly, by considering $e^{-\epsilon t}u(t,x)$ with $e^\epsilon = 1+\eta$, we find that $u(t,x) \le (1+\eta)\tilde{u}(t,x)$ and so $F(g) \le (1+\eta)B(g)$ if $g \in X_+$ and $\|g\| \le \delta$ with an appropriate $\delta > 0$. □

Let $w \in C^2(\bar{\Omega})$ be the solution of

$$(17.17)\qquad -\Delta w = 1 \text{ on } \Omega, \qquad w = 0 \text{ on } \partial\Omega.$$

By the maximum principle $w(x) > 0$ for all $x \in \Omega$.

Further let $G : (0,\infty) \times \bar{\Omega}^2 \to \mathbb{R}_+$ be the Green's function associated with the differential operator $\partial_t - \Delta$ and Dirichlet boundary conditions. The following result has been proved in [**194**, L.6.4].

LEMMA 17.8. *For every $t > 0$ there exists some $c_t > 1$ such that*

$$(1/c_t)w(x)w(y) \le G(t,x,y) \le c_t w(x)w(y), \qquad x,y \in \bar{\Omega}.$$

We define $w_0 \in C_+(\bar{\Omega})$ by

$$(17.18)\qquad w_0(x) = \varphi(x,1,1)w(x), \qquad x \in \bar{\Omega}.$$

By Assumption 17.3 (d), $w_0 \not\equiv 0$.

PROPOSITION 17.9. *There exists some $c > 0$ such that, for $i = 1,2$ and all $x \in \bar{\Omega}$ and $g = (g_1, g_2) \in X_+$,*

$$B_i(g)(x), F_i(g)(x)$$
$$\leq cw(x)\Big(p_i(x)\int_\Omega w(y)g_i(y)dy + \beta_i(x)\varphi(x,1,1)\max_{j=1,2}\int_\Omega w(y)g_j(y)dy\Big).$$

Further there exists some $\epsilon > 0$ such that, for $i = 1,2$ and all $x \in \bar{\Omega}$ and $g = (g_1, g_2) \in X_+$,

$$B_i(g)(x) \geq \epsilon w(x)\Big(p_i(x)\int_\Omega w(y)g_i(y)dy + \beta_i(x)\varphi(x,1,1)\min_{j=1,2}\int_\Omega w(y)g_j(y)dy\Big).$$

Finally, for any $g = (g_1, g_2) \in X_+$, there exists some $\epsilon_g > 0$ such that, for $i = 1,2$ and all $x \in \bar{\Omega}$,

$$F_i(g)(x) \geq \epsilon_g w(x)\Big(p_i(x)\int_\Omega w(y)g_i(y)dy + \beta_i(x)\varphi(x,1,1)\min_{j=1,2}\int_\Omega w(y)g_j(y)dy\Big).$$

where $\epsilon_g > 0$ can be chosen uniformly for g in bounded subsets of X_+.

PROOF. For the solutions of (17.2) with Dirichlet boundary conditions and initial values g_i,

$$u_i(1,x) \leq \int_\Omega G(d_i, x, y)g_i(y)dy, \qquad x \in \Omega.$$

By Lemma 17.8, for some $c > 0$,

$$u_i(1,x) \leq cw(x)\int_\Omega w(y)g_i(y)dy, \qquad x \in \Omega.$$

Since $\varphi(x,\cdot)$ is order-preserving and homogeneous, by (17.8) and Lemma 4.24 (a), the upper estimates for $F_i(g)$ and $B_i(g)$ follow.

Let $g \in X_+$. By the maximum principle, $u_i(t,x) \leq \|g\|$ for all $t \in [0,1]$, $x \in \bar{\Omega}$.

By the properties of μ_i, there exists some $\tilde{c} > 0$ which only depends on $\|g\|$ such that

$$\mu_i(t,x,u(t,x)) \leq \tilde{c}, \qquad t \in [0,1], \quad x \in \bar{\Omega}$$

By the properties of a Green's function,

$$u_i(1,x) \geq \int_\Omega G(d_i, x, y)e^{-\tilde{c}}g_i(y)dy.$$

By Lemma 17.8, there exists some $\epsilon_g > 0$ that only depends on $\|g\|$ such that

$$u_i(1,x) \geq \epsilon_g w(x)\int_\Omega w(y)g_i(y)dy.$$

As for the solutions $\tilde{u}_i$ with mortality rates $\mu_i(t,x,0)$, we obtain

$$\tilde{u}_i(1,x) \geq \epsilon w(x)\int_\Omega w(y)g_i(y)dy,$$

with $\epsilon > 0$ not depending on g. Since $\varphi(x,\cdot)$ is homogeneous and order-preserving and β_i bounded away from 0, by (17.8) and Lemma 4.24 (a) the lower estimates for $F_i(g)$ and $B_i(g)$ follow. □

PROPOSITION 17.10. *Let $\mathcal{T}_0 = \mathbf{r}(B) > 1$. Then the population persists uniformly weakly in the following sense: There exists some $\epsilon > 0$ such that $\limsup_{n\to\infty} \|F^n(g)\| \geq \epsilon$ for every $g = (g_1, g_2) \in X_+$ with $g_1 \not\equiv 0$ and $g_2 \not\equiv 0$.*

PROOF. We check the assumptions of Proposition 14.28.

B is a lower order-derivative of F by Proposition 17.7.

We define $\rho(g) = \min_{j=1,2} \int_\Omega w(y) g_j(y) dy$ for $g = (g_1, g_2) \in X_+$. Since $w(y) > 0$ for all $y \in \Omega$, $\rho(g) > 0$ if and only if $g_1 \not\equiv 0$ and $g_2 \not\equiv 0$.

By Proposition 17.7 and Theorem 9.10, there exists some $v = (v_1, v_2) \in \dot{X}_+$ such that $B(v) = rv$ with $r = \mathbf{r}(B) > 1$. Set $\theta(g) = [g]_v$ for $g \in X_+$. By Remark 11.13, $\theta(B(g)) \geq r\theta(g)$ for all $x \in X_+$ and assumption (b) of Proposition 14.28 is satisfied.

By Proposition 17.9, B is uniformly $\tilde{v}$-bounded with $\tilde{v} = (\tilde{v}_1, \tilde{v}_2)$,

$$\tilde{v}_i(x) = w(x) p_j(x) + \varphi(x, 1, 1) w(x) \beta_j(x), \qquad i = 1, 2, \quad x \in \bar{\Omega}.$$

See Definition 6.1. Notice that $\tilde{v} \in X_+$ and that v is $\tilde{v}$-bounded (Definition 2.59). By Proposition 17.9, if $\rho(g) > 0$, then there exists some $\epsilon > 0$ such that $F(g) \geq \epsilon \tilde{v}$ and $\rho(F(g)) > 0$ and $\theta(F(g)) > 0$. Recall Assumption 17.3 (d).

This implies the properties (a) and (b) of Proposition 14.28, and the assertion follows. □

COROLLARY 17.11. *If $\mathcal{T}_0 = \mathbf{r}(B) > 1$, there exists some $\epsilon > 0$ such that, for all $g = (g_1, g_2) \in X_+$, $g_1 \not\equiv 0$, $g_2 \not\equiv 0$,*

$$\limsup_{n \to \infty} \max_{j=1,2} \int_\Omega w(y) g_{n,i}(y) dy \geq \epsilon,$$

where $F^n(g) = (g_{n,1}, g_{n,2})$, $n \in \mathbb{N}$.

PROOF. By Proposition 17.9, since $u \in X_+$ is bounded,

$$\|F(g)\| \leq \tilde{c} \max_{i=1,2} \int_\Omega w(y) g_i(y) dy, \qquad f \in C_+(\bar{\Omega}).$$

Apply Proposition 17.10. □

PROOF OF THEOREM 17.4. The first result in Theorem 17.4 now follows from Proposition 17.7 and from Theorem 14.13. The second result follows from Proposition 17.7 and from Theorem 14.22. The third result follows from Corollary 17.11. □

PROOF OF THEOREM 17.5. We only prove the second equation. Fix $f, m \in C_+(\bar{\Omega})$ with $f \not\equiv 0$, $m \not\equiv 0$. Using the upper and lower estimates in Proposition 17.9, there exists some $c > 0$ (which depends on f and m) such that

$$B(g) \leq c\|g\| B(f, m), \qquad g \in X_+.$$

Since B is order-preserving and homogeneous, so is B^n for $n \in \mathbb{N}$ and

$$B^n(g) \leq c\|g\| B^n(f, m), \qquad g \in X_+, \quad n \in \mathbb{N}.$$

Again by Proposition 17.9, after modifying $c > 0$,

$$B^{n+1}(g)(x) \leq c\|g\| \int_\Omega [f_n(y) + m_n(y)] dy, \qquad g \in X_+, \quad n \in \mathbb{N},$$

where $(f_n, m_n) = B^n(f, m)$. By definition of the operator norm, (4.1),

$$\|B^{n+1}\| \leq c \int_\Omega [f_n(y) + m_n(y)] dy, \qquad n \in \mathbb{N}.$$

We take roots,

$$\|B^{n+1}\|^{1/n} \leq c^{1/n}\Big(\int_\Omega [f_n(y) + m_n(y)]dy\Big)^{1/n}, \qquad n \in \mathbb{N},$$

and the limit as $n \to \infty$, (see (5.1)),

$$\mathbf{r}(B) \leq \liminf_{n\to\infty} \Big(\int_\Omega [f_n(y) + m_n(y)]dy\Big)^{1/n}.$$

In turn,

$$\int_\Omega [f_n(y) + m_n(y)]dy \leq 2\|B^n(f,m)\| \, |\Omega| \leq 2\|B^n\| \, \|(f,m)\| \, |\Omega|,$$

with $|\Omega|$ denoting the Lebesgue measure of Ω; so equality follows by taking roots and then limits. □

17.3. Basic turnover number and critical domain size

We consider the special case

$$B_i(g)(x) = p_i\tilde{u}_i(1,x) + \beta_i\varphi(\tilde{u}(x)), \tag{17.19}$$

where $\tilde{u}_1$ and $\tilde{u}_2$ are the solutions of

$$(\partial_t - d_i\Delta)\tilde{u}_i = -\mu_i(t,0)\tilde{u}_i, \qquad 0 \leq t \leq 1, \quad x \in \Omega, \tag{17.20}$$

with Dirichlet boundary conditions $\tilde{u}_i(t,x) = 0, \quad x \in \partial\Omega$, and initial conditions

$$\tilde{u}_i(0,x) = g_i(x) \qquad x \in \bar{\Omega}. \tag{17.21}$$

where $g = (g_1, g_2)$.

To find $\mathbf{r}(B)$ and an associated positive eigenfunction of B, we choose $v \in C^2_+(\Omega)$ and $\lambda = \lambda_\Omega > 0$ as the principal eigenfunction-eigenvalue-pair of the Laplace operator with Dirichlet boundary conditions,

$$\begin{aligned} \Delta v =& -\lambda v \qquad \text{on } \Omega,\\ v =& \, 0 \qquad \text{on } \partial\Omega. \end{aligned} \tag{17.22}$$

This eigenfunction exists under our assumptions and is strictly positive on Ω. λ and v are uniquely determined up to normalization of v. They contain the essential information about the size and the shape of the habitat Ω of the population and whether the habitat is suitable for population survival.

We try to determine $\tilde{u}_i$ by separation of variables,

$$\tilde{u}_i(t,x) = \gamma_i(t)v(x). \tag{17.23}$$

We substitute this ansatz into (17.20) and obtain the following initial-value-problem for γ_i:

$$\begin{aligned} \gamma_i' =& -(d_i\lambda + \mu_i(t,0))\gamma_i,\\ \gamma_i(0) =& \, w_i, \qquad w = (w_1, w_2), \end{aligned} \tag{17.24}$$

with $w \in (0,\infty)^2$ to be determined. This decoupled system is uniquely solved by

$$\gamma_i(t) = w_i e^{-d_i\lambda t}\exp\Big(-\int_0^t \mu_i(s,0)ds\Big). \tag{17.25}$$

We substitute this result into (17.23),

$$\tilde{u}_i(1,x) = q_i w_i v(x), \qquad q_i = e^{-d_i\lambda} \exp\Big(- \int_0^1 \mu_i(s,0)ds\Big). \tag{17.26}$$

We fit these formulas into (17.19) and (17.21) and use the homogeneity of φ,

$$\begin{aligned} g(x) =& v(x)w, \\ B_i(g)(x) =& p_i q_i w_i v(x) + \beta_1 v(x)\varphi(q_1 w_1, q_2 w_2), \qquad i = 1, 2. \end{aligned} \tag{17.27}$$

So, with the separation of variables ansatz (17.23), the infinite-dimensional eigenvector problem $rg = B(g)$ reduces to the two-dimensional eigenvector problem

$$r w_j = p_j\, q_j w_j + \beta_j\, \varphi(q_1 w_1, q_2 w_2), \qquad j = 1, 2. \tag{17.28}$$

By setting $q_j w_j = \tilde{w}_j$, we obtain the equivalent eigenvalue problem

$$r \tilde{w}_j = p_j\, q_j \tilde{w}_j + q_j \beta_j\, \varphi(\tilde{w}), \qquad j = 1, 2, \tag{17.29}$$

which we have already dealt with in Section 15.1.2.

17.3.1. Basic reproduction number. By Theorem 17.5, there is no other positive eigenvalue associated with a nonnegative eigenfunction for (17.29) than the basic turnover number $r = \mathcal{T}_0 = \mathbf{r}(B)$. This implies that $\mathcal{T}_0 - 1$ has the same sign as $\mathcal{R}_0 - 1$ (Theorem 15.3), where

$$\mathcal{R}_0 = \varphi\Big(\frac{q_1\beta_1}{1 - p_1 q_1}, \frac{q_2\beta_2}{1 - p_2 q_2}\Big).$$

Recall (17.26),

$$q_i = e^{-\lambda d_i}\zeta_i, \qquad \zeta_i = \exp\Big(- \int_0^1 \mu_i(s,0)ds\Big),$$

with $\lambda = \lambda_\Omega > 0$ given by (17.22). $\mathcal{R}_0$ can be viewed as *basic reproduction number* of the population. If

$$\varphi\Big(\frac{\zeta_1\beta_1}{1 - \zeta_1 p_1}, \frac{\zeta_2\beta_2}{1 - \zeta_2 p_2}\Big) \le 1,$$

then, whatever the size and shape of the domain, $\mathcal{R}_0 < 1$ and the population cannot persist at low densities. If this number is strictly greater than 1, $\mathcal{R}_0$ is a strictly decreasing function of $\lambda > 0$ that tends to zero as $\lambda \to \infty$ so that there is a critical $\lambda^\sharp > 0$ such that $\mathcal{R}_0 = 1$ which is a measure of the critical domain size for the population [**32, 147**].

If $p_j = 0$ for $j = 1, 2$ (in the case of a semelparous population) and the diffusion rates $d_i = d$ are the same for both sexes, then the critical value of λ at which $\mathcal{R}_0 = 1$ is

$$\lambda^\sharp = (1/d) \ln \varphi(\zeta_1\beta_1, \zeta_2\beta_2).$$

See [**127**]. It is well known that if $\Omega_1 \subseteq \Omega_2$, then $\lambda_{\Omega_1} \ge \lambda_{\Omega_2}$ with strict inequality if $\Omega_2 \setminus \overline{\Omega_1}$ is a nonempty open set. So $\lambda^\sharp$ measures the *critical critical size* of the domain (or habitat or patch) for population persistence versus population extinction at small population sizes.

17.3.2. Critical domain size. In order to illustrate how the principal eigenvalue λ of the Laplace operator is related to the size and shape of the domain Ω, we consider a rectangle with length L and height H,

$$\Omega = (0, L) \times (0, H), \tag{17.30}$$

and

$$v(x_1, x_2) = \sin(\pi x_1/L)\sin(\pi x_2/H). \tag{17.31}$$

v is strictly positive on Ω and zero on $\partial\Omega$,

$$\Delta v = -\lambda v, \qquad \lambda = (\pi/L)^2 + (\pi/H)^2. \tag{17.32}$$

It is well known that if $\Omega_1 \subseteq \Omega_2$, then $\lambda_{\Omega_1} \geq \lambda_{\Omega_2}$ with strict inequality if $\Omega_2 \setminus \overline{\Omega_1}$ is a nonempty open set. See Corollary 2.3 in [**33**]. Thus, $\mathcal{R}_0$ increases with domain size.

17.4. More on persistence

We additionally assume that $\mu_j(t,x,z) \geq \breve{\mu}_j(t)$ for all $t \in [0,1]$, $x \in \bar{\Omega}$, and $z \in \mathbb{R}^2_+$ and that

$$\sup_{x\in\Omega} p_i(x) e^{-\int_0^1 \breve{\mu}_i(t)dt} := \alpha_i < 1, \qquad j = 1, 2. \tag{17.33}$$

By the maximum principle, for the solutions $u = (u_1, u_2)$ of (17.2) and (17.3)

$$\|u_i(1,\cdot)\|_\infty \leq e^{-\int_0^1 \breve{\mu}_i(t)dt} \|u_i(0,\cdot)\|_\infty. \tag{17.34}$$

By (17.8),

$$F_i(g)(x) \leq \alpha_i \|g_i\|_\infty + \beta_i(x)\varphi(x, u(1,x)), \qquad g \in X_+. \tag{17.35}$$

By Lemma 17.8, there is some $c > 0$ such that

$$F_i(g)(x) \leq \alpha_i\|g_i\|_\infty + \beta_i(x) c w(x) \varphi\Big(x, \int_\Omega w(y) g_1(y)dy, \int_\Omega g_2(y)dy\Big). \tag{17.36}$$

We define $\rho : X_+ \to \mathbb{R}_+$ by

$$\begin{aligned} \rho(g) = &\sup_{x\in\Omega} \varphi\Big(x, \int_\Omega w(y)g_1(y)dy, \int_\Omega w(y) g_2(y)dy\Big), \\ &g = (g_1, g_2) \in X_+, \end{aligned} \tag{17.37}$$

and

$$\alpha = \max_{i=1,2} \alpha_i \quad \text{and} \quad \tilde{c} = c \max_{i=1,2} \sup \beta_i(\Omega) \sup w(\Omega).$$

Then $\alpha \in (0,1)$, ρ is homogeneous, continuous and order-preserving and

$$\|F(g)\| \leq \alpha\|g\| + \tilde{c}\rho(g), \qquad g \in X_+. \tag{17.38}$$

By induction,

$$\|F^n(g)\| \leq \alpha^n \|g\| + \tilde{c}\sum_{j=0}^{n-1} \alpha^j \rho\big(F^{n-1-j}(g)\big), \qquad n \in \mathbb{N}, \quad g \in X_+. \tag{17.39}$$

The subsequent result now follows from Proposition 17.10.

THEOREM 17.12. *Let $\mathcal{T}_0 > 1$ and the Assumption* (17.33) *be satisfied. Then the semiflow induced by F is uniformly weakly ρ-persistent: There exists some $\epsilon > 0$ such that*

$$\limsup_{n\to\infty} \rho(F^n(g)) \ge \epsilon$$

for all $g \in X_+$ with $\rho(g) > 0$.

PROOF. Suppose the assertion is false. Let $\epsilon > 0$. Then there exists some $g \in X_+$ with $\rho(g) > 0$ such that

$$\limsup_{n\to\infty} \rho(F^n(g)) \le \epsilon.$$

Now we can apply Fatou's lemma to (17.39), using the counting measure,

$$\limsup_{n\to\infty} \|F^n(g)\| \le \frac{\tilde{c}}{1-\alpha} \limsup_{n\to\infty} \rho(F^n(g)) \le \frac{\tilde{c}}{1-\alpha}\epsilon.$$

This is a contradiction to Proposition 17.10. Notice that $\rho(g) > 0$ if and only if $g_1 \not\equiv 0$ and $g_2 \not\equiv 0$. □

17.5. Point-dissipativity

Assume that there are continuous functions $\nu_1, \nu_2 : \mathbb{R}^2_+ \to \mathbb{R}_+$ such that

$$\mu_i(t,x,y) \ge \nu_i(t,y_i), \qquad t \in [0,1], \quad x \in \Omega, \quad y = (y_1,y_2) \in \mathbb{R}^2_+, \tag{17.40}$$

and

$$\int_0^1 \nu_i(t,\xi)dt \to \infty, \qquad \xi \to \infty. \tag{17.41}$$

Assume that the functions ν_i satisfy a local Lipschitz condition: For any $c > 0$, there exists some Lipschitz constant $\Lambda > 0$ such that

$$|\nu_i(t,z) - \nu_i(t,\tilde{z})| \le \Lambda|z-\tilde{z}|, \qquad z,\tilde{z} \in \mathbb{R}_+, \quad z+\tilde{z} \le c, \quad t \in [0,1]. \tag{17.42}$$

Recall the solutions u_i of

$$\begin{aligned} (\partial_t - d_i\Delta)u_i &= -\mu_i(t,x,u)u_i, \qquad 0 < t \le 1, \quad x \in \Omega, \quad i = 1,2, \\ u &= (u_1,u_2), \\ u_i(t,x) &= 0, \qquad x \in \partial\Omega, \\ u_i(0,x) &= g_i(x), \qquad x \in \bar{\Omega}, \quad i = 1,2. \end{aligned} \tag{17.43}$$

Consider the ordinary differential equations

$$U_i'(t) = -\nu_i(t,U_i(t))U_i(t). \tag{17.44}$$

It follows from (17.40) and a standard comparison argument that

$$u_i(t,x) \le U_i(t), \qquad t \in [0,1], \quad x \in \Omega, \tag{17.45}$$

where U_i solves (17.44) with

$$U_i(0) = \sup g_i(\bar{\Omega}). \tag{17.46}$$

Assume that

$$\varphi(x,1,s) + \varphi(x,s,1) \le c_\varphi, \qquad s \ge 0, \quad x \in \Omega. \tag{17.47}$$

This is a plausible assumption: If there is only one female or one male, then the number of pairs being formed is bounded however many members of the other sex are present.

Assumption (15.25) and the other properties of φ imply that

$$\varphi(x, y) \le c_\varphi \min\{y_1, y_2\}, \qquad y = (y_1, y_2) \in \mathbb{R}^2_+. \tag{17.48}$$

Let

$$G_i(z) = \big(\sup p_i(\Omega) + \sup \beta_i(\Omega) c_\varphi\big) U_i(1), \tag{17.49}$$

where U_i solves (17.44) with $U_i(0) = z \in \mathbb{R}_+$. By (17.8) and (17.44)–(17.49),

$$F_i(g)(x) \le G_i(\sup g_i(\Omega)), \qquad x \in \bar\Omega, \quad i = 1, 2, \quad g = (g_1, g_2) \in X_+. \tag{17.50}$$

Since (17.44) is a scalar ODE, $U_i(1)$ is an increasing function of $U_i(0)$. By (17.49), G is increasing. By induction,

$$\begin{aligned} F^n(g)(x) \le \big(G_1^n(\sup g_1(\Omega)), G_2^n(\sup g_2(\Omega))\big), \\ n \in \mathbb{N}, \quad g = (g_1, g_2) \in X_+. \end{aligned} \tag{17.51}$$

Indeed, if $n \in \mathbb{N}$ and (17.51) holds for n, then

$$F^{n+1}(g)(x) = F^n(F(g))(x) \le (G_1^n(\sup F_1(g)(\Omega), G_2^n(\sup F_2(g)(\Omega)).$$

Since the G_j are increasing, by (17.50),

$$\begin{aligned} F^{n+1}(g)(x) &\le \big(G_1^n(G_1(\sup g_1(\Omega)), G_2^n(G_2(\sup g_2(\Omega)))\big) \\ &= \big(G_1^{n+1}(\sup g_1(\bar\Omega)), G_2^{n+1}(\sup g_2(\Omega))\big), \end{aligned}$$

and (17.51) holds for $n + 1$.

A similar proof as for Proposition 15.7 shows that the G_i in (17.49) are point-dissipative. By (17.51), F inherits point-dissipativity from G_1 and G_2.

THEOREM 17.13. *Assume in addition that there is some $c_\varphi > 0$ such that $\varphi(x, 1, s) + \varphi(x, s, 1) \le c_\varphi$ for all $s \ge 0$ and $x \in \Omega$ and that* (17.40)–(17.42) *are satisfied.*

Then $F : X_+ \to X_+$ is point-dissipative and compact, and the semiflow induced by F has a compact attractor of points.

PROOF. Recall the solutions $u = (u_1, u_2)$ of (17.2) with Dirichlet boundary conditions and initial conditions $u(0) = g$, $g \in X_+$. It follows from the smoothing properties of parabolic differential equations that the map $X_+ \ni g \mapsto u(1, \cdot) \in X_+$ is compact and so is the composition $F(g) = H(u(1, \cdot))$ with H in (17.5). Since F is point-dissipative, the induced semiflow has a compact attractor of points by [**187**, Thm.2.28]. □

17.6. Order permanence

Recall the homogeneous continuous order-preserving functional ρ in (17.37),

$$\rho(g) = \sup_{x \in \Omega} \varphi\Big(x, \int_\Omega w(y) g_1(y) dy, \int_\Omega w(y) g_2(y) dy\Big), \qquad g = (g_1, g_2) \in X_+.$$

Combine Theorem 17.13, Theorem 17.12 and [**187**, Thm.4.13] to get the following result.

THEOREM 17.14. *Let $\mathcal{T}_0 > 1$ and the Assumption* (17.33) *be satisfied.*

Assume in addition that there is some $c_\varphi > 0$ such that $\varphi(x, 1, s) + \varphi(x, s, 1) \le c_\varphi$ for all $s \ge 0$ and $x \in \Omega$ and that (17.40)–(17.42) *are satisfied.*

Then the semiflow induced by F is uniformly ρ-persistent: There exists some $\epsilon > 0$ such that

$$\liminf_{n\to\infty} \rho(F^n(g)) \geq \epsilon$$

for all $g \in X_+$ with $\rho(g) > 0$.

We define

$$w_i(x) = p_i(x) + \beta_i(x)\varphi(x, 1, 1)w(x), \qquad x \in \Omega, \quad i = 1, 2, \tag{17.52}$$

with w in (17.17).

THEOREM 17.15. *Let $\mathcal{T}_0 > 1$ and equation (17.33) be satisfied.*

Assume in addition that there is some $c_\varphi > 0$ such that $\varphi(x, 1, s) + \varphi(x, s, 1) \leq c_\varphi$ for all $s \geq 0$ and $x \in \Omega$ and that (17.40)–(17.42) are satisfied.

(a) *Then the semiflow induced by F is order-permanent:*

There exist some $c > \epsilon > 0$ such that for any sequence (g_n) in X_+ with $g_n = (g_{n1}, g_{n2})$ and $g_n = F^n(g_1)$, $g_1 = (g_{11}, g_{12}) \in X_+$, $g_1 \not\equiv 0$ and $g_2 \not\equiv 0$, there exists some $m \in \mathbb{N}$ such that

$$\epsilon w_i(x) \leq g_{ni}(x) \leq c w_i(x), \qquad x \in \bar{\Omega}, \quad n \in \mathbb{N}, \quad n \geq m, \quad i = 1, 2. \tag{17.53}$$

(b) *There exist a fixed point $F(g) = g \in X_+$, $g = (g_1, g_2)$, and some $c > \epsilon > 0$ such that $\epsilon w_i(x) \leq g_i(x) \leq c g_i(x)$ for all $x \in \bar{\Omega}$, $i = 1, 2$.*

PROOF. (a) The right inequality in (17.53) follows from Theorem 17.13 and Proposition 17.9.

As for the left inequality in (17.53), we apply [**187**, Cor.4.22] in combination with Theorem 17.13 and the fact that F is compact. Let $(g_n)_{n\in\mathbb{Z}}$ be a sequence in X_+ with $g_n = F(g_{n-1})$ and $g_n = (g_{n1}, g_{n2})$ for all $n \in \mathbb{Z}$. Assume that the closure of $\{g_n; n \in \mathbb{Z}\}$ is a compact subset of X_+ and $\inf_{n\in\mathbb{Z}} \rho(g_n) > 0$. Let

$$\tilde{\rho}(g) = \min\Big\{ \int_\Omega w(y) g_{n1}(y) dy, \int_\Omega w(y) g_{n1}(y) dy \Big\}, \qquad g = (g_1, g_2) \in X_+. \tag{17.54}$$

Then $\tilde{\rho}$ is lower semi-continuous, even continuous. By (17.48),

$$\inf_{n\in\mathbb{Z}} \tilde{\rho}(g_n) > 0.$$

By [**187**, Cor.4.22], there exists some $\epsilon_0 > 0$ such that

$$\liminf_{n\to\infty} \tilde{\rho}(g_n) \geq \epsilon_0$$

for all $g \in X_+$ with $\rho(g) > 0$.

By Proposition 17.9 and (17.52), there exists some $\epsilon > 0$ such that for any sequence (g_n) with $g_n = (g_{n1}, g_{n2})$ in X_+ and $g_n = F^n(g)$, $n \in \mathbb{N}$, with $g = (g_1, g_2) \in X_+$, $g_1 \not\equiv 0$ and $g_2 \not\equiv 0$, there exists some $m \in \mathbb{N}$ such that

$$g_{ni}(x) \geq \epsilon w_i(x), \qquad x \in \bar{\Omega}, \quad n \geq m, \quad i = 1, 2.$$

(b) We apply [**187**, Thm.6.2] with $\tilde{\rho} : X_+ \to X_+$ in (17.54) as the ρ and X in [**187**, Thm.6.2]. Then $\tilde{\rho}$ is continuous and concave. It follows from the proof in (a) that (the semiflow Φ induced by) F is uniformly $\tilde{\rho}$-persistent. By Theorem 17.13, Φ has a compact attractor of points and F is compact. The hypothesis (H1) in [**187**, Thm.6.2] follows from Proposition 17.9 and the strict positivity of the w.

By [**187**, Thm.6.2], there exists a fixed point $F(g) = g \in X_+$ with $\tilde{\rho}(g) > 0$. A similar proof as in (a) provides the desired estimates. □

CHAPTER 18

Nonlocal spatial spread of semelparous two-sex populations

If the females and males in a spatially distributed two-sex population spread by diffusion as in the previous Chapter 17, both sexes easily spread over the whole habitat and have no problems in finding each other to mate and reproduce. If they spread in different (nonlocal) ways, males and females may concentrate in different parts of the habitat and mating and reproduction may not happen sufficiently often for the population to persist. Dispersal is sex-biased in many animal populations; for instance, members of one sex may tend to travel longer distances than members of the other sex ([**162**] and the references therein).

For simplicity, to explore the impact of sex-dependent dispersal on population survival, we consider a semelparous population where individuals reproduce only once in their life-time. We look at the case of very short reproduction seasons. Differently from the two-sex models in Chapter 15 and 17, because of the semelparity, it is possible to keep track only of the spatial density of neomates (newborn individuals) in every year. This requires that the population count is taken immediately after the reproductive season (post-reproductive census [**78**]). (Before the reproductive season, one would need to count both males and females. For the importance of census times see [**152**].)

This chapter highlights parts of and adds to [**205**]. Nonlocal spatial spread of iteroparous population models without sex distinction is considered in [**206**].

18.1. A recursive equation for population densities

The habitat of the population is represented by a metric space S. Our mathematical tool for modeling the dispersal of females and males are Feller kernels $\kappa_j : \mathcal{B} \times S \to \mathbb{R}_+$, $j = 1, 2$, where $\mathcal{B}$ is the σ-algebra of Borel subsets of S. See Chapter 13, in particular Section 13.3.

Let $C^b(S)$ denote the Banach space of bounded continuous real-valued functions on S with the supremum norm $\|f\|_\infty = \sup_{s\in S} |f(s)|$ and $C^b_+(S)$ the closed cone of nonnegative functions in $C^b(S)$.

If $h \in C^b(S)$,

$$(K_j h)(s) = \int_S h(t)\kappa_j(dt, s), \qquad s \in S, \tag{18.1}$$

provides a function $K_j h \in C^b(S)$. So, the Feller kernels κ_j induce bounded linear operators K_j on $C^b(S)$ that are positive, i.e., they map the cone $C^b_+(S)$ into itself. If we choose h as the constant function with value 1 in (18.1), we obtain that

$$\kappa_j(S, \cdot) \in C^b_+(S), \qquad j = 1, 2. \tag{18.2}$$

As explained after Definition 13.16, $\kappa_j(T, \cdot)$ is a Borel measurable function for any $T \in \mathcal{B}$. Thus, the Feller kernels κ_j also induce positive bounded linear operators on $M^b(S)$ that are denoted by K_j as well. Recall that $M^b(S)$ denotes the Banach space of bounded Borel measurable functions with values in $\mathbb{R}$ under the supremum norm, while $M^b_+(S)$ denotes the closed cone of those functions that are nonnegative.

It will be useful, indeed, to have K_j operate on $M^b(S)$ as well, as

$$\kappa_j(T,s) = (K_j\,\chi_T)(s), \qquad s \in S, \quad T \in \mathcal{B}. \tag{18.3}$$

Here χ_T is the *characteristic (indicator) function* of the subset T of S,

$$\chi_T(s) = 1, \quad s \in T; \qquad \chi_T(s) = 0, \quad s \in S \setminus T. \tag{18.4}$$

Anyway, whether the linear maps K_j operate on $C^b_+(S)$ or $M^b_+(S)$, they map the spatial density of neonates after the reproductive season (irrespective of their sex) to the respective spatial densities of female and male adults after one year.

To model reproduction, let $b(s, f, m)$ denote the amount of offspring produced at $s \in S$ during the short reproductive season by f females and m males. We call b the *mating and birth function.*

The *yearly population turnover operator* (Section 1.1) is given by

$$F(h)(s) = b\big(s, (K_1h)(s), (K_2h)(s)\big), \qquad s \in S, \quad h \in C^b_+(S). \tag{18.5}$$

F maps this year's spatial density of neonates to next year's spatial density of neonates (immediately after the reproductive season). Since the population is semelparous, the neonates have no overlapping generations, and F coincides with the *next generation operator.*

Traditionally, the recursive equation $h_n = F(h_{n-1})$, $n \in \mathbb{N}$, with given h_0, is called a *difference equation* though there is no difference taken in this formula. Since (18.5) involves the integral operators K_j, it is more accurately referred to as an *integro-difference equation* [**134, 135, 151, 152, 165, 171, 192, 221**]. It is also called an *integral projection model* (see [**74, 78, 79, 89**] and the references therein) because F gives a projection of next year's spatial density based on this year's spatial density.

However the equations are called, the Feller kernels therein often are of the form

$$\kappa_j(T,s) = \int_T k_j(t,s)\nu_j(dt), \qquad s \in S, \quad T \in \mathcal{B}, \tag{18.6}$$

with $k_j \in C^b_+(S \times S)$ and nonnegative measures ν_j on $\mathcal{B}$. See [**130**, Sec.8], for instance. Feller kernels are not only more general, but also allow for more succinct mathematical formulations. Sometimes, the special case (18.6) is unavoidable, though e.g., if traveling wave solutions and spreading speeds are considered [**116, 125, 148, 149**]. We highlight [**162**], which is also a two-sex model with homogeneous mating function.

18.1.1. Some remarks on interpretation. If S is an open nonempty subset of $\mathbb{R}^n$, the Lebesgue measure induces a natural master measure on S, and a function $h \in C^b_+(S)$ can be interpreted as spatial density of neomates, adult females and adults males, respectively, by $\int_T h(s)ds$ giving the number of neonates located in the Borel subset T of S. If S is an arbitrary metric space, there may be no natural master measure, and $h \in C^b_+(S)$ is to be interpreted as the measure-independent spatial density of neomates, for example, as follows.

Let μ be a nonnegative finite measure on $\mathcal{B}$. Then $\int_T h d\mu$ is the amount of neonates in the set T measured by μ. Depending on μ, this could be the biomass of neonates in T or the food eaten by them in a day. Let $s \in S$ and $\mathcal{B}_\mu = \{T \in \mathcal{B};\ \mu(T) > 0\}$. Since h is continuous at s,

$$\frac{1}{\mu(T)} \int_T h d\mu \to h(s), \qquad s \in T \in \mathcal{B}_\mu, \quad \Delta(T) \to 0.$$

Here $\Delta(T) = \sup\{d(t, \tilde{t});\ t, \tilde{t} \in T\}$ is the diameter of T.

If the population state space is not $C_+^b(S)$ but $\mathcal{M}_+(S)$, the cone of nonnegative measures, the Feller kernels can have a very natural interpretation. See Section 13.2 and 19.1.

It is difficult, however, to interpret the Feller kernels κ_j that describe the dispersal of the two sexes in the state space $C_+^b(S)$. Let S be an open subset of $\mathbb{R}^n$ and $\kappa_j(T, s) = \int_T k_j(t, s) dt$. Then $\int_U \kappa_j(t, s) ds$ is the probability that a neonate that is in t at the beginning of the year will be still alive and in the set $U \in \mathcal{B}$ at the end of the year. But κ_j does not seem to have an interpretation itself.

An interpretation of κ_j even becomes more difficult if certain proportions $p_j \in [0, 1]$ of females or males do not spread at all and

$$\kappa_j(T, s) = (1 - p_j) \int_T k_j(t, s) dt + p_j \chi_T(s), \qquad T \in \mathcal{B}, \quad s \in S,$$

though each ingredient of the right-hand side has a clear interpretation. Here χ_T is the indicator function of T defined in (18.4).

Consider dispersal kernels κ_j in separable form

$$\kappa_j(T, s) = \gamma_j(s) \mu_j(T), \qquad T \in \mathcal{B}, \quad s \in S,$$

with continuous $\gamma_j : S \to [0, 1]$ and probability measures μ_j with values in $[0, 1]$. In this simple case, $\mu_j(T)$ can be interpreted as the probabilities of having been located in set T at the beginning of the year, and $\gamma_j(s)$ can be interpreted as the proportion of neonates that have the appropriate sex and are still alive and are located at $s \in S$ at the end of the year.

For further interpretative issues we refer to [**205**, Sec.1.1].

18.2. Reproduction and mating

The amount of offspring produced by males and females is related to the number of mated pairs they form during the reproductive season. Without competitive or cooperative interference between pairs, the amount of offspring should be proportional to the number of pairs.

Let $\varphi(s, f, m)$ be the amount of offspring produced at $s \in S$ during the short reproductive season by f females and m males if there is no competitive or cooperative interference between pairs but the only competition is between singles looking for mates. We assume that

$$\varphi(s, f, m) > 0, \qquad s \in S, \quad f, m \in (0, \infty),$$

$$\frac{b(s, f, m)}{\varphi(s, f, m)} \to 1, \qquad f, m \to 0, \tag{18.7}$$

with the convergence being uniform for $s \in S$.

Recall that $b(s, f, m)$ denotes the amount of offspring produced at $s \in S$ during the short reproductive season by f females and m males. We call b the *mating and birth function.*

If $\beta(s)$ is the number of offspring per mated pair at location s without competitive or cooperative interference by other pairs, then $\varphi(s, f, m)/\beta(s)$ is the number of mated pairs formed at s by f females and m males. Notice that this interpretation of φ is somewhat different from Sections 1.3, 4.5, and 17.1.2 and Chapter 15. In this chapter, we will refer to φ as the *basic mating and birth function.*

An extensive literature about pair formation assumes that, for $s \in S$, the mating function $(f, m) \mapsto \varphi(s, f, m)/\beta(s)$ is homogeneous, i.e.,

$$\varphi(s, \alpha f, \alpha m) = \alpha\varphi(s, f, m), \qquad \alpha, f, m \in \mathbb{R}_+, \quad s \in S. \tag{18.8}$$

Examples are variants of the harmonic mean, geometric mean, and the minimum function

$$\left.\begin{aligned} \varphi(s, f, m) &= \frac{fm}{\beta_1(s)f + \beta_2(s)m} \\ \varphi(s, f, m) &= a(s)f^{q(s)}m^{1-q(s)} \\ \varphi(s, f, m) &= \min\{\beta_1(s)f, \beta_2(s)m\} \end{aligned}\right\} s \in S, \quad f, m \in \mathbb{R}_+.$$

Here $q : S \to (0, 1)$ and $\beta_j, a : S \to (0, \infty)$, $j = 1, 2$. See [**83, 108, 109, 117, 118, 130, 161, 202**], the literature therein, and Section 4.5 where you can learn how to construct your favorite mating function (Remark 4.25). Since we assume that the mating function is homogeneous, Allee effects are excluded which may occur when males and females have difficulties finding each other at low densities and $\varphi(f, m)$ is proportional to fm as $f, m \to 0$. For such a mating function, the turnover operator F would be (Frechet) differentiable at the extinction state 0, the derivative at 0 would be the zero operator and the extinction state would be locally stable.

Equation (18.8) implies that the operator $B : C_+^b(S) \to \mathbb{R}^S$

$$(B(h))(s) = \varphi\big(s, (K_1h)(s), (K_2h)(s)\big), \qquad s \in S, \quad h \in C_+^b(S), \tag{18.9}$$

is homogeneous,

$$B(\alpha h) = \alpha B(h), \qquad \alpha \in \mathbb{R}_+, \quad h \in C_+^b(S). \tag{18.10}$$

Our interpretation of φ suggests that B is the *basic turnover operator* associated with the turnover operator F. It follows from (18.7) that B is the *order-derivative* of F (Section 14.2): For any $\epsilon \in (0, 1)$, there exists some $\delta > 0$ such that

$$(1 - \epsilon)B(h) \le F(h) \le (1 + \epsilon)B(h), \qquad h \in C_+^b(S), \quad \|h\|_\infty \le \delta. \tag{18.11}$$

In general, F is not Frechet-differentiable at $0 = F(0)$ because it is not additive on X_+ (Remark 14.8).

18.3. The basic turnover operator

Throughout this chapter, we assume that, for $j = 1, 2$, $\kappa_j : \mathcal{B} \times S \to \mathbb{R}_+$ are Feller kernels.

ASSUMPTION 18.1. *The birth function b in* (18.5) *has the following properties:*

(a) $b : S \times \mathbb{R}_+^2 \to \mathbb{R}_+$ *is continuous.*
(b) $b(s, f, m) = 0$ *if* $f = 0$ *or* $m = 0$, $s \in S$,
(c) $b(s, f, m) > 0$, $s \in S$, $f > 0$, $m > 0$.

(d) *If* $c > 0$, $b(S \times [0,c]^2)$ *is bounded.*

Since b is continuous by Assumption 18.1, we can define the Nemytskii (superposition) operator $N_b : C_+^b(S) \times C_+^b(S) \to C_+^b(S)$ associated with b [**55**, Exp.7.5],

$$N_b(h,\tilde{h})(s) = b\big(s, h(s), \tilde{h}(s)\big), \qquad s \in S, \quad h, \tilde{h} \in C_+^b(S). \tag{18.12}$$

The population turnover operator F can be written as

$$F(h) = N_b(K_1 h, K_2 h), \qquad h \in C_+^b(S), \tag{18.13}$$

and we see that $F : C_+^b(S) \to C_+^b(S)$. We can also consider N and F as maps on $M_+^b(S)$. Notice that

$$(F(\chi_T))(s) = b\big(s, \kappa_1(T,s), \kappa_2(T,s)\big), \qquad s \in S, \quad T \in \mathcal{B}, \tag{18.14}$$

where χ_T is the characteristic (indicator) function of T, (18.4).

Rewriting a system for females and males of a semelparous population as one equation for offspring using the operator F above may require scaling and absorbing parameters into κ_j. See [**162**, system (7)].

ASSUMPTION 18.2. *The function* $\varphi : S \times \mathbb{R}_+^2 \to \mathbb{R}_+$ *in* (18.7) *has the same properties as the function* b *in Assumption* 18.1,

$\varphi(s,\cdot)$ *is homogeneous and order-preserving on* $\mathbb{R}_+^2$ *for all* $s \in S$, *and*

$$\begin{aligned} &\varphi(s,f,m) > 0, \qquad s \in S, \quad f, m > 0, \\ &\frac{b(s,f,m)}{\varphi(s,f,m)} \to 1, \qquad 0 < f, m \to 0 \text{ uniformly for } s \in S. \end{aligned} \tag{18.15}$$

We define the basic turnover operator (in this case, also the basic next generation operator) by

$$B(h)(s) = \varphi\big(s, (K_1 h)(s), (K_2 h)(s)\big), \qquad s \in S, \quad h \in C_+^b(S). \tag{18.16}$$

Here, K_j are given by (18.1). We see that B maps $C_+^b(S)$ into itself.

We can also consider B as an operator on $M_+^b(S)$,

$$B(\chi_T)(s) = \varphi\big(s, \kappa_1(T,s), \kappa_2(T,s)\big), \qquad s \in S, \quad T \in \mathcal{B}, \tag{18.17}$$

where χ_T is the characteristic (indicator) function of the set T. See (18.4).

The spectral radius of B is the same for both cones, and $\mathcal{T}_0 = \mathbf{r}(B)$ can be interpreted as the basic population turnover number (Section 1.2).

THEOREM 18.3. *Let* φ *satisfy the Assumptions* 18.2. *Then* $B : C_+^b(S) \to C_+^b(S)$ *is continuous, homogeneous, and order-preserving and the order-derivative of* F. B *is uniformly* u_0*-bounded for*

$$u_0(s) = \varphi\big(s, \kappa_1(S,s), \kappa_2(S,s)\big), \qquad s \in S.$$

If $\mathcal{T}_0 = \mathbf{r}(B) > 0$, *there exists a bounded, homogeneous, order-preserving eigenfunctional* $\theta : C_+^b(S) \to \mathbb{R}_+$ *with* $\theta \circ B = \mathcal{T}_0 B$ *and* $\theta(u_0) > 0$.

Recall Definition 6.1.

REMARK 18.4.

(a) If, for the basic birth function φ, $\varphi(s,\cdot)$ is concave on $\mathbb{R}_+^2$ for each $s \in S$ (as it is typically the case, see Section 4.5), then the eigenfunctional θ is concave on $C_+^b(S)$.

(b) Since we can consider the basic turnover operator B also on $M_+^b(S)$, (18.16), the eigenfunctional θ can be extended to $M_+^b(S)$,

$$\theta\Big(\varphi(\cdot,\kappa_1(T,\cdot),\kappa_2(T,\cdot))\Big) = \mathcal{T}_0\,\varphi(\cdot,\kappa_1(T,\cdot),\kappa_2(T,\cdot)), \qquad T \in \mathcal{B}.$$

PROOF. By (18.1), $K_j(h)(s) \le \|h\|_\infty \kappa_j(S,s)$, for all $s \in S$. By (18.16), since φ is order-preserving and homogeneous,

$$B(h)(s) \le \varphi\big(s, \|h\|_\infty\kappa_1(S,s), \|h\|_\infty\kappa_2(S,s)\big) = \|h\|_\infty u_0(s).$$

So B is uniformly u_0-bounded (Definition 6.1).

The statement and the remark concerning the eigenfunctional follow from Theorem 11.14. □

THEOREM 18.5. *Consider B given by* (18.16) *on $X_+ = C_+^b(S)$ or $X_+ = M_+^b(S)$. Then*

$$\mathbf{r}(B) = \lim_{n\to\infty} \|B^n(u)\|_\infty^{1/n},$$

where u is the constant function with value 1.

PROOF. Apply Theorem 6.14. Notice that $\mathbf{r}_+(B)$ is used in [**199**, Thm.12.10] where this result is also proved. □

Considering B on $M_+^b(S)$, we have the following estimates. Recall the lower and upper Collatz–Wielandt bound of B, (5.30) and (6.6) and the upper Collatz–Wielandt spectral radius (Section 5.5).

COROLLARY 18.6. *Let u be the constant function with value one. Then*

$$\begin{aligned}\sup_{s\in S}\varphi\big(s,\kappa_1(S,s),\kappa_2(S,s)\big) &\ge CW_u(B) = \eta^u(B) = \gamma_u(B) = \mathbf{r}(B) = \mathcal{T}_0\\ &= cw(B) \ge \sup_{\emptyset\ne T\in\mathcal{B}}\ \inf_{s\in T}\varphi\big(s,\kappa_1(T,s),\kappa_2(T,s)\big).\end{aligned}$$

PROOF. For the first inequality, use Theorem 6.15.

For the last inequality, apply Theorem 5.31 with $X_B = M_+^b(S) = X_+$. In (5.30), specialize $x = \chi_T$, $\emptyset \ne T \in \mathcal{B}$. Here χ_T is the indicator function of the set T described in (18.4). The other inequalities and equalities follow from Theorem 9.21. □

Recall the concepts of Feller kernels that are tight and that are uniform (Sections 13.4 and 13.6).

PROPOSITION 18.7. *Let φ satisfy Assumption* 18.2 *and let $\{\varphi(s,\cdot); s\in S\}$ be equicontinuous on $\mathbb{R}_+^2$.*

Let the κ_i be tight uniform Feller kernels and $\mathcal{T}_0 = \mathbf{r}(B) > 0$. Then there exists some $h \in \dot{C}_+^b(S)$ such that $B(h) = \mathcal{T}_0 h$. Further,

$$CW_u(B) = \eta^u(B) = \gamma_u(B) = \mathbf{r}(B) = cw(B).$$

The equicontinuity assumption means that for any $f_0, m_0 \in \mathbb{R}_+$ and any $\epsilon > 0$ there is some $\delta > 0$ such that $|\varphi(s,f,m) - \varphi(s,f_0,m_0)| < \epsilon$ for all $s \in S$ and for all $f, m \in \mathbb{R}_+$ with $|f - f_0| + |m - m_0| < \delta$.

PROOF. $C^b(S)$ with the supremum norm is an ordered Banach space with closed normal solid cone $C_+^b(S)$. B is uniformly u_0-bounded for u_0 in Theorem 18.3 and for any $u \in \breve{C}_+^b(S)$. We apply Theorem 9.23.

By [**205**, Thm.7.5], B^2 is a compact operator and B is pseudo-compact by Corollary 9.28. (In order to get a taste how the assumption that the κ_i are uniform tight Feller kernels get into play, see Proposition 13.44 and Proposition 13.46 (b) and (c).)

The assumptions of Theorem 9.23 are satisfied and existence of h follows and implies $\mathbf{r}(B) \le cw(B)$ by definition of the lower Collatz–Wielandt bound (equation (5.30)). The equalities now follow from Corollary 18.6. □

18.4. Stability and instability results

We first consider the subthreshold case in which the basic turnover number is smaller than 1. $\mathcal{T}_0 = \mathbf{r}(B) < 1$.

18.4.1. Local stability and global extinction results.

THEOREM 18.8. *Let b satisfy the Assumption 18.1 and φ the Assumption 18.2. Let the κ_j be Feller kernels.*

Assume that $\mathcal{T}_0 = \mathbf{r}(B) < 1$.

(a) *Then the extinction state is locally asymptotically stable in the following sense:*

For each $\alpha \in (\mathcal{T}_0, 1)$, there exist some $\delta_\alpha > 0$ and $M_\alpha \ge 1$ such that,

$$\|F^n(f)\|_\infty \le M_\alpha \alpha^n \|f\|_\infty, \qquad n \in \mathbb{N},$$

if $f \in C_+^b(S)$ with $\|f\|_\infty \le \delta_\alpha$.

(b) *If $b(s, f, m) \le \varphi(s, f, m)$ for all $s \in S$ and $f, m \in \mathbb{R}_+$, the extinction state is globally stable in the following sense:*

For each $\alpha \in (\mathcal{T}_0, 1)$ there exists some $M_\alpha \ge 1$ such that $\|F^n(f)\|_\infty \le M_\alpha \alpha^n \|f\|_\infty$ for all $n \in \mathbb{N}$ and all $f \in C_+^b(S)$.

PROOF. (a) Apply Theorem 14.13.

(b) Since B is order-preserving, for all $n \in \mathbb{N}$ and $f \in C_+^b(S)$, $F^n(f) \le B^n(f)$. Since $\mathcal{T}_0 = \mathbf{r}(B) < 1$, $\|F^n(f)\|_\infty \le \|B^n\| \|f\|_\infty \to 0$, as $n \to \infty$. □

THEOREM 18.9. *Let b satisfy the Assumption 18.1 and φ the Assumption 18.2. Assume that the κ_j are Feller kernels.*

Let $T \in \mathcal{B}$, $T \ne \emptyset$, and $r \in (0,1)$ such that

$$\varphi\big(s, \kappa_1(T,s), \kappa_2(T,s)\big) \le r, \qquad s \in T,$$
$$\kappa_1(T,s)\kappa_2(T,s) = 0, \qquad s \in S \setminus T.$$

Then, for all $\alpha \in (r,1)$, there exists some $\delta_\alpha > 0$ such that the following holds for all $h \in C_+^b(S)$ with $h(s) = 0$ for $s \in S \setminus T$ and $\|h\|_\infty \le \delta_\alpha$:

$\|F^n(h)\|_\infty \le \alpha^n \delta_\alpha \to 0$ as $n \to \infty$ and $F^n(h)(s) = 0$ for all $n \in \mathbb{N}$ and $s \in S \setminus T$.

PROOF. Apply Theorem 14.10 with $X = M^b(S)$ and $X_+ = M_+^b(S)$ and $v = \chi_T$, the indicator (characteristic) function of T described in (18.4). We endow X with the supremum norm which is order-preserving on X_+, $\|v\|_\infty = 1$. Recall that $\varphi(s,f,m) = 0$ if $fm = 0$, which implies that $B(v) \le rv$.

For $\alpha \in (r,1)$, choose $\delta_\alpha > 0$ as guaranteed by Theorem 14.10. Let $h \in M_+^b(S)$, $h(s) = 0$ for $s \in S \setminus T$ and $\|h\|_\infty \le \delta_\alpha$. Then $h \le \delta_\alpha v$ and $\|F^n(h)\|_\infty \le \alpha^n \delta\alpha$ for all $n \in \mathbb{N}$. □

18.4.2. An instability result.

THEOREM 18.10. *Let b satisfy the Assumption 18.1 and φ the Assumption 18.2. Let the κ_j be Feller kernels.*

Then, if $\mathcal{T}_0 = \mathbf{r}(B) > 1$, there exists some $\epsilon_0 > 0$ such that

$$\sup_{n \in \mathbb{Z}_+} \|F^n(h)\|_\infty \ge \epsilon_0$$

for any $h \in C_+^b(S)$ that is bounded away from zero.

In particular, the zero function (the extinction equilibrium) is unstable.

PROOF. We combine Theorems 18.3 and 14.11 with u_0 being given as in Theorem 18.3. Since $\theta(u_0) > 0$ and θ is homogeneous and order-preserving, $\theta(h) > 0$ for any u_0-positive h, i.e., for any h with an $\epsilon > 0$ such that $h(s) \ge \epsilon u_0(s)$ for all $s \in S$.

Since u_0 is bounded, $\theta(h) > 0$ for any $h \in C_+^b(S)$ that is bounded away from zero. □

18.5. Uniform weak persistence

Recall the concepts of sustaining Feller kernels (Definition 13.63) and of a finite collection of jointly topologically irreducible Feller kernels (Definition 13.68).

DEFINITION 18.11. *A Feller kernel $\kappa : \mathcal{B} \times S \to \mathbb{R}_+$ is called* stringent *if there exist a compact subsets W of S and some $\delta \in (0,1)$ such that $\kappa(W,s) \ge \delta\kappa(S,s)$ for all $s \in S \setminus W$.*

REMARK 18.12. A Feller kernel $\kappa : \mathcal{B} \times S \to \mathbb{R}_+$ is *stringent* if and only if there exist a compact subset W of S and some $c \in (0,\infty)$ such that $\kappa(S \setminus W, s) \le c\,\kappa(W,s)$ for all $s \in S \setminus W$.

A Feller kernel is trivially stringent if S is compact. It is also stringent if $\kappa(S,\cdot)$ is bounded away from 0 and the kernel κ is tight (Definition 3.16).

The following result may help to prove by contradiction that a Feller kernel is stringent.

LEMMA 18.13. *Suppose that (W_n) is an increasing sequence of compact subsets of S and that κ is not stringent. Then, for any sequence (c_n) in $(0,\infty)$, there exists a sequence (s_n) with $s_n \in S \setminus W_n$ and $\kappa(S \setminus W_n, s_n) \ge c_n \kappa(W_n, s_n)$ for all $n \in \mathbb{N}$.*

The subsequent result and its proof are the center of this section.

THEOREM 18.14. *Let the κ_j, $j = 1,2$, be stringent sustaining Feller kernels that are jointly topologically irreducible. Let φ satisfy the Assumption 18.2 and b the Assumption 18.1 and $\mathcal{T}_0 = \mathbf{r}(B) > 1$.*

Then the population is uniformly weakly norm-persistent: There is some $\epsilon > 0$ such that $\limsup_{n\to\infty} \|F^n(h)\|_\infty \ge \epsilon$ for all $h \in \dot{C}_+^b(S)$.

PROOF. We apply Theorem 14.29. We consider the ordered Banach space $X = C^b(S)$ with the supremum norm $\|h\| = \|h\|_\infty = \sup_{s\in S} |h(s)|$ and the closed cone of nonnegative functions in X.

Assumption (a) of Theorem 14.29 follows from Theorem 13.69.

We now prove assumption (b) and (c) of Theorem 14.29. Since the κ_i, $i = 1, 2$, are stringent, by Definition 18.11 there exist compact subsets W_i of W and numbers $\delta_i > 0$ such that

$$\kappa_i(W_i, s) \geq \delta_i \kappa_i(S, s), \qquad s \in S \setminus S_i, \quad i = 1, 2.$$

Set $W = W_1 \cup W_2$ and $\delta_0 = \min\{\delta_1, \delta_2\}$. Then W is a compact subset of S and

$$\kappa_i(W, s) \geq \delta_0 \kappa_i(S, s), \qquad s \in S \setminus W, \quad i = 1, 2. \tag{18.18}$$

Let $h \in \dot{C}_+^b(S)$. Since W is compact, there exists some $m \in \mathbb{N}$ such that $W \subseteq S_n(h)$ for all $n \geq m$. Here the S_n are as in Theorem 13.69 but with B replacing F.

For all $n \geq m$, since W is compact and $B^n(h)$ is continuous and strictly positive on $S_n(h)$, there exists some $\epsilon_n > 0$ such that $B^n(h)(t) \geq \epsilon_n$ for all $t \in W$. If $t \in S \setminus W$,

$$(K_i B^n(h))(t) \geq \epsilon_n \kappa_i(W, t) \geq \epsilon_n \delta_0 \kappa_i(S, t).$$

So, there exists some $\tilde{\epsilon}_n > 0$ such that $(K_i B^n(h))(s) \geq \tilde{\epsilon}_n \kappa_i(S, s)$ for all $s \in S$. Since $\varphi(s, \cdot)$ is order-preserving and homogeneous,

$$B^{n+1}(h)(s) \geq \varphi\big(s, \tilde{\epsilon}_n \kappa_1(S, s), \tilde{\epsilon}_n \kappa_2(S, s)\big) = \tilde{\epsilon}_n u_0(s), \qquad s \in S,$$

with u_0 from Theorem 18.3. This shows that $B^{n+1}(h)$ is u_0-positive for all $n \geq m$.

Theorem 18.3 also provides an order-preserving homogeneous bounded eigenfunctional θ of B that satisfies assumption (b) and (c) of Theorem 14.29, which in turn implies that there is some $\epsilon > 0$ such that $\limsup_{n\to\infty} \|F^n(h)\|_\infty \geq \epsilon$ for all $h \in \dot{C}_+^b(S)$. □

We refer to [**205**], Sections 3.5, 3.6, 8 and 9, for results concerning uniform and order persistence.

In Theorem 18.14, the Feller kernels are assumed to have so many properties that some example is in place.

Let $\tilde{S}$ be a metric space under the metric $\tilde{d}$. Let S be an open subset of $\tilde{S}$ such that $\bar{S}$ is a compact subset of $\tilde{S}$. S is a metric space under the restriction d of $\tilde{d}$ to $S \times S$.

We assume that $\tilde{\mu}$ is a finite nonnegative measure on the $\tilde{\mathcal{B}}$, the Borel σ-algebra of $\tilde{S}$, $\tilde{\mu} \in \mathcal{M}_+(\tilde{S})$. Since S is an open subset of $\tilde{S}$, the restriction of $\tilde{\mu}$ is also a measure on the Borel σ-algebra of S, $\mathcal{B}$. Actually,

$$\mathcal{B} \subseteq \tilde{\mathcal{B}}, \qquad \mathcal{B} = \{S \cap T; T \in \tilde{\mathcal{B}}\}.$$

We assume that $\tilde{\mu}(U) > 0$ for any nonempty open subset of S.

For instance, let $\tilde{S} = \mathbb{R}^N$, $\tilde{\mu}$ the Lebesgue measure on $\mathbb{R}^n$, and S be an open bounded subset of $\mathbb{R}^N$.

For $\tilde{s} \in \tilde{S}$ and $\delta > 0$, let

$$\tilde{U}_\delta(\tilde{s}) = \{t \in \tilde{S}; \tilde{d}(t, \tilde{s}) < \delta\}$$

be the δ-neighborhood of $\tilde{s}$ in $\tilde{S}$. $U_\delta(s)$ denotes the δ-neighborhood of $s \in S$ in S.

LEMMA 18.15. *Let (W_n) be an increasing sequence of subsets of S the union of which is S, $W_n \in \tilde{\mathcal{B}}$. Then, for any $\delta > 0$,*

$$\lim_{n\to\infty} \inf_{s \in S} \tilde{\mu}\big(W_n \cap U_\delta(s)\big) > 0.$$

PROOF. Notice that $\inf_{s\in\bar S}\tilde\mu(W_n\cap U_\delta(s))$ are terms of an increasing sequence. Suppose that the statement is false. Then there exists some $\delta>0$ such that for all $n\in\mathbb{N}$ there exist some $s_n\in S$ such that

$$\tilde\mu(W_n\cap\tilde U_\delta(s_n))\to 0,\qquad n\to\infty. \tag{18.19}$$

Since $\bar S$ is a compact subset of $\tilde S$, after choosing a subsequence we can assume that $s_n\to s$ as $n\to\infty$ with some $s\in\bar S$ and (18.19) still holds. There exists some $N\in\mathbb{N}$ such that $d(s_n,s)<\delta/2$ for $n\ge N$. By the triangle inequality, for $n\ge N$, $\tilde U_{\delta/2}(s)\subseteq\tilde U_\delta(s_n)$. So, for $n\ge N$,

$$\tilde\mu(W_n\cap\tilde U_\delta(s_n))\ge\tilde\mu(W_n\cap\tilde U_{\delta/2}(s))\overset{n\to\infty}{\longrightarrow}\tilde\mu(S\cap\tilde U_{\delta/2}(s))>0. \tag{18.20}$$

The last inequality holds because $s\in\bar S$ and, by Lemma A.21, $U=S\cap\tilde U_{\delta/2}(s)$ is a nonempty subset of S that is open as intersection of two open sets; so $\tilde\mu(U)>0$ by assumption. We obtain a contradiction between (18.20) and (18.19). □

EXAMPLE 18.16. Let $\tilde S, S,\tilde\mu$ be as before. Let $\kappa:\mathcal{B}\times S\to\mathbb{R}_+$ be a Feller kernel. Further let $\delta>0$ such that

$$\kappa(T,s)\ge\kappa(S,s)\,\tilde\mu(T),\qquad s\in S,\quad T\in\mathcal{B},\quad T\subseteq U_\delta(s). \tag{18.21}$$

Then κ is stringent. If $\kappa(S,s)>0$ for all $s\in S$, κ is a colonization kernel.

PROOF. Set

$$W_n=\{t\in S; D(t)\ge 1/n\},\quad n\in\mathbb{N},$$

where $D(t)=d(t,\tilde S\setminus S)$ is the distance from t to the closed set $\tilde S\setminus S$, (2.36). Since $\bar S$ is a compact subset of $\tilde S$ by assumption, W_n is a compact subset of S, (W_n) is increasing and its union is S. By (18.21),

$$\kappa(s,W_n)\ge\kappa\big(s,W_n\cap U_\delta(s)\big)\ge\kappa(S,s)\,\tilde\mu\big(W_n\cap U_\delta(s)\big),\qquad n\in\mathbb{N},\quad s\in S. \tag{18.22}$$

The assertion now follows from Lemma 18.15. As for κ being a colonization kernel, recall that $\tilde\mu(U)>0$ for all nonempty open subsets of S and apply (18.21). □

For more elementary examples, see the next section.

18.6. Population survival and sex-biased dispersal

In this chapter, we have given conditions that establish the spectral radius $\mathcal{T}_0$ of the basic population turnover operator (the homogeneous first order approximation of the population turnover operator) as threshold between population extinction and population survival. The Assumptions 18.1 and 18.2 are these conditions and are imposed throughout this section (cf. [**205**, Sec.10]).

While $\mathcal{T}_0>1$ is sufficient to make the extinction state, the zero vector, unstable, it is not sufficient for uniform weak persistence. For instance, it is possible that

$$\mathcal{T}_0\ge\inf_{s\in S}\varphi(s,\kappa_1(S,s),\kappa_2(S,s))>1$$

(Corollary 18.6) and that there is some nonempty open subset U of S such that

$$\kappa_1(U,s)\kappa_2(U,s)=0,\qquad s\in S. \tag{18.23}$$

Then, a semelparous population the newborns of which are concentrated in U will be extinct after one generation because there are no male and female born in U that will meet at the same place during the mating season. See (18.14).

The latter is avoided if the kernels κ_1 and κ_2 are sustaining (Definition 13.63) and jointly topologically irreducible (Definition 13.68). If this is the case, the influence of sex-biased dispersal on population persistence is still difficult to appreciate because the spectral radius is difficult to determine or to estimate except in special cases. See Corollary 18.6.

One may like to compare the forthcoming considerations with the findings in [**162**] as to how sex-biased dispersal influences the success of invading species.

18.6.1. Separable dispersal. We assume that the dispersal Feller kernels have the form

$$\kappa_j(T,s) = \gamma_j(s)\mu_j(T), \qquad s \in S, \quad T \in \mathcal{B}, \quad j = 1,2, \tag{18.24}$$

with $\gamma_j \in \dot{C}_+^b(S)$ and $\mu_j \in \dot{\mathcal{M}}_+(S)$, $j = 1,2$. Actually, μ_j is a probability measure where $\mu_j(T)$ is the probability that a neonate of the respective sex was born in the set T; $\gamma_j(s) \in (0,1]$ is the birthplace-independent probability that an individual of the respective sex gets to s.

The next year turnover operator has the form

$$F(h)(s) = b\Big(s, \gamma_1(s)\int_S h d\mu_1, \gamma_2(s)\int_S h d\mu_2\Big), \qquad s \in S, \quad h \in C_+^b(S). \tag{18.25}$$

Since $b(s,f,m) = 0$ if $fm = 0$ by Assumption 18.1, F is the zero operator if $\gamma_1(s)\gamma_2(s) = 0$ for all $s \in S$, i.e., if females and males move to two disjoint parts of the habitat. This means that the population dies out after one generation whatever the initial densities are.

Define functionals $\theta_j : C_+^b(S) \to \mathbb{R}_+$ by

$$\begin{aligned} \theta_1(h)(s) &= \min\Big\{\int_S h d\mu_1, \int_S h d\mu_2\Big\}. \\ \theta_2(h)(s) &= \max\Big\{\int_S h d\mu_1, \int_S h d\mu_2\Big\}. \end{aligned} \tag{18.26}$$

The θ_j are bounded, order-preserving and homogeneous. By the properties of φ (Assumption 18.2), for all $s \in S$,

$$\begin{aligned} B(h)(s) &\le \varphi(s, \gamma_1(s)\theta_2(h), \gamma_2(s)\theta_2(h)) = u(s)\theta_2(h) =: B_2(h)(s), \\ u(s) &= \varphi(s, \gamma_1(s), \gamma_2(s)). \end{aligned}$$

Similarly, for all $s \in S$,

$$B(h)(s) \ge u(s)\theta_1(h) =: B_1(h)(s).$$

The operators B_j on $C_+^b(S)$ are continuous, order-preserving, homogeneous and of rank one. By Example 5.1, $\mathbf{r}(B_j) = \theta_j(u)$, $j = 1,2$. By Corollaries 6.24 and 6.25,

$$\theta_1(u) \le \mathbf{r}(B) = \mathcal{T}_0 \le \theta_2(u).$$

If $\mu_1 = \mu_2$, the estimates become sharp and

$$\mathcal{T}_0 = \mathbf{r}(B) = \int_S \varphi(s, \gamma_1(s), \gamma_2(s))\mu_1(ds).$$

$\mathcal{T}_0$ is a threshold parameter between local stability and instability of the extinction state in the sense of Theorems 18.8 and 18.10. For $\mathcal{T}_0 > 1$ to imply uniform weak persistence (Theorem 18.14), we have assumed that the kernels κ_1 and κ_2 are sustaining, jointly topologically irreducible and stringent. In our special case, this means that the $\gamma_j(s) > 0$ for all $s \in S$ and $j = 1,2$ and that $\mu_j(T) > 0$ for every nonempty open subset T of S and $j = 1, 2$. The kernel κ_j is stringent if μ_j is tight which is the case if the metric space is complete and separable.

18.6.2. The strategy of no dispersal. Since the two sexes may end up in disjoint parts of the habitat during the mating season and reproduction fails leading to extinction of the population after one generation (Section 18.6.1), one dispersal strategy may be not to disperse. The corresponding dispersal kernels are

$$\kappa_j(T, s) = q_j(s)\chi_T(s), \qquad s \in S, \quad T \in \mathcal{B}, \tag{18.27}$$

with continuous functions $q_j : S \to [0, 1]$ and indicator functions χ_T (see (18.4)). These Feller kernels are not topologically irreducible (unless S is a singleton set). In this special case, we have the interpretation that

$$\kappa_j(S, s) = q_j(s) \tag{18.28}$$

is the probability that that a member of sex j is alive and located at $s \in S$ at the end of the year (the beginning of the mating season). By (18.27),

$$(K_j h)(s) = q_j(s)h(s), \qquad s \in S, \quad h \in M^b(S).$$

Since φ is homogeneous, by (18.16),

$$(B(h))(s) = \varphi\big(s, q_1(s), q_2(s)\big)\, h(s), \quad s \in S, \quad h \in C_+^b(S). \tag{18.29}$$

LEMMA 18.17. $\mathcal{T}_0 = \mathbf{r}(B) = \sup_{s \in S} \varphi\big(s, q_1(s), q_2(s)\big)$.

PROOF. By (18.29), B is a additive multiplication operator,

$$\begin{aligned}(B(h))(s) =& \gamma(s)h(s), \qquad s \in S, \quad h \in C_+^b(S),\\ \gamma(s) =& \varphi\big(s, q_1(s), q_2(s)\big).\end{aligned}$$

Set $\mathcal{T}_0 = \sup \gamma(S)$. By induction, for all $n \in \mathbb{N}$,

$$(B^n(h))(s) = (\gamma(s))^n h(s), \qquad s \in S, \quad h \in C_+^b(S).$$

By (4.1),

$$\|B^n\| \le \sup_{s \in S} (\gamma(s))^n = \mathcal{T}_0^n.$$

The constant nonnegative functions are elements of $C_+^b(S)$, and

$$\|B^n h\|_\infty = \mathcal{T}_0^n \, \|h\|_\infty, \qquad n \in \mathbb{N},$$

if h is constant and nonnegative. In combination, $\|B^n\| = \mathcal{T}_0^n$. By (1.7) or (5.1), $\mathbf{r}(B) = \mathcal{T}_0$. □

Since these dispersal kernels are not topologically irreducible, $\mathcal{T}_0 > 1$ only guarantees instability of the extinction state (Theorem 18.10) but no persistence. Indeed, if

$$\inf_{s \in S} \varphi(s, q_1(s), q_2(s)) < 1,$$

the population may die out if it is located in a part $T \in \mathcal{B}$ satisfying

$$\sup_{s \in T} \varphi(s, q_1(s), q_2(s)) < 1.$$

See Theorem 18.9. For the population to persist, some males and females would need to move to a more favorable part of the habitat.

18.6.3. Combination dispersal strategies. To avoid the risks of dispersal and of non-dispersal in the previous sections, the two sexes may choose mixed strategies where some but not all individuals of each sex disperse. Let us assume that each individual makes the extreme choice of either staying put or dispersing indiscriminately over the habitat. Let $p_j \in [0,1]$ be the probabilities of the respective sexes to stay put (also called non-dispersal probabilities). We choose dispersal kernels

$$\kappa_j(T,s) = \vartheta_j\big(p_j \chi_T(s) + (1-p_j)\gamma(s)\mu_j(T)\big), \qquad s \in S, \quad T \in \mathcal{B}. \tag{18.30}$$

Here μ_j is a probability measure where $\mu_j(T)$ is the probability that a dispersing neonate was born in the set T.

The numbers $\vartheta_j \in [0,1]$ are the probabilities of having the appropriate sex and of surviving the year, and $\gamma(s) \in (0,1]$ is the sex and birthplace independent probability at which a dispersing individual gets to s. Assume $\gamma \in \dot C^b_+(S)$.

By (18.1) and (18.30),

$$(K_j h)(s) = \vartheta_j\Big(p_j h(s) + (1-p_j)\gamma(s)\int_S h\, d\mu_j\Big), \qquad s \in S, \quad h \in C^b(S).$$

This formula suggests that γ is an eigenfunction of K_j, $j = 1,2$,

$$(K_j\gamma)(s) = \vartheta_j\Big(p_j\gamma(s) + (1-p_j)\gamma(s)\int_S \gamma d\mu_j\Big) = \lambda_j\gamma(s),$$

with eigenvalues

$$\lambda_j = \vartheta_j\Big(p_j + (1-p_j)\int_S \gamma\, d\mu_j\Big). \tag{18.31}$$

Since φ is homogeneous, by (18.16),

$$B(\gamma)(s) = \varphi\big(s, \lambda_1, \lambda_2\big)\,\gamma(s), \qquad s \in S. \tag{18.32}$$

If $\varphi(s,\cdot)$ is independent of s, then γ is an eigenfunction of B associated with the eigenvalue $\varphi(\lambda_1,\lambda_2)$.

THEOREM 18.18. *Let $\varphi(s,\cdot)$ be independent of s.*

(a) *The spectral radius of B is*

$$\mathcal{T}_0 = \mathbf{r}(B) = \varphi(\lambda_1,\lambda_2), \tag{18.33}$$

with λ_j from (18.31)

(b) *Let γ be strictly positive on S. For $j = 1,2$, let $p_j < 1$, μ_j be tight and $\mu_j(U) > 0$ for any nonempty open subset U of S. Let φ satisfy the Assumption 18.2 and b the Assumption 18.1.*

Then the κ_j are stringent sustaining jointly top-irreducible kernels, and uniform weak persistence holds provided that $\varphi(\lambda_1,\lambda_2) > 1$.

PROOF. By our assumptions, γ is an eigenfunction of B associated with $\varphi(\lambda_1,\lambda_2)$.

(a) By Theorem 5.31 and (5.30), $\mathbf{r}(B) \ge cw(B) \ge \varphi(\lambda_1,\lambda_2)$.

We first assume that γ is bounded away from zero, i.e., γ is an interior point of $C^b_+(S)$ and a uniform order unit for $C^b_+(S)$. See (2.34) and Proposition 2.73. Then $\mathbf{r}(B) \le CW_\gamma(B) \le \varphi(\lambda_1,\lambda_2)$ by Theorem 6.15 because B is uniformly order-bounded with order bound γ.

Otherwise, for $n \in \mathbb{N}$, we define

$$\gamma_n(s) = \gamma(s) + (1/n), \qquad s \in S,$$

and let B_n be the homogeneous bounded operator with γ_n replacing γ. Then γ_n is a uniform order unit for $C_+^b(S)$. By our preceding considerations,

$$\begin{aligned}\mathbf{r}(B_n) &= \varphi(\lambda_1(n), \lambda_2(n)),\\ \lambda_j(n) &= \vartheta_j\Big(p_j + (1-p_j)\int_S \gamma_n d\mu_j\Big)\\ &= \vartheta_j\Big(p_j + (1-p_j)\Big[\int_S \gamma d\mu_j + \frac{1}{n}\mu_j(S)\Big]\Big).\end{aligned}$$

Since $\gamma \le \gamma_n$ and φ is order-preserving, $B(h) \le B_n(h)$ in $C_+^b(S)$ for all $h \in C_+^b(S)$ and all $n \in \mathbb{N}$. Since B is order-preserving and $C_+^b(S)$ a normal cone, by Corollary 6.25,

$$\varphi(\lambda_1, \lambda_2) \le \mathbf{r}(B) \le \mathbf{r}(B_n) = \varphi(\lambda_1(n), \lambda_2(n)).$$

Since $\lambda_j(n) \to \lambda_j$ as $n \to \infty$ and φ is continuous, $\mathbf{r}(B) = \varphi(\lambda_1, \lambda_2)$.

(b) See Theorem 18.14 and the end of Section 18.6.1. □

REMARK 18.19. If $\varphi(s, \cdot)$ depends on $s \in S$, part (b) still holds if $\mathcal{T}_1 := \inf_{s\in S} \varphi(s, \lambda_1, \lambda_2) > 1$. By (18.32),

$$B(\gamma)(s) \ge \mathcal{T}_1 \gamma(s), \quad s \in S.$$

By Theorem 5.31 and (5.30), $\mathbf{r}(B) \ge cw(B) \ge \mathcal{T}_1 > 1$.

REMARK 18.20. Let $\phi(s, \cdot)$ depend on $s \in S$ and set

$$\mathcal{T}_2 = \sup_{s\in S} \varphi(s, \lambda_1, \lambda_2).$$

By (18.32),

$$B(\gamma)(s) \le \mathcal{T}_2 \gamma(s), \qquad s \in S.$$

Assume that γ is bounded away from 0, $\inf \gamma(S) > 0$. Then γ is comparable to u with $u(s) = 1$ for all $s \in S$. Since B is uniformly u-bounded, $CW_u(B) \le \mathcal{T}_2$ by (6.6) and $\mathbf{r}(B) \le CW_u(B)$ by Theorem 6.15. In particular, Theorem 18.8 holds if $\mathcal{T}_2 < 1$.

We continue to assume that $\varphi(s, \cdot)$ is independent of s.

By our assumptions, $\int_S \gamma d\mu_j \in (0, 1]$. λ_j is independent of p_j if $\int_S \gamma d\mu_j = 1$. If $\int_S \gamma d\mu_j < 1$, λ_j depends on p_j in a strictly increasing way. By (18.33), the basic turnover number $\mathcal{T}_0$ depends on p_j in an increasing way. This is remarkable, because the kernels (18.30) and thus the operator B do not necessarily depend on p_j in an increasing way. See (18.17) and recall Section 6.5.

If $p_j = 1$ for $j = 1$ or $j = 2$, however, $\mathcal{T}_0$ partially looses its importance as a threshold parameter. Let us look at the case that all individuals of one sex stay where they are and all individuals of the other sex disperse, say $p_1 = 0$ and $p_2 = 1$. By (18.31),

$$\mathcal{T}_0 = \varphi\Big(\vartheta_1, \vartheta_2 \int_S \gamma\, d\mu_2\Big). \tag{18.34}$$

Here it is convenient to use that F and B are also maps on $M^b_+(S)$, the closed cone of $M^b(S)$, the Banach space of bounded Borel measurable functions with the supremum norm. Then

$$B(\chi_T)(s) = \varphi\big(\vartheta_1 \chi_T(s)), \vartheta_2 \gamma(s)\mu_2(T)\big), \qquad s \in S.$$

Since $\varphi(f,m) = 0$ if $f = 0$ or $m = 0$,

$$B(\chi_T)(s) = \begin{cases} \varphi\big(\vartheta_1, \vartheta_2\gamma(s)\mu_2(T)\big), & s \in T, \\ 0, & s \in S \setminus T. \end{cases}$$

Since φ is order-preserving and $\gamma(s) \le 1$ for all $s \in S$,

$$B(\chi_T) \le \lambda_T \chi_T, \qquad \lambda_T = \varphi\big(\vartheta_1, \vartheta_2\, \mu_2(T)\big). \tag{18.35}$$

If T is a sufficiently small subset of S with nonempty interior such that $\mu_2(T)$ is small enough, it can happen by (18.34) and (18.35) that $\lambda_T < 1 < \mathcal{T}_0$. Then, by Theorem 14.10, there is some $\delta > 0$ such that

$$F^n(h) \to 0, \quad n \to \infty,$$

for any $h \in C^b_+(S)$ with $h(s) = 0$ for all $s \in S \setminus T$ and $\|h\|_\infty \le \delta$.

In fact, since γ is an eigenfunction of B, we have a bistable situation as there is some $\epsilon > 0$ such that $\sup_{n \in \mathbb{Z}_+} \|F^n(h)\|_\infty \ge \epsilon$ for all $h \in C^b_+(S)$ that are γ-positive (Theorem 18.10).

We refer to [**205**, Sec.10.3] for a comparison of the semelparous two-sex population to corresponding one-sex populations.

In view of our discussion of the continuity of the spectral radius (Chapters 7 and 10), we mention that $\mathcal{T}_0$ in (18.33) continuously depends on all model ingredients.

18.6.4. Continuity of the spectral radius. The last observation raises the question whether the spectral radius of B, $\mathcal{T}_0 = \mathbf{r}(B)$, depends continuously on the model ingredients if $\varphi(s,\cdot)$ depends on $s \in S$. For simplicity, we single out the non-dispersal probabilities p_1 and p_2, which seem to be the most interesting in the context of our discussion.

This is a special case of the continuous dependence of the basic turnover $\mathcal{T}_0$ number on the dispersal kernels κ_1 and κ_2 which can be addressed similarly. The continuous dependence of $\mathcal{T}_0$ on the mating and birth function φ is discussed in [**208**].

We plan on applying Theorem 10.6.

Let us assume that φ satisfies a Lipschitz condition: There is some $\Lambda_\varphi > 0$, such that

$$|\varphi(s,f,m) - \varphi(s,\tilde f,\tilde m)| \le \Lambda_\varphi \max\big\{|f-\tilde f|, |m - \tilde m|\big\}, \qquad s \in S, \quad f,\tilde f, m, \tilde m \ge 0. \tag{18.36}$$

Then B is uniformly continuous on bounded sets. Further, $\|B - \tilde B\| \to 0$ if $(p_1,p_2) \to (\tilde p_1, \tilde p_2)$ and $\tilde B$ is obtained from B by replacing (p_1, p_2) by $(\tilde p_1, \tilde p_2)$.

By Theorem 7.1, $\mathbf{r}(B)$ is an upper semicontinuous function of (p_1,p_2):

If $p_1, p_2 \in [0,1]$ and $\epsilon > 0$, there is some $\delta > 0$ such that $\mathbf{r}(\tilde B) < \mathbf{r}(B) + \epsilon$ for all

$$\tilde p_1, \tilde p_2 \in [0,1], \qquad |\tilde p_1 - p_1| + |\tilde p_2 - p_2| < \delta.$$

See Appendix A.4.

To find conditions for the lower semicontinuity of $\mathbf{r}(B)$ in (p_1, p_2), we will apply Theorem 10.6. Notice that B is not compact and, apparently, has no compact power so that we cannot apply the results in [**146**].

We assume that γ_j, $j = 1, 2$, and $\varphi(\cdot, 1, 1)$ are not only bounded on S but also bounded away from 0 on S, $\inf \gamma_j(S) > 0$, $\inf_{s\in S} \varphi(s, 1, 1) > 0$. Let u be the constant function on S with value 1. Then $\|\cdot\|_u = \|\cdot\|_\infty$.

Then $\|B(h) - \tilde{B}(h)\|_u \le \|B - \tilde{B}\| \, \|h\|$ for all $h \in C^b_+(S)$.

Further, B is uniformly u-bounded and, if $p_j < 1$ for $j = 1, 2$, pointwise u-positive. Indeed, since $\varphi(s, \cdot)$ is homogeneous and order-preserving for all $s \in S$,

$$\begin{aligned}(Bh)(s) &\ge \varphi\Big(s, \, (1-p_1)\gamma(s)\int_S h d\mu_1, \, (1-p_2)\gamma(s)\int_S h d\mu_1\Big) \\ &\ge \varphi(s,1,1)\gamma(s) \min_{j=1,2}(1-p_j)\int_S h d\mu_j.\end{aligned}$$

This implies that $\mathbf{r}(B) > 0$. It remains to find conditions for $cw(B) \ge \mathbf{r}(B)$.

We aim at applying Theorem 9.2.

To this end, we define the so-called Nemytskii (or superposition) operator for φ [**55**, Exp.7.5],

$$N_\varphi(h_1, h_2)(s) = \varphi(s, h_1(s), h_2(s)), \qquad h_1, h_2 \in C^b_+(S), \quad s \in S. \tag{18.37}$$

N_ϕ inherits Lipschitz continuity on $C^b_+(S) \times C^b_+(S)$ from φ, (18.36).

We also introduce the bounded linear operators $L, \tilde{L} : C^b(S) \to C^b(S) \times C^b(S)$,

$$\begin{aligned} &Lh = (\vartheta_1 p_1 h, \vartheta_2 p_2 h), \\ &\tilde{L}h = \Big(\vartheta_1(1-p_1)\Big(\int_S h\, d\mu_1\Big)\gamma, \vartheta_2(1-p_2)\Big(\int_S h\, d\mu_2 \gamma\Big)\gamma\Big), \\ &h \in C^b_+(S). \end{aligned} \tag{18.38}$$

Then

$$B = N_\varphi(L + \tilde{L}).$$

By Theorems 5.14 and 5.33,

$$\begin{aligned} \mathbf{r}(B) &= \mathbf{r}(LN_\varphi + \tilde{L}N_\varphi), \\ cw(B) &= cw(LN_\varphi + \tilde{L}N_\varphi), \end{aligned} \tag{18.39}$$

To show that $\mathbf{r}(B) \le cw(B)$, it is sufficient to show that $\mathbf{r}(B)$ is an eigenvalue of $LN_\phi + \tilde{L}N_\phi$. See Remark 10.10. To this end, we use Theorem 9.2. Now $\tilde{L}$ is a compact homogeneous operator and so is $\tilde{L}N_\varphi$, while LN_φ is Lipschitz continuous on $C^b_+(S) \times C^b_+(S)$ because we have assumed that φ is Lipschitz continuous. If we use the norm $\|(h_1, h_2)\| = \max_{j=1,2} \|h_j\|_\infty$, $h_j \in C^b_+(S)$, $\tilde{L}N_\varphi$ has Lipschitz constant $\max_{j=1,2} \vartheta_j p_j \Lambda$ from (18.36). Recall Remark 18.19 which gives a lower estimate of $\mathbf{r}(B)$.

THEOREM 18.21. *Let*

$$\Lambda \max_{j=1,2} \vartheta_j p_j^\circ < \inf_{s\in S} \varphi(s, \lambda_1, \lambda_2) \tag{18.40}$$

with λ_j *given by* (18.31),

$$\lambda_j = \vartheta_j \Big(p_j^\circ + (1 - p_j^\circ) \int_S \gamma \, d\mu_j \Big).$$

Then the spectral radius of B *as a function of* (p_1, p_2) *is continuous at* (p_1°, p_2°).

PROOF. The left-hand side of (18.40) is a Lipschitz constant of $\tilde{L}N_\varphi$ while the right-hand side is a lower estimate of the spectral radius of $\mathbf{r}(B)$. By Theorem 9.2, $\mathbf{r}(B)$ is an eigenvalue of $LN_\phi + \tilde{L}N_\phi$ and so $\mathbf{r}(B) \leq cw(B)$ by (18.39). □

EXAMPLE 18.22. Let

$$\varphi(s, f, m) = \beta(s) \min\{f, m\}, \qquad s \in S, \quad f, m \geq 0.$$

Here $\beta(s) > 0$ is the per pair fertility (number of offspring) at $s \in S$. Then the Lipschitz constant of φ in (18.36) is $\Lambda_\varphi = \sup \beta(S)$. By Theorem 18.21, the spectral radius of B as a function of (p_1, p_2) is continuous at (p_1°, p_2°) if

$$\sup \beta(S) \max_{j=1,2} \vartheta_j p_j^\circ < \inf \beta(S) \min_{j=1,2} \vartheta_j \Big(p_j^\circ + (1 - p_j^\circ) \int_S \gamma \, d\mu_j \Big).$$

CHAPTER 19

Populations with measure-valued structural distributions

The measurable space S with σ-algebra $\mathcal{B}$ represents the state space of individual structural characteristics of a structured population. A measure $\mu : \mathcal{B} \to \mathbb{R}_+$ has a natural interpretation as the structural distribution of the population with $\mu(T)$ being the number (or biomass) of individuals with structural characteristics in the set $T \in \mathcal{B}$. See Chapter 3. Actually, the specific formulation of the population model below suggests that the structure is given by spatial location, with S representing the habitat of the population. See Section 19.4 for a more general model framework that may include physiological structures given by age, body size, viral load, and so on.

19.1. A recursive equation for measures

We consider a difference equation on $\mathcal{M}_+(S)$, the cone of nonnegative finite measures on $\mathcal{B}$,

$$\mu_n(T) = F(\mu_{n-1})(T), \qquad n \in \mathbb{N}, \quad T \in \mathcal{B}, \tag{19.1}$$

with $\mu_0 \in \mathcal{M}_+(S)$ and $F : \mathcal{M}_+(S) \to \mathcal{M}_+(S)$ given by

$$F(\mu)(T) = \int_S \Big[\int_T g\Big(s, t, \int_S \xi(s,t,z)\mu(dz)\Big) P(dt, s) \Big] \mu(ds), \tag{19.2}$$

for $\mu \in \mathcal{M}_+(S)$ and $T \in \mathcal{B}$.

The difference equation describes the year-to-year development of a spatially distributed population with a short reproductive season. The population census is taken in such a way that the reproductive season is at the end of the year (post-reproductive census). The measure μ_n is the spatial distribution at the beginning of the nth year, after the end of the reproduction season of the $(n-1)^{th}$ year.

The measure kernel $P : \mathcal{B} \times S \to \mathbb{R}_+$ describes spatial migration and survival without intraspecific competition: $P(T, s)$ is the probability that an individual that is located at $s \in S$ at the beginning of the year does not die from non-competitive causes of death during the year and is located within the set $T \in \mathcal{B}$ at the end of the year.

$g = g_1 + g_2$ has two additive components. $g_1 : S^2 \times \mathbb{R}_+ \to [0,1]$ describes how survival is lowered by intraspecific competition with u being the force of competition. $g_2 : S^2 \times \mathbb{R}_+ \to \mathbb{R}_+$ describes per capita reproduction and how it is affected by intraspecific competition.

$g_1(s,t,u) \in [0,1]$ is the chance of an individual that is at s at the beginning of the year and would be at t at the end of year not to die due to intraspecific competition if u is the force of competition that is exerted on it. If the species is semelparous, $g_1 = 0$; if not, $g_1(s,t,0) = 1$ for all $s, t \in S$.

$g_2(s,t,u)$ is the amount of offspring of an individual that is at s at the beginning of the year and would be at t at the end of the year, born at the end of year, if u is the force of competition exerted on that individual. Notice that this assumes that the offspring has the same location that the parent has immediately before given birth. This is why the model is well suited for spatial distribution but not so well suited for structural distributions for which the structural characteristic of the offspring is different from the structural characteristic of the parent (body size, e.g.).

$u(s,t) = \int_S \xi(s,t,x)\mu(dx)$ is the force of competition exerted by a population of distribution $\mu \in \mathcal{M}_+(S)$ on an individual that is at $s \in S$ at the beginning of the year and at $t \in S$ at the end of the year. In the time-discrete formulation of the model, it is ignored which way the individual gets from s to t. $\xi(s,t,x)$ is the competitive influence of an individual that is at x at the beginning of the year on an individual that is at s at the beginning of the year and at t at the end of the year.

19.2. Habitats that are just measurable spaces

If S is a measurable space with σ-algebra $\mathcal{B}$, we choose the variation norm $\mathcal{M}(S)$, denoted by $\|\cdot\|$ or the equivalent supremum norm $\|\mu\|_\infty = \sup_{T\in\mathcal{B}} |\mu(T)|$. See Section 3.1. We make the following assumptions.

ASSUMPTION 19.1. *These are the assumptions for g.*

(i) *$g : S^2 \times \mathbb{R}_+ \to \mathbb{R}_+$ is bounded and measurable where $S^2 \times \mathbb{R}_+$ is equipped with the product σ-algebra.*
(ii) *For any $c > 0$, $\inf g\big(S^2 \times [0,c]\big) > 0$.*
(iii) *$g(s,t,u) \to g(s,t,0)$ as $u \to 0$ uniformly for $s,t \in S$.*

ASSUMPTION 19.2. *$\xi : S^3 \to \mathbb{R}_+$ is bounded and measurable where S^3 is equipped with the product σ-algebra.*

ASSUMPTION 19.3.

(i) *$P : \mathcal{B} \times S \to \mathbb{R}_+$ is a measure kernel.*
(ii) *$0 < P(S,s) \le 1$ for all $s \in S$.*

The linear first order approximation for F, A, will be given by

$$(A\mu)(T) = \int_S \Big(\int_T g(s,t,0)P(dt,s)\Big)\mu(ds). \tag{19.3}$$

Notice that A is induced by the measure kernel

$$\kappa(T,s) = \int_T g(s,t,0)P(dt,s), \qquad T \in \mathcal{B}, \quad s \in S. \tag{19.4}$$

We also consider the operator $\tilde{A}$ induced by P,

$$(\tilde{A}\mu)(T) = \int_S P(T,s)\mu(ds). \tag{19.5}$$

PROPOSITION 19.4. *Assume Assumptions 19.1, 19.2 and 19.3. Then F is an operator from $\mathcal{M}_+(S)$ to $\mathcal{M}_+(S)$ and A and $\tilde{A}$ are linear bounded positive operators on $\mathcal{M}(S)$. Further, the restriction of the linear bounded map A to $\mathcal{M}_+(S)$ is the order-derivative on F.*

Finally, if $\mu \in \dot{\mathcal{M}}_+(S)$, also $F(\mu) \in \dot{\mathcal{M}}_+(S)$.

PROOF. The first assertions follow by standard measure theory. To check that A is the order-derivative of F, let $\epsilon \in (0,1)$. By Assumption 19.1 (ii) and (iii), there exists some $\tilde{\delta} > 0$ such that $(1-\epsilon)g(s,t,0) \le g(s,t,u) \le (1+\epsilon)g(s,t,0)$ for all $u \in [0,\tilde{\delta}]$ and $s,t \in S$. Since ξ is bounded by Assumption 19.2, there exists some $\delta > 0$ such that $\xi(s,t,z)\delta \le \tilde{\delta}$ for all $s,t,z \in S$. Now assume that $\|\mu\|_\flat \le \delta$. Then $\int_S \xi(s,t,z)\mu(dz) \le \sup_{z\in S} \xi(s,t,z)\delta \le \tilde{\delta}$ for all $s,t \in S$. This implies that $(1-\epsilon)A(\mu) \le F(\mu) \le (1+\epsilon)A(\mu)$.

As for the last statement, notice that

$$F(\mu)(S) = \int_S \Big(\int_S w(s,t)P(dt,s) \Big) \mu(ds),$$

where $w(s,t) = g(s,t,v(s,t))$ with $v(s,t) = \int_S \xi(s,t,z)\mu(dz)$, $s,t \in S$. Since v is bounded by Assumption 19.2, $w(s,t) > 0$ for all $s,t \in S$ by Assumption 19.1. Since $P(S,s) > 0$ for all $s \in S$ by Assumption 19.3, $\int_S w(s,t)P(dt,s) > 0$ for all $s \in S$ and so $F(\mu)(S) > 0$ because $\mu(S) > 0$. □

Since A in (19.3) as the order-derivative of the turnover operator F is the basic turnover operator, we expect the spectral radius of A,

$$\mathcal{T}_0 = \mathbf{r}(A), \tag{19.6}$$

the basic turnover number, to be the threshold parameter between population extinction and population persistence. We can also view $\mathcal{T}_0$ as the spectral radius of the kernel κ in (19.4),

$$\mathcal{T}_0 = \mathbf{r}(\kappa). \tag{19.7}$$

See Section 13.2.2.

19.2.1. Local (global) asymptotic stability of the extinction state in the subthreshold case. Since $\mathcal{M}(S)$ with the variation norm is an ordered Banach space and $\mathcal{M}_+(S)$ a normal closed generating (and non-flat) cone, the following local stability result follows from Theorem 14.13.

THEOREM 19.5. *Assume Assumptions* 19.1, 19.2 *and* 19.3 *and* $\mathcal{T}_0 = \mathbf{r}(A) = \mathbf{r}(\kappa) < 1$.

Then the origin (extinction fixed point) is locally asymptotically stable in the following sense:

For each $\alpha \in (r,1)$, *there exist some* $\delta_0 > 0$ *and* $M \ge 1$ *such that* $\|F^n(x)\| \le M\alpha^n\|x\|$ *for all* $n \in \mathbb{N}$ *and all* $x \in X_+$ *with* $\|x\| \le \delta_0$.

If, in addition, $g(s,t,u) \le g(s,t,0)$ *for all* $u \ge 0$, $s,t \in S$, *then the origin is globally stable in the sense that for each* $\alpha \in (r,1)$ *there exists some* $M \ge 1$ *such that* $\|F^n(x)\| \le M\alpha^n\|x\|$ *for all* $n \in \mathbb{N}$ *and all* $x \in X_+$.

THEOREM 19.6. *Assume Assumptions* 19.1, 19.2 *and* 19.3 *and* $\mathcal{T}_0 = \mathbf{r}(A) = \mathbf{r}(\kappa) = 1$.

Assume that $g(s,t,u) \le g(s,t,0)$ *for all* $s,t \in S$ *and* $u \ge 0$ *and* $\mathbf{r}(A) = 1$. *Further, assume that for any* $\delta > 0$ *there exists some* $\epsilon \in (0,1)$ *such that* $g(s,t,u) \le \epsilon g(s,t,0)$ *for all* $s,t \in S$ *and* $u \ge \delta$.

Finally, assume that $\inf \xi(S^3) > 0$.

Then $\liminf_{n\to\infty} F^n(\mu)(S) = 0$ *for any* $\mu \in \mathcal{M}_+(S)$.

PROOF. We apply Theorem 14.23. Since $g(s,t,u) \le g(s,t,0)$, $F(\mu) \le A\mu$ for all $\mu \in \mathcal{M}_+(S)$.

Let $\delta > 0$ and $\mu \in \mathcal{M}_+(S)$ with $\mu(S) \ge \delta$. Then $\int_S \xi(s,t,z)\mu(dz) \ge \delta \inf \xi(S^3) =: \tilde{\delta} > 0$ for all $s,t \in S$. By assumption, there exists some $\epsilon \in (0,1)$ such that $g(s,t,\int_S \xi(s,t,z)\mu(dz)) \le \epsilon g(s,t,0)$ for all $s,t \in S$. By (19.2) and (19.3), $F(\mu) \le \epsilon A\mu$.

All assumptions of the relevant part of Theorem 14.23 are verified and $\liminf_{n\to\infty} F^n(\mu)(S) = 0$ follows for all $\mu \in \mathcal{M}(S)$. □

19.2.2. Dissipativity of the induced semiflow.

THEOREM 19.7. *Assume Assumptions* 19.1, 19.2 *and* 19.3.

Further assume that $g(s,t,u) \to 0$ as $u \to \infty$ uniformly for $s,t \in S$ and that $\inf \xi(S^3) > 0$.

Then the following hold:

(a) $\frac{\|F(\mu)\|}{\|\mu\|} \to 0$ *as* $\|\mu\| \to \infty$.

(b) *For any bounded subset B of X_+, there exists a bounded convex subset $\tilde{B}$ of X_+ such that $F^n(B) \subseteq \tilde{B}$ for all $n \in \mathbb{N}$.*

(c) *There exists a bounded convex subset B_0 of X_+ such that for each $x \in X_+$ there exists some $m \in \mathbb{N}$ such that $F^n(x) \in B_0$ for all $n \ge m$.*

PROOF. (a) For $\mu \in \mathcal{M}_+(S)$, by Assumption 19.3 and (19.2),

$$\frac{F(\mu)(S)}{\mu(S)} \le \sup_{s,t\in S} g\Big(s,t,\int_S \xi(s,t,z)\mu(dz)\Big).$$

Further

$$\int_S \xi(s,t,z)\mu(dz) \ge \inf \xi \ \mu(S).$$

Since $\inf \xi > 0$ by assumption, as $\mu(S) \to \infty$, $\int_S \xi(s,t,z)\mu(dz) \to \infty$ uniformly in $s,t \in S$. By our additional assumption for g, $F(\mu)(S)/\mu(S) \to 0$ as $\mu(S) \to \infty$.

(b)–(c) Apply Theorem 14.15 to $X = \mathcal{M}(S)$ equipped with the variation norm and recall that $\|\mu\| = \mu(S) =: \theta(\mu)$ for $\mu \in \mathcal{M}_+(S)$. □

19.3. Habitats that are metric spaces

Let S be a metric space and $\mathcal{B}$ the Borel σ-algebra. Recall the operators F, A, $\tilde{A}$ in (19.2), (19.3), and (19.4).

PROPOSITION 19.8. *Assume Assumptions* 19.3, 19.1 (i) *and* 19.2 (i). *Then F, A and $\tilde{A}$ are operators from $\mathcal{M}_+^s(S)$ to $\mathcal{M}_+^s(S)$.*

If P is even tight, then F, $\tilde{A}$, and A are compact from $\mathcal{M}_+(S) \to \mathcal{M}_+^t(S)$ with respect to the flat norm.

PROOF. By Proposition 19.4, F is an operator from $\mathcal{M}_+(S)$ to $\mathcal{M}_+(S)$ and A and $\tilde{A}$ are additive bounded operators on $\mathcal{M}_+(S)$.

We only show the proof of the additional statements for F. The ones for A and $\tilde{A}$ are similar, even easier.

Let $K \in \mathcal{B}$ and $\mu \in \mathcal{M}_+(S)$. By Assumption 19.1 (i),

$$F(\mu)(S\setminus K) \le \sup g \int_S P(S\setminus K, s)\mu(ds) = \sup g\ (\tilde{A}\mu)(S\setminus K).$$

By Theorem 13.22, $\tilde{A}$ maps $\mathcal{M}_+^s(S)$ into itself and so does F.

Let $\mathcal{N}$ be a bounded subset of $\mathcal{M}_+(S)$. For all $\mu \in \mathcal{N}$, by the inequality above,

$$F(\mu)(S \setminus K) \le \sup g \sup_{x \in S} P(S \setminus K, x) \sup_{\mu \in \mathcal{N}} \mu(S).$$

With $K = \emptyset$, we see that $\{F(\mu)(S); \mu \in \mathcal{N}\}$ is bounded. For compact subsets K of S, we see that $F(\mathcal{N})$ inherits tightness from P. By Theorem 3.30, $F(\mathcal{N})$ is compact with respect to the flat norm. □

19.3.1. Instability of the extinction state in the superthreshold case.

THEOREM 19.9. *Make Assumptions* 19.1, 19.2 *and* 19.3.

Assume that $g(\cdot, 0)$ *is continuous on* S^2 *and that* P *is a tight Feller kernel. Let* $\mathcal{T}_0 = \mathbf{r}(A) = \mathbf{r}(\kappa) > 1$.

Then there exists some $\nu \in \mathcal{M}_+^t(S)$ *such that* $A\nu = \mathcal{T}_0\nu$, *and there is some* $\epsilon > 0$ *such that* $\sup_{n \in \mathbb{Z}_+} \|F^n(\mu)\| \ge \epsilon$ *for all* ν*-positive* $\mu \in \mathcal{M}_+^s(A)$.

PROOF. We apply Corollary 14.12. By Proposition 19.4, A is the order-derivative of F. By Corollary 13.36, the measure kernel κ in (19.4) is a tight Feller kernel. By Theorem 13.39, there exists some $\nu \in \dot{\mathcal{M}}_+^s(S)$ such that $A\nu = \mathcal{T}_0\nu$. Since κ is tight, $\nu \in \mathcal{M}_+^t(S)$. By Assumption 19.1(ii) and Assumption 19.2, for any $\mu \in \mathcal{M}_+(S)$, there exists some $\delta > 0$ such that $F(\mu) \ge \delta A\mu$.

The assumptions of Corollary 14.12 have been verified, and it implies the assertion. □

19.3.2. Uniform weak persistence in the superthreshold case.

We make stronger assumptions for the migration kernel P and obtain instability of the extinction state in the stronger form of uniform weak persistence.

THEOREM 19.10. *Assume Assumptions* 19.1, 19.2 *and* 19.3.

Assume that $g(\cdot, 0)$ *is continuous on* S^2 *and* $P = P_1 + P_2$ *where the* P_j *are tight Feller kernels,* P_2 *is a uniform Feller kernel and* P *is a topologically irreducible kernel.*

Finally assume that $\mathcal{T}_0 = \mathbf{r}(A) = \mathbf{r}(\kappa) > 1 \ge \mathbf{r}(\kappa_1)$ *where* κ_1 *is the kernel given by*

$$\kappa_1(T, s) = \int_T g(s, t, 0) P_1(dt, s), \qquad T \in \mathcal{B}, \quad s \in S.$$

Then, the population persists uniformly weakly: There exists some $\epsilon > 0$ *such that* $\limsup_{n \to \infty} F^n(\mu)(S) \ge \epsilon$ *for all* $\mu \in \dot{\mathcal{M}}_+(S)$.

PROOF. We apply Theorem 14.29.

By Proposition 19.4, A is the order-derivative of F and $F^n(\mu) \in \dot{\mathcal{M}}_+(S)$ for all $n \in \mathbb{N}$ and $\mu \in \dot{\mathcal{M}}_+(S)$.

By Proposition 13.49, the kernels $\kappa_j(T, s) = \int_T g(s, t, 0) P_j(dt, s)$ are tight Feller kernels and the kernel κ_2 is a uniform Feller kernel. Notice that $\kappa_1 + \kappa_2$ induces the map A_* on $M^b(S)$.

By Theorem 13.52, there is some $f \in \dot{C}^b(S)$ such that $\mathcal{T}_0 f = A_* f$ with $\mathcal{T}_0 = \mathbf{r}(A_*) = \mathbf{r}(\kappa) > 1$. Since P is a topologically irreducible kernel and $g(s, t, 0) > 0$ for all $s, t \in S$, also $\kappa_1 + \kappa_2$ is a topologically irreducible kernel by Proposition 13.62.

By Proposition 13.58, $f(s) > 0$ for all $s \in S$. The definition $\theta(\mu) = \int_S f d\mu$ provides a continuous linear positive functional θ on $\mathcal{M}(S)$ with $\theta(\mu) > 0$ for all $\mu \in \dot{\mathcal{M}}_+(S)$ and $\theta(A\mu) = \int_S (A_* f) d\mu = \mathcal{T}_0 \theta(\mu)$.

The assumptions of Theorem 14.29 have been verified, and the assertion follows. □

19.3.3. Population extinction in the threshold case.

THEOREM 19.11. *Assume Assumptions* 19.1, 19.2 *and* 19.3 *and* $\mathbf{r}(A) = 1$.

Assume that $g(s,t,u) \le g(s,t,0)$ *for all* $s,t \in S$ *and* $u \ge 0$ *and that* $g(\cdot,0)$ *is continuous on* S^2 . *Further, assume that for any* $\delta > 0$ *there exists some* $\epsilon \in (0,1)$ *such that* $g(s,t,u) \le \epsilon g(s,t,0)$ *for all* $s,t \in S$ *and* $u \ge \delta$.

Moreover, assume that $\inf \xi(S^3) > 0$.

Assume that $P = P_1 + P_2$ *where the* P_j *are tight Feller kernels and* P_2 *is a uniform Feller kernel and* $\mathbf{r}(A_{1*}) < 1$ *where* A_{1*} *is the map induced by* P_1 *on* $M^b(S)$.

Finally assume that $\mathcal{T}_0 = \mathbf{r}(A) = \mathbf{r}(\kappa) = \mathbf{r}(A_*) > 1 > \mathbf{r}(\kappa_1)$ *where* κ_1 *is the kernel given by* $\kappa_1(T,s) = \int_T g(s,t,0) P_1(dt,s)$, $T \in \mathcal{B}$.

Then there exists some $f \in \dot{C}^b_+(S)$ *with* $A_* f = f$ *such that* $\inf_S f dF(\mu) \le \int_S f d\mu$ *and* $\int_S f dF^n(\mu) \to 0$ *as* $n \to \infty$ *for all* $\mu \in \mathcal{M}_+(S)$. *If* P *is a topologically irreducible kernel,* $f(s) > 0$ *for all* $s \in S$.

PROOF. The existence of the eigenfunction f follows as in the proof of Theorem 19.10. Let $\mu \in \mathcal{M}_+(S)$.

By Theorem 19.6, $\liminf_{n\to\infty} F^n(\mu)(S) = 0$. So $\liminf_{n\to\infty} \int_S f dF^n(\mu) = 0$. Since $\int_S f dF(\mu) \le \int_S f dA(\mu) = \int_S A_* f d\mu = \int_S f d\mu$, the sequence $\big(\int_S f dF^n(\mu)\big)$ is decreasing and the assertion follows. □

19.3.4. Stronger results with stronger assumptions. Among other things, we would like to establish uniform persistence of the population instead of uniform weak persistence, i.e., replace lim sup by lim inf in Theorem 19.10.

For clarity, we present a collection of the new assumptions, which are stronger than the previous ones.

ASSUMPTION 19.12. *These are the assumptions for* g.

(i) $g : S^2 \times \mathbb{R}_+ \to \mathbb{R}_+$ *be bounded and continuous and* $g(x,y,\cdot)$ *be uniformly continuous, uniformly for* $x,y \in S$.
(ii) *For any* $c > 0$, $\inf g(S^2 \times [0,c]) > 0$.
(iii) $g(s,t,u) \to g(s,t,0)$ *as* $u \to 0$ *uniformly for* $s,t \in S$.
(iv) $g(s,t,u) \to 0$ *as* $u \to \infty$ *uniformly for* $s,t \in S$.

ASSUMPTION 19.13.

(i) $\xi : S^3 \to \mathbb{R}_+$ *is bounded and uniformly continuous.*
(ii) $\inf \xi(S^3) > 0$.

REMARK 19.14. Assumption 19.13 (i) could be replaced by ξ being continuous and $\{\xi(s,t,\cdot); s,t \in S\}$ begin uniformly equicontinuous: For any $\epsilon > 0$, there exists $\delta > 0$ such that $|\xi(s,t,y) - \xi(s,t,z)| < \epsilon$ for all $s,t,y,z \in S$ with $d(y,z) < \delta$.

ASSUMPTION 19.15.

(i) $P : \mathcal{B} \times S \to \mathbb{R}_+$ *is a tight Feller kernel of separable measures.*
(ii) $0 < P(S,s) \le 1$ *for all* $s \in S$.

PROPOSITION 19.16. *Assume Assumptions* 19.15, 19.12 (i) *and* 19.13 (i). *Then* F *and* A *are continuous operators from* $\mathcal{M}_+^s(S)$ *to itself and* κ *is a Feller kernel.*

PROOF. We only show that F is continuous. The proof for A is similar, actually easier.

Let $\mu \in \mathcal{M}_+^s(S)$ and (μ_n) a sequence in $\mathcal{M}_+^s(S)$ such that $\|\mu_n - \mu\|_\flat \to 0$ as $n \to \infty$. Let $f : S \to [0,1]$ and $|f(s) - f(t)| \le d(s,t)$. By Propositions 3.23 and 19.8, it is sufficient to show that $\int_S f\,dF(\mu_n) \to \int_S f\,dF(\mu)$ as $n \to \infty$. By definition of F,

$$\begin{aligned}\Big|\int_S f\,dF(\mu_n) - \int_S f\,dF(\mu)\Big| \\ \le \Big| \int_S \Big[\int_S f(t) g\big(s,t,v_n(s,t)\big) P(dt,s)\Big] \mu_n(ds) \\ - \int_S \Big[\int_S f(t) g\big(s,t,v(s,t)\big) P(dt,s)\Big] \mu(ds)\Big|.\end{aligned}$$

Here

$$v_n(s,t) = \int_S \xi(s,t,z)\mu_n(dz), \qquad v(s,t) = \int_S \xi(s,t,z)\mu(dz). \tag{19.8}$$

By the triangle inequality,

$$\begin{aligned}\Big|\int_S f\,dF(\mu_n) - \int_S f\,dF(\mu)\Big| \\ \le \Big| \int_S \Big[\int_S f(t) g\big(s,t,v_n(s,t)\big) P(dt,s)\Big] \mu_n(ds) \\ - \int_S \Big[\int_S f(t) g\big(s,t,v(s,t)\big) P(dt,s)\Big] \mu_n(ds)\Big| \\ + \Big| \int_S \Big[\int_S f(t) g\big(s,t,v(s,t)\big) P(dt,s)\Big] \mu_n(ds) \\ - \int_S \Big[\int_S f(t) g\big(s,t,v(s,t)\big) P(dt,s)\Big] \mu(ds)\Big|.\end{aligned} \tag{19.9}$$

By Assumption 19.15 (ii), the first term on the right-hand side can be estimated by

$$\sup f \sup_{s,t\in S} \Big| g\big(s,t,v_n(s,t)\big) - g\big(s,t,v(s,t)\big)\Big| \mu_n(S).$$

In order to show that this expression tends to 0 as $n \to \infty$, it is sufficient to show that

$$\sup_{s,t\in S} \Big| g\big(s,t,v_n(s,t)\big) - g\big(s,t,v(s,t)\big)\Big| \stackrel{n\to\infty}{\longrightarrow} 0. \tag{19.10}$$

Since the continuity of $\xi(s,t,\cdot)$ is uniform, uniformly for $s,t \in S$ and ξ is bounded by Assumption 19.13 (i),

$$v_n(s,t) \to v(s,t), \qquad n \to \infty,$$

uniformly for $s,t \in S$ by Proposition 3.15 and (19.8). Since the continuity of $g(s,\cdot)$ is uniform, uniformly for $s \in S$ by Assumption 19.12 (i), (19.10) follows and the first term on the right-hand side of inequality (19.9) converges to 0 as $n \to \infty$.

To show that the second term on the right-hand side of inequality (19.9) converges to 0 as $n \to \infty$, it is sufficient to show that

$$\int_S h(s,t)P(dt,s) = \tilde{h}(s)$$

is a continuous function of $s \in S$ (Theorem 3.5) where $h : S^2 \to \mathbb{R}_+$ is given by

$$h(s,t) = f(t)g\big(s,t,v(s,t)\big), \qquad s,t \in S,$$

with v in (19.8). Since ξ is continuous and bounded by Assumption 19.13 (i), v is a continuous function by Lebesgue's theorem of dominated convergence and (19.8). Since $g : S^2 \times \mathbb{R}_+ \to \mathbb{R}_+$ is continuous and bounded and f is continuous and bounded, h is continuous and bounded. By Corollary 13.36, $\tilde{h} \in C^b(S)$ because P is a tight Feller kernel. □

THEOREM 19.17. *Make the Assumptions* 19.12 (i) *and* (iv), 19.13 *and* 19.15.
Then the semiflow induced by F *on* $\mathcal{M}_+^s(S)$ *has a compact attractor of bounded sets.*

PROOF. By Proposition 19.8 and [**187**, Rem.2.26b], the semiflow induced by F is asymptotically smooth. By Theorem 19.7, Proposition 19.16 and Theorem 14.16, the semiflow induced by F has a compact attractor of bounded sets. □

THEOREM 19.18. *Make the Assumptions* 19.12, 19.13 *and* 19.15.
Assume that the Feller kernel P *is tight and that* $\mathcal{T}_0 = \mathbf{r}(\kappa) = \mathbf{r}(A) > 1$.
Then F *has a nonzero fixed point in* $\mathcal{M}_+^s(S)$.

PROOF. Combine Propositions 19.16, 19.8 and 19.4, and Corollary 14.26. □

THEOREM 19.19. *Make Assumptions* 19.12, 19.13 *and* 19.15.
Assume that $P = P_1 + P_2$ *where the* P_j *are tight Feller kernels and* P_2 *is a uniform Feller kernel and* P *is a topologically irreducible kernel.*
Finally, assume that $\mathcal{T}_0 = \mathbf{r}(\kappa) = \mathbf{r}(A) > 1 \geq \mathbf{r}(\kappa_1)$ *where* κ_1 *is the kernel given by* $\kappa_1(T,s) = \int_T g(s,t,0)P_1(dt,s)$, $T \in \mathcal{B}$.
Then there is some $\epsilon > 0$ *such that* $\liminf_{n\to\infty} F^n(\mu)(S) \geq \epsilon$ *for any* $\mu \in \dot{\mathcal{M}}_+^s(S)$.

PROOF. Combine Theorems 19.10 and 19.17, Proposition 19.4 and [**187**, Thm.4.5] with $\rho(\mu) = \mu(S)$. □

THEOREM 19.20. *Assume Assumptions* 19.12, 19.13 *and* 19.15.
Assume that $g(s,t,u) \leq g(s,t,0)$ *for all* $s,t \in S$ *and* $u \geq 0$ *and* $g(\cdot,0)$ *is continuous on* S^2 . *Further, assume that for any* $\delta > 0$ *there exists some* $\epsilon \in (0,1)$ *such that* $g(s,t,u) \leq \epsilon g(s,t,0)$ *for all* $s,t \in S$ *and* $u \geq \delta$.
Moreover, assume that $\inf \xi(S^3) > 0$.
Assume that $P = P_1 + P_2$ *where the* P_j *are tight Feller kernels,* P_2 *is a uniform Feller kernel and* P *is a topologically irreducible kernel.*
Finally assume that $\mathcal{T}_0 = \mathbf{r}(A) = \mathbf{r}(\kappa) = 1 > \mathbf{r}(\kappa_1)$ *where* κ_1 *is the kernel given by* $\kappa_1(T,s) = \int_T g(s,t,0)P_1(dt,s)$, $T \in \mathcal{B}, s \in S$.
Then $F^n(\mu)(S) \to 0$ *as* $n \to \infty$ *uniformly for* μ *in bounded subsets of* $\mathcal{M}_+(S)$.

PROOF. By Theorem 19.17, the semiflow induced by F has a compact attractor $\mathcal{A}$ of bounded sets. Since $F(0) = 0$, $0 \in \mathcal{A}$. We claim $\mathcal{A} = \{0\}$, which implies the assertion.

We apply [**187**, Thm.2.52] with $\tilde{A} = \{0\}$.

By Theorem 19.11, there exists some $f \in \dot{C}^b_+(S)$, $f(s) > 0$ for all $s \in S$, such that $\int_S f dF(\mu) \le \int_S f d\mu$ and $\int_S f dF^n(\mu) \to 0$ as $n \to \infty$ for all $\mu \in \mathcal{M}_-(S)$.

Then $L(\mu) = \int_S f d\mu$ is a Lyapunov function for F on $\mathcal{M}^s_+(S)$ and $L(0) = 0$. Let $(\mu_n)_{n\in\mathbb{Z}}$ satisfy $\mu_{n+1} = F(\mu_n)$ for all $n \in \mathbb{Z}$ and $L(\mu_n) = L(\mu_0)$ for all $n \in \mathbb{Z}$. By the properties of f, $L(\mu_n) \to 0$ as $n \to 0$. So $L(\mu_n) = 0$ for all $n \in \mathbb{Z}$. Since $f(s) > 0$ for all $s \in S$, $\mu_n = 0$ for all $n \in \mathbb{Z}$.

So, the assumptions of Theorem [**187**, Thm.2.52] are satisfied which implies that $\mathcal{A} \subseteq \tilde{\mathcal{A}} = \{0\}$. $\square$

19.3.5. Completely metrizable habitat. In this section, it is assumed that the metric space S can be made complete by switching to a topologically equivalent metric (one that generates the same open, closed, and compact sets).

We consider $\mathcal{M}(S)$ and $\mathcal{M}^s(S)$ with the flat norm associated with this complete metric.

LEMMA 19.21. *Assume that $\xi : S^3 \to \mathbb{R}_+$ is continuous and bounded, (μ_n) a sequence in $\mathcal{M}^s_+(S)$ and $\mu \in \mathcal{M}^s_+(S)$ and $\|\mu_n - \mu\|_\flat \to 0$ as $n \to \infty$. Then*

$$\int_S \xi(s,t,z)\mu_n(dz) \to \int_S \xi(s,t,z)\mu(dz), \qquad n \to \infty,$$

uniformly for s, t in any compact subset K of S^2.

PROOF. Suppose not. Then there exist some $\epsilon > 0$ and a strictly increasing sequence (n_k) and sequences (s_k), (t_k) with $(s_k, t_k) \in K$ such that

$$\Big| \int_S \xi(s_k,t_k,z)\mu_{n_k}(dz) - \int_S \xi(s_k,t_k,z)\mu(dz) \Big| \ge \epsilon, \qquad k \in \mathbb{N}. \tag{19.11}$$

After choosing subsequences, $s_k \to s$ and $t_k \to t$ as $k \to \infty$ for some $s,t \in S$, $(s,t) \in K$. By the triangle inequality,

$$\begin{aligned}
&\Big| \int_S \xi(s_k,t_k,z)\mu_{n_k}(dz) - \int_S \xi(s_k,t_k,z)\mu(dz) \Big| \\
&\quad \le \int_S |\xi(s_k,t_k,z) - \xi(s,t,z)|\mu_{n_k}(dz) \\
&\quad + \Big| \int_S \xi(s,t,z)\mu_{n_k}(dz) - \int_S \xi(s,t,z)\mu(dz) \Big| \\
&\quad + \int_S |\xi(s,t,z) - \xi(s_k,t_k,z)|\mu(dz).
\end{aligned} \tag{19.12}$$

Since $\xi(s,t,\cdot)$ is continuous on S, $\Big| \int_S \xi(s,t,z)\mu_{n_k}(dz) - \int_S \xi(s,t,z)\mu(dz) \Big| \to 0$ as $n \to \infty$ by Theorem 3.5. So

$$\begin{aligned}
&\limsup_{n\to\infty} \Big| \int_S \xi(s_k,t_k,z)\mu_{n_k}(dz) - \int_S \xi(s_k,t_k,z)\mu(dz) \Big| \\
&\quad \le \limsup_{n\to\infty} \int_S |\xi(s_k,t_k,z) - \xi(s,t,z)|\mu_{n_k}(dz) \\
&\quad + \limsup_{n\to\infty} \int_S |\xi(s,t,z) - \xi(s_k,t_k,z)|\mu(dz).
\end{aligned} \tag{19.13}$$

We will show that $\int_S |\xi(s_k, t_k, z) - \xi(s, t, z)|\mu_{n_k}(dz) \to 0$ as $k \to \infty$. A similar proof shows that $\int_S |\xi(s_k, t_k, z) - \xi(s, t, z)|\mu(dz) \to 0$ as $k \to \infty$. This provides a contradiction to (19.11).

Since $\|\mu_n - \mu\|_\flat \to 0$ and S is a complete metric space, $\{\mu_n; n \in \mathbb{N}\}$ is tight by Proposition 3.38. So, there exists a sequence (K_j) of compact subsets of S such that $\mu_n(S \setminus K_j) \to 0$ as $j \to \infty$ uniformly for $n \in \mathbb{N}$. By the triangle inequality, for each $j \in \mathbb{N}$,

$$\begin{aligned}&\int_S |\xi(s_k, t_k, z) - \xi(s, t, z)|\mu_{n_k}(dz) \\ &\qquad \le \int_{K_j} |\xi(s_k, t_k, z) - \xi(s, t, z)|\mu_{n_k}(dz) + 2\sup \xi(S^3)\mu_{n_k}(S \setminus K_j).\end{aligned}$$

For each $j \in \mathbb{N}$, since K_j is compact and ξ is continuous on S^3, $|\xi(s_k, t_k, z) - \xi(s, t, z)| \to 0$ as $k \to \infty$, uniformly for $z \in K_j$ and

$$\limsup_{k\to\infty} \int_S |\xi(s_k, t_k, z) - \xi(s, t, z)|\mu_{n_k}(dz) \le 2\sup \xi(S^3) \sup_{n\in\mathbb{N}} \mu_n(S \setminus K_j).$$

Since this holds for any $j \in \mathbb{N}$, we can take the limit as $j \to \infty$ and obtain $\limsup_{k\to\infty} \int_S |\xi(s_k, t_k, z) - \xi(s, t, z)|\mu_{n_k}(dz) = 0$. □

For clarity, we present a collection of the new assumptions, which can be a little weaker than the preceding ones because of the completeness assumption.

ASSUMPTION 19.22. *These are the assumptions for g.*

(i) *$g : S^2 \times \mathbb{R}_+ \to \mathbb{R}_+$ be bounded and continuous.*
(ii) *For any $c > 0$, $\inf g(S^2 \times [0, c]) > 0$.*
(iii) *$g(s, t, u) \to g(s, t, 0)$ as $u \to 0$ uniformly for $s, t \in S$.*
(iv) *$g(s, t, u) \to 0$ as $u \to \infty$ uniformly for $s, t \in S$.*

ASSUMPTION 19.23.

(i) *$\xi : S^3 \to \mathbb{R}_+$ is bounded and continuous.*
(ii) *$\inf \xi(S^3) > 0$.*

ASSUMPTION 19.24.

(i) *$P : \mathcal{B} \times S \to \mathbb{R}_+$ is a tight Feller kernel of separable measures.*
(ii) *$0 < P(S, s) \le 1$ for all $s \in S$.*

PROPOSITION 19.25. *Assume Assumptions* 19.24, 19.22 (i) *and* 19.23 (i). *Then F and A are continuous operators from $\mathcal{M}_+^s(S)$ to itself and κ is a Feller kernel.*

PROOF. We only show that F is continuous. The proof for A is similar, actually easier, and so is the proof that κ is a Feller kernel.

Let $\mu \in \mathcal{M}_+^s(S)$ and (μ_n) a sequence in $\mathcal{M}_+^s(S)$ such that $\|\mu_n - \mu\|_\flat \to 0$ as $n \to \infty$. Then $\{\mu_n; n \in \mathbb{N}\}$ is tight by Proposition 3.38.

Let $f : S \to [0, 1]$ and $|f(s) - f(t)| \le d(s, t)$. By Propositions 3.23 and 19.8, it is sufficient to show that $\int_S f\, dF(\mu_n) \to \int_S f\, dF(\mu)$. By definition of F,

$$\begin{aligned}&\Big|\int_S f\, dF(\mu_n) - \int_S f\, dF(\mu)\Big| \\ &\quad \le \Big|\int_S \Big[\int_S f(t) w_n(s, t) P(dt, s)\Big]\mu_n(ds) - \int_S \Big[\int_S f(t) w(s, t) P(dt, s)\Big]\mu(ds)\Big|.\end{aligned}$$

Here

$$w_n(s,t) = g\big(s,t,v_n(s,t)\big), \qquad w(s,t) = g\big(s,t,v(s,t)\big)$$

$$v_n(s,t) = \int_S \xi(s,t,z)\mu_n(dz), \qquad v(s,t) = \int_S \xi(s,t,z)\mu(dz). \tag{19.14}$$

By the triangle inequality,

$$\begin{aligned}\Big|\int_S f\,dF(\mu_n) - \int_S f\,dF(\mu)\Big| \\ \le \int_S \Big[\int_S f(t)|w_n(s,t) - w(s,t)|P(dt,s)\Big]\mu_n(ds) \\ + \Big|\int_S h\,d\mu_n - \int_S h\,d\mu\Big|\end{aligned} \tag{19.15}$$

with

$$h(s) = \int_S f(t)w(s,t)P(dt,s).$$

To show that the second term on the right-hand side of inequality (19.15) converges to 0 as $n \to \infty$, it is sufficient to show that h is a continuous function on S (Theorem 3.5). Since ξ is continuous and bounded by Assumption 19.13 (i), v is a continuous function by Lebesgue's theorem of dominated convergence and (19.8). Since $g : S^2 \times \mathbb{R}_+ \to \mathbb{R}_+$ is continuous and bounded, w is continuous and bounded. f is continuous and bounded as well. By Corollary 13.36, $h \in C^b(S)$ because P is a tight Feller kernel.

So

$$\begin{aligned}\limsup_{n\to\infty}\Big|\int_S f\,dF(\mu_n) - \int_S f\,dF(\mu)\Big| \\ \le \limsup_{n\to\infty}\int_S \Big[\int_S f(t)|w_n(s,t) - w(s,t)|P(dt,s)\Big]\mu_n(ds).\end{aligned} \tag{19.16}$$

Since $\{P(\cdot,s); s \in S\}$ and $\{\mu_n; n \in \mathbb{N}\}$ are tight, there exists a sequence (K_j) of compact subsets of S such that

$$P(S\setminus K_j, s) \stackrel{j\to\infty}{\longrightarrow} 0, \qquad \mu_n(S\setminus K_j) \stackrel{j\to\infty}{\longrightarrow} 0, \quad \text{uniformly for } s \in S \text{ and } n \in \mathbb{N}. \tag{19.17}$$

For any $j \in \mathbb{N}$,

$$\begin{aligned}\limsup_{n\to\infty}\Big|\int_S f\,dF(\mu_n) - \int_S f\,dF(\mu)\Big| \\ \le \limsup_{n\to\infty}\int_{K_j}\Big[\int_{K_j} f(t)|w_n(s,t) - w(s,t)|P(dt,s)\Big]\mu_n(ds) \\ + \limsup_{n\to\infty}\int_{S\setminus K_j}\Big[\int_S f(t)(w_n(s,t) + w(s,t))P(dt,s)\Big]\mu_n(ds) \\ + \limsup_{n\to\infty}\int_S\Big[\int_{S\setminus K_j} f(t)(w_n(s,t) + w(s,t))P(dt,s)\Big]\mu_n(ds).\end{aligned} \tag{19.18}$$

By Assumption 19.15 (ii), the first term on the right-hand side of the inequality can be estimated by

$$\sup f \sup_{s,t\in K_j}\Big|w_n(s,t) - w(s,t)\Big| \sup_{n\in\mathbb{N}} \mu_n(S).$$

In order to show that this expression tends to 0 as $n \to \infty$, it is sufficient to show that

$$\sup_{s,t\in K_j} \Big| g\big(s,t,v_n(s,t)\big) - g\big(s,t,v(s,t)\big)\Big| \stackrel{n\to\infty}{\longrightarrow} 0. \tag{19.19}$$

By Lemma 19.21,

$$v_n(s,t) \to v(s,t), \qquad n \to \infty,$$

uniformly for $s,t \in K_j^2$. Since g is uniformly continuous on $K_j^2 \times [0,a]$ for any $a \in (0,\infty)$, (19.19) follows and the first term on the right-hand side of inequality (19.18) equals 0.

Thus, by (19.18), for any $j \in \mathbb{N}$,

$$\begin{aligned} &\limsup_{n\to\infty} \Big| \int_S f\,dF(\mu_n) - \int_S f\,dF(\mu)\Big| \\ &\quad \le (\sup f)2(\sup g)\Big[\sup_{n\in\mathbb{N}} \mu_n(S\setminus K_j) + \sup_{s\in S} P(S\setminus K_j)\sup_{n\in\mathbb{N}}\mu_n(S)\Big]. \end{aligned} \tag{19.20}$$

Since this holds for any $j \in \mathbb{N}$, by (19.17) we can take the limit as $j \to \infty$ and obtain

$$\limsup_{n\to\infty} \Big| \int_S f\,dF(\mu_n) - \int_S f\,dF(\mu)\Big| = 0. \qquad \square$$

The same proof as for Theorem 19.19 provides the following result.

THEOREM 19.26. *Make Assumptions* 19.22, 19.23 *and* 19.24.

Assume that $P = P_1 + P_2$ where the P_j are tight Feller kernels, P_2 is a uniform Feller kernel and P is a topologically irreducible kernel.

Finally, assume that $\mathcal{T}_0 = \mathbf{r}(\kappa) = \mathbf{r}(A) > 1 \ge \mathbf{r}(\kappa_1)$ where κ_1 is the kernel given by $\kappa_1(T,s) = \int_T g(s,t,0)P_1(dt,s)$, $T \in \mathcal{B}, s \in S$.

Then there is some $\epsilon > 0$ such that $\liminf_{n\to\infty} F^n(\mu)(S) \ge \epsilon$ for any $\mu \in \dot{\mathcal{M}}_+^s(S)$.

We now assume that P is strongly topologically irreducible (Section 13.7.3).

THEOREM 19.27. *Make Assumptions* 19.22, 19.23 *and* 19.24.

Assume that $P = P_1 + P_2$ where the P_j are tight Feller kernels and P_2 is a uniform Feller kernel and

$$\mathbf{r}(\kappa_1) \le 1 < \mathbf{r}(\kappa).$$

Finally, assume that P is a strongly topologically irreducible kernel.

Then, for any $f \in \dot{C}_+^b(S)$, there exists some $\epsilon_f > 0$ such that

$$\liminf_{n\to\infty} \int_S f\, dF^n(\mu) \ge \epsilon_f$$

for any $\mu \in \dot{\mathcal{M}}_+^s(S)$.

PROOF. Let $f \in \dot{C}_+^b(S)$. We apply [**187**, Cor.4.22] with $X = \mathcal{M}_+^s(S)$ endowed with the flat norm and with $\rho(\mu) = \mu(S)$ and $\tilde\rho(\mu) = \int_S f d\mu$ for $\mu \in \mathcal{M}_+^s(S)$. Then ρ and $\tilde\rho$ are continuous on $\mathcal{M}_+^s(S)$.

In the language of [**187**, Cor.4.22], the semiflow induced by F is uniformly ρ-persistent by Theorem 19.26.

Let $(\mu_n)_{n\in\mathbb{Z}}$ be total F-trajectory, i.e., a sequence in $\mathcal{M}_+^s(S)$ such that

$$\mu_{n+1} = F(\mu_n), \qquad n \in \mathbb{Z}, \tag{19.21}$$

and $\{\mu_n; n \in \mathbb{N}\}$ has compact closure and

$$\delta := \inf_{n\in\mathbb{Z}} \mu_n(S) > 0. \tag{19.22}$$

There exists some $c > 0$ such that $\mu_n(S) \le c$ for all $n \in \mathbb{Z}$. Then

$$\int_S \xi(s,t,z) d\mu_n(z) \le \sup \xi(S^3)c, \qquad n \in \mathbb{Z}, \quad s,t \in S.$$

By Assumptions 19.23 and 19.22, there is some $\delta_0 > 0$ such that

$$g\Big(s,t,\int_S \xi(s,t,z)d\mu_n(z)\Big) \ge \delta_0, \qquad n \in \mathbb{Z}, \quad s,t \in S.$$

By (19.2), $F(\mu_n) \ge \delta_0 \tilde{A}(\mu_n)$ for all $n \in \mathbb{Z}$, where $\tilde{A}$ is the map on $\mathcal{M}_+^s(S)$ induced by P. By induction and (19.21), $F^k(\mu_n) \ge \delta_0^k \tilde{A}^k(\mu_n)$ for all $k \in \mathbb{N}$ and $n \in \mathbb{Z}$.

Now

$$\int_S f \, dF^k(\mu_n) \ge \delta_0^k \int_S f \, d\tilde{A}^k(\mu_n) = \delta_0^k \int_S (\tilde{A}_*^k f) \, d\mu_n, \tag{19.23}$$

where $\tilde{A}_*$ is the map induced on $C^b(S)$ by the kernel P.

By Theorem 19.17, the semiflow induced by F has a compact attractor of bounded set and so is point-dissipative and asymptotically smooth [**187**, Thm.2.33].

Since $\{\mu_n; n \in \mathbb{Z}\}$ has compact closure and S is complete, $\{\mu_n; n \in \mathbb{Z}\}$ is tight by Proposition 3.38. So there exist a compact subset K of S such that $\mu_n(S \setminus K) < \delta/2$ and $\mu_n(K) \ge \delta/2$ for all $n \in \mathbb{Z}$. Recall (19.22). Since K is compact and P is strongly topologically irreducible, by Lemma 13.71, there exists some $k \in \mathbb{N}$ such that $(\tilde{A}_*^k f)(s) > 0$ for all $s \in K$. For this k, since $\tilde{A}_*^k f$ is continuous and K compact, $\inf(\tilde{A}_*^k f)(K) > 0$. By (19.23),

$$\int_S f \, dF^k(\mu_n) \ge \delta_0^k \inf(\tilde{A}_*^k f)(K)\mu_n(K) \ge \delta_0^k \inf(\tilde{A}_*^k f)(\delta/2) =: \delta_1.$$

Since $F^k(\mu_n) = \mu_{k+n}$, this implies that $\int_S f \, d\mu_{k+n} \ge \delta_1 > 0$ for all $n \in \mathbb{Z}$. So $\tilde{\rho}(\mu_0) = \int_S f \, d\mu_{k-k} \ge \delta_1$.

All assumptions of [**187**, Cor4.22] have been verified which implies that there is some $\epsilon_f > 0$ such that

$$\liminf_{n\to\infty} \tilde{\rho}(F_n(\mu)) = \liminf_{n\to\infty} \int_S f \, dF^n(\mu) \ge \epsilon_f$$

for all $\mu \in \mathcal{M}_+^s(S)$ with $\rho(\mu) = \mu(S) > 0$. □

19.4. A more general framework

In this chapter, we have considered a turnover operator that is particular suited for populations that are spatially distributed. Another population structure may require a more general framework as it is presented in [**203, 204**]. There, we consider yearly turnover operators F on $\mathcal{M}_+(S)$ of the form

$$F(\mu)(T) = \int_S \kappa^\mu(T,s)\mu(ds), \qquad \mu \in \mathcal{M}_+(S), \quad T \in \mathcal{B}. \tag{19.24}$$

Here $\{\kappa^\mu; \mu \in \mathcal{M}(S)\}$ is a family of measure kernels $\kappa^\mu : \mathcal{B} \times S \to \mathbb{R}_+$. If μ is the zero measure, we write κ^o. In this chapter,

$$\kappa^\mu(T,s) = \int_T g\Big(s,t,\int_S \xi(s,t,z)\mu(dz)\Big)P(dt,s), \qquad T \in \mathcal{B}, \quad s \in S.$$

The results are analogous to those obtained in this chapter. If a population is physiologically structured, one of the main challenges in applying the framework in [**203, 204**] is to come up with reasonable assumptions that make the kernel κ^o topologically irreducible or even strongly topologically irreducible.

19.5. A homogeneous operator on the cone of nonnegative measures

It is the purpose of this section to convey a feeling for the difficulties in dealing with two-sex population models in the state space of measures. Also see [**101**, Sec.6.4].

Let $(S, \mathcal{B})$ be a metric space and $X = \mathcal{M}(S)$ be the vector lettuce of finite measures on the Borel σ-algebra $\mathcal{B}$ of S with order cone $X_+ = \mathcal{M}_+(S)$. $\mathcal{M}(S)$ is a Banach space under the variation norm. $\mathcal{M}(S)$ also carries the flat norm without being a Banach space under the flat norm. $\mathcal{M}_+(S)$ is closed under either norm. See Chapter 3.

If S represents the habitat of a spatially distributed population with two sexes and the nonnegative measures μ_1 and μ_2 describe the spatial distributions of females and males respectively, it does not seem straightforward to find expressions for the spatial distribution of resulting mated pairs. One reasonable possibility is taking the infimum $\mu_1 \wedge \mu_2$ given by

$$(\mu_1 \wedge \mu_2)(T) = \inf \big\{\mu_1(U) + \mu_2(T \setminus U);\ U \in \mathcal{B},\, U \subseteq T\big\}, \qquad T \in \mathcal{B}, \tag{19.25}$$

which is the special case of (3.3) for a set of two measures. This is motivated by the following fact.

REMARK 19.28. If $\nu \in \mathcal{M}_+(S)$ and $\mu_j(T) = \int_T f_j d\nu$ for all $T \in \mathcal{B}$, then

$$(\mu_1 \wedge \mu_2)(T) = \int_T \min\{f_1(s), f_2(s)\}\, \nu(ds), \qquad T \in \mathcal{B}. \tag{19.26}$$

Recall that taking the minimum is one of the possible mating functions (Section 4.5).

If $\varphi : \mathbb{R}^2_+ \to \mathbb{R}$ is a general homogeneous function, the formula

$$(\mu_1 \wedge \mu_2)(T) = \int_T \varphi\big(f_1(s), f_2(s)\big)\, \nu(ds), \qquad T \in \mathcal{B}, \tag{19.27}$$

seems difficult to extend from a pair of measures that are dominated by a common measure ν to a pair of general measures.

We consider a semelparous population with two sexes where individuals reproduce only once in their life-time. We look at the case of very short reproduction seasons. We let the year start with the end of the reproductive season, and a measure μ describes the spatial distribution of neonates (newborn individuals) at the beginning year.

Let $\kappa_1, \kappa_2 : \mathcal{B} \times S \to \mathbb{R}_+$ be measures kernels (Definition 13.8) and A_1, A_2 the induced linear operators on $\mathcal{M}(S)$,

$$(A_j\mu)(T) = \int_S \kappa_j(T, s)\mu(ds), \qquad \mu \in \mathcal{M}(S), \quad T \in \mathcal{B}.$$

$\kappa_j(T, s)$ is the respective probability that a neonate that is at $s \in S$ at the beginning of the year has the appropriate sex and is still alive and is located in the set T at the end of the year. So the A_j map the spatial distribution of neonates at the beginning of the year to the respective distributions of females and males at the end of the year. The infimum models the formation of pairs as explained above. Further, we assume that offspring has the same location as the pair that produces it and that $\beta(s)$ is the amount of surviving neonates produced by a pair that is located at $s \in S$. We assume that $\beta \in M_+^b(S)$.

$B : \mathcal{M}_+(S) \to \mathcal{M}_+(S)$ is defined by

$$B(\mu)(T) = \int_T \beta(s) \left[(A_1\mu) \wedge (A_2\mu)\right](ds), \qquad \mu \in \mathcal{M}_+(S), \quad T \in \mathcal{B}.$$

If μ is the spatial distribution of the neonates at the beginning of a given year, $B(\mu)$ is the spatial distribution of the neonates at the beginning of the next year (immediately after the reproductive season).

PROPOSITION 19.29. *B is a bounded homogeneous operator.*

(a) *If κ_1 or κ_2 are tight, then B is compact with respect to the flat norm and maps $\mathcal{M}_+(S)$ into $\mathcal{M}_+^t(S)$.*
(b) *If both κ_1 and κ_2 are uniform Feller kernels, then B is continuous from $\mathcal{M}_+(S)$ with respect to the flat norm to $\mathcal{M}_+(S)$ with the variation norm.*

Recall that the minimum operation is not continuous with respect to the flat norm (Corollary 3.14). For the concepts of tight measure kernels and uniform Feller kernels see Section 13.4 and Definition 13.45.

PROOF. Define $\tilde{B}(\mu) = (A_1\mu) \wedge (A_2\mu)$, $\mu \in \mathcal{M}_+(S)$. Then

$$B(\mu)(T) = \int_T \beta \, d\tilde{B}(\mu), \qquad \mu \in \mathcal{M}_+(S), \quad T \in \mathcal{B}. \tag{19.28}$$

$\tilde{B}$ is a bounded homogeneous operator by Example 4.1, and B inherits this property by (19.28).

(a) Let $\mathcal{N}$ be a bounded subset of $\mathcal{M}_+(S)$. By symmetry, we can assume that κ_2 is tight. Then $A_2(\mathcal{N})$ is a tight bounded subset of $\mathcal{M}_+^t(S)$. Let $\epsilon > 0$. Then there exists a compact subset K_2 of S such the $(A_2\mu)(S \setminus K_2) \le \epsilon$ for all $\mu \in \mathcal{N}$. For all $\mu \in \mathcal{N}$,

$$\begin{aligned}\tilde{B}(\mu)(S \setminus K) =& \inf\left\{(A_1\mu)(T) + (A_2\mu)((S \setminus K_2) \setminus T); T \in \mathcal{B}\right\} \\ \le& (A_1\mu)(\emptyset) + (A_2\mu)(S \setminus K_2) \le \epsilon.\end{aligned}$$

This proves that $\tilde{B}(\mathcal{N})$ is a bounded tight set of measures. By (19.28), $B(\mathcal{N})$ is tight and bounded. By Theorem 3.30, $B(\mathcal{N})$ is a compact set with respect to the flat norm.

(b) By Proposition 13.46, the operators A_j are continuous from $\mathcal{M}_+(S)$ with the flat norm to $\mathcal{M}_+(S)$ with the variation norm. Since $\wedge : \mathcal{M}_+(S) \times \mathcal{M}_+(S) \to \mathcal{M}_+(S)$ is continuous with respect to the variation norm, $\tilde{B}$ is continuous from $\mathcal{M}_+(S)$ with the flat norm to $\mathcal{M}_+(S)$ with the variation norm. By (19.28), B is continuous from $\mathcal{M}_+(S)$ with the flat norm to $\mathcal{M}_+(S)$ with the variation norm. □

THEOREM 19.30. *Let κ_1 and κ_2 be uniform Feller kernels and let at least one of them be tight. Let A_1, A_2 be the induced linear operators on $\mathcal{M}(S)$, $\beta \in M_+^b(S)$, and $B : \mathcal{M}_+(S) \to \mathcal{M}_+(S)$ defined by*

$$B(\mu)(T) = \int_T \beta(s)[(A_1\mu) \wedge (A_2\mu)](ds), \qquad \mu \in \mathcal{M}_+(S), \quad T \in \mathcal{B}.$$

Then, if $\mathcal{T} = \mathbf{r}(B) > 0$, *there exists some* $\mu \in \mathcal{M}_+^t(S)$, $\mu(S) = 1$, *such that* $B(\mu) = \mathcal{T}\mu$.

PROOF. By Proposition 19.29, B is compact and continuous on $\mathcal{M}_+(S)$ with the flat norm and maps $\mathcal{M}_+(S)$ into $\mathcal{M}_+^t(S)$. Existence of the eigenvector μ now follows from Theorem 9.10. □

Exercises.

19.5.1. Prove Remark 19.28.
Hint: The Radon–Nikodym Theorem may come in handy [**9**, Sec.13.6].

APPENDIX A

Some tools from real analysis

A.1. Metric spaces

The concept of a metric space is fundamental for this monograph. The population state spaces are special metric spaces, namely normed vector spaces, and the state spaces of individual characteristics in Chapters 18 and 19 are metric spaces. The following primer on metric spaces has been adapted from [**187**, Sec.1.1] and expanded to the purposes of this monograph.

DEFINITION A.1. *A* metric space (S,d) *is a nonempty set* S *together with a function* $d : S \times S \to \mathbb{R}$ *which satisfies the following axioms:*

(1) $d(s,t) = d(t,s)$ *for all* $s,t \in S$. *[symmetry]*
(2) $d(s,u) \le d(s,t) + d(t,u)$ *for all* $s,t,u \in S$. *[triangle inequality]*
(3) $d(s,s) = 0$ *for all* $s \in S$.
(4) $d(s,t) \neq 0$ *for all* $s,t \in S$ *with* $s \neq t$.

$d(s,t)$ *measures the distance between* s *and* t. d *is called a* metric *on* S. *If* (4) *does not hold,* d *is called a* semimetric.

EXAMPLE A.2 (discrete metric). Let S be an arbitrary nonempty set. Define

$$d(s,t) = \begin{cases} 0, & s = t, \\ 1, & s \neq t. \end{cases}$$

Then d is a metric on S, called the *discrete metric* or *trivial metric*.

PROOF. The only nontrivial property is the triangle inequality.

Case 1: $s = u$.
Then $d(s,u) = 0 = 0 + 0 \le d(s,t) + d(t,s)$.

Case 2: $s \neq u$.
Then $s \neq t$ or $t \neq u$. So $d(s,t) + d(t,u) \ge 1 = d(s,u)$. □

This metric is useless for applications but great for finding counterexamples.

DEFINITION A.3. *A metric space* (S,d) *is called* uniformly discrete *if there is some* $\epsilon > 0$ *such that* $d(s,t) > \epsilon$ *for all* $s,t \in S$ *with* $s \neq t$ [**113**].

Obviously, a nonempty set becomes a uniformly discrete metric space under the discrete metric.

PROPOSITION A.4. *Let* d *be a semimetric on the nonempty set* S. *Then*

$$|d(s,t) - d(t,u)| \le d(s,u) \qquad \forall s,t,u \in S.$$

In particular, $d(s,u) \ge 0$ *for all* $s,u \in S$.

PROOF. Obviously, the first statement implies the second. Let $s, t, u \in S$. Then, by the triangle inequality,

$$d(s,t) \leq d(s,u) + d(t,u).$$

By subtraction,

$$d(s,t) - d(t,u) \leq d(s,u).$$

Again, by the triangle inequality,

$$d(t,u) \leq d(t,s) + d(s,u).$$

By subtraction and symmetry of the semimetric,

$$-(d(s,t) - d(t,u)) = d(t,u) - d(t,s) \leq d(s,t).$$

We combine the two inequalities and obtain the first statement. Recall that the absolute value $|\alpha|$ of a real number α is either α or $-\alpha$. □

DEFINITION A.5 (Congergence). *Let (S, d) be a metric space and (s_j) a sequence in S, $s \in S$. We say that (s_j)* converges to s *if, for any $\epsilon > 0$, there exists an $N \in \mathbb{N}$ such that*

$$d(s_j, s) < \epsilon \qquad j \in \mathbb{N}, \quad j > N.$$

The point $s \in S$ to which the sequence (s_j) converges is called the limit *of the sequence, and we write*

$$s = \lim_{j \to \infty} s_j$$

or

$$s_j \to s, \qquad j \to \infty.$$

REMARK A.6. A sequence (s_j) converges towards a point s in a metric space (S, d) if and only if $d(s_j, s) \to 0$ as $j \to \infty$ in $\mathbb{R}$ with the absolute value.

The following relation will be needed in the future.

LEMMA A.7. *Let S be a metric space and d its metric.*

If (s_n), (t_n) are sequences in S and $s_n \to s \in S$, $t_n \to t \in S$ as $n \to \infty$, then $d(s_n, t_n) \to d(s,t)$ as $n \to \infty$.

PROOF. By Proposition A.4,

$$\big|d(s,t) - d(t,u)\big| \leq d(s,u) \qquad \forall s, t, u \in S.$$

Let (s_n), (t_n) be sequences in S and $s, t \in S$. Then

$$\big|d(s_n, t_n) - d(s,t)\big| = \big|d(s_n, t_n) - d(t_n, s) + d(t_n, s) - d(s,t)\big|.$$

By the triangle inequality for the absolute value and the inequality above,

$$\begin{aligned}\big|d(s_m, t_n) - d(s,t)\big| &\leq \big|d(s_n, t_n) - d(t_n, s)\big| + \big|d(t_n, s) - d(s,t)\big| \\ &\leq d(s_n, s) + d(t_n, t).\end{aligned}$$

If $t_n \to t$ and $s_n \to s$, then $d(t_n, t) \to 0$ and $d(s_n, s) \to 0$, so $d(s_n, t_n) - d(s,t) \to 0$ as $n \to \infty$. □

DEFINITION A.8 (distance from a point to a set). *Let $\emptyset \neq T \subset S$ and $s \in S$. Then the* distance from the point s to the set T *is defined as*

$$d(s, T) := \inf\{d(s,t); t \in T\}.$$

LEMMA A.9. *Let* $\emptyset \neq T \subseteq S$. *Then*

$$|d(s,T) - d(u,T)| \leq d(s,u), \qquad s,u \in S.$$

PROOF. Let $s,u \in S$ and $t \in T$. By the triangle inequality,

$$d(s,t) \leq d(s,u) + d(u,t).$$

By definition of the infimum,

$$d(s,T) \leq d(s,u) + d(u,t).$$

By subtraction

$$d(s,T) - d(s,u) \leq d(u,t).$$

Since $t \in T$ was arbitrary, this inequality holds for all $t \in T$ and so, by definition of the infimum,

$$d(s,T) - d(s,u) \leq d(u,T).$$

We reorganize,

$$d(s,T) - d(u,T) \leq d(s,u).$$

By symmetry, this inequality also holds, if s and u are interchanged, and the inequality follows. □

A.1.1. Compact sets.

DEFINITION A.10. *The subset T of a metric space (S,d) is said to be* compact *if every infinite sequence (t_n) in T has a subsequence which has a limit in S.*

In a topolcgical Hausdorff space, this property is called *sequential compactness.*

LEMMA A.11. *Let $T \subset S$, $s \in S$, and let T be nonempty and compact. Then*

$$d(s,T) = \min\{d(s,t); t \in T\}.$$

PROOF. By definition of the infimum, there exists a sequence (t_n) in T such that $d(s,t_n) \to d(s,T)$ as $n \to \infty$. Since T is compact, there exists a subsequence (t_{n_j}) of (t_n) and some $t \in T$ such that $t_{n_j} \to t$ as $j \to \infty$. By Lemma A.7, $d(s,t_{n_j}) \to d(s,t)$ as $j \to \infty$; but also $d(s,t_{n_j}) \to d(s,T)$ as $j \to \infty$. Hence $d(s,T) = d(s,t)$ with $t \in T$. □

A.1.2. Limit points, closures, and closed sets.

DEFINITION A.12. *Let T be a subset of a metric space (S,d).*

- *A point $s \in S$ is called a* limit point *of T if there is a sequence $(t_n)_{n=1}^{\infty}$ in T such that $t_n \to s$ as $n \to \infty$.*

  Notice that every $t \in T$ is a limit point of T.
- *A point $s \in S$ is called an* accumulation point *of T, if s is a limit point of T and the sequence (t_n) can be chosen such that $t_n \neq s$ for all $n \in \mathbb{N}$.*

DEFINITION A.13. *Let T be the subset of a metric space (S,d).*

- *The* closure *of T, denoted by $\bar{T}$, is the set of all limit points of S.*

  $\bar{T}$ is the union of T and the set of accumulation points of T. Since every $t \in T$ is a limit point of T, $T \subset \bar{T}$.
- *T is called* closed *if $T = \bar{T}$, i.e., if T contains all its limit points.*

Exercises.

A.1.1. Let S be a metric space and T a compact subset of S. Show that T is closed.

A.1.2. Let S be a metric space and T be a subset of S.

Show that the closure of T is a compact subset of S if (and only if) every sequence in T has a subsequence that converges in S.

A.1.3. Dense sets.

DEFINITION A.14. *Let (S,d) be a metric space. Let $U \subseteq T \subseteq S$. Then U is called dense in T if $T \subseteq \bar{U}$. In other words: U is dense in T if and only if, for every $t \in T$ there exists a sequence (u_n) in U such that $u_n \to t$ as $n \to \infty$.*

EXAMPLE A.15. $\mathbb{Q}$ and $\mathbb{R} \setminus \mathbb{Q}$ are dense in $\mathbb{R}$.

DEFINITION A.16. *A subset T of a metric space S is called* precompact *if its closure $\bar{T}$ is compact.*

A.1.4. Interior points and open sets. The *open neighborhood* of $s \in S$, centered at s with radius r, is defined as

$$U_r(s) = \{t \in S; d(t,s) < r\}, \tag{A.1}$$

while the *closed neighborhood* is defined as

$$\bar{U}_r(s) = \{t \in S; d(t,s) \leq r\}. \tag{A.2}$$

DEFINITION A.17. *Let U be the subset of a metric space S. A point $s \in U$ is called an* interior point *of U if there exists some $\epsilon > 0$ such that $U_\epsilon(s) \subseteq U$. The set of interior points of U is called the* interior *of U and is denoted by $\breve{U}$ or Int U.*

If $s \in S$ and $U \subseteq S$, then U is called a neighborhood of s *if s is an interior point of U.*

DEFINITION A.18. *Let U be the subset of a metric space S with metric d. U is called* open *if $U = \breve{U}$, i.e., every point in U is an interior point of U.*

Warning: While, in a normed vector space, the closed neighborhood is the closure of the open neighborhood and the open neighborhood is the interior of the closed neighborhood, this is not necessarily true in an arbitrary metric space.

Many metric spaces are subsets of normed vectors spaces over a field $\mathbb{K}$, $\mathbb{K} = \mathbb{R}$ or $\mathbb{K} = \mathbb{C}$.

DEFINITION A.19. *Let X be a vector space over $\mathbb{K}$. A function $\xi : X \to \mathbb{R}$ is called a* seminorm *on X, if the following hold:*

1. $\xi(x+y) \leq \xi(x) + \xi(y)$ *for all $x, y \in X$.* *[triangle inequality]*
2. $\xi(\alpha x) = |\alpha|\xi(x)$ *for all $x \in X$, $\alpha \in \mathbb{K}$.* *[homogeneity]*

ξ is called a norm *on X if, in addition,*

3. $\xi(x) \neq 0$ *if $x \in X$, $x \neq 0$.*

In this case, we often write $\|x\|$ instead of $\xi(x)$ and call $(X, \|\cdot\|)$ a normed vector space. The norm induces a metric on X by $d(x,y) = \|x-y\|$. If the vector space is finite-dimensional, we will sometimes use $|\cdot|$ for the norm.

REMARK A.20.

(a) Let (S,d) be a metric space, $x,y \in X$, $r > 0$. Then $d(y,\bar{U}_r(x)) \geq d(y,x) - r$.
(b) Let $(V,\|\cdot\|)$ be a normed vector space and d the metric induced by $\|\cdot\|$. Then $d(y,U_r(x)) = d(y,\bar{U}_r(x)) = d(y,x) - r$ for all $x \in X$, $r > 0$ and $y \in X \setminus U_r(x)$.
(c) Give an example of a metric space where $d(y,\bar{U}_r(x)) > d(y,x) - r$ for appropriate $x \in X$, $r > 0$, $y \in X \setminus \bar{U}_r(x)$.

LEMMA A.21. *Let S be the subset of a metric space X. Then $x \in X$ is a limit point of S if and only if $S \cap U_\epsilon(x) \neq \emptyset$ for all $\epsilon > 0$.*

PROOF. Suppose that x is a limit point of S. Then there exists a sequence (x_n) in S such that $x_n \to x$ as $n \to \infty$. Let $\epsilon > 0$. Then there exists some $N \in \mathbb{N}$ such that $d(x_n, x) < \epsilon$ for all $n \in \mathbb{N}$ with $n > N$, i.e., $x_n \in U_\epsilon(x)$ for $n > N$. Hence $U_\epsilon(x) \cap S \neq \emptyset$.

Now assume that $x \in X$ and $U_\epsilon(x) \cap S \neq \emptyset$ for all $\epsilon > 0$. Then, for each $n \in \mathbb{N}$, there exists an element $x_n \in S \cap U_{1/n}(x)$. This implies that $d(x_n, x) < 1/n$ and $S \ni x_n \to x$ as $n \to \infty$. Hence x is a limit point of S by Definition A.12. □

Exercises.

A.1.3. Let S be metric space. Show that the set of open sets as defined in Definition A.18 defines a topology on S (Definition 3.41).

A.1.5. Bounded sets, diameter.

DEFINITION A.22 (diameter, bounded set). *The* diameter *of a subset T of the metric space S is defined as*

$$\operatorname{diam} T = \sup\{d(s,t); s,t \in T\}. \tag{A.3}$$

A subset T of S is called bounded *if it has a finite diameter, i.e., there is some $c \in (0,\infty)$ such that $d(s,t) \leq c$ for all $s,t \in T$.*

Let T be a subset of a metric space S and $\epsilon > 0$. The ϵ-neighborhood of T is defined as the set of points whose distance to T is less than ϵ,

$$U_\epsilon(T) := \{s \in S; d(s,T) < \epsilon\}. \tag{A.4}$$

Notice that

$$U_\epsilon(T) = \{s \in S; \exists t \in T \text{ such that } d(s,t) < \epsilon\} = \bigcup_{t \in T} U_\epsilon(t). \tag{A.5}$$

The letter U traditionally reminds us of the German word "Umgebung" which means "neighborhood."

Exercises.

A.1.4. Let T be a subset of a metric space S and $\epsilon > 0$. Recall that the ϵ-neighborhood of T is defined as the set

$$U_\epsilon(T) := \{s \in S; d(s,T) < \epsilon\}.$$

Show:

(a) $U_\epsilon(T)$ is an open set.
(b) If C is a compact subset of S and D a closed subset of S with $C \cap D = \emptyset$, then there exists some $\epsilon > 0$ such that $U_\epsilon(C) \cap D = \emptyset$.

A.1.6. Totally bounded and separable sets. For checking that a set is compact, the following can be useful. Proof of the upcoming results can be found in many analysis books ([**159**, Chap. 7.9], e.g.).

DEFINITION A.23. *Let T be the subset of a metric space S with metric d. T is called* totally bounded *if, for each $\epsilon > 0$, there are finitely many points $s_1, \dots, s_n \in S$ such that*

$$S \subseteq \bigcup_{j=1}^{n} U_\epsilon(s_j).$$

In other words, for each $t \in T$, there exists some $j \in \{1, \dots, n\}$ such that $d(t, s_j) < \epsilon$. The set $\{s_1, \dots, s_n\}$ is called an ϵ-net.

REMARK A.24.

(a) It is immediate that every subset of a totally bounded set is totally bounded itself.
(b) The ϵ-net in Definition A.23 can be chosen as a subset of T.

PROPOSITION A.25. *Let T be the subset of a metric space (S, d). If T is compact, then T is totally bounded.*

To obtain a converse we need the following concepts.

DEFINITION A.26. *A sequence (s_j) in a metric space (S, d) is called* a Cauchy sequence *(or fundamental) if, for every $\epsilon > 0$, there exists some $N \in \mathbb{N}$ such that*

$$d(s_j, s_k) < \epsilon, \qquad j, k > N.$$

DEFINITION A.27. *A subset T of metric space (S, d) is* complete *if every Cauchy sequence in T has a limit in T. A normed vector space that is complete under the metric induced by the norm is called a Banach space.*

THEOREM A.28. *Let T be the subset of a metric space (S, d). Then T is totally bounded if and only if every sequence in T has a subsequence that is a Cauchy sequence.*

See [**72**, Thm.A.11] for a proof.

REMARK A.29. That proof also provides the following statement:
T is not totally bounded if and only if there exist some $\epsilon > 0$ and a sequence (t_n) in T such that $d(t_n, t_m) \geq \epsilon$ for all $n, m \in \mathbb{N}$ with $n \neq m$.

By using Theorem A.28, the next result follows from the respective definitions.

THEOREM A.30. *Let T be the subset of a metric space (S, d). Then T is compact if and only if T is complete and T is a totally bounded subset of S.*

For an alternative proof see [**159**, Thm.7.7].

Exercises.

A.1.5. Let T be the subset of a metric space S. Show that, if T is totally bounded, so is $\bar{T}$.

A.1.6. Let T be the subset of a complete metric space. Show that, if T is totally bounded, $\bar{T}$ is compact.

DEFINITION A.31. *Let S be a metric space. A subset T of S is called* separable *if it contains a countable set M such that $M \subseteq T \subseteq \bar{M}$.*

In other words, T is separable if it has a countable dense subset (Definition A.14).

Recall that a nonempty set M is countable if there exists a function $f : \mathbb{N} \to M$ such that $M = f(\mathbb{N})$ (in other words if M is the range of a sequence). [**159**, Sec.1.3].

Actually, the countable set M does not need to be a subset of S but only of X.

PROPOSITION A.32. *Let S be a metric space. A subset T of S is separable if there exists a countable subset M of S such that $T \subseteq \bar{M}$.*

COROLLARY A.33. *In a metric space, the subset of a separable set is also separable.*

PROPOSITION A.34. *In a metric space, a totally bounded subset is separable.*

Exercises.

A.1.7. Let S be a metric space. Show that the countable union of separable sets is separable: If $\{S_n; n \in \mathbb{N}\}$ is a countable family of separable sets S_n in S, then $\bigcup_{n\in\mathbb{N}} S_n$ is a separable subset of S.

Hint: Countable unions of countable sets are countable [**159**, Prop.1.10].

A.1.8. Let S be a nonempty set with the discrete metric (Example A.2).

Show:

(a) All subsets of S are open, closed, complete and bounded.
(b) A subset of S is compact if and only if it is finite.
(c) A subset of S is totally bounded if and only if it is finite.
(d) A subset of S is separable if and only if it is countable.
(e) The metric space (S, d) is locally compact: For any open subset U of S and any $s \in U$, there exist an open set V and a compact set W such that $s \in V \subseteq W \subseteq U$ [**159**, Prop.7.22].

A.1.7. Continuous functions. Equivalence of metrics.

DEFINITION A.35. *A function $f : S_1 \to S_2$ from a metric space (S_1, d_1) to a metric space (S_2, d_2) is* continuous *at a point $s_1 \in S_1$, if for each $\epsilon > 0$ there exists some $\delta > 0$ such that, for all $s \in S_1$,*

$$d_1(s, s_1) < \delta \implies d_2(f(s), f(s_1)) < \epsilon.$$

The function f is continuous *if f is continuous at every $s_1 \in S_1$.*

Two metrics d_1 and d_2 on a nonempty set S are called topologically equivalent *if the identity map $s \mapsto s$ is continuous from (S, d_1) to (S, d_2) and vice versa.*

EXAMPLE A.36. Let V be a vector space, $V \neq \{\mathbb{O}\}$, and $\|\cdot\|$ an arbitrary norm on V. Then the metric induced by $\|\cdot\|$ is not topologically equivalent to the discrete metric, denoted by d. More precisely, while any map from (V, d) to $(V, \|\cdot\|)$ is continuous, the identify map from $(V, \|\cdot\|)$ to (V, d) is not continuous at any point of V.

Definition A.37. *A function $f : S_1 \to S_2$ from a metric space (S_1, d_1) to a metric space (S_2, d_2) is called* uniformly continuous *on a subset D of S_1 if for every $\epsilon > 0$ there exists some $\delta > 0$ such that, for all $s, t \in D$,*

$$d_1(s,t) < \delta \implies d_2(f(s), f(t)) < \epsilon.$$

f is called uniformly continuous *if it is uniformly continuous on S_1.*

Definition A.38. *Let $f : S_1 \to S_2$ be a function from a metric space (S_1, d_1) to a metric space (S_2, d_2) and $D \subseteq S_1$. f is called* Lipschitz continuous *on D if there exists a constant $\Lambda > 0$ such that*

$$d_2(f(s), f(t)) \le \Lambda d_1(s,t) \quad \textit{for all } s, t \in D.$$

*Λ is called **a*** Lipschitz constant *for f on D.*

*If D contains at least two points, **the*** Lipschitz constant *for f on D is defined as*

$$\Lambda(f, D) = \sup\left\{ \frac{d_2(f(s), f(t))}{d_1(s,t)}; s, t \in D, s \ne t \right\}.$$

If $D = S_1$, we say that f is Lipschitz continuous.

Two metrics d_1 and d_2 on the same nonempty set S are called equivalent *if the identity map is Lipschitz continuous from (S, d_1) to (S, d_2) and from (S, d_2) to (S, d_1).*

Lemma A.39. *Every Lipschitz continuous function is uniformly continuous. Every uniformly continuous function is continuous.*

Example A.40. Let S be a metric space with metric d. Let t be a fixed but arbitrary element in S and define $f : S \to \mathbb{R}$ by $f(s) = d(s,t)$, $s \in S$. Then f is Lipschitz continuous with $\Lambda = 1$ as a Lipschitz constant.

This follows directly from Proposition A.4 and the symmetry of the metric.

Example A.41. Let S be a metric space with metric d. Let T be a nonempty subset of S and define $f : S \to \mathbb{R}$ by $f(s) = d(s, T)$, $s \in S$. Then f is Lipschitz continuous with $\Lambda = 1$ as a Lipschitz constant.

This follows directly from Lemma A.9.

Example A.42 (Minimum function). Equip $\mathbb{R}^2$ with the max-norm and consider $f : \mathbb{R}^2 \to \mathbb{R}$,

$$f(x,y) = \min\{x, y\}, \qquad x, y \in \mathbb{R}.$$

Then f is Lipschitz continuous with Lipschitz constant 1. (This can be seen by checking many cases.)

Exercises.

A.1.9. Let X be a vector space and $\|\cdot\|_1$ and $\|\cdot\|_2$ be norms on X. Show: The induced metrics are equivalent if and only if there exist constants $\Lambda_2, \Lambda_1 > 0$ such that

$$\|x\|_1 \le \Lambda_2 \|x\|_2 \quad \text{and} \quad \|x\|_2 \le \Lambda_1 \|x\|_1 \quad \text{for all } x \in X. \tag{A.6}$$

If one and then both of the two equivalent statements hold, we say that *the two norms are equivalent.*

Exercises: Sequential characterization of (uniform) continuity.

A.1.10. Let (S_1, d_1) and (S_2, d_2) be metric spaces and $f : S_1 \to S_2$, $s \in S_1$. Show f is continuous at s if and only if $f(s_n) \to f(s)$ as $n \to \infty$ for every sequence (s_n) in S_1 with $s_n \to s$ as $n \to \infty$.

A.1.11. Let S be a nonempty set endowed with two metrics d_1 and d_2. Show that d_1 and d_2 are topologically equivalent if and only if any sequence that converges in one of the two metric spaces also converges in the other.

In other words, two metrics are topologically equivalent if and only if convergence does not depend on which of the two metrics is chosen.

So, the closure of a set taken with respect to one metric is the same as the closure taken with respect to the other metric.

Further, a subset is closed (compact) with respect to one metric if (and only if) it is closed (compact) with respect to the other metric.

A.1.12. Let (S_1, d_1) and (S_2, d_2) be metric spaces and $f : S_1 \to S_2$. Show that f is uniformly continuous if and only if $d_2(f(s_n), f(t_n)) \to 0$ as $n \to \infty$ for every two sequences (s_n) and (t_n) in S_1 with $d_1(s_n, t_n) \to 0$ as $n \to \infty$.

A.1.8. Compact operators. Let S_1 and S_2 be metric spaces, $T \subseteq S_1$ and $f : T \to S_2$. Recall the definition of a bounded set, Definition A.22.

DEFINITION A.43. *The operator f is called* compact *if, for any bounded subset D of T, $f(D)$ is contained in a compact subset of S_2 (i.e., the closure of $f(D)$ is a compact subset of S_2).*

Exercises.

A.1.13. Show: The operator f is compact if and only if, for any bounded sequence (s_n) in T, the sequence $(f(s_n))$ has a subsequence that converges in S_2.

Hint: For $\Leftarrow$ use Exercise A.1.2.

A.1.14. Show:

(a) If f is compact and D is a bounded subset of T, then $f(D)$ is a totally bounded subset of S_2.
(b) If S_2 is complete and if, for any bounded subset D of T, $f(D)$ is a totally bounded subset of S_2, then f is compact.

A.2. The extended real numbers

Let $\tilde{\mathbb{R}} = \mathbb{R} \cup \{-\infty, \infty\}$ and $\mathbb{R}_\infty = \mathbb{R} \cup \{\infty\}$.

Define a metric d on $\tilde{\mathbb{R}}$ by $d(x, y) = |\phi(x) - \phi(y)|$ for $x, y \in \tilde{\mathbb{R}}$ with a function $\phi : \tilde{\mathbb{R}} \to [-1, 1]$ given by

$$\phi(x) = \begin{cases} \frac{x}{1+|x|}, & x \in \mathbb{R}, \\ 1, & x = \infty, \\ -1, & x = -\infty. \end{cases}$$

Notice that ϕ is continuous at every $x \in \mathbb{R}$, ϕ strictly increasing on $\mathbb{R}$, and $\phi(x) \to \pm 1$ as $x \to \pm\infty$. One readily checks that d is a metric on $\tilde{\mathbb{R}}$ and has the following properties:

- If $x \in \mathbb{R}$ and (x_n) is a sequence in $\mathbb{R}$, then $d(x_n, x) \to 0$ if and only if $|x_n - x| \to 0$;
- If (x_n) is a sequence in $\mathbb{R}$, then $d(x_n, \pm\infty) \to 0$ if and only if for every $c > 0$ there is some $N \in \mathbb{N}$ such that $\pm x_n > c$ for all $n \geq N$.

It follows that $(\tilde{\mathbb{R}}, d)$ is a compact metric space.

We extend the order on $\mathbb{R}$ to $\tilde{\mathbb{R}}$ by

$$-\infty < x < \infty, \qquad x \in \mathbb{R}.$$

Intervals are now defined as usually, e.g., $(a, \infty] = \{x \in \tilde{\mathbb{R}}; a < x \leq \infty\}$.

Multiplication is extended by

$$0 \cdot x = 0, \quad x \in \tilde{\mathbb{R}}, \qquad \pm\infty \cdot x = \pm\infty, \quad x > 0, \qquad \pm\infty \cdot x = \mp\infty, \quad x < 0.$$

Addition is extended by

$$\begin{aligned} x + \infty &= \infty = \infty + x, && x \in \mathbb{R}_\infty, \\ x + (-\infty) &= -\infty = -\infty + x, && x \in \mathbb{R} \cup \{-\infty\}. \end{aligned}$$

$\infty + (-\infty)$ is left undefined.

Notice that no longer $x < y$ implies $x + z < y + z$ if $z = \pm\infty$, but $x \leq y$ implies $x + z \leq y + z$ whenever both sums are defined.

Notions as the supremum (least upper bound) and infimum (largest lower bound) of subsets of $\tilde{R}$ are to be understood with respect to this extended order.

A.3. Sequences

If (a_n) is a sequence in $\mathbb{R}_\infty$, then, for all $m \in \mathbb{N}$, $\sum_{n=1}^m a_n$ is defined and an element in $\mathbb{R}_\infty$ and

$$\sum_{n=1}^{\infty} a_n := \lim_{m \to \infty} \sum_{n=1}^{m} a_n,$$

provided that the limit exists in $\tilde{\mathbb{R}}$ with metric d.

In particular, this means that none of the terms of the series is $-\infty$; further, if at least one term is ∞, the series is ∞.

A.3.1. Limit superior and limit inferior.

Let (x_n) be a sequence in $[-\infty, \infty]$. We define

$$x_\infty = \liminf_{n \to \infty} x_n = \sup_{m \in \mathbb{N}} \Big(\inf_{n \geq m} x_n \Big)$$

and

$$x^\infty = \limsup_{n \to \infty} x_n = \inf_{m \in \mathbb{N}} \Big(\sup_{n \geq m} x_n \Big).$$

Notice that

$$\begin{aligned} \limsup_{n \to \infty} (-x_n) &= -\liminf_{n \to \infty} x_n \\ -\limsup_{n \to \infty} x_n &= \liminf_{n \to \infty} (-x_n) \end{aligned} \tag{A.7}$$

Exercises. The first exercise is to get familiar with the notions of limit inferior and limit superior.

A.3.1. Show:

(a) $x_\infty = \lim_{m\to\infty} \Big(\inf_{n\geq m} x_n \Big)$.

(b) $x^\infty = \lim_{m\to\infty} \Big(\sup_{n\geq m} x_n \Big)$.

(c) $x_\infty \leq x^\infty$.

(d) $x_\infty = x^\infty$ if and only if (x_n) converges.

The following exercise is related to Section 1.3 and the formulation of uniform weak and uniform persistence using the limit superior and the limit inferior.

A.3.2. Let (x_n) be a sequence in $\mathbb{R}$. Let $\epsilon \in \mathbb{R}$. Then the following hold:

(a) If $x_\infty > \epsilon$, then $x_n > \epsilon$ for all but finitely many $n \in \mathbb{N}$.

(b) If $x_n \geq \epsilon$ for all but finitely many $n \in \mathbb{N}$, then $x_\infty \geq \epsilon$.

(c) If $x^\infty > \epsilon$, then $x_n > \epsilon$ for infinitely many $n \in \mathbb{N}$.

(d) If $x_n \geq \epsilon$ for infinitely many $n \in \mathbb{N}$, then $x^\infty \geq \epsilon$.

For more see [**159**, Sec.2.2].

A.3.2. Towards the definition of the spectral radius. The next result is known since a long time in its logarithmic version, [**172**, p.17, problem 98]. See also [**174**, p.425].

LEMMA A.44. *Let $(a_n)_{n\geq k}$, $k \in \mathbb{Z}_+$, be a sequence in $\mathbb{R}_+$ such that*

$$a_{n+m} \leq a_n a_m, \qquad n, m \geq k.$$

Then $a_n^{1/n} \to \inf_{m\geq k} a_m^{1/m}$ as $n \to \infty$.

PROOF. Let $\inf_{m=k}^\infty a_m^{1/m} =: \alpha < c$.

Then there exists some $m \in \mathbb{Z}_+$ with $m \geq k$ such that $a_m^{1/m} < c$.

Any number $n \in \mathbb{N}$ with $n \geq 2m$ has a unique representation $n = pm + q$ with $p = p_n \in \mathbb{N}$ and $m \leq q < 2m$. Then, for $n \geq 2m$,

$$a_n \leq a_m^p a_q \leq c^{mp} a_q.$$

and

$$a_n^{1/n} \leq c^{p_n m/n} \beta^{1/n}, \qquad \beta = \max_{q=m}^{2m} a_q.$$

We can assume that $\beta > 0$, otherwise $a_n = 0$ for all $n \geq 2m$ and the claim holds. As $n \to \infty$, $p_n m/n \to 1$ and $\beta^{1/n} \to 1$ and $\limsup_{n\to\infty} a_n^{1/n} \leq c$. Since c was any number larger than the infimum, the limes superior and inferior coincide and the limit exists and equals the infimum. □

LEMMA A.45. *Let $(b_n)_{n\geq k}$, $k \in \mathbb{Z}_+$, be a sequence in $(0, \infty)$ such that*

$$b_{n+m} \geq b_n b_m, \qquad n, m \geq k.$$

Then $b_n^{1/n} \to \sup_{m\geq k} b_m^{1/m}$ as $n \to \infty$.

PROOF. Set $a_n = 1/b_n$. By the previous result,

$$a_n \to \inf_{m\geq k} a_m^{1/m} = \Big(\sup_{m\geq k} b_m^{1/m} \Big)^{-1}.$$

This implies the assertion. □

LEMMA A.46. *Let $(a_n)_{n\geq k}$ be a sequence in $\mathbb{R}_+$ and $L \in \mathbb{R}_+$ such that $a_n^{1/n} \to L$ as $n \to \infty$. Then, for any $m \in \mathbb{N}$, $a_{nm}^{1/n} \to L^m$ as $n \to \infty$.*

PROOF. Since subsequences of a convergent sequence converge to the same limit,

$$L = \lim_{n\to\infty} a_{nm}^{1/(nm)}.$$

By the limit rules,

$$L^m = \lim_{n\to\infty} a_{nm}^{m/(nm)} = \lim_{n\to\infty} a_{nm}^{1/n}.$$

□

LEMMA A.47 (Bonsall [**27**]). *Let (a_n) be an unbounded sequence in $\mathbb{R}_+$. Then there exists a sequence (n_j) in $\mathbb{N}$ such that $n_j \to \infty$ and*

$$a_{n_j} \to \infty, \quad j \to \infty, \qquad a_k \leq a_{n_j}, \qquad k = 1, \dots, n_j, \quad j \in \mathbb{N}.$$

PROOF. For each $j \in \mathbb{N}$, choose $n_j \leq j$ such that $a_{n_j} = \max\{a_1, \dots, a_j\}$. Since (a_n) is unbounded, $n_j \to \infty$ and $a_{n_j} \to \infty$ as $j \to \infty$. By construction, $a_k \leq a_{n_j}$ for $k = 1, \dots, j \geq n_j$. □

A.4. Semicontinuous functions

Since semicontinuity of functions with values in $\mathbb{R}$ plays an unusually important role in this book, it seems appropriate to develop that subject in some detail. While continuity can be defined for functions between arbitrary metric spaces (Appendix A.1.7), semicontinuity is mainly defined for functions with range in $\mathbb{R}$.

Let S be a metric space with metric d (Appendix A.1).

A function $f : S \to \mathbb{R}$ is called *lower semicontinuous* at $s \in S$ if for any $\epsilon > 0$ there exists some $\delta > 0$ such that $f(t) > f(s) - \epsilon$ for all $t \in S$ with $d(t, s) < \delta$.

It is called *upper semicontinuous* at $s \in S$ if for any $\epsilon > 0$ there exists some $\delta > 0$ such that $f(t) < f(s) + \epsilon$ for all $t \in S$ with $d(t, s) < \delta$.

f is continuous at $s \in S$ if and only if it is lower and upper semicontinuous at s.

f is (upper, lower semi-) continuous (on S) if f is (upper, lower semi-) continuous at every $s \in S$.

Cf. [**112**, II.7].

REMARK A.48. Let $f, g : S \to \mathbb{R}$, $s \in S$.

If f is lower semicontinuous at s, then $-f$ is upper semicontinuous at s.

If f is upper semicontinuous at s, then $-f$ is lower semicontinuous at s.

If f, g are lower semicontinuous at s, so is $f + g$.

If f, g are upper semicontinuous at s, so is $f + g$

If f is continuous at s and g is lower semicontinuous at s, then $f - g$ is upper semicontinuous at s.

If f is continuous at s and g is upper semicontinuous at s, then $f - g$ is lower semicontinuous at s.

PROOF. As for the first two statements, multiply the inequalities in the definitions above by -1. The other statements follow from standard ϵ&δ arguments. □

LEMMA A.49. *Let $f : S \to \mathbb{R}$, $s \in S$. The following are equivalent:*

(a) *f is lower semicontinuous at $s \in S$.*

(b) *For any $\epsilon > 0$, s is an interior point of the set $f^{-1}(s - \epsilon, \infty)$ (Definition A.17).*

(c) *For any sequence (s_n) in S with limit s,*

$$\liminf_{n\to\infty} f(s_n) \geq f(s). \tag{A.8}$$

PROOF. The equivalence of (a) and (b) follows from the definition of an interior point.

(a) $\Rightarrow$ (c): Let f be lower semicontinuous at $s \in S$. Let $\epsilon > 0$. Then there exists some $\delta > 0$ such that some $f(t) > f(s) - \epsilon$ for all $t \in S$ with $d(t, s) < \delta$. Let (s_n) be a sequence in S with limit $s \in S$. Then there exist some $m \in \mathbb{N}$ such that $d(s_n, s) < \delta$ for all $n \geq m$ and so $f(s_n) \geq f(s) - \epsilon$ for all $n \geq m$. Thus

$$\inf_{n\geq m} f(s_n) \geq f(s) - \epsilon.$$

By definition of the limit inferior (Appendix A.3.1),

$$\liminf_{n\to\infty} f(s_n) \geq f(s) - \epsilon.$$

Since $\epsilon > 0$ has been arbitrary, (A.8) follows.

(c) $\Rightarrow$ (a): We prove the contrapositive. Assume that f is not lower semicontinuous. Then there exists some $\epsilon > 0$ such that for all $n \in \mathbb{N}$ there exists some $s_n \in S$ with $d(s_n, s) < 1/n$ and $f(s_n) \leq f(s) - \epsilon$. Then $s_n \to s$ as $n \to \infty$. For every $m \in N$, $\inf_{n\geq m} f(s_n) \leq f(s) - \epsilon$ and (A.8) does not hold by definition of the limit inferior. □

The next result follows from this one by Remark A.48 and (A.7).

LEMMA A.50. *Let $f : S \to \mathbb{R}$, $s \in S$. The following are equivalent:*

(a) *f is upper semicontinuous at $s \in S$.*

(b) *For any $\epsilon > 0$, s is an interior point of the set $f^{-1}(-\infty, s + \epsilon)$.*

(c) *For any sequence $s_n)$ in S with limit s,*

$$\limsup_{n\to\infty} f(s_n) \leq f(s).$$

The following characterization of semicontinuous functions is used as a definition for general topological spaces [**71**, Def.10.2].

PROPOSITION A.51. *Let $f : S \to R$.*

(a) *Then, f is lower semicontinuous if and only if $f^{-1}(r, \infty)$ is an open subset of S for any $r \in \mathbb{R}$.*

(b) *Further, f is upper continuous if and only if $f^{-1}(-\infty, r)$ is open subset of S for any $r \in \mathbb{R}$.*

PROOF. This follows from the previous results and the fact that a subset U of S is open if any $s \in S$ is an interior point of S. Recall Definitions A.17 and A.18. □

LEMMA A.52. *Let $\mathcal{F}$ be a set of functions $f : S \to \mathbb{R}$ such that $\{f(s); f \in \mathcal{F}\}$ is bounded below for any $s \in S$. Then, if $s \in S$ and f is upper semicontinuous at s for any $f \in \mathcal{F}$, the function f_1 given by*

$$f_1(t) := \inf\{f(t), f \in \mathcal{F}\}, \qquad t \in S,$$

is upper semicontinuous at $s \in S$.

PROOF. Let $s \in S$ and f be upper semicontinuous at s for any $f \in \mathcal{F}$. Let $\epsilon > 0$. Since the infimum is the largest lower bound, $f_1(s)+(\epsilon/2)$ is no a lower bound of $\{f(s); f \in \mathcal{F}\}$. Thus there exists some $f \in \mathcal{F}$ such that $f(s) < f_1(s) + (\epsilon/2)$. Since f is upper semicontinuous, there is some $\delta > 0$ such that $f(t) < f(s) + (\epsilon/2)$ for all $t \in S$ with $d(t,s) < \delta$. So, $f_1(t) \le f(t) < f_1(s) + \epsilon$ for all $t \in S$ with $d(t,s) < \delta$. □

PROOF. ([**71**, 10.4]). Notice that, for any $r \in \mathbb{R}$,

$$f_1^{-1}(-\infty, r) = \bigcup_{f \in \mathcal{F}} f^{-1}(-\infty, r)$$

and use that arbitrary unions of open sets are open. □

LEMMA A.53. *Let $\mathcal{F}$ be a set of functions $f : S \to \mathbb{R}$ such that $\{f(s); f \in \mathcal{F}\}$ is bounded above for any $s \in S$. Then, if f is lower semicontinuous at s for any $f \in \mathcal{F}$, the function f_2 given by*

$$f_2(t) := \sup\{f(t), f \in \mathcal{F}\} =: \big(\sup_{f \in \mathcal{F}} f\big)(t), \qquad t \in S,$$

is lower semicontinuous at $s \in S$.

PROOF. Let $\mathcal{F}$ be a set of functions $f : S \to \mathbb{R}$ such that $\{f(s); f \in \mathcal{F}\}$ is bounded above for any $s \in S$. Then $\{-f(s); f \in \mathcal{F}\}$ is bounded below for all $s \in S$. Let $s \in S$ and f be lower semicontinuous at s for any $f \in \mathcal{F}$. By Remark A.48, $-f$ is upper semicontinuous at s for any $f \in \mathcal{F}$.

By Lemma A.52, $\inf_{f \in \mathcal{F}}(-f)$ is upper semicontinuous at s. But $\inf_{f \in \mathcal{F}}(-f) = -\sup_{f \in cF} f$. Again by Remark A.48, $f_2 = \sup_{f \in \mathcal{F}} f$ is lower semicontinous. □

PROOF. ([**71**, 10.4]). Notice that, for any $r \in \mathbb{R}$,

$$f_1^{-1}(r, \infty) = \bigcup_{f \in \mathcal{F}} f^{-1}(r, \infty)$$

and use that arbitrary unions of open sets are open. □

EXAMPLE A.54.

(a) Let S be a metric space, $\emptyset \ne U \subseteq S$.

Then U is an open subset of S if and only if χ_U is lower semicontinuous.

Further, U is a closed subset of S if and only if χ_U is upper semicontinuous.

(b) Let X be an ordered normed vector space with closed cone X_+ and $u \in X_+$. Then $\|\cdot\|_u$ is lower semicontinuous on $X_u \cap X_+$ with respect to the norm on X.

(c) Let X be an ordered normed vector space with closed cone X_+ and $u \in \dot{X}_+$. Then $[\cdot]_u$ is upper semicontinuous on X_+ with respect to the norm on X.

(d) Under relatively mild conditions, the spectral radius is an upper semicontinuous function of bounded homogeneous operators, while the conditions for lower semicontinuity are quite restrictive.

PROOF. (a) Let $\emptyset \neq U \subset S$. Then, for $r \in \mathbb{R}$, we have the trichotomy

$$\chi_U^{-1}(r,\infty) = \begin{cases} \emptyset, & r \geq 1, \\ U, & 0 \leq r < 1, \\ S, & r < 0. \end{cases}$$

By Proposition A.51, χ_U is lower semicontinuous if and only if U is open.

The following chain of equivalences holds:

$$\begin{aligned} U \text{ is closed} &\iff S \setminus U \text{ is open} \\ &\iff \chi_{S\setminus U} = 1 - \chi_U \text{ is lower semicontinuous (part (a))} \\ &\iff -\chi_U \text{ is lower semicontinuous (easy to see)} \\ &\iff \chi_U \text{ is upper semicontinuous (Remark A.48).} \end{aligned}$$

(b) See Proposition 2.70.

(c) See Lemma 4.21. □

Bibliography

[1] A. S. Ackleh, R. M. Colombo, P. Goatin, S. Hille, and A. Muntean (guest editors), *Mathematical modeling with measures*, Math. Biosci. Engin. **17** (2020), Special issue.

[2] A. S. Ackleh, R. M. Colombo, S. C. Hille, and A. Muntean, *Preface to "Modeling with measures"*, Math. Biosci. Eng. **12** (2015), no. 2, i–ii, DOI 10.3934/mbe.2015.12.2i. MR3327921

[3] A. S. Ackleh, V. K. Chellamuthu, and K. Ito, *Finite difference approximations for measure-valued solutions of a hierarchically size-structured population model*, Math. Biosci. Eng. **12** (2015), no. 2, 233–258, DOI 10.3934/mbe.2015.12.233. MR3327922

[4] A. S. Ackleh, J. Cleveland, and H. R. Thieme, *Population dynamics under selection and mutation: long-time behavior for differential equations in measure spaces*, J. Differential Equations **261** (2016), no. 2, 1472–1505, DOI 10.1016/j.jde.2016.04.008. MR3494403

[5] A. S. Ackleh, B. G. Fitzpatrick, and H. R. Thieme, *Rate distributions and survival of the fittest: a formulation on the space of measures*, Discrete Contin. Dyn. Syst. Ser. B **5** (2005), no. 4, 917–928, DOI 10.3934/dcdsb.2005.5.917. MR2170216

[6] A. S. Ackleh, R. J. Sacker, and P. Salceanu, *On a discrete selection-mutation model*, J. Difference Equ. Appl. **20** (2014), no. 10, 1383–1403, DOI 10.1080/10236198.2014.933819. MR3240393

[7] A. S. Ackleh and P. L. Salceanu, *Competitive exclusion and coexistence in an n-species Ricker model*, J. Biol. Dyn. **9** (2015), no. suppl. 1, 321–331, DOI 10.1080/17513758.2015.1020576. MR3360881

[8] M. Akian, S. Gaubert, and R. D. Nussbaum, *A Collatz–Wielandt characterization of the spectral radius of order-preserving homogeneous maps on cones*, `arXiv:1112.5968v1`, 2011.

[9] C. D. Aliprantis and K. C. Border, *Infinite Dimensional Analysis. A Hitchhiker's Guide*, 3rd ed., Springer, Berlin, 2006. MR2378491

[10] C. D. Aliprantis and R. Tourky, *Cones and duality*, Graduate Studies in Mathematics, vol. 84, American Mathematical Society, Providence, RI, 2007, DOI 10.1090/gsm/084. MR2317344

[11] L. J. S. Allen, E. J. Allen, and C. B. Jonsson, *The impact of environmental variation on hantavirus infection in rodents*, Mathematical studies on human disease dynamics, Contemp. Math., vol. 410, Amer. Math. Soc., Providence, RI 2006, pp. 1–15, DOI 10.1090/conm/410/07717. MR2277878

[12] L. J. S. Allen and P. van den Driessche, *The basic reproduction number in some discrete-time epidemic models*, J. Difference Equ. Appl. **14** (2008), no. 10-11, 1127–1147, DOI 10.1080/10236190802332308. MR2447189

[13] L. Ambrosio, N. Gigli, and G. Savaré, *Gradient flows in metric spaces and in the space of probability measures*, Lectures in Mathematics ETH Zürich, Birkhäuser Verlag, Basel, 2005. MR2129498

[14] T. Andô, *On fundamental properties of a Banach space with a cone*, Pacific J. Math. **12** (1962), 1163–1169. MR150572

[15] J. Appell, E. De Pascale, and A. Vignoli, *Nonlinear spectral theory*, De Gruyter Series in Nonlinear Analysis and Applications, vol. 10, Walter de Gruyter & Co., Berlin, 2004, DOI 10.1515/9783110199260. MR2059617

[16] W. Arendt, *Positive semigroups of kernel operators*, Positivity **12** (2008), no. 1, 25–44, DOI 10.1007/s11117-007-2137-z. MR2373131

[17] W. Arendt, P. R. Chernoff, and T. Kato, *A generalization of dissipativity and positive semigroups*, J. Operator Theory **8** (1982), no. 1, 167–180. MR670183

[18] W. Arendt and R. Nittka, *Equivalent complete norms and positivity*, Arch. Math. (Basel) **92** (2009), no. 5, 414–427, DOI 10.1007/s00013-009-3190-6. MR2506943

[19] B. S. Attili and S. F. Mallak, *Existence of limit cycles in a predator-prey system with a functional response of the form* $\arctan(ax)$, Commun. Math. Anal. **1** (2006), no. 1, 33–40. MR2250379

[20] N. Bacaër and E. H. Ait Dads, *On the biological interpretation of a definition for the parameter R_0 in periodic population models*, J. Math. Biol. **65** (2012), no. 4, 601–621, DOI 10.1007/s00285-011-0479-4. MR2966664

[21] M. F. Barnsley, *Fractals everywhere*, 2nd ed., Academic Press Professional, Boston, MA, 1993. Revised with the assistance of and with a foreword by Hawley Rising, III. MR1231795

[22] H. Bauer, *Probability theory and elements of measure theory*, Second edition of the translation by R. B. Burckel from the third German edition, Probability and Mathematical Statistics, Academic Press, Inc. [Harcourt Brace Jovanovich, Publishers], London-New York, 1981. MR636091

[23] G. Birkhoff, *Uniformly semi-primitive multiplicative processes*, Trans. Amer. Math. Soc. **104** (1962), 37–51, DOI 10.2307/1993931. MR146100

[24] V. I. Bogachev, *Measure theory. Vol. I, II*, Springer-Verlag, Berlin, 2007, DOI 10.1007/978-3-540-34514-5. MR2267655

[25] E. Bohl, *Monotonie: Lösbarkeit und Numerik bei Operatorgleichungen* (German), Springer Tracts in Natural Philosophy, Vol. 25, Springer-Verlag, Berlin-New York, 1974. MR636414

[26] F. F. Bonsall, *Endomorphisms of partially ordered vector spaces*, J. London Math. Soc. **30** (1955), 133–144, DOI 10.1112/jlms/s1-30.2.133. MR68753

[27] F. F. Bonsall, *Linear operators in complete positive cones*, Proc. London Math. Soc. (3) **8** (1958), 53–75, DOI 10.1112/plms/s3-8.1.53. MR92938

[28] F. F. Bonsall, *Lectures on some fixed point theorems of functional analysis*, Tata Institute of Fundamental Research, Bombay, 1962. Notes by K. B. Vedak. MR198173

[29] C. J. Briggs, H. C. J. Godfray, *The dynamics of insect-pathogen interactions in stage-structured populations*, Amer. Nat. **145** (1995), 855-887

[30] K. Bryan and T. Leise, *The* $25,000,000,000 *eigenvector: the linear algebra behind Google*, SIAM Rev. **48** (2006), no. 3, 569–581, DOI 10.1137/050623280. MR2278443

[31] J.-E. Busse, P. Gwiazda, and A. Marciniak-Czochra, *Mass concentration in a nonlocal model of clonal selection*, J. Math. Biol. **73** (2016), no. 4, 1001–1033, DOI 10.1007/s00285-016-0979-3. MR3545528

[32] R. S. Cantrell and C. Cosner, *Effects of domain size on the persistence of populations in a diffusive food-chain model with Beddington-DeAngelis functional response*, Natur. Resource Modeling **14** (2001), no. 3, 335–367. MR1871914

[33] R. S. Cantrell and C. Cosner, *Spatial ecology via reaction-diffusion equations*, Wiley Series in Mathematical and Computational Biology, John Wiley & Sons, Ltd., Chichester, 2003, DOI 10.1002/0470871296. MR2191264

[34] J. A. Carillo, P. Gwiazda, K. Kropielnicka, and A. Marciniak-Czochra, *The escalator boxcar train method for a system of age-structured equations in the space of measures*, SIAM J. Num. Anal. **11** (2019), 123–143.

[35] B. H. Carlson and H. R. Thieme, *Can infectious diseases that do not confer immunity eradicate host species?*, J. Biol. Systems **32** (2024), DOI 10.1142/S0218339023400016.

[36] H. Caswell, *Matrix population models*, Sinauer Associates Inc., Sunderland, 1989.

[37] H. Caswell, *Matrix population models – Construction, analysis, and interpretation*, Sinauer Associates Inc., Sunderland, 2000.

[38] J. Cleveland, *Basic stage structure measure valued evolutionary game model*, Math. Biosci. Eng. **12** (2015), no. 2, 291–310, DOI 10.3934/mbe.2015.12.291. MR3327925

[39] J. Cleveland and A. S. Ackleh, *Evolutionary game theory on measure spaces: well-posedness*, Nonlinear Anal. Real World Appl. **14** (2013), no. 1, 785–797, DOI 10.1016/j.nonrwa.2012.08.002. MR2969873

[40] L. Collatz, *Einschliessungssatz für die Eigenwerte von Integralgleichungen* (German), Math. Z. **47** (1941), 395–398, DOI 10.1007/BF01180971. MR7086

[41] L. Collatz, *Einschliessungssatz für die charakteristischen Zahlen von Matrizen* (German), Math. Z. **48** (1942), 221–226, DOI 10.1007/BF01180013. MR8590

[42] J. M. Cushing, *An introduction to structured population dynamics*, CBMS-NSF Regional Conference Series in Applied Mathematics, vol. 71, Society for Industrial and Applied Mathematics (SIAM), Philadelphia, PA, 1998, DOI 10.1137/1.9781611970005. MR1636703

[43] J. M. Cushing, *Matrix models and population dynamics*, Mathematical biology, IAS/Park City Math. Ser., vol. 14, Amer. Math. Soc., Providence, RI, 2009, pp. 49–149, DOI 10.1090/pcms/014/04. MR2522049

[44] J. M. Cushing, *On the relationship between r and R_0 and its role in the bifurcation of stable equilibria of Darwinian matrix models*, J. Biol. Dyn. **5** (2011), no. 3, 277–297, DOI 10.1080/17513758.2010.491583. MR2818158

[45] J. M. Cushing, *Discrete time darwinian dynamics and semelparity versus iteroparity*, Math. Biosci. Eng. **16** (2019), no. 4, 1815–1835, DOI 10.3934/mbe.2019088. MR4030661

[46] J. M. Cushing, *Difference equations as models of evolutionary population dynamics*, J. Biol. Dyn. **13** (2019), no. 1, 103–127, DOI 10.1080/17513758.2019.1574034. MR3907772

[47] J. M. Cushing, R. F. Costantino, B. Dennis, R. A. Desharnais, and S. M. Henson, *Chaos in ecology: Experimental nonlinear dynamics*, Theoretical Ecology Series, vol. 1, Academic Press (Elsevier Science), New York, 2003.

[48] J. M. Cushing and A. P. Farrell, *A bifurcation theorem for nonlinear matrix models of population dynamics*, J. Difference Equ. Appl. **26** (2020), no. 1, 25–44, DOI 10.1080/10236198.2019.1699916. MR4065246

[49] J. M. Cushing and K. Stefanko, *A Darwinian dynamic model for the evolution of post-reproduction survival*, J. Biol. Systems **29** (2021), no. 2, 433–450, DOI 10.1142/S0218339021400088. MR4274354

[50] J. M. Cushing and Y. Zhou, *The net reproductive value and stability in matrix population models*, Nat. Res. Mod. **8** (1994), 297–333.

[51] R. Datko, *Extending a theorem of A. M. Liapunov to Hilbert space*, J. Math. Anal. Appl. **32** (1970), 610–616, DOI 10.1016/0022-247X(70)90283-0. MR268717

[52] R. O. Davies, *A non-Prokhorov space*, Bull. London Math. Soc. **3** (1971), 341–342, DOI 10.1112/blms/3.3.341. MR293579

[53] N. V. Davydova, O. Diekmann, and S. A. van Gils, *Year class coexistence or competitive exclusion for strict biennials?*, J. Math. Biol. **46** (2003), no. 2, 95–131, DOI 10.1007/s00285-002-0167-5. MR1963068

[54] N. V. Davydova, O. Diekmann, and S. A. van Gils, *On circulant populations. I. The algebra of semelparity*, Linear Algebra Appl. **398** (2005), 185–243, DOI 10.1016/j.laa.2004.12.020. MR2121350

[55] K. Deimling, *Nonlinear functional analysis*, Springer-Verlag, Berlin, 1985, DOI 10.1007/978-3-662-00547-7. MR787404

[56] P. De Leenheer, Z. Gregg, and J. McCaslin, *Some limitations on the use of the basic reproduction number*, J. Difference Eqs. Appl. (2024), 1–25, DOI 10.1080/10236198.2024.2308110.

[57] A. de Roos and L. Persson, *Population and community ecology of ontogenic development*, Princeton University Press, Princeton, NJ, 2013.

[58] O. Diekmann, N. Davydova, and S. van Gils, *On a boom and bust year class cycle*, J. Difference Equ. Appl. **11** (2005), no. 4-5, 327–335, DOI 10.1080/10236190412331335409. MR2151678

[59] O. Diekmann, M. Gyllenberg, H. Huang, M. Kirkilionis, J. A. J. Metz, and H. R. Thieme, *On the formulation and analysis of general deterministic structured population models. II. Nonlinear theory*, J. Math. Biol. **43** (2001), no. 2, 157–189, DOI 10.1007/s002850170002. MR1860461

[60] O. Diekmann, M. Gyllenberg, J. A. J. Metz, and H. Thieme, *The "cumulative" formulation of (physiologically) structured population models*, Evolution equations, control theory, and biomathematics (Han sur Lesse, 1991), Lecture Notes in Pure and Appl. Math., vol. 155, Dekker, New York, 1994, pp. 145–154. MR1254894

[61] O. Diekmann, M. Gyllenberg, J. A. J. Metz, and H. R. Thieme, *On the formulation and analysis of general deterministic structured population models. I. Linear theory*, J. Math. Biol. **36** (1998), no. 4, 349–388, DOI 10.1007/s002850050104. MR1624188

[62] O. Diekmann, H. Heesterbeek, and T. Britton, *Mathematical tools for understanding infectious disease dynamics*, Princeton Series in Theoretical and Computational Biology, Princeton University Press, Princeton, NJ, 2013. MR3015083

[63] O. Diekmann, J. A. P. Heesterbeek, and J. A. J. Metz, *On the definition and the computation of the basic reproduction ratio R_0 in models for infectious diseases in heterogeneous populations*, J. Math. Biol. **28** (1990), no. 4, 365–382, DOI 10.1007/BF00178324. MR1057044

[64] O. Diekmann and R. Planqué, *The winner takes it all: how semelparous insects can become periodical*, J. Math. Biol. **80** (2020), no. 1-2, 283–301, DOI 10.1007/s00285-019-01362-3. MR4062821

[65] O. Diekmann and S. A. van Gils, *Invariance and symmetry in a year-class model*, Bifurcation, symmetry and patterns (Porto, 2000), Trends Math., Birkhäuser, Basel, 2003, pp. 141–150. MR2014364

[66] O. Diekmann and S. A. van Gils, *On the cyclic replicator equation and the dynamics of semelparous populations*, SIAM J. Appl. Dyn. Syst. **8** (2009), no. 3, 1160–1189, DOI 10.1137/080722734. MR2551259

[67] O. Diekmann, Y. Wang, and P. Yan, *Carrying simplices in discrete competitive systems and age-structured semelparous populations*, Discrete Contin. Dyn. Syst. **20** (2008), no. 1, 37–52, DOI 10.3934/dcds.2008.20.37. MR2350059

[68] A. Ducrot, Q. Griette, Z. Liu, and P. Magal, *Differential equations and population dynamics II: Advanced approaches*, Lecture Notes on Mathematical Modelling in the Life Sciences, Springer Cham, In preparation.

[69] R. M. Dudley, *Convergence of Baire measures*, Studia Math. **27** (1966), 251–268, DOI 10.4064/sm-27-3-251-268. MR200710

[70] R. M. Dudley, *Real analysis and probability*, Cambridge Studies in Advanced Mathematics, vol. 74, Cambridge University Press, Cambridge, 2002. Revised reprint of the 1989 original, DOI 10.1017/CBO9780511755347. MR1932358

[71] J. Dugundji, *Topology*, Allyn and Bacon, Inc., Boston, MA, 1966. MR193606

[72] C. Düll, P. Gwiazda, A. Marciniak-Czochra, and J. Skrzeczkowski, *Spaces of measures and their applications to structured population models*, Cambridge Monographs on Applied and Computational Mathematics, vol. 36, Cambridge University Press, Cambridge, 2022. MR4309603

[73] N. Dunford and J. T. Schwartz, *Linear operators. Part I*, Wiley Classics Library, John Wiley & Sons, Inc., New York, 1988. General theory; With the assistance of William G. Bade and Robert G. Bartle; Reprint of the 1958 original; A Wiley-Interscience Publication. MR1009162

[74] E. A. Eager, R. Rebarber, and B. Tenhumberg, *Modeling and analysis of a density-dependent stochastic integral projection model for a disturbance specialist plant and its seed bank*, Bull. Math. Biol. **76** (2014), no. 7, 1809–1834, DOI 10.1007/s11538-014-9978-y. MR3249937

[75] S. Elaydi, *An introduction to difference equations*, 3rd ed., Undergraduate Texts in Mathematics, Springer, New York, 2005. MR2128146

[76] S. N. Elaydi, *Discrete chaos*, 2nd ed., Chapman & Hall/CRC, Boca Raton, FL, 2008. With applications in science and engineering; With a foreword by Robert M. May; With 1 CD-ROM (Windows, Macintosh and UNIX). MR2364977

[77] S. N. Elaydi, J. M. Cushing, and A.-A. Yakubu, *Discrete mathematical biology*, Springer, New York, To appear.

[78] S. P. Ellner, D. Z. Childs, and M. Rees, *Data-driven modelling of structured populations*, Lecture Notes on Mathematical Modelling in the Life Sciences, Springer, [Cham], 2016. A practical guide to the integral projection model, DOI 10.1007/978-3-319-28893-2. MR3496931

[79] S. P. Ellner and M. Rees, *Stochastic stable population growth in integral projection models: theory and application*, J. Math. Biol. **54** (2007), no. 2, 227–256, DOI 10.1007/s00285-006-0044-8. MR2284066

[80] K. E. Emmert and L. J. S. Allen, *Population persistence and extinction in a discrete-time, stage-structured epidemic model*, J. Difference Equ. Appl. **10** (2004), no. 13-15, 1177–1199, DOI 10.1080/10236190410001654151. MR2100721

[81] K.-J. Engel and R. Nagel, *One-parameter semigroups for linear evolution equations*, Graduate Texts in Mathematics, vol. 194, Springer-Verlag, New York, 2000. With contributions by S. Brendle, M. Campiti, T. Hahn, G. Metafune, G. Nickel, D. Pallara, C. Perazzoli, A. Rhandi, S. Romanelli and R. Schnaubelt. MR1721989

[82] J. H. M. Evers, S. C. Hille, and A. Muntean, *Mild solutions to a measure-valued mass evolution problem with flux boundary conditions*, J. Differential Equations **259** (2015), no. 3, 1068–1097, DOI 10.1016/j.jde.2015.02.037. MR3342408

[83] A. P. Farrell, J. P. Collins, A. L. Greer, and H. R. Thieme, *Do fatal infectious diseases eradicate host species?*, J. Math. Biol. **77** (2018), no. 6-7, 2103–2164, DOI 10.1007/s00285-018-1249-3. MR3881875

[84] A. P. Farrell and H. R. Thieme, *Predator—prey/host—parasite: a fragile ecoepidemic system under homogeneous infection incidence*, Discrete Contin. Dyn. Syst. Ser. B **26** (2021), no. 1, 217–267, DOI 10.3934/dcdsb.2020328. MR4191544

[85] H. Federer, *Geometric measure theory*, Die Grundlehren der mathematischen Wissenschaften, Band 153, Springer-Verlag New York, Inc., New York, 1969. MR257325

[86] H. Federer, *Colloquium lectures on geometric measure theory*, Bull. Amer. Math. Soc. **84** (1978), no. 3, 291–338, DOI 10.1090/S0002-9904-1978-14462-0. MR467473

[87] K.-H. Förster and B. Nagy, *On the Collatz-Wielandt numbers and the local spectral radius of a nonnegative operator*, Proceedings of the Fourth Haifa Matrix Theory Conference (Haifa, 1988), Linear Algebra Appl. **120** (1989), 193–205, DOI 10.1016/0024-3795(89)90378-9. MR1010057

[88] R. Fortet and E. Mourier, *Convergence de la répartition empirique vers la répartition théorique* (French), Ann. Sci. École Norm. Sup. (3) **70** (1953), 267–285. MR61325

[89] D. Franco, C. Guiver, H. Logemann, and J. Perán, *Boundedness, persistence and stability for classes of forced difference equations arising in population ecology*, J. Math. Biol. **79** (2019), no. 3, 1029–1076, DOI 10.1007/s00285-019-01388-7. MR3987188

[90] E. Franco, M. Gyllenberg, and O. Diekmann, *One dimensional reduction of a renewal equation for a measure-valued function of time describing population dynamics*, Acta Appl. Math. **175** (2021), Paper No. 12, 67, DOI 10.1007/s10440-021-00440-3. MR4323318

[91] E. Franco, O. Diekmann, and M. Gyllenberg, *Modelling physiologically structured populations: renewal equations and partial differential equations*, J. Evol. Equ. **23** (2023), no. 3, Paper No. 46, 62, DOI 10.1007/s00028-023-00880-4. MR4600253

[92] H. I. Freedman and P. Moson, *Persistence definitions and their connections*, Proc. Amer. Math. Soc. **109** (1990), no. 4, 1025–1033, DOI 10.2307/2048133. MR1012928

[93] I. M. Gel'fand, *Normierte ringe*, Mat. Sbornik N.S. **9** (1941), 3–24.

[94] X. Gong and B. Piccoli, *A measure model for the spread of viral infections with mutations*, Netw. Heterog. Media **17** (2022), no. 3, 427–442, DOI 10.3934/nhm.2022015. MR4421533

[95] G. Gripenberg, *On the definition of the cone spectral radius*, Proc. Amer. Math. Soc. **143** (2015), no. 4, 1617–1625, DOI 10.1090/S0002-9939-2014-12375-6. MR3314074

[96] C. Guiver, *On the strict monotonicity of spectral radii for classes of bounded positive linear operators*, Positivity **22** (2018), no. 4, 1173–1190, DOI 10.1007/s11117-018-0566-5. MR3843462

[97] P. Gwiazda, J. Jabłoński, A. Marciniak-Czochra, and A. Ulikowska, *Analysis of particle methods for structured population models with nonlocal boundary term in the framework of bounded Lipschitz distance*, Numer. Methods Partial Differential Equations **30** (2014), no. 6, 1797–1820, DOI 10.1002/num.21879. MR3267354

[98] P. Gwiazda, K. Kropielnicka, and A. Marciniak-Czochra, *The escalator boxcar train method for a system of age-structured equations*, Netw. Heterog. Media **11** (2016), no. 1, 123–143, DOI 10.3934/nhm.2016.11.123. MR3461738

[99] P. Gwiazda, T. Lorenz, and A. Marciniak-Czochra, *A nonlinear structured population model: Lipschitz continuity of measure-valued solutions with respect to model ingredients*, J. Differential Equations **248** (2010), no. 11, 2703–2735, DOI 10.1016/j.jde.2010.02.010. MR2644146

[100] P. Gwiazda and A. Marciniak-Czochra, *Structured population equations in metric spaces*, J. Hyperbolic Differ. Equ. **7** (2010), no. 4, 733–773, DOI 10.1142/S021989161000227X. MR2746205

[101] P. Gwiazda, A. Marciniak-Czochra, and H. R. Thieme, *Measures under the flat norm as ordered normed vector space*, Positivity **22** (2018), no. 1, 105–138, DOI 10.1007/s11117-017-0503-z. MR3764634

[102] P. Gwiazda, P. Orlinski, and A. Ulikowska, *Finite range method of approximation for balance laws in measure spaces*, Kinet. Relat. Models **10** (2017), no. 3, 669–688, DOI 10.3934/krm.2017027. MR3591128

[103] M. Gyllenberg, J. Jiang, and L. Niu, *Chaotic attractors in Atkinson-Allen model of four competing species*, J. Biol. Dyn. **14** (2020), no. 1, 440–453, DOI 10.1080/17513758.2020.1779828. MR4115411

[104] M. Gyllenberg, J. Jiang, L. Niu, and P. Yan, *Permanence and universal classification of discrete-time competitive systems via the carrying simplex*, Discrete Contin. Dyn. Syst. **40** (2020), no. 3, 1621–1663, DOI 10.3934/dcds.2020088. MR4063939

[105] K. P. Hadeler, *Pair formation in age-structured populations*, Acta Appl. Math. **14** (1989), no. 1-2, 91–102, DOI 10.1007/BF00046676. Evolution and control in biological systems (Laxenburg, 1987). MR990038

[106] K. P. Hadeler, *Periodic solutions of homogeneous equations*, J. Differential Equations **95** (1992), no. 1, 183–202, DOI 10.1016/0022-0396(92)90049-S. MR1142283

[107] K. P. Hadeler, *Pair formation*, J. Math. Biol. **64** (2012), no. 4, 613–645, DOI 10.1007/s00285-011-0454-0. MR2886028

[108] K. P. Hadeler, *Topics in mathematical biology*, Lecture Notes on Mathematical Modelling in the Life Sciences, Springer, Cham, 2017, DOI 10.1007/978-3-319-65621-2. MR3752189

[109] K. P. Hadeler, R. Waldstätter, and A. Wörz-Busekros, *Models for pair formation in bisexual populations*, J. Math. Biol. **26** (1988), no. 6, 635–649, DOI 10.1007/BF00276145. MR973342

[110] H. J. A. M. Heijmans, *Markov semigroups and structured population dynamics*, Aspects of positivity in functional analysis (Tübingen, 1985), North-Holland Math. Stud., vol. 122, North-Holland, Amsterdam, 1986, pp. 199–208. MR859728

[111] Ph. Clément, H. J. A. M. Heijmans, S. Angenent, C. J. van Duijn, and B. de Pagter, *One-parameter semigroups*, CWI Monographs, vol. 5, North-Holland Publishing Co., Amsterdam, 1987. MR915552

[112] E. Hewitt and K. Stromberg, *Real and abstract analysis*, Springer, Berlin Heidelberg, 1956.

[113] S. C. Hille and D. T. H. Worm, *Embedding of semigroups of Lipschitz maps into positive linear semigroups on ordered Banach spaces generated by measures*, Integral Equations Operator Theory **63** (2009), no. 3, 351–371, DOI 10.1007/s00020-008-1652-z. MR2491036

[114] S. C. Hille and D. T. H. Worm, *Continuity properties of Markov semigroups and their restrictions to invariant L^1-spaces*, Semigroup Forum **79** (2009), no. 3, 575–600, DOI 10.1007/s00233-009-9176-7. MR2564065

[115] F. Hoppensteadt, *Mathematical theories of populations: demographics, genetics and epidemics*, Regional Conference Series in Applied Mathematics, Society for Industrial and Applied Mathematics, Philadelphia, PA, 1975. MR526771

[116] S.-B. Hsu and X.-Q. Zhao, *Spreading speeds and traveling waves for nonmonotone integrodifference equations*, SIAM J. Math. Anal. **40** (2008), no. 2, 776–789, DOI 10.1137/070703016. MR2438786

[117] M. Iannelli, M. Martcheva, and F. A. Milner, *Gender-structured population modeling*, Frontiers in Applied Mathematics, vol. 31, Society for Industrial and Applied Mathematics (SIAM), Philadelphia, PA, 2005. Mathematical methods, numerics, and simulations, DOI 10.1137/1.9780898717488. MR2134254

[118] H. Inaba, *The Malthusian parameter and R_0 for heterogeneous populations in periodic environments*, Math. Biosci. Eng. **9** (2012), no. 2, 313–346, DOI 10.3934/mbe.2012.9.313. MR2897081

[119] H. Inaba, *On a new perspective of the basic reproduction number in heterogeneous environments*, J. Math. Biol. **65** (2012), no. 2, 309–348, DOI 10.1007/s00285-011-0463-z. MR2944513

[120] H. Inaba, *Age-structured population dynamics in demography and epidemiology*, Springer, Singapore, 2017, DOI 10.1007/978-981-10-0188-8. MR3616174

[121] H. Inaba, *The basic reproduction number R_0 in time-heterogeneous environments*, J. Math. Biol. **79** (2019), no. 2, 731–764, DOI 10.1007/s00285-019-01375-y. MR3982710

[122] V. S. Ivlev, *Experimental ecology of the feeding of fishes*, Yale University Press, New Haven, 1955.

[123] P. Jagers, *The deterministic evolution of general branching populations*, State of the art in probability and statistics (Leiden, 1999), IMS Lecture Notes Monogr. Ser., vol. 36, Inst. Math. Statist., Beachwood, OH, 2001, pp. 384–398, DOI 10.1214/lnms/1215090079. MR1836571

[124] A. D. Jassby, T. Platt, *Mathematical formulation of the relationship between photosynthesis and light for phytoplankton*, Limnol. Oceanogr. **21** (1976), 540-547

[125] L. Jiang, T. Yi, and X.-Q. Zhao, *Propagation dynamics for a class of integro-difference equations in a shifting environment*, J. Differential Equations **380** (2024), 491–515, DOI 10.1016/j.jde.2023.10.053. MR4667626

[126] W. Jin, *Persistence of Discrete Dynamical Systems in Infinite Dimensional State Spaces*, ProQuest LLC, Ann Arbor, MI, 2014. Thesis (Ph.D.)–Arizona State University. MR3232336

[127] W. Jin, H. L. Smith, and H. R. Thieme, *Persistence and critical domain size for diffusing populations with two sexes and short reproductive season*, J. Dynam. Differential Equations **28** (2016), no. 3-4, 689–705, DOI 10.1007/s10884-015-9434-1. MR3537352

[128] W. Jin, H. L. Smith, and H. R. Thieme, *Persistence versus extinction for a class of discrete-time structured population models*, J. Math. Biol. **72** (2016), no. 4, 821–850, DOI 10.1007/s00285-015-0898-8. MR3459168

[129] W. Jin and H. R. Thieme, *Persistence and extinction of diffusing populations with two sexes and short reproductive season*, Discrete Contin. Dyn. Syst. Ser. B **19** (2014), no. 10, 3209–3218, DOI 10.3934/dcdsb.2014.19.3209. MR3327899

[130] W. Jin and H. Thieme, *An extinction/persistence threshold for sexually reproducing populations: the cone spectral radius*, Discrete Contin. Dyn. Syst. Ser. B **21** (2016), no. 2, 447–470, DOI 10.3934/dcdsb.2016.21.447. MR3432306

[131] S. Karlin, *Positive operators*, J. Math. Mech. **8** (1959), 907–937, DOI 10.1512/iumj.1959.8.58058. MR114138

[132] T. Kato, *Perturbation theory for linear operators*, 2nd ed., Grundlehren der Mathematischen Wissenschaften, Band 132, Springer-Verlag, Berlin-New York, 1976. MR407617

[133] A. Klenke, *Probability theory*, Translation from the German edition. Universitext, Springer, London, 2014. A comprehensive course, DOI 10.1007/978-1-4471-5361-0. MR3112259

[134] M. Kot and W. M. Schaffer, *Discrete-time growth-dispersal models*, Math. Biosci. **80** (1986), no. 1, 109–136, DOI 10.1016/0025-5564(86)90069-6. MR849698

[135] M. A. Krasnosel'skij, *Positive solutions of operator equations*, Noordhoff, Groningen, 1964.

[136] M. A. Krasnosel'skij, Je. A. Lifshits, and A. V. Sobolev, *Positive linear systems: The method of positive operators*, Heldermann Verlag, Berlin, 1989.

[137] M. A. Krasnosel'skii and P. P. Zabreiko, *Geometrical methods of nonlinear analysis*, Springer, Berlin Heidelberg, 1984.

[138] U. Krause, *Positive dynamical systems in discrete time*, De Gruyter Studies in Mathematics, vol. 62, De Gruyter, Berlin, 2015. Theory, models, and applications, DOI 10.1515/9783110365696. MR3309532

[139] M. Krein, *Sur les opérations linéaires transformant un certain ensemble conique en lui-même* (French), C. R. (Doklady) Acad. Sci. URSS (N.S.) **23** (1939), 749–752. MR2018

[140] M. G. Kreĭn and M. A. Rutman, *Linear operators leaving invariant a cone in a Banach space* (Russian), Uspehi Matem. Nauk (N.S.) **3** (1948), no. 1(23), 3–95. MR27128

[141] T. Lant and H. R. Thieme, *Markov transition functions and semigroups of measures*, Semigroup Forum **74** (2007), no. 3, 337–369, DOI 10.1007/s00233-006-0636-z. MR2321572

[142] A. Lasota and M. C. Mackey, *Chaos, fractals, and noise*, 2nd ed., Applied Mathematical Sciences, vol. 97, Springer-Verlag, New York, 1994. Stochastic aspects of dynamics, DOI 10.1007/978-1-4612-4286-4. MR1244104

[143] A. Lasota, J. Myjak, and T. Szarek, *Markov operators with a unique invariant measure*, J. Math. Anal. Appl. **276** (2002), no. 1, 343–356, DOI 10.1016/S0022-247X(02)00457-2. MR1944355

[144] B. Lemmens, B. Lins, R. Nussbaum, and M. Wortel, *Denjoy-Wolff theorems for Hilbert's and Thompson's metric spaces*, J. Anal. Math. **134** (2018), no. 2, 671–718, DOI 10.1007/s11854-018-0022-2. MR3771496

[145] B. Lemmens and R. Nussbaum, *Nonlinear Perron-Frobenius theory*, Cambridge Tracts in Mathematics, vol. 189, Cambridge University Press, Cambridge, 2012, DOI 10.1017/CBO9781139026079. MR2953648

[146] B. Lemmens and R. Nussbaum, *Continuity of the cone spectral radius*, Proc. Amer. Math. Soc. **141** (2013), no. 8, 2741–2754, DOI 10.1090/S0002-9939-2013-11520-0. MR3056564

[147] M. A. Lewis and B. Li, *Spreading speed, traveling waves, and minimal domain size in impulsive reaction-diffusion models*, Bull. Math. Biol. **74** (2012), no. 10, 2383–2402, DOI 10.1007/s11538-012-9757-6. MR2978784

[148] M. A. Lewis, N. G. Marculis, and Z. Shen, *Integrodifference equations in the presence of climate change: persistence criterion, travelling waves and inside dynamics*, J. Math. Biol. **77** (2018), no. 6-7, 1649–1687, DOI 10.1007/s00285-018-1206-1. MR3881861

[149] B. Li, M. A. Lewis, and H. F. Weinberger, *Existence of traveling waves for integral recursions with nonmonotone growth functions*, J. Math. Biol. **58** (2009), no. 3, 323–338, DOI 10.1007/s00285-008-0175-1. MR2470192

[150] C.-K. Li and H. Schneider, *Applications of Perron-Frobenius theory to population dynamics*, J. Math. Biol. **44** (2002), no. 5, 450–462, DOI 10.1007/s002850100132. MR1908132

[151] F. Lutscher, *Integrodifference equations in spatial ecology*, Interdisciplinary Applied Mathematics, vol. 49, Springer, Cham, [2019] ©2019. With a foreword by Mark Lewis, DOI 10.1007/978-3-030-29294-2. MR3971004

[152] F. Lutscher and S. V. Petrovskii, *The importance of census times in discrete-time growth-dispersal models*, J. Biol. Dyn. **2** (2008), no. 1, 55–63, DOI 10.1080/17513750701769899. MR2401580

[153] P. Magal, *Global stability for differential equations with homogeneous nonlinearity and application to population dynamics*, Discrete Contin. Dyn. Syst. Ser. B **2** (2002), no. 4, 541–560, DOI 10.3934/dcdsb.2002.2.541. MR1921237

[154] P. Magal and X.-Q. Zhao, *Global attractors and steady states for uniformly persistent dynamical systems*, SIAM J. Math. Anal. **37** (2005), no. 1, 251–275, DOI 10.1137/S0036141003439173. MR2172756

[155] D. K. Mallawaarachchi, L. J. S. Allen, and C. Carey, *Stability and permanence in gender- and stage-structured models for the boreal toad*, J. Biol. Dyn. **5** (2011), no. 1, 1–26, DOI 10.1080/17513751003777515. MR2787170

[156] J. Mallet-Paret and R. D. Nussbaum, *Eigenvalues for a class of homogeneous cone maps arising from max-plus operators*, Discrete Contin. Dyn. Syst. **8** (2002), no. 3, 519–562, DOI 10.3934/dcds.2002.8.519. MR1897866

[157] J. Mallet-Paret and R. D. Nussbaum, *Generalizing the Krein-Rutman theorem, measures of noncompactness and the fixed point index*, J. Fixed Point Theory Appl. **7** (2010), no. 1, 103–143, DOI 10.1007/s11784-010-0010-3. MR2652513

[158] J. Mallet-Paret and R. D. Nussbaum, *Asymptotic fixed point theory and the beer barrel theorem*, J. Fixed Point Theory Appl. **4** (2008), no. 2, 203–245, DOI 10.1007/s11784-008-0095-0. MR2465551

[159] J. N. McDonald and N. A. Weiss, *A course in real analysis*, Academic Press, Inc., San Diego, CA, 1999. Biographies by Carol A. Weiss. MR1680810

[160] R. E. Mickens, *Difference equations*, 2nd ed., Van Nostrand Reinhold Co., New York, 1990. Theory and applications. MR1158461

[161] T. E. X. Miller and B. D. Inouye, *Confronting two-sex demographic models with data*, Ecology **92** (2011), 2141–2151.

[162] T. E. X. Miller, A. K. Shaw, B. D. Inouye, and M. G. Neubert, *Sex-biased dispersal and the speed of two-sex invasions*, Amer. Nat. **177** (2011), DOI: 10.1086/659628.

[163] V. Müller and A. Peperko, *On the Bonsall cone spectral radius and the approximate point spectrum*, Discrete Contin. Dyn. Syst. **37** (2017), no. 10, 5337–5354, DOI 10.3934/dcds.2017232. MR3668365

[164] G. J. Murphy, *Continuity of the spectrum and spectral radius*, Proc. Amer. Math. Soc. **82** (1981), no. 4, 619–621, DOI 10.2307/2043782. MR614889

[165] J. Musgrave and F. Lutscher, *Integrodifference equations in patchy landscapes II: population level consequences*, J. Math. Biol. **69** (2014), no. 3, 617–658, DOI 10.1007/s00285-013-0715-1. MR3248070

[166] H. Neunzert, *An introduction to the nonlinear Boltzmann-Vlasov equation*, Kinetic theories and the Boltzmann equation (Montecatini, 1981), Lecture Notes in Math., vol. 1048, Springer, Berlin, 1984, pp. 60–110, DOI 10.1007/BFb0071878. MR740721

[167] R. D. Nussbaum, *Eigenvectors of nonlinear positive operators and the linear Kreĭn-Rutman theorem*, Fixed point theory (Sherbrooke, Que., 1980), Lecture Notes in Math., vol. 886, Springer, Berlin, 1981, pp. 309–330. MR643014

[168] R. D. Nussbaum, *Hilbert's projective metric and iterated nonlinear maps*, Mem. Amer. Math. Soc. **75** (1988), no. 391, iv+137, DOI 10.1090/memo/0391. MR961211

[169] R. D. Nussbaum, *Iterated nonlinear maps and Hilbert's projective metric. II*, Mem. Amer. Math. Soc. **79** (1989), no. 401, iv+118, DOI 10.1090/memo/0401. MR963567

[170] R. D. Nussbaum, *Eigenvectors of order-preserving linear operators*, J. London Math. Soc. (2) **58** (1998), no. 2, 480–496, DOI 10.1112/S0024610798006425. MR1668148

[171] C. Poetzsche, *Numerical dynamics of integrodifference equations: global attractivity in a C^0-setting*, SIAM J. Numer. Anal. **57** (2019), no. 5, 2121–2141, DOI 10.1137/18M1214469. MR4000220

[172] G. Poólya and G. Szegö, *Aufgaben und Lehrsätze aus der analysis*, I, Springer, Berlin, 1925.

[173] A. Quarteroni, R. Sacco, and F. Saleri, *Numerical mathematics*, 2nd ed., Texts in Applied Mathematics, vol. 37, Springer-Verlag, Berlin, 2007, DOI 10.1007/b98885. MR2265914

[174] F. Riesz and B. Sz.-Nagy, *Functional analysis*, Dover Books on Advanced Mathematics, Dover Publications, Inc., New York, 1990. Translated from the second French edition by Leo F. Boron; Reprint of the 1955 original. MR1068530

[175] W. Rudin, *Real and complex analysis*, McGraw-Hill Book Co., New York-Toronto-London, 1966. MR210528

[176] P. L. Salceanu and H. L. Smith, *Persistence in a discrete-time, stage-structured epidemic model*, J. Difference Equ. Appl. **16** (2010), no. 1, 73–103, DOI 10.1080/10236190802400733. MR2589558

[177] H. Schaefer, *Positive Transformationen in lokalkonvexen halbgeordneten Vektorräumen* (German), Math. Ann. **129** (1955), 323–329, DOI 10.1007/BF01362375. MR70965

[178] H. Schaefer, *Halbgeordnete lokalkonvexe Vektorräume. II* (German), Math. Ann. **138** (1959), 259–286, DOI 10.1007/BF01342907. MR106402

[179] H. Schaefer, *Some spectral properties of positive linear operators*, Pacific J. Math. **10** (1960), 1009–1019. MR115090

[180] H. H. Schaefer, *Topological vector spaces*, The Macmillan Company, New York; Collier Macmillan Ltd., London, 1966. MR193469

[181] H. H. Schaefer, *Banach lattices and positive operators*, Die Grundlehren der mathematischen Wissenschaften, Band 215, Springer-Verlag, New York-Heidelberg, 1974. MR423039

[182] G. Seo and G. S. K. Wolkowicz, *Existence of multiple limit cycles in a predator-prey model with* $\arctan(ax)$ *as functional response*, Commun. Math. Anal. **18** (2015), no. 1, 64–68. MR3365174

[183] G. Seo and G. S. K. Wolkowicz, *Sensitivity of the dynamics of the general Rosenzweig-MacArthur model to the mathematical form of the functional response: a bifurcation theory approach*, J. Math. Biol. **76** (2018), no. 7, 1873–1906, DOI 10.1007/s00285-017-1201-y. MR3788219

[184] V. M. Shurenkov, *On the relationship between spectral radii and Perron Roots*, Chalmers Univ Tech and Göteborg Univ, Preprint.

[185] J. G. Skellam, *Random dispersal in theoretical populations*, Biometrika **38** (1951), 196–218, DOI 10.1093/biomet/38.1-2.196. MR43440

[186] H. L. Smith, *Monotone dynamical systems*, Mathematical Surveys and Monographs, vol. 41, American Mathematical Society, Providence, RI, 1995. An introduction to the theory of competitive and cooperative systems. MR1319817

[187] H. L. Smith and H. R. Thieme, *Dynamical systems and population persistence*, Graduate Studies in Mathematics, vol. 118, American Mathematical Society, Providence, RI, 2011, DOI 10.1090/gsm/118. MR2731633

[188] H. L. Smith and H. R. Thieme, *Persistence and global stability for a class of discrete time structured population models*, Discrete Contin. Dyn. Syst. **33** (2013), no. 10, 4627–4646, DOI 10.3934/dcds.2013.33.4627. MR3049095

[189] J. Smoller, *Shock waves and reaction-diffusion equations*, Grundlehren der Mathematischen Wissenschaften, vol. 258, Springer-Verlag, New York-Berlin, 1983. MR688146

[190] K. Taira, *Semigroups, boundary value problems and Markov processes*, Springer Monographs in Mathematics, Springer-Verlag, Berlin, 2004, DOI 10.1007/978-3-662-09857-8. MR2019537

[191] A. L. J. Thieme, *Wanderings with Lady M: a happy threesome*, Math. Biosci. Eng. **7** (2010), no. 1, iv–ix, DOI 10.3934/mbe.2010.7.1iv. MR2654276

[192] H. R. Thieme, *On a class of Hammerstein integral equations*, Manuscripta Math. **29** (1979), no. 1, 49–84, DOI 10.1007/BF01309313. MR539788

[193] H. R. Thieme, *Well-posedness of physiologically structured population models for Daphnia magna. How biological concepts can benefit by abstract mathematical analysis*, J. Math. Biol. **26** (1988), no. 3, 299–317, DOI 10.1007/BF00277393. MR949095

[194] H. R. Thieme, *Asymptotic proportionality (weak ergodicity) and conditional asymptotic equality of solutions to time-heterogeneous sublinear difference and differential equations*, J. Differential Equations **73** (1988), no. 2, 237–268, DOI 10.1016/0022-0396(88)90107-6. MR943942

[195] H. R. Thieme, *Mathematics in population biology*, Princeton Series in Theoretical and Computational Biology, Princeton University Press, Princeton, NJ, 2003. MR1993355

[196] H. R. Thieme, *Spectral bound and reproduction number for infinite-dimensional population structure and time heterogeneity*, SIAM J. Appl. Math. **70** (2009), no. 1, 188–211, DOI 10.1137/080732870. MR2505085

[197] H.R. Thieme, *Eigenvectors and eigenfunctionals of homogeneous order-preserving maps*, `arXiv:1302.3905v1`, 2013.

[198] H. R. Thieme, *Comparison of spectral radii and Collatz–Wielandt numbers for homogeneous maps, and other applications of the monotone companion norm on ordered normed vector spaces*, `arXiv:1406.6657v2`, 2016.

[199] H. R. Thieme, *Spectral radii and Collatz-Wielandt numbers for homogeneous order-preserving maps and the monotone companion norm*, Ordered structures and applications, Trends Math., Birkhäuser/Springer, Cham, 2016, pp. 415–467, DOI 10.1007/978-3-319-27842-1_26. MR3708356

[200] H. R. Thieme, *Eigenfunctionals of homogeneous order-preserving maps with applications to sexually reproducing populations*, J. Dynam. Differential Equations **28** (2016), no. 3-4, 1115–1144, DOI 10.1007/s10884-015-9463-9. MR3537366

[201] H. R. Thieme, *Eigenvectors of homogeneous order-bounded order-preserving maps*, Discrete Contin. Dyn. Syst. Ser. B **22** (2017), no. 3, 1073–1097, DOI 10.3934/dcdsb.2017053. MR3639156

[202] H. R. Thieme, *From homogeneous eigenvalue problems to two-sex population dynamics*, J. Math. Biol. **75** (2017), no. 4, 783–804, DOI 10.1007/s00285-017-1114-9. MR3687220

[203] H. R. Thieme, *Discrete-time population dynamics on the state space of measures*, Math. Biosci. Eng. **17** (2020), no. 2, 1168–1217, DOI 10.3934/mbe.2020061. MR4060809

[204] H. R. Thieme, *Persistent discrete-time dynamics on measures*, Progress on difference equations and discrete dynamical systems, Springer Proc. Math. Stat., vol. 341, Springer, Cham, [2020] ©2020, pp. 59–100, DOI 10.1007/978-3-030-60107-2_4. MR4219146

[205] H. R. Thieme, *Discrete-time population dynamics of spatially distributed semelparous two-sex populations*, J. Math. Biol. **83** (2021), no. 2, Paper No. 18, 40, DOI 10.1007/s00285-021-01649-4. MR4292766

[206] H. R. Thieme, *Discrete-time dynamics of structured populations via Feller kernels*, Discrete Contin. Dyn. Syst. Ser. B **27** (2022), no. 2, 1091–1119, DOI 10.3934/dcdsb.2021082. MR4361674

[207] H. R. Thieme, *Reproduction number versus turnover number in structured discrete-time population models*, Advances in discrete dynamical systems, difference equations and applications, Springer Proc. Math. Stat., vol. 416, Springer, Cham, [2023] ©2023, pp. 495–539, DOI 10.1007/978-3-031-25225-9_23. MR4606694

[208] H. R. Thieme, *Continuity of the spectral radius, applied to structured semelparous two-sex population models*, J. Difference Eqs. Appl. (2023), DOI 10.1080/10236198.2023.224912.

[209] H. R. Thieme and J. Yang, *An endemic model with variable re-infection rate and applications to influenza*, Math. Biosci. **180** (2002), 207–235, DOI 10.1016/S0025-5564(02)00102-5. John A. Jacquez memorial volume. MR1950755

[210] A. C. Thompson, *On certain contraction mappings in a partially ordered vector space*, Proc. Amer. Math. Soc. **14** (1963), 438–443, DOI 10.2307/2033816. MR149237

[211] A. C. Thompson, *On the eigenvectors of some not-necessarily-linear transformations*, Proc. London Math. Soc. (3) **15** (1965), 577–598, DOI 10.1112/plms/s3-15.1.577. MR182882

[212] A. J. Tromba, *The beer barrel theorem*, Functional differential equations and approximation of fixed points (Proc. Summer School and Conf., Univ. Bonn, Bonn, 1978), Lecture Notes in Math., vol. 730, Springer, Berlin, 1979, pp. 484–488. A new proof of the asymptotic conjecture in fixed point theory. MR548002

[213] A. Ulikowska, *Structured population models in metric spaces*, Dissertation, Warsaw University, 2013.

[214] P. van den Driessche and J. Watmough, *Reproduction numbers and sub-threshold endemic equilibria for compartmental models of disease transmission*, Math. Biosci. **180** (2002), 29–48, DOI 10.1016/S0025-5564(02)00108-6. John A. Jacquez memorial volume. MR1950747

[215] A. W. van der Vaart and J. A. Wellner, *Weak convergence and empirical processes*, Springer Series in Statistics, Springer-Verlag, New York, 1996. With applications to statistics, DOI 10.1007/978-1-4757-2545-2. MR1385671

[216] J. S. Vandergraft, *Spectral properties of matrices which have invariant cones*, SIAM J. Appl. Math. **16** (1968), 1208–1222, DOI 10.1137/0116101. MR244284

[217] R. von Mises and H. Pollaczek-Geiringer, *Praktische Verfahren der Gleichungsauflösung*, Zeitschrift für Angewandte Mathematik und Mechanik **9** (1929), 58–77, 152–164.

[218] X. Wang and X.-Q. Zhao, *Target reproduction numbers for reaction-diffusion population models*, J. Math. Biol. **81** (2020), no. 2, 625–647, DOI 10.1007/s00285-020-01523-9. MR4135694

[219] F.-B. Wang, L. Zhang, and X.-Q. Zhao, *Basic reproduction ratios for time-periodic homogeneous evolution systems*, SIAM J. Appl. Math. **83** (2023), no. 5, 1806–1831, DOI 10.1137/22M1531865. MR4642254

[220] H. Wielandt, *Unzerlegbare, nicht negative Matrizen* (German), Math. Z. **52** (1950), 642–648, DOI 10.1007/BF02230720. MR35265

[221] R. Wu and X.-Q. Zhao, *Propagation dynamics for a spatially periodic integrodifference competition model*, J. Differential Equations **264** (2018), no. 10, 6507–6534, DOI 10.1016/j.jde.2018.01.039. MR3770057

[222] K. Wysocki, *Some ergodic theorems for solutions of homogeneous differential equations*, SIAM J. Math. Anal. **24** (1993), no. 3, 681–702, DOI 10.1137/0524042. MR1215432

[223] K. Yosida, *Functional analysis*, 2nd ed., Springer, Berlin Heidelberg, 1965–1968.

[224] P. P. Zabreĭko, M. A. Krasnosel′skiĭ, and Ju. V. Pokornyĭ, *A certain class of positive linear operators* (Russian), Funkcional. Anal. i Priložen. **5** (1971), no. 4, 9–17. MR291838

[225] Y. Zhang and X.-Q. Zhao, *Bistable travelling waves in competitive recursion systems*, J. Differential Equations **252** (2012), no. 3, 2630–2647, DOI 10.1016/j.jde.2011.10.005. MR2860634

[226] X.-Q. Zhao, *Dynamical systems in population biology*, 2nd ed., CMS Books in Mathematics/Ouvrages de Mathématiques de la SMC, Springer, Cham, 2017, DOI 10.1007/978-3-319-56433-3. MR3643081

Index

Selected Published Titles in This Series

281 **Horst R. Thieme,** Discrete-Time Dynamics of Structured Populations and Homogeneous Order-Preserving Operators, 2024

277 **George Metcalfe, Francesco Paoli, and Constantine Tsinakis,** Residuated Structures in Algebra and Logic, 2023

276 **Balázs Bárány, Károly Simon, and Boris Solomyak,** Self-similar and Self-affine Sets and Measures, 2023

275 **Alekos Vidras and Alain Yger,** Multidimensional Residue Theory and Applications, 2023

274 **Willi Freeden and M. Zuhair Nashed,** Recovery Methodologies: Regularization and Sampling, 2023

273 **Gennadiy Feldman,** Characterization of Probability Distributions on Locally Compact Abelian Groups, 2023

272 **Tadashi Ochiai,** Iwasawa Theory and Its Perspective, Volume 1, 2023

271 **Hideto Asashiba,** Categories and Representation Theory, 2022

270 **Leslie Hogben, Jephian C.-H. Lin, and Bryan L. Shader,** Inverse Problems and Zero Forcing for Graphs, 2022

269 **Ralph S. Freese, Ralph N. McKenzie, George F. McNulty, and Walter F. Taylor,** Algebras, Lattices, Varieties, 2022

268 **Ralph S. Freese, Ralph N. McKenzie, George F. McNulty, and Walter F. Taylor,** Algebras, Lattices, Varieties, 2022

267 **Dragana S. Cvetković Ilić,** Completion Problems on Operator Matrices, 2022

266 **Kate Juschenko,** Amenability of Discrete Groups by Examples, 2022

265 **Nabil H. Mustafa,** Sampling in Combinatorial and Geometric Set Systems, 2022

264 **J. Scott Carter and Seiichi Kamada,** Diagrammatic Algebra, 2021

263 **Vugar E. Ismailov,** Ridge Functions and Applications in Neural Networks, 2021

262 **Ragnar-Olaf Buchweitz,** Maximal Cohen–Macaulay Modules and Tate Cohomology, 2021

261 **Shiri Artstein-Avidan, Apostolos Giannopoulos, and Vitali D. Milman,** Asymptotic Geometric Analysis, Part II, 2021

260 **Lindsay N. Childs, Cornelius Greither, Kevin P. Keating, Alan Koch, Timothy Kohl, Paul J. Truman, and Robert G. Underwood,** Hopf Algebras and Galois Module Theory, 2021

259 **William Heinzer, Christel Rotthaus, and Sylvia Wiegand,** Integral Domains Inside Noetherian Power Series Rings, 2021

258 **Pramod N. Achar,** Perverse Sheaves and Applications to Representation Theory, 2021

257 **Juha Kinnunen, Juha Lehrbäck, and Antti Vähäkangas,** Maximal Function Methods for Sobolev Spaces, 2021

256 **Michio Jimbo, Tetsuji Miwa, and Fedor Smirnov,** Local Operators in Integrable Models I, 2021

255 **Alexandre Boritchev and Sergei Kuksin,** One-Dimensional Turbulence and the Stochastic Burgers Equation, 2021

254 **Karim Belabas and Henri Cohen,** Numerical Algorithms for Number Theory, 2021

253 **Robert R. Bruner and John Rognes,** The Adams Spectral Sequence for Topological Modular Forms, 2021

252 **Julie Déserti,** The Cremona Group and Its Subgroups, 2021

251 **David Hoff,** Linear and Quasilinear Parabolic Systems, 2020

250 **Bachir Bekka and Pierre de la Harpe,** Unitary Representations of Groups, Duals, and Characters, 2020

For a complete list of titles in this series, visit the
AMS Bookstore at **www.ams.org/bookstore/survseries/**.